ROUTLEDGE HANDBOOK OF FOOD AS A COMMONS

From the scientific and industrial revolution to the present day, food – an essential element of life – has been progressively transformed into a private, transnational, mono-dimensional commodity of mass consumption for a global market. But over the last decade there has been an increased recognition that this can be challenged and reconceptualized if food is regarded and enacted as a commons.

This Handbook provides the first comprehensive review and synthesis of knowledge and new thinking on how food and food systems can be thought, interpreted and practiced around the old/new paradigms of commons and commoning. The overall aim is to investigate the multiple constraints that occur within and sustain the dominant food and nutrition regime and to explore how it can change when different elements of the current food systems are explored and re-imagined from a commons perspective. Chapters do not define the notion of commons but engage with different schools of thought:

- the economic approach, based on rivalry and excludability;
- the political approach, recognizing the plurality of social constructions and incorporating epistemologies from the South;
- the legal approach that describes three types of proprietary regimes (private, public and collective) and different layers of entitlement (bundles of rights); and
- the radical-activist approach that considers the commons as the most subversive, coherent and history-rooted alternative to the dominant neoliberal narrative.

These schools have different and rather diverging epistemologies, vocabularies, ideological stances and policy proposals to deal with the construction of food systems, their governance, the distributive implications and the socio-ecological impact on Nature and Society.

The book sparks the debate on food as a commons between and within disciplines, with particular attention to spaces of resistance (food sovereignty, de-growth, open knowledge, transition town, occupations, bottom-up social innovations) and organizational scales (local food, national policies, South–South collaborations, international governance and multi-national agreements). Overall, it shows the consequences of a shift to the alternative paradigm of food as a commons in terms of food, the planet and living beings.

Jose Luis Vivero-Pol is a Research Fellow on food governance and agri-food transitions at the Centre for the Philosophy of Law and the Earth and Life Institute, Catholic University of Louvain, Belgium. An agronomist and development worker with 20 years of experience in food security and nutritional policies, right to food, food sovereignty movements and biodiversity conservation schemes, mostly working in the Global South (Latin America, Africa and Asia). At present, he is working with the United Nations World Food Programme in Asia.

Tomaso Ferrando is Lecturer in Law at the University of Bristol Law School, UK. He has been Visiting Professor at the Universita' di Torino, Italy, and Universidad Externado de Colombia, Colombia, and Resident Fellow at the Institute for Global Law and Policy at Harvard Law School, USA. He is an active member of the Legal Action Committee of the Global Legal Action Network and the Extraterritorial Obligation Consortium.

Olivier De Schutter is Professor at the Catholic University of Louvain, Belgium, SciencesPo Paris, France, and the College of Europe, Natolin, Poland. He has been Visiting Professor at Yale, UCLA and Columbia Universities in the USA. He is a member of the Committee on Economic, Social and Cultural Rights of the United Nations and Co-Chair of IPES-Food, the International Panel of Experts on Sustainable Food Systems. Formerly, he was UN Special Rapporteur on the right to food and Chair of the EU Network of Independent Experts on Fundamental Rights.

Ugo Mattei is Alfred and Hannah Fromm Professor in International Law at the University of California, Hastings College of the Law, USA, and Professor of Civil Law at the University of Turin, Italy. He is the Academic Coordinator of the International University College of Turin and in 2017 he received the Elinor Ostrom Award for his scholarship on the commons. Previously, he was Professor at the University of Trento, Italy and Visiting Professor at University of Montpellier, France; Berkeley College, Yale University, USA; and University of Cambridge, UK.

ROUTLEDGE HANDBOOK OF FOOD AS A COMMONS

Edited by Jose Luis Vivero-Pol, Tomaso Ferrando, Olivier De Schutter and Ugo Mattei

Routledge
Taylor & Francis Group
LONDON AND NEW YORK

earthscan
from Routledge

First published 2019 by Routledge

2 Park Square, Milton Park, Abingdon, Oxfordshire OX14 4RN
52 Vanderbilt Avenue, New York, NY 10017

Routledge is an imprint of the Taylor & Francis Group, an informa business

First issued in paperback 2020

British Library Cataloguing-in-Publication Data
A catalogue record for this book is available from the British Library

Library of Congress Cataloging-in-Publication Data
A catalog record for this title has been requested

ISBN: 978-1-138-06262-7 (hbk)
ISBN: 978-0-367-62856-7 (pbk)

Typeset in Bembo
by Deanta Global Publishing Services, Chennai, India

CONTENTS

FIGURES

CO-WRITERS

Molly D. Anderson is the William R. Kenan, Jr. Professor of Food Studies at Middlebury College in Vermont, USA, where she teaches and coordinates a Food Studies Program. She is especially interested in multi-actor collaborations for sustainable food systems, sustainability metrics and assessment, food system resilience, human rights in the food system, food security and the right to food in the US and other industrialized countries. She is involved in food system planning and analysis of solutions at scales from the local to the international.

Peter Andrée is Associate Professor in the Department of Political Science at Carleton University, Canada. His research focuses on the politics of food and the environment. He practices, and teaches, community-based, participatory research methods. He is co-editor of *Globalization and Food Sovereignty: Global and Local Change in the New Politics of Food* (University of Toronto Press, 2014), which examines social movements the world over seeking to build more sustainable and just food systems.

Bálint Balázs is a senior researcher and Director of the Environmental Social Science Research Group, Budapest, Hungary. With a background in sociology and experience in local food systems, policy analysis and participatory action research, he is interested in interactions between actors and policies in sustainability transitions. He is a member of the Community-Supported Agriculture Research Group and the research network Environment and Society (RN12) of the European Sociological Association.

Christian Barrère teaches economics at the University of Reims, France. He mainly works on edible goods, including characteristics related to taste such as gastronomy, wine, tourism and fashion, and their links to cultural and institutional heritages.

Michel Bauwens is a key theorist in the emerging field of P2P. He is the founder of the P2P Foundation and is working with a global community of researchers and activists to explore P2P and the Commons as emerging paradigms of economic, political and social organization.

Petra Benyei is a predoctoral researcher affiliated to the Institute of Environmental Science and Technology (ICTA-UAB) and part of the core development team of the CONECT-e project (www.conecte.es). Her main research interest is Traditional Agroecological Knowledge through innovative approaches such as Citizen Science, under a commons framework.

Patrick Bond is a Distinguished Professor of Political Economy at the University of Witwatersrand's School of Governance, South Africa. He is presently researching economic crisis and the dynamics of uneven and combined development, political ecology (energy, water and climate change), social mobilization, public policy and geopolitics.

Laura Calvet-Mir (Ph.D. in Environmental Science) is a research fellow at IN3-UOC (Barcelona, Spain) and affiliated researcher at ICTA-UAB (Barcelona). Her research focuses on urban and rural agriculture, ecosystem services, political ecology, agroecological transitions and traditional ecological knowledge. In studying local social-ecological systems, she combines ethnographic approaches with quantitative methods.

Marina Chang is Research Associate at the Centre for Agroecology, Water and Resilience at Coventry University, UK. Taking food as an entry point, her research focuses on the theory and practice of the commons as an alternative to the existing dominant capitalism. She endeavors to integrate research, education, public policy, community development and enterprise into a coherent force for change.

Brendan Coolsaet is Assistant Professor at the European School of Political and Sciences (ESPOL, Lille Catholic University, France) and a member of the Global Environmental Justice group (University of East Anglia, UK). He works on environmental justice, biodiversity conservation and food politics. At ESPOL, he teaches on the Master of Food Policy program.

Ivan Cucco is a Contract Researcher at the University of Naples "l'Orientale" for the project *Empowering civil society and workers in Vietnam*. He regularly works as a consultant for FAO and IFAD with a focus on food and nutrition security, family resilience and food systems analysis. He has recently co-authored *Social Network Analysis for the territorial assessment and Mapping of Food Security and Nutrition Systems: A Methodological Approach* (FAO, 2018) and "Cooperatives and alternative food networks in Italy" (*Journal of Rural Studies*, 53, 2017).

Olivier De Schutter is Professor at the Catholic University of Louvain, SciencesPo Paris, France, and the College of Europe, Natolin, Poland. He has been Visiting Professor at Yale, UCLA and Columbia Universities in the USA. He is a member of the Committee on Economic, Social and Cultural Rights of the United Nations and Co-Chair of IPES-Food, the International Panel of Experts on Sustainable Food Systems. Formerly, he was UN Special Rapporteur on the right to food and Chair of the EU Network of Independent Experts on Fundamental Rights.

Tomaso Ferrando is Lecturer in Law at the University of Bristol Law School, UK. He has been Visiting Professor at the Universita' di Torino, Italy, and Universidad Externado de Colombia, Colombia, and Resident Fellow at the Institute for Global Law and Policy at Harvard Law School, USA. He is an active member of the Legal Action Committee of the Global Legal Action Network and the Extraterritorial Obligation Consortium.

Maria Fonte is Associate Professor at the University of Naples Federico II, Italy, Adjunct Professor at the American University of Rome, Italy, and Project Professor at the University of Kyoto, Japan. Her publications include works on agro-biotechnology, local agro-food systems and rural development. Publications include many articles in international journal and a book on *Organismi geneticamente modificati* (Franco Angeli, 2004) and *Naming Food after Places. Food Relocalisation and Knowledge in Rural Development*, co-edited with Apostolos Papadopoulos (Routledge, 2010).

Christine Frison is an FWO post-doctoral researcher at the Law Faculty of the University of Antwerp and associate researcher at the Centre for Philosophy of Law of the Université catholique de Louvain, Belgium. She worked as Legal Research Fellow with the Centre for International Sustainable Development Law (McGill University, Canada) where she specialized in the field of international environmental law, biodiversity, biosafety and agrobiodiversity international law. Her latest book is *The Commons, Plant Breeding and Agricultural Research* (Routledge, 2018).

Mary Galvin is Associate Professor at the University of Johannesburg's Centre for Social Change, South Africa. She has worked on water and sanitation issues in South Africa as a researcher, development practitioner and social justice activist since 1992. Her work focuses on local water governance and citizen engagement, political ecology and new approaches to sanitation.

Eric Holt-Giménez is the Executive Director of the Institute for Food and Development Policy/Food First, USA. An agroecologist and political economist holding a PhD, he lectures at universities in North America, Latin America, and Europe. Author of several books and dozens of academic articles on agroecology, food systems, and social movements, he has worked closely with farmers organizations in the Americas for over 40 years.

Tara Kenny has a BSc in International Development and Food Policy, a Master's in Public Health and is currently in the final phase of a PhD focused on food insecurity in Ireland with the Department of Human Geography at University College Cork, Ireland. Her research is funded by an Irish Research Council Government of Ireland Postgraduate scholarship.

George Kent retired from the University of Hawai'i and currently serves as Adjunct Professor with universities in Australia and California. He is Deputy Editor of the online journal *World Nutrition*. His most recent books are *Caring about Hunger* (Irene Publishing, 2016) and *Governments Push Infant Formula* (Irene Publishing, 2017).

Hugo Martorell studies food policy and governance in Canada as a PhD student in food in Carleton University, Canada. His Master's Thesis at Concordia University focused on regional food planning in Montreal, Quebec. He continues to collaborate with civil society organizations and academic partnerships as a researcher and advocate.

Ugo Mattei is Alfred and Hannah Fromm Professor in International Law at the University of California, Hastings College of the Law, USA, and Professor of Civil Law at the University of Turin, Italy. He is the Academic Coordinator of the International University College of Turin and in 2017 he received the Elinor Ostrom Award for his scholarship on the commons. Previously, he was Professor at the University of Trento, Italy and Visiting Professor at University of Montpellier, France; Berkeley College, Yale University, USA; and University of Cambridge, UK.

Chris Maughan is a British Academy Research Postdoctoral Fellow based at the Centre for Agroecology, Water and Resilience, Coventry University, UK. His current research explores the meaning and practice of 'innovation' in European grassroots agroecology. For the last ten years he has been directly involved in a variety of community food growing projects in the UK.

John O'Neill is Professor of Political Economy at the University of Manchester, UK. He has written widely on philosophy, political economy and environmental policy. His books include *Markets, Deliberation and Environment* (Routledge, 2007), *The Market: Ethics, Knowledge and Politics* (Routledge, 1998) and *Ecology, Policy and Politics: Human Well-Being and the Natural World* (Routledge, 1993). He is co-author of *Environmental Values* (Routledge, 2008).

Alex Pazaitis is a core member of the P2P Lab research collective, spin off of the Tallinn University of Technology, Estonia and the P2P Foundation. He writes and lectures on the political economy of the commons, innovation policy, open cooperativism and the blockchain.

Giacomo Pettenati is post-doctoral researcher in Political and Economic Geography at the Department of Cultures, Politics and Society of the University of Turin, Italy. His main research interests are food geographies, mountain areas society and development, and landscape management.

Victoria Reyes-García (PhD in Antropology, University of Florida, USA) is ICREA Research Professor at the Institut de Ciència i Tecnologia Ambientals (Universitat Autònoma de Barcelona, Spain). Her research centers around the benefits generated by local ecological knowledge and the dynamic nature of these knowledge systems, including traditional agricultural knowledge.

Peter M. Rosset is Professor of Agroecology at El Colegio de la Frontera Sur (ECOSUR), Mexico, where he is a member of the Research Group on Scaling Up Agroecology. He is also CAPES-FUNCAP Visiting Professor (BPV) at the Crateús Education Faculty (FAEC) and the Post-Graduate Program on Sociolofy (PPGS) of the Universidade Estadual do Ceará (UECE) in Brazil. He is co-coordinator of the Land Research Action Network (LRAN).

Colin Sage has worked at University College Cork, Ireland for two decades conducting research on food systems and sustainability, most recently in Tasmania. He is the author of *Environment and Food* (Routledge, 2011) and co-editor of *Food Transgressions* (Routledge, 2013) and *Transdisciplinary Perspectives on Transitions to Sustainability* (Routledge, 2016). He serves as Chair of the Cork Food Policy Council.

Cristian Timmermann studied philosophy and political science at the Ludwig-Maximilian University, Germany, and Wageningen University, The Netherlands. He is currently a post-doctoral researcher at Universidad de Chile, after holding previously research positions at the Universidad Nacional Autónoma de México and the Ben-Gurion University of the Negev, Israel. His research areas are agricultural ethics, global justice, science policy, resource governance and property theories.

Cristina Tirado-von der Pahlen is Director of International Climate Initiatives at the Center for Urban Resilience at Loyola Marymount University and Faculty at the Environmental Sciences Program, USA. She serves at the Secretariat of the Mediterranean Cities Climate Consortium and moderates the UN Standing Committee of Nutrition e-group on Climate and Nutrition. She is a lead author of the next IPCC AR6. Formerly she was WHO Regional Adviser in Europe and Latin America, Coordinator of the WHO Surveillance Program in Berlin, Director of the PHI Center for Climate and Health in Oakland and UCLA affiliate.

Alessia Toldo is post-doctoral researcher in Political and Economic Geography at the Department of Cultures, Politics and Society of the University of Turin, Italy. Her main research interests are food geographies and urban food policies, social inclusion and justice.

Valentín Val is an anthropologist, PhD candidate and member of the Research Group on Scaling Up Agroecology at El Colegio de la Frontera Sur (ECOSUR) in Chiapas, Mexico. Along with the Koman Ilel collective (www.komanilel.org), he co-produced a video documentary and helped create the Multimedia Peasant School about the "campesino-a-campesino" agroecology movement in Cuba (www.agreocologia.espora.org).

Penny Van Esterik is Professor Emerita of Anthropology, retired from York University, Toronto, Canada, where she taught nutritional anthropology, advocacy anthropology and feminist theory. Her fieldwork was primarily in Southeast Asia.

Ilja van Lammeren is a Master's student at the Norwegian University of Life Sciences, the Department of International Environment and Development Studies (NORAGRIC). Her studies center around agrarian political economy, development politics and political ecology. From June 2016 to February 2017, she was a research-intern at the Institute for Food and Development Policy/Food First.

Jose Luis Vivero-Pol is a Research Fellow on food governance and agri-food transitions at the Centre for the Philosophy of Law and the Earth and Life Institute, Catholic University of Louvain, Belgium. An agronomist and development worker with 20 years of experience in food security and nutritional policies, right to food, food sovereignty movements and biodiversity conservation schemes, mostly working in the Global South (Latin America, Africa and Asia). At present, he is working with the United Nations World Food Programme in Asia.

Noah Zerbe is Professor of Politics at Humboldt State University in California, USA. He is the author of numerous articles on the politics of food and the commodification processes. His current research focuses on the political economy of alternative food systems.

PREFACE

Human society is developing a troublesome relationship with food. Nowadays, the food regime is such that many eat poorly and badly so that others can have access to all the food they desire and can pay for. Moreover, industrial production and global distribution of food are major driving forces in pushing the environment beyond the planet's ecological boundaries. Within that scenario, the idea that it is feasible to reach the right to adequate food (an entitlement) and food and nutrition security (a Global Public Good) by means of food as a commodity (a for-profit private good) under conditions of extreme inequality inevitably crumbles. The aim of this book is to investigate the multiple enclosures that occur within and sustain the dominant food and nutrition regime and to explore how it would change if food were regarded and enacted as a commons.

From the industrial revolution to the present, food – an essential element of life – has been progressively transformed into a private, transnational, mono-dimensional commodity of mass consumption in a global market. As in many other areas of people's livelihoods, the mechanisms of enclosure of common resources through legislation, excessive pricing and patents have obviously played a major role in limiting the production, transformation and consumption of food as a commons. The social construct of food as a commodity, in fact, denies its non-economic attributes (healthy, sustainable, cultural determinant, vital fuel for our bodies, human right, natural resource) in favour of its tradable features, namely durability, appearance and the standardization of naturally diverse food products, leading to a neglect of nutrition-related properties of food, alongside an emphasis on cheap calories.

The change in the food paradigm, when it happens, will be driven by the dialectic clash between the hegemonic economic epistemology (the economic understanding of the commons with the tragedy of the commons, the absolute primacy of private property and market exchanges) and non-dominant alternatives (political, historical, legal and radical-activist approaches to the commons) that are gaining legitimacy in recent decades. In this book, we explore the multiple approaches to food from different schools of thought: the economic approach, which is rather ontological and based on two theoretical features (rivalry and excludability); the political approach, which recognizes the plurality of social constructions around the consideration of food and incorporates epistemologies from the South; the legal approach, which describes three types of proprietary regimes (private, public and collective) and different layers of entitlement (bundles of rights); and the radical-activist approach, which considers the commons as the most subversive, coherent and history-rooted alternative to the dominant neoliberal narrative and praxis. These schools have different and rather diverging epistemologies, vocabularies, ideological stances and policy proposals to deal with food systems and their governance, entitlements and externalities.

This book aims to enrich the debate on food as a commons between and within disciplines, niches of resistance (transition towns, food sovereignty, de-growth, open knowledge, commons) and organizational scales (local food, national policies, South–South collaborations, and international governance and agreements). The goal is to answer a straightforward question: What would be the consequence in terms of food policies of the shift to an alternative paradigm of food as a commons?

The consideration of food as a commodity can be questioned and modified, and it is actually happening in many contemporary grassroots movements and customary indigenous traditions all over the world. Small-scale farmers, peasants, fishermen, sensitized urban consumers, food security activists, legal and political academics and human rights advocates, amongst others, are re-constructing a different food paradigm in multiple loci (urban and rural areas in the Global South and North) by defending the public nature of many food-producing resources, such as seeds, water, land and agricultural knowledge. The de-commodification and commoning of food (and of the whole food system as the broader objective) will open up the transition towards a new food regime in which primacy rests in its absolute need for human beings (nutrition + culture + community) and not in profit maximization, and in which the different dimensions of food are properly valued.

1
INTRODUCTION
The food commons are coming …

Jose Luis Vivero-Pol, Tomaso Ferrando, Olivier De Schutter and Ugo Mattei

Seeing with new eyes

Since the fall of the Berlin Wall, only one vision has become hegemonic worldwide. The marginalization of any alternative to the single thought, also known as the end of history (Fukuyama, 1989; IUC, 2009), has quickly generated what is known as neoliberalism, the new form of hybridization between public sovereignty and private corporations that has come to dominate contemporary structures of global governance (Harvey, 2007). This arrangement, with a crucial role for the military industrial complex, has not only produced new forms of world disorders. It has also disrupted the fundamental understanding of modernity, that of a neat distinction between a public and a private sector. The new hybrid corporate power, the current form of capital accumulation, now runs the world within a logic of global sovereignty that defeats every form of democratic control. Every single aspect of human life has been attracted within this bio-political machinery so that the very human being is now commodified like every other aspect of nature. The most tangible manifestation of this process is in the domain of two of the fundamental building blocks of human life: water and food. These two essential components of life are now almost entirely transformed into commodities, leading to forms of domination and subordination that are difficult to overestimate. The consequences of the current extractive system are so deep as to produce a new geological era, the so-called Anthropocene (Crutzen, 2006; Purdy, 2015) or Capitalocene (Moore, 2017), which is likely to destroy the very conditions of life and human civilization (Brown, 2008; Capra and Mattei, 2015).

It is as a reaction to the massive abuses visited upon nature and community by the imperatives of reproduction of the dominant structure of power that the commons have re-emerged. This notion has the ambition to ground a counter-narrative and a political and institutional organization capable of shifting our pattern of development from an extractive and individual into a generative and collective mode. It is not, however, a new notion, as the commons have long constituted one way to organize and govern the relationship between society and nature resources (Sahlins, 1972; Mauss, 2002; De Moor, 2011; Ferrando and Vivero-Pol, 2017). The re-birth of the commons as an alternative, generative vision against neoliberal corporate plunder started with two heroic battles in the Global South. One emerged in 1994 in Chiapas (Mexico) with the Movimiento Zapatista as a reaction to the entry of the North American Free Trade Agreement (NAFTA) into force. It was ignited by the impossibility for local farmers to survive with dignity in the global corporate food system (in its broader sense). The other arose in Cochabamba (Bolivia) in 2001, triggered by the need to defend water against corporate privatization by an American company with the support of the national government. In both cases, the commons were invoked to defend local communities against governments transformed into cronies of global corporate interests. Food and water, components of our very physical existence, have therefore been at the origins of the re-birth of the commons as a strategy of defence and (hopefully) of transformation: defence against ongoing commodification of commons still owned or governed collectively, and transformation to re-invent or design de novo forms to use, steward and share resources important for the community outside the market and state logic. Interestingly, however, while the notion of water as a commons is now widely recognized and has grounded many battles even in the Global North (Barlow and Clarke, 2017; Bieler, 2017), food as a commons has not been a conscious target of political battles and civic claims (Ferrando, 2016); indeed, it has not even been a paradigm of research (Vivero-Pol, 2017a).

Yet, we believe food can also be valued and governed as a commons and that approaching it under this intellectual framework offers important insights into a possible alternative vision coherent with the needs of reproduction of life rather than of capital. This book aims to open that discussion in the belief that we can obtain for food at least some of the (though partial) successes that we have been able to obtain with water.

Valuing food as a commodity is at odds with human history

Capitalism has been thriving and reproducing a troublesome relationship with food and food systems. The contemporary food regime of corporations and financial investors is such that while many eat poorly and badly, others have access to all the food they desire: purchasing power is what separates the two. Moreover, industrial production and global distribution of food are major driving forces in pushing the environment beyond its planetary and ecological boundaries, mortgaging the livelihood of future generations. This scenario is characterized by extreme inequality and power imbalances. At its centre is the idea that food is an object for sale (a commodity) and the food system is nothing but an opportunity to extract private value. In such a context, achieving the universal right to adequate food (a legal entitlement), food and nutrition security (a global public good) or food justice and food sovereignty inevitably appears a long-term vision at best, a utopian goal at worst. It is therefore essential to broaden political imagination: to explore and practice alternative paradigms of food and visions of food systems capable of overcoming the normative, technical, political lock-ins the industrial food system has created (IPES-Food, 2016). The paradigm of food as a commons, as a way to value food and to govern its production and allocation, will unlock our imagination, encouraging us to design other types of policies and legal frameworks for the food system that have been so far disallowed because they were not aligned to the dominant narratives of capitalism (Wright, 2013).

The aim of this book is to investigate the multiple enclosures at the basis of the dominant industrial food regime and to explore how such enclosures could be challenged by re-describing and re-conceptualizing food as a commons. As in many other areas of people's livelihoods, enclosures, plunder and exclusions have occurred through legislation, pricing, patents, discourses and public violence (Mattei and Nader, 2008). As a consequence, the opportunities for the production, transformation and consumption of food as a commons have been marginalized and repressed. The social construction of food as a commodity, in fact, denies its non-economic attributes (as vital fuel for our bodies, as a human right, as a product of Nature, or as an element of our culture) in favour of an exclusive focus on its tradable features, such as its external appearance and packaging, taste, or shelf-life, but first and foremost, its price and calorie content.

Inevitably, this leads to neglect social and relational properties of food, alongside an emphasis on cheap calories and the dismissal of the ecological role of food systems in stewarding biodiversity and nature's inherent connection with society and the organization of the economy (Díaz et al., 2018; Moore, 2017).

From the scientific and industrial revolutions of the 18th century to the present day, capitalist thinking and its practices have increasingly transformed food – an essential element of life – into a private, mono-dimensional commodity for mass consumption in a globalized market. Over the last decade, however, there has been an increased recognition that this view of food as a commodity, as a social construct, can be challenged. Food can be re-conceptualized differently: it can be valued and governed as a commons, and it is constructed as such in a range of initiatives in all world regions.

A subversion of the food paradigm that sustains the current mainstream food system, when it happens, will shed light on the conflict between the hegemonic economic epistemology (an epistemology in which the commons lead to the "tragedy" of overexploitation, and in which private property and allocation through market mechanisms predominate) and the non-dominant alternatives (political, historical, legal and radical–activist approaches to the commons), which have been gaining legitimacy in recent decades.

The framing of food as a commodity, the production of which responds to price signals and the allocation of which depends on purchasing power, is increasingly being challenged. Alternative framings have been proposed, often implicitly, by a range of grassroots movements and customary indigenous traditions all over the world. Small-scale farmers, peasants and fisherfolk, farmworkers, conscious eaters and regulators, food security activists, academics and human rights advocates, among others, are developing alternative food paradigms in multiple loci (urban and rural areas in the Global South and North) by defending the public nature of many food-producing resources such as seeds, water, land and agricultural knowledge. The de-commodification and commoning of food (and of the whole food system as the broader objective) will open up a transition towards a plurality of new food regimes. As a result, features other than exchange value shall be given greater recognition: food, under these competing paradigms, is re-conceptualized as essential to the satisfaction of a human need (nutrition + culture + community), with justice, democracy and the inherent recognition of the ecological limits and moral obligations as pivotal elements. Food systems will emerge in various forms, and the individual freedom to extract nature and maximize profits will be deemed incompatible with the common good of people and the planet (Patel and Moore, 2018).

This book aims to enrich the debate on food as a commons between and within disciplines, niches of resistance (transition towns, food sovereignty, de-growth, open knowledge, commons) and organizational scales (local food systems and national policies, South–South collaborations and international governance and agreements). It asks two questions: What would food policies look like, once we shift to the paradigm of food as a commons? And how do we get there?

The thriving commons as a civic counter-movement to the global food crises

The commons are back … if they were ever gone. The multiple crises the world has faced in the last decades have prompted scholars, policy makers and activists to seek solutions that enable us to live a satisfying, fair and sustainable life within planetary boundaries. The reappearance of the commons represents a promising transformative pathway to replace the neoliberal model. Historically, the commons have been associated with a record of resilience, collective governance and sustainability. They provide an inspirational narrative based on solid moral grounds. Commons thinking offers a counter-claim to the idea that society is and should be composed of atomized individuals, acting as rational agents seeking to maximize their individual utility and competing against other individuals in order to thrive as a separate individual rather than as a member of an ecological collectivity.[1] However, the narrative of the commons was marginalized in the 20th century by the ascent of possessive individualism (Macpherson, 1971), rational choice (Schelling, 1984), the diffusion of the individualistic ethos and domination proper of colonialism, the objectification of nature, social Darwinism (Leonard, 2009) and the famous fable of the "tragedy of the commons" (Hardin, 1968). Unlike these, the commons discourse recognizes that people shall live their lives as aware individuals deeply embedded in, and not acting against, social relationships and the environment. Moreover, individuals' active participation is essential to realizing collective and personal goals, moving away from a purely individual rights-based, market-based and private-property worldview.

From a historical perspective, treating food as a pure commodity devoid of other important dimensions is an anomaly. For centuries, food was cultivated in common and considered a mythological or sacred item; it was allocated according to need, rather than on the basis of the ability to pay. In different times and geographies, food shaped civilizations and socio-economic transformations. Often, it was considered so important in terms of culture, religion and survival that its production and distribution were governed by non-market rules; production, distribution and consumption were collective activities, done in common rather than alone or within the nuclear family (Diamond, 1997; Fraser and Rimas, 2011; Montanori, 2006). Food-producing commons were ubiquitous in the world, and history records are full of commons-based food production systems ranging from the early Babylonian Empire (Renger, 1995), ancient India (Gopal, 1961), the Roman Empire (Jones, 1986), Medieval Europe (Linebaugh, 2008) and early modern Japan (Brown, 2011). Food was considered a commons as well as a public tool, with diverse and certainly evolving proprietary schemes ranging from a private good given for free to idle Temple priests, a resource levied by kings and feudal lords as well as a public tool used by Roman emperors, Mayan dignitaries and the British government to prevent disturbances and appease the revolting crowds (Jones, 1986; Schuftan, 2015; Kent, 2015). Food always carried many dimensions, and, except in recent history, it was never reduced to a tradeable priced good.

However, in the Western context, the idea of the commons was gradually abandoned: the enclosures movement, which started in England in the 16th century, and the abolition of the poor laws by the Poor Law Amendment Act in 1834 symbolize this shift (Polanyi, [1944] 2001). The commons re-entered the political and social agenda only in the 1980s, as a countermovement to – as society's self-defense against – the commodification process that accelerated in the last quarter of the 20th century (Appadurai, 1986). For decades, the commons have been dismissed as a failed system of governance and resource management (Bloemen and Hammerstein, 2015). They have now been gradually rehabilitated in the legal, political and economic domains, especially in the environmental and knowledge realms (Benkler, 2013; Capra and Mattei, 2015). Today, there is a growing recognition that the hegemonic market–state duet, with their capitalist system and individualistic ethos, is inadequate to tackle the global and multiple disruptions that living beings and the planet confront on a daily basis.

All over the world, socio-economic imaginations are regaining ground as alternative narratives and praxis to the hegemonic neoliberal version of capitalism (e.g., happiness, de-growth, *buen vivir*, resilience, transition, sharing economy, peer-production). Moreover, innovative commons-based initiatives are mushrooming, with examples ranging from the local level (e.g., the maintenance of communal forests owned by parishes in Galicia villages), to the national level (e.g., the path-breaking initiative promoted by the government of Ecuador to collectively design public policies that can support knowledge commons [Vila-Viñas and Barandiaran, 2015]), to the regional level (e.g., the first European Citizens' Initiative, which demanded that water be treated as a public good and commons [Bieler, 2017] or the European commoners establishing a European Commons Assembly). Some commoners organize to defend old commons from current modes of enclosure and commodification (e.g., land grabbing or privatization of municipal water services), while others are inventing new commons in the knowledge domain (Creative Commons Licenses, online services and digital content) and in the cities (food councils, commoning disused public squares and abandoned buildings, sharing meals with neighbours and the broader community). In all these cases, the theory and praxis of the commons (Quarta and Ferrando, 2015) operate as counter-hegemonic or alter-hegemonic, gathering around a diffused dissent and the desire for new forms of imagination (Vivero-Pol, 2017a).

The multiplicity of commons: different vocabularies, understandings and practices

Before embarking on the reading of this volume, it is important to highlight that the commons continue to have different readings (Mattei, 2013), each with its different trajectories and implications. Legal, political, economic, cultural and ecological approaches talk about commons and inform knowledge and ideologies, which are then reflected in the creation of different schools of thought and vocabularies that examine, interpret and influence our understanding of the nature of the commons. As resources that are important for human beings, commons have "multiple personalities" (Wall, 2014) and therefore multiple phenomenologies (Mattei, 2012) and vocabularies to describe them. This is not an anomaly but rather a characteristic of societies already highlighted, among other theories, by legal pluralism (Engle-Merry, 1988) and institutional diversity (Ostrom, 1990). The plurality of definitions of the commons in the public and academic discourses renders it difficult to reach a consensus on which resources, situations and policy decisions are deemed to be considered as commons or for the common good. This situation affects food directly, with its consideration as a commons strongly contested in academic and political domains (Vivero-Pol, 2017b). One source of discrepancy of understanding the commons stems from the fact that collective ethical notions of what a commons is according to different communities (commons as a social construct) have developed in parallel with theoretical approaches proposed by influential thinkers (in particular among economists of the institutionalist branch) and with political decisions made by elites (experimenting with a political approach to commons).

Different academic disciplines have addressed the commons by relying on the epistemologies (cognitive tools and accumulated knowledge) that characterize each, be that economics, law, history or political science. These epistemologies have been blended with dominant ideologies and politics, as academia is often influenced by the ruling elites (Wallerstein, 2016). Other versions of the commons emerge from grassroots activists, the "commoners" who develop a range of practices questioning, mostly implicitly, the dominant understanding of food as a commodity. These varied approaches to a complex, place-based and multi-faceted theme have shaped the different meanings and implications of the commons that we have at present. These understandings have evolved into an interdisciplinary approach (Laerhoven and Berge, 2011) that now seeks to expand beyond the academic walls to incorporate the meanings of commoners, combining different sources of knowledge in a transdisciplinary perspective (Dedeurwaerdere, 2014). However, various definitions of the commons still co-exist: the debate today is not only between an individualistic approach, in which the allocation of goods occurs through a combination of the state and the market, and an approach that makes room for the commons; it is also a debate within the community of scholars and activists who rely on the commons as to how to define the commons, how to govern them and which political implications follow from this counter-hegemonic paradigm (Benkler, 2013; Hess and Ostrom, 2007).

Therefore, with such a rich array of proponents and practitioners, the academic theory of the commons cannot be considered uniform, coherent or consolidated. However, diversity should not be perceived as a threat. On the contrary, the existence of colliding theoretical approaches underlines tensions and fault-lines, revealing the different epistemic regards to resources and practices that are essential to human societies and individuals.

The different meanings of the commons to economists and policy makers

Commons as public goods

In its most widespread and general meaning, a common good describes a specific resource that is shared with and benefits all or most members of a given community. Commons, owned in common or shared within the community, satisfy needs that go unmet by either markets or institutions. However, in the economic and political parlance, commons are identified (and named) as public goods in some cases or as common-pool resources in others. On the one hand, political scholars define public goods as those material and immaterial goods deemed to be desirable by the public (Hampson and Hay, 2004) because of the utilities they generate in favour of the society (Ver Eecke, 1999). Although their nature as public good does not automatically imply their open accessibility to all, goods like water, pollination, soil fertility and sunlight are often considered commons and public goods as they are fundamental to the idea that life is not for sale (Shiva, 2005; Patel, 2007). On the other hand, the notion of commons (or common-pool resources, as termed by Elinor Ostrom) is different from public goods in neoclassical economics parlance. The term commons is often utilized to define a large set of human and natural systems that is

> sufficiently large that it is difficult, but not impossible, to define recognized users and exclude other users altogether. Further, each person's use of such resources subtracts benefits that others might enjoy. Fisheries and forests are two common-pool resources that are of great concern in this era of major ecological challenges. Others include irrigation systems, groundwater basins, pastures and grazing systems, lakes, oceans, and the Earth's atmosphere.
>
> *(Ostrom, 2009)*

Throughout the world, natural fisheries, common grazing pastures, forests and biodiversity are examples of open-access resources prone to the tragedy of the commons, a fable that was proposed by Garrett Hardin (1968) and gained ample support at the end of 20th century. However, Hardin's generalized postulates were not based on sufficient evidence. When such evidences were gathered and analyzed by Elinor Ostrom (1990), the weaknesses and inconsistencies of the tragedy were exposed and debunked, leading Hardin to recognize the limits of his approach (Hardin, 1994). Unfortunately, however, Hardin's tragedy, with all its limits of reductionism, proved to be of exceptional predictive power if the commons idea is opened enough to include our whole world. Global corporations today roam it to satisfy the unlimited short-term profit motive that is codified in their DNA, the corporate charter. They act exactly like the self-interested farmers in Hardin's parable, enjoying a space of *no law* because the national legal systems are (captured and) ill-equipped to limit corporate power and its exceptional volatility. Land grabbing, water privatization and many other issues directly connected to food plunder cannot be understood outside of such clear perception (Ferrando, 2017). Ostrom's critique of Hardin, by theoretically denying the tragedy, may in practice shield the corporate entities that as a matter of fact cause it in the global arena. This is perhaps the main reason why the commons should not be approached as a positive object that can be defined ontologically. The political dimension cannot be overlooked. This is why the economic approach, as that of other social sciences, provides little understanding and no political agency. A true genuinely phenomenological social theory capable of developing a *total critique* is needed (Unger and Crawford, 1996).

Although the literature on public goods and common-pool resources is extensive and diverse, such literature typically relies on the standard economic definition of public goods, which is based on the two ontological characteristics of non-rivalry and non-excludability (Samuelson, 1954; Buchanan, 1965; Musgrave, 1959). A public good is a good that is both non-excludable and non-rivalrous, in that individuals cannot be effectively excluded from its use and use by one individual does not reduce availability to others. A pure public good is an extreme case of a positive externality. There is, in general, no profit motivation to lead private firms to supply a socially efficient quantity of such goods (in other terms, markets undersupply such goods). In many cases, markets for public goods will not even exist (i.e., clean air). Private goods, however, cannot be enjoyed simultaneously by many people, and individuals can be prevented from using them either by physical means or by property rights (including intellectual property rights such as patents). Pure public goods provided by the government are usually financed from tax revenues. Different funding options result in different economic outcomes in terms of the distribution of the cost burden between taxpayers and users of the good or service.

Commons: a political construct to govern resources or to radically transform the system?

Because of their non-excludable character, public goods result in a collective action problem: all those who benefit from the provision of a local public good find it costly to contribute and would prefer others to pay for the good instead. If everyone follows the selfish dominant strategy, hoping to freely ride on the contributions of others, then the good is not provided or is under-provided. Yet, everyone would be better off if everyone contributed. Institutions allow for the overcoming of such collective action problems by imposing compliance with formal or informal rules with the aim of producing socially optimal outcomes (Ostrom, 2005). Another problem that has gained particular relevance in the recent period is that "public" no longer means the communities who manage their local resources but rather the central governing authority that controls these resources. In theory, public still means people; in practice, public means government decoupled from the people's social/ecological rights to their common goods (Quilligan, 2012).

Very often, public goods and commons are used as interchangeable terms, the former mostly used in the economic and political realms and the latter predominant in the social and environmental sciences domains. However, in both economic and political terms, food could be considered as an essential resource that requires management as a social mandate in order to guarantee the right to food for all: due to its vital role in allowing people to lead active and healthy lives, its access cannot be made conditional on purchasing power (De Schutter and Pistor, 2015). As such, considering and governing food as a commons would simply mean recognizing food for what it is. Some authors, like Giacomo Pettenati et al. (this volume), have also gone as far as claiming that the uniqueness of food is such that the whole food system should be re-imagined as a commons. Others, like Cristian Timmermann (this volume), have also suggested the condition of food and nutrition security (FNS) should be considered a global commons or a global public good as it is beneficial for the community, the nations and the planet in general. FNS as a state of affairs is not rivalrous (my own food and nutrition security does not prevent you from having yours), but it is definitely excludable (as we can see at present, with over 800 million people with no food security at all), although ethically abominable.

The transformative and imaginative potential of the commons has been synthetized by the idea of the commons as a political tool and horizon. Such understanding of the commons is currently adopted by two different intellectual streams, which differ from each other on the basis of the primary subject of analysis: the resource or the governing community.

Those who focus on the properties of the resource recognize that rivalry and excludability can be molded by societal norms and technology but at the same time accept that commons are defined by these two features (Kaul et al., 1999; Kaul et al., 2003). Actually, it is not rare to find scholars using the terms public goods and commons interchangeably, especially when dealing with global public goods and global commons (Buck, 1998; Brousseau et al., 2012). For this stream, global commons are resources that provide benefits that are strongly universal in terms of countries, or whose benefits extend to all population groups and generations (Hjorth Agerskov, 2005); they have been the building blocks of different civilizations (Wolf, 2012). Examples range from clean air, weather data collection or internet, to stable currencies or standardized norms (e.g., ISO system). This understanding of global commons requires little more than forms of intergovernmental cooperation, voluntary guidelines to corporate actors and minor adjustments in policies and international law. Moreover, the transformative power of collective arrangements by people or communities outside the market and state duopoly is not contemplated here. Resource-based commons can co-exist with neoliberal markets, given their focus on non-appropriable resources (those termed as market failures) and the benefits they provide. That explains why global commons–global public goods have been increasingly embraced by the "institutional mainstream", as they can easily fit the dominant narrative of capitalism (Birdsall and Diofasi, 2015). In the European Commission, global public goods are now the subject of a thematic programme of the Development Cooperation Instrument.

For scholars and activists in the second stream, commons are not about the nature of a good but rather the way in which societies organize around essential goods that are produced, reproduced and managed collectively (Workshop on Governing Knowledge Commons, 2014). By commons, they do not mean things (rivers, forests, land, etc.), information or knowledge content or places defined by their material properties. They mean a way of doing things together in order to strengthen democratic self-determination. In this view, commons are self-regulated social arrangements to govern material and immaterial resources deemed essential for all and are place- and time-restricted and vary according to different societies, circumstances and technological developments. Commons can be distinguished from non-commons by the institutionalized sharing of resources among members of a community (Madison et al., 2010), what is often known as "commoning". It is "commoning" together that confers on a material, or non-material, common resource its commons consideration (Dardot and Laval, 2014). Commoning is about human/nature relationships (Bollier and Helfrich, 2015), and therefore the human-made consideration of what a commons is requires a specification for each place in our own time (Friedmann, 2015).

Commoning, as a form of governance, differs from the market allocation mechanism based on individual profit maximization and state governance based on command and control. It demands new institutions, goal setting and forms of interaction, thereby forming the bedrock to support a new moral narrative, a new transition pathway, a new economic model and a new relationship with nature and the planet Earth. Commons are a system of decision-making, collective ownership and value-based purposes that challenge the for-profit ethos of the market and the state's pretense to a monopoly on the definition of the common good and to acting "parens patriae" in the name of the whole polity. Commons are not about maximizing individual utilities, selfish individualism or legitimizing the use of force but rather collective decisions, institutions, property and shared goals to maximize everybody's wellbeing.

Transformative-wise, those two streams present diverging characteristics: the resource-based scholars see the commons as self-regulated forms of governance that can co-exist with current forms of free-market and capital accumulation of private-property regimes and absolute sovereign states (e.g., see a critique of the approaches defended by neo-institutionalists or neo-hardiniens in Caffentzis, 2012). Conversely, the governance-based proponents conceive of the commons as a transformative narrative, rooted in history but innovative enough to challenge the hegemonic duopoly formed by the neoliberal market and the state (Dardot and Laval, 2014; Wall, 2014; Capra and Mattei, 2015) This stream directly collides with the basic foundations of capitalism, such as the primacy of individual property over other rights, the sovereignty of the individual consumer over collective wellbeing, the lack of limits to resource accumulation and competition as the main driver of progress rather than cooperation (McCarthy, 2005; Hardt and Negri, 2009; Jeffrey et al., 2012; Verhaegen, 2015). Commons hold different values, goals, narratives, ethical principles and functioning from the capitalist market. From the very moment that we accept that the community has an instituting power to create a commons (resource, property regime, governing institution and purpose), we accept that the community is bestowed with legal and political powers to regulate the resources important to it, making commoning transformational and counter-hegemonic, since the state aims to retain those instituting powers to issue policies and enact laws and the market aims to retain its supremacy to allocate and govern scarce resources.

The charter to navigate the chapters

This book presents a different normative view of food, as a commons instead of a commodity, based on the recognition of the multi-dimensional nature of food and its essential role for humans, as well as on the praxis carried out by customary food practices, rural or indigenous, and contemporary civic food actions, urban or consumer-driven. The different understandings of food as a commons are place- and time-situated, with meanings and governing institutions created specifically by each polity, and thus there is no one interpretation of that concept. They converge, however, in their refusal to treat food exclusively as a monetized commodity. Far from being concerned about the lack of homogeneity, the editors value those discrepancies positively as there cannot be just one monolithic narrative about the polysemic concepts of food and commons.

With their diversity of approaches and their multiplicity of angles, the various chapters enrich the debate on food as a commons between and within disciplines, niches of resistance (transition towns, food sovereignty, de-growth, open knowledge, commons) and organizational scales (local food, national policies, South–South collaborations and international governance and agreements), exploring the different dimensions that reframe food as a commons and deploying a wide array of practical initiatives in rural and urban settings, in the Global South and the Global North, that actually materialize this narrative.

It is not our intention to provide an academic definition of what we consider "food as a commons" (although some authors in this book have already provided their own understandings). In that sense, we defend those understandings of "food as a commons" that are related to food democracy, food justice, food sovereignty or right to food practices. Said otherwise, we believe that valuing food as a commons informs the idea that communities should invent new ways of guaranteeing access to adequate and preferred food for all by setting up social innovations of various sorts, "de-commodifying" food and creating in the process a sort of "socio-diversity" of food alternatives that create multiple food systems that value food differently. And yet, all those alternatives oppose and deny the mono-dimensional valuation of food as a for-profit commodity.

For us, it is essential to understand that the concept of commons is socially and environmentally relational and cannot thus be understood without the particular value-based relations between the community and nature and within the community itself (Bollier, 2016; Verhaegen, 2015). Commons encompass networking, bond-creation, social learning among citizens, empowerment, caring and emancipatory meanings through community praxis. Actually, as historian Peter Linebaugh (2008) said, the concept of commons is best understood as a verb, and commoning can be understood as a means to rediscover the embeddedness of the individual in society and nature (Clausen, 2016). As a matter of fact, people, communities, activists, scholars and practitioners all over the world engage with commons on a daily basis. They live in both urban and rural settings and they protect, produce and imagine conceptions of the world that go beyond the dominant paradigm of privatization and exclusion (Walljasper, 2010). In that sense, we agree that "each commons is also somebody else's commons" (Shiva, 2005) and that in the web of life what is connected to a certain community is always connected to others (both human and non-human) beyond that community.

The choice of the chapters and the authors was not an easy task. It was certainly influenced by our networks and positioning as academics from the Global North. For this reason, there is no pretension of exhaustivity, but rather the desire to see analogous projects thriving elsewhere. Furthermore, chapters do not represent all the existing debates around food and food systems as a commons, and they are inherently contextual and inspired by the histories and experiences of their authors. We are aware that much more can and must be said about the intellectual, practical and methodological shift that is brought by the de-commodification of food. We hope that this collection can help to open up spaces and carve cracks in the mainstream, presenting other ways of engaging with food and food systems.

Rebranding food and alternative narratives of transition

The first part of the book sets the stage. Its five chapters directly challenge the commodity-based nature of the mainstream narrative around food and food systems and invite the readers to imagine alternative scenarios. Here, the authors explore different theoretical approaches to normative views of food, as a commons or as a public good, that reject the absolute commodification of food, understood as the hegemonic cultural narrative that impinges the mainstream food system and the productivist paradigm. Those approaches are based on the multiple dimensions of food; the non-Christian cosmologies; the de-commodification of food by also de-commodifying the components that produce that food; the open-source, peer-to-peer ethos and the sharing economy; and the emergent political construction of global public goods.

In the opening piece, José Luis Vivero-Pol departs from the multiple understandings of food to underline the reductionism resulting from the consideration of food as a commodity: such a framing, he argues, obscures other non-economic dimensions of food quite relevant to humans. For him, it is not enough to say that food is not a commodity, but it is essential to discuss its role as life enabler, natural resource, human right, cultural determinant, tradeable good and public good. All these dimensions must be taken into consideration if we are to radically shift the terms of the debate around food as a commons, but none of them is visible when we accept the mono-dimensional valuation of food as a commodity.

In the second contribution, Giacomo Pettenati, Alessia Toldo and Tomaso Ferrando engage in a dialogue with the idea of food as a commons presented in the introductory chapter but offer an additional provocative twist. In their opinion, it is not enough to focus on food as the product of the food system. On the contrary, the de-commodifying power of the commons must redesign the entirety of the food system and, as such, redefine each single element that composes it. In their eyes, food cannot be dissociated from the deeper and broader socio-economic–ecological food system that generates it. Therefore, land, seeds, gender, energy, labour, landscape, the convivial act of eating, food waste and all other components of the food system must be re-thought, re-imagined and practiced according to the radical and ecological paradigm of commoning and the commons. Otherwise, no real transformation can be achieved.

On a similar line, Marina Chang's chapter refines the idea of food system as a commons and enriches it with insights from critical feminism and non-Western traditions. In her chapter, she constructs a holistic, interconnected and intersectional idea of care as the core of growing a commons food regime in order to create synergistic outcomes in a world held together by an array of disciplines, organizations, institutions, movements and forms of discursive power, and at a multitude of sites across the social domain. Growing a care-based commons food regime, she concludes, is like entering a new epoch of history: the pattern is not written, but we make history by living, experiencing, generating, reproducing and protecting the food commons towards ecological and just food systems.

In their chapter, Alex Pazaitis and Michel Bauwens converge on food through their thinking about prefigurative social order, technological innovation and commons-based peer production. In the context of a productive civil society of contributors with an ethical market economy and an enabling partner state, they claim, a set of policies that target the empowerment of social production may lead to an open-source agricultural revolution. Through the construction of an integrated ecosystem and the enactment of specific policies that favour the transition, the different dynamics of Commons-Based Peer Production and the emerging political economy could thus be brought together and facilitate the construction of a commons-based sustainable agricultural system. Contrary to the mainstream food system in which resource accumulation, heavy subsidies for unsustainable and unhealthy practices and exploitation for profit without including the true account of food becomes the norm, a commons-based food system revolves around collective governance, rational utilization of natural resources (considering the livelihood of future generations) and a fair distribution of revenues and food products.

In the last contribution of the first section, Cristian Timmermann closes the circle of narratives of transition by focusing on food security as one of the most debated and – often – abused concepts in the domain of food systems studies. For Timmermann, food security brings a number of benefits to humanity from which nobody can be excluded and which can be simultaneously enjoyed by all. As such, an innovative understanding of food as a commons must be accompanied by an innovative understanding of food security as a public good that can be deployed to assess policies and decisions affecting food production, distribution and access. The author offers a five-fold theory of food security as a public good based on normative rationale and political implications, unfolding one of the multiple dimensions of food (as posited by Vivero-Pol in this volume). He also highlights the advantages of a shifting paradigm with regards to not only food but also the broader intellectual and policy framework.

Exploring the multiple dimensions of food

The second part of the book explores the multiple dimensions of food and how they have been constructed through continuous interaction with and clashes between nature, authority, market, history and communities. Recasting food as a commons enables us to better value and protect the multi-dimensionality of food and thereby to reverse the mono-dimensional approach to food as a commodity that still prevails. The various dimensions of food explored in these chapters in no way preclude or restrain other dimensions of food that could go beyond the ones presented here. Actually, Cristina Tirado (this volume) already proposes a seventh dimension of food as a medicine to be added to the six dimensions mentioned by Jose Luis Vivero-Pol (this volume).

In Chapter 7, John O'Neill approaches these interactions through the lenses of the conflict between conceptions of food as a vital human need and food as a commodity. In response to the consolidation of the "new" moral economy of the market and the paradigm of food as a commodity, egalitarian forms of mutual aid were developed and grounded in the acknowledgement of mutual dependence and common neediness. He explores how the first theorists of the market economy obscured the claims of need and replaced mutual dependence with individual competition. Today, although often invisible, the practices of mutual aid in working-class communities and the arguments for universal social protection remind us of the possibility of other readings of food that are rooted in the acknowledgment of the vulnerability that characterizes states of dependency as those that every human has with regard to food: we all need to eat food every day.

In Chapter 8, George Kent infuses his studies on community-based food systems with the notion of food as a commons and highlights the benefits that can be derived if we organize communities in ways that facilitate positive social interaction, minimize exploitation and indifference, and encourage caring for the others, whether your relatives, neighbours or more distant humans. By setting up community-level food projects and treating food as a commons, he claims, food systems can facilitate people's working and playing together and, in that way, support their caring about one another's wellbeing. In a world made up of strong local communities with strong local food systems, we can grow a global food system that works well for both living beings and the planet. His approach is certainly bottom-up, departing from local communities, and then networking with other similar caring niches. However, this can only occur once we realize that the food system is not a terrarium that can be objectified and studied but rather a complex set of socio-ecological relations in permanent flux that shapes communities and the space around them, at the same time that it is shaped in turn by these communities.

Departing from the recent initiatives of infant and young child feeding in emergencies (IYCF-E) and the SafelyFed scheme of communal support for breastfeeding mothers in situations of need, Penny Van Esterik offers in Chapter 9 a reflection on food as a cultural core. In a society that tends to donate industrial infant formula, purchases breastmilk for profit and proposes individualistic solutions to infant food security, she claims, the creation of collective spaces for mothers and the satisfaction of their needs represents a paradigm change that has significant implications on both society and individuals. More importantly, the discourse of food as a commodity makes culture in the global food system invisible and devalues nurturing practices such as postpartum care, home cooking, regional food preservation techniques, gardening, food sharing through feasting and commensality. Whatever has a value but is not priced by the market gets obscured. On the contrary, food and food systems as a commons make culture and diversity visible, away from standardization and homogenization. Van Esterik makes a call for ethnographies of community-based food commons, which would make visible how the commons work in different cultural settings and the link between food and societies.

Finally, Noah Zerbe's contribution in Chapter 10 provides the reader with a genealogy of the idea of food as a commodity, another food dimension worth exploring because it became hegemonic in the global food system of the 21st century. In order to better understand the need for transition and where the possibilities lie, Zerbe traces the commodification of food in political and economic terms from the colonial food regime, through the rise of the United States, and then to the consolidation of the current neoliberal food regime. Through a combination of legal, political and economic elements, he shows how the strengthening and global expansion of neoliberal capitalism, with its associated narratives of enclosing the commons, absolute proprietary rights, individualism and the moral supremacy of market rules over other allocation mechanisms, has fundamentally modified the global food regime, resulting in the transformation of food from a vital component of life into an instrument for speculative investment and profit maximization. In the industrial, neoliberal food system, food is produced to earn profit and not to feed people adequately. It is only by knowing the premises and processes that shaped the narrative behind the dominant food regime, he claims, that alternative imaginations and new forms of resistance can be organized.

Food-related elements considered as commons

Policy makers and academics are moving from the stringent and binary division of the world into public and private goods to a looser but more practical definition of the circumstances that take into consideration utility rather than ownership, as highlighted by the example of the so-called global commons, which would remain undersupplied in the absence of robust cooperation mechanisms. This move is nothing but a reflection of the multiple experiences on the ground by grassroots organizations, civic collective actions and customary societies that value food in its multiple dimensions and not just based on its market price. Regarding food and its system of production, some material and non-material elements are already considered, although only to a certain extent and in certain contexts, to fall beyond the public/private division and are associated with the ideas of commons, while the status of others is contested (genetic resources, wild foods and water) or generally regarded as private goods (agro-chemical inputs, labour, etc.). This section presents immaterial knowledge commons (traditional agricultural knowledge, public science and gastronomy) that are considered and practiced as a commons in current food systems. Moreover, two material food producing inputs, the normative valuation of which is quite contested by the neoliberal hegemonic narrative, namely genetic resources and water, are also discussed in detail, with cases studies on South Africa, Germany and the International Treaty on Plant Genetic Resources for Food and Agriculture. The aim of this part is to contribute to an expansive understanding of food as a commons that departs from the reductionist idea of food as an object and connects multiple layers and scales.

The first chapter in this part is authored by Victoria Reyes-García, Petra Benyei and Laura Calvet-Mir, three experts of traditional agricultural knowledge (TAK). Their contribution engages with the idea that TAK can be governed as a commons. They understand commons as resources used by a group of people who have self-designed a set of rules to manage the social dilemmas derived from their collective use. Knowledge commons in this case illustrate well the political construction of commons, regardless of the nature of the resource, by people's instituting power. To illustrate the governance of TAK under the commons framework, they present two case studies in which TAK is shared by communities of users who operate at different scales, local and global (through a web-based platform). Valuing TAK as a commons, they conclude, is not just an intellectual exercise but a political stand against the commodification of knowledge by close intellectual property rights (e.g., seed patents).

Chapter 12, by Molly Anderson, further explores the links between food, knowledge and commons. She challenges the ongoing privatization of food and agriculture scientific knowledge, highlighting the fact that the private sector has been assuming a greater proportion of research funding and, as a consequence, is taking advantage of the strengthening of intellectual property rights to recoup its investments. The chapter explores those mechanisms as ways to commodify knowledge. These trends, she claims, are dangerous because they limit the quality and scope of scientific knowledge about food and agriculture, which not only rests upon millennia of uncompensated public participation but also helps the public to adapt to changing environmental conditions, caused in large part by private sector activities and externalization of costs. However, she concludes, these trends are not inevitable, and shifts in public policies and investment can build on existing models of knowledge commons to allow scientific knowledge of food and agriculture to be recognized and governed as a global public good.

A third food-related element discussed in this section is gastronomy, as the way in which food is combined and presented as an object of aesthetic and culinary consumption. In light of the increased spectacularization of food, Christian Barrère posits in Chapter 13 that modern Western societies present themselves as democratic and, along those lines, pretend to export worldwide their model of gastronomy, even in countries that have mainly been characterized by very different gastronomic trajectories. However, the combination between gastronomy and commodification makes contemporary highly marketed gastronomy anything but democratic. On the contrary, it is based on an aristocratic framework that under-values popular gastronomies and celebrates sophistication of recipes, scarcity and high value of foodstuffs, richness of setting, etc. It is thus time to imagine a new pathway for multiple gastronomies that breaks with joint market–elitist gastronomy and recognize the popular, open-knowledge and shared bases of gastronomy and cuisine. A possible solution, Barrère concludes, may reside in the mix of recipes and cultures that accompanies multi-culturalism and cross-boundaries dialogues. Circulation and coexistence of popular gastronomies, as much as the people who create them, become therefore the pillars on which to build a new model of gastronomy, more democratic, ecological and pluralist.

In Chapter 14, Christine Frison and Brendan Coolsaet enrich the conversation with a discussion of the possibility of governing plant and animal genetic resources for food and agriculture as commons. With the help of two case studies, the Global Seed Commons established under the International Treaty on Plant Genetic Resources for Food and Agriculture and the reintroduction and "commonification" of a traditional pig breed by a local community enterprise in Schwäbisch Hall, Germany, the authors conclude that innovative legal frameworks and governance arrangements inspired by the philosophy of the commons can facilitate access to and sharing of genetic resources for food and agriculture, hence helping to ensure the transition towards more ecological and just agri-food systems.

With their chapter on water, food and climate commoning in South Africa, Patrick Bond and Mary Galvin push the reader to think about food in close connection with water, climate change and bottom-up forms of organization, tensions and resistance. Using the case of South Africa's most deprived urban areas as an example, the authors show that commoning is not simply a matter of technicist collective resource management but rather a political ideology in which socio-ecological contradictions inevitably emerge. In particular, the illegal reconnection of water pipes by poor households and the support to those unable to pay for water that took place in South Africa during the period of the most intense drought, combined with pressure to commercialize water resources and its accompanying social contestation, lead them to reflect on the strong potential for commoning as a catalyst of self-regulated collective action, social contestation and the making of new rules from bottom up.

Commoning from below: current examples of commons-based food systems

Although the almost complete commodification of food has pervaded most national food systems and the global dynamics, there are still numerous examples where the underlying narrative about food is not based in its commodity properties or the value-in-exchange only. Those examples range from customary indigenous food systems that are resisting the privatization waves of the globalizing neoliberal doctrine to the contemporary civic movements that are trying to regain control of decision-making in local, urban and regional food systems. In this book, we have called those examples "commoning from below", i.e., contemporary examples of food systems that are based on a non-commodified understanding of food. These national examples prove the existence of narratives of food transition other than the productivist discourse of commodified food, and how these narratives are being constructed and revised by a dialectical process between governmental policies and civic collective actions. The examples from Cuba, Canada, Ireland and Hungary show that, like any other social process, this commoning from below is not exempted from power tensions, inequalities and flaws. Although limited and at times contradictory, the four experiences reveal that alternative considerations of food are possible and already practiced, although in some cases with less transformative implications than imagined. Nevertheless, all of them share two important features: the valuation of the multiple meanings of food to people and the questioning of the balance of power in the food system, where the market and the state are no longer seen as the two only actors. People organizing themselves to produce, transform and consume food outside of market-driven and state-driven structures emerge as the third pillar of a tricentric food system where healthy and fair food is guaranteed to every human being. Throughout the world, rural and urban communities are constructing and performing forms of social innovation where food is not only an object of consumption but is recognized in its multiple dimensions.

Peter M. Rosset and Valentín Val take the readers to Cuba in Chapter 16. They present the way in which the "campesino a campesino" agro-ecological movement may be strengthened by the adoption of the methodology of and assumptions about food as a commons. They analyze the horizontal, peasant-to-peasant learning and sharing methodology through the lenses of its communal and collective visions of food. Their conclusion, which opens to dialogue and recognition of the common struggles of food sovereignty and "food-as-a-commons" movements, is that a commons-based vision of food and the food systems are more effective at achieving food sovereignty than conventional practices based on more individual and capitalist views of food.

In Chapter 17, Hugo Martorell and Peter Andrée change geography and approach to present the case of the national food policy in Canada. In their account, we discover that networks and coalitions of civil society organizations are actively working towards integrating values of food as a commons and a public good, with a focus on strengthening their role in food governance, from local urban policy councils to national institutions. They thus draw on some of the experiences of the commoning of food governance that have been instituted in different provinces and territories and reason on the opportunities and tensions that emerge when a polycentric and self-organized commons-based governance is combined with the role of public authorities as facilitators. In their conclusions, they propose that a Canadian food policy should build on provincial and territorial food security networks and existing governance arrangements in order to increase the population's access to healthy food. However, scaling these diverse arrangements at a federal level would bring into play ideological and operational tensions and new challenges to be addressed.

Then, we move to Ireland and a different topic in Chapter 18. Tara Kenny and Colin Sage deal with a theme of extreme topicality and relevance for both public and private actors involved in the food system in Europe: the commodification of food surplus as charitable provision. Through the analysis of some initiatives undertaken in Ireland, the authors discuss the implications and hurdles that charitable food provisioning may interpose to the transition towards a commons-based food system. Without dismissing the importance of feeding people and addressing hunger at a time of austerity, the authors highlight the intrinsic inequality and unsustainability that characterize a system based on excesses and volatile solutions to hunger, using the left-overs of an industrialized food system. A radical transition, they conclude, would rather require moving beyond the current two-tiered food system and its schizophrenia. The paradigm of commons and its focus on multi-dimensional, multi-stakeholder, local and resilience-enhancing systems would thus represent an ally in this shift.

In Chapter 19, the final chapter of this part, Bálint Balázs describes the thriving community-based food self-provisioning in Central and Eastern European countries as socially inclusive practices that involve all strata of society and are deeply rooted in customary traditions. Based primarily on bartering and gifting relations between families, relatives and neighbours, these emerging food systems build and strengthen communities, at the same time saving money and empowering households by not just playing the consumer's role but also self-producing part of its food needs. These practices are based on inherited traditions and have become an important non-market source of local food that reflects the principles of sustainability and preferred local gastronomies (two dimensions of food not always valued in monetary terms). The "re-commonification" of food systems in Central and Eastern European countries, Balázs concludes, has a solid foundation and promising future, as it is propelling high proportions of the population along a sustainable pathway towards new food regimes.

Dialogue of alternative narratives of transition

The 2008 and 2011 food price peaks were two important events that positioned food at the very top of political agendas at national and international levels. Concerns about the food supply required to feed a growing population with diminishing natural resources under highly unpredictable climatic conditions have triggered thousands of events, debates, innovative actions and policies aimed at securing more and better food for all. Yet, hunger is still prevalent and obesity is rampant, in both the Global North and the Global South. How to transit from our unsustainable and unfair industrial food system towards a better one for the people and the planet is nowadays a major topic for politicians and citizens alike.

The fifth part of this volume engages with alternative scenarios and imaginations and explores the convergences, current and potential synergies and elements of tension and possible conflict between the food commons narrative and other relevant counter- and alter-hegemonic narratives that currently confront the industrial food system, such as the food sovereignty movement, the urban food initiatives, the anti-land grabbing constituency or the climate and health constituencies, since the multiple crises (i.e., food, climate, biodiversity, health, energy) seem to be strongly interconnected. Since the food system is the most important transformer on Earth, the way we regard food is linked to possible solutions to all other planetary crises. The aim here is to stress the links between competing narratives about food and existing struggles and attempts to imagine just and ecological food systems. The editors' hope is that the vocabulary and imaginary of food as a commons will help strengthen the actions of movements and individuals who are already deploying intellectual and practical tools to challenge the contradictions and socio-environmental injustices of the dominant food system. That is why this dialogue of alternatives of transition is deemed so relevant: only through a convergence of constituencies, recognizing the diversity of approaches but the unicity of goals, can the mainstream food system, which is both unsustainable and unfair, be changed into an alternative system that guarantees food for all within the planet's boundaries. Of all the possible interlocutors, we have chosen three. However, we believe that this volume, as much as the rationale of commons and commoning, must be seen as a continuous and dynamic process that is constantly enriched, redefined and strengthened by dialogues with other collectives and constituencies combating the inequalities of the current dominant industrial food system.

The first dialogue, contained in Chapter 20, is to do with food justice and food sovereignty. There, Eric Holt-Giménez and Ilja van Lammeren engage with the question of whether food as a commons can advance food sovereignty. In their response, the authors recognize that the link between a global call for food commons and the struggle for food sovereignty may seem straightforward. However, they conclude this is true only when they are superficially analyzed and that both concepts are highly complicated on the ground. In their conclusions, they suggest that a nuanced approach to understanding the commons as a contested terrain of struggle is needed to help determine whether and to what extent a food commons as a strategy for food sovereignty can serve not only as a utopic beacon but also as an effective form of transformative resistance. It is thus up to the advocates of food and food systems as a commons to think about the practical and political implications that the paradigmatic shift may produce. As editors, we welcome the invitation and look forward to building collectively a better understanding of the concrete opportunities and limits that lie behind the ideas proposed in this volume, and to engaging with food sovereignty activists and scholars on how to further develop the links between both narratives.

Then, in Chapter 21, Chris Maughan and Tomaso Ferrando look at ongoing struggles for land as a commons in the United Kingdom and Italy to make the case that the fight for food as a commons cannot be detached from the struggle for a de-commodification of all the elements that compose food systems. In this contribution, they explore concrete examples in which the paradigm of the commons has been utilized to support the struggle for land and soil as key components in the creation of ecological and democratic food systems. In their analysis, civil society–led processes that aim to regain land for the collectivity may thus provide important connective tissue between the radical outliers of food commoning and broad-based support for food systems that nourish the collective, rather than enriching the few.

In the third conversation (Chapter 22), Maria Fonte and Ivan Cucco use the aspirational paradigm of the commons to engage with the potential and limits of local food systems. On the one hand, localism can help with transitioning towards a more equitable, ethical and sustainable agro-food system. However, the idea of localism can also support protectionism and neo-ruralist ideologies that reinforce bounded, defensive and spatial strategies. A true emancipation, they claim, can only take place when food ceases to be perceived as a commodity and is understood in its multi-dimensional value, namely natural and economic resource, right, culture and place-based identity. In their reading, food as a commons plays a crucial political role in the construction of a real utopian project to achieve an aspirational and inspirational fair and sustainable food system. Re-thought and re-imagined, food regains its multi-dimensional value and becomes the basis of heterogeneous ecosystems and communities of people and nature, in which social justice and democratic powers may prevail and where a non-capitalist or post-capitalist economy is achievable.

In the final contribution of this part (Chapter 23), Cristina Tirado-von der Pahlen explores how climate change impacts the multiple dimensions of food, proposing a new conceptual health-related dimension to add to the theoretical approach to food dimensions presented in this volume: food can also be valued as a medicine. Moreover, departing from the consequences of climate change effects over human health, nutrition and food security, she highlights the relevant role the industrial food system has in global warming and the obesity pandemic that is ravaging all countries, either in high-income Western nations or the impoverished Global South. The current way of producing and consuming food, including food waste and high meat consumption, is the biggest contributor to greenhouse gas emissions and is also the biggest user of water resources, biodiversity destruction and soil pollution. As the main goal of the global food system shall be to nourish everybody adequately, respecting the limits of natural renewable resources and stewarding the food-producing resources, there is a need to shift the normative consideration of food from an only-for-profit good to a sustainable resource that delivers healthy diets for all without mortgaging the planet. At the end of the chapter, Cristina proposes multiple leverages to transit from the current unsustainable and unhealthy food system towards a food commons system, establishing a dialogue between the most progressive policy and legal ideas from the academic mainstream with the most palatable proposals from the commoners' side.

Un-common exploration of food commons

Through history, with differences in time and space, food has been transformed from the common concern of a community into the individualized concern of human consumers. This is a process of transformation of commons into capital that was already studied by Karl Polanyi. In his book, *The Great Transformation* ([1944], 2001), Polanyi analyzed the commodification of three former commons, namely labour, money and land, and identified "disembedded" capitalism as the root cause of the tensions between markets and democracy. This decoupling generated, through plunder and exploitation, a deeply internalized "extractive" vision of the legal order (Capra and Mattei, 2015; De Schutter and Pistor, 2015). Polanyi then proposed a pathway to "re-embed" markets within society. In the last thirty years, neoliberalism has all but precluded every alternative to a few global extractive giants entrusted with feeding the world with obscene profits and completely anti-ecological practices. Reversing this trend is a matter of survival of life on its planet and must become perhaps the single most important matter of discussion in public conversation in the decades to come.

Yet it is not. The chapters included in this collection are all efforts to think collectively about this fundamental question: How should we change the system in order to transform the excessive accumulation of capital into revamped, sustainable commons (Mattei and Quarta, 2018)? The scholarly community has the duty and responsibility to develop better alternatives to the current disasters and not to consider natural or normal the situation we have inherited. Political choices are open, and we believe the contributors of this collection have offered some important materials to inform them.

Note

1 This idea is epitomized by the Latin sentence "*Homo homini lupus*", created by Plautus (254–184 BC) and rendered popular by Thomas Hobbes (1588–1679). The opposite narrative of cooperation, collectivism and solidarity is, however, defended by authors such as de Waal (2006, 3), Bowles and Gintis (2013) or Kropotkin (1902).

Bibliography

Appadurai, A. 1986. Introduction: Commodities and the politics of value. In Appadurai, A., ed. *The social life of things: Commodities in cultural perspective*, 3–63. Cambridge, UK: Cambridge University Press.

Barlow, M., and Clarke, T. 2017. *Blue gold: The battle against corporate theft of the world's water*. London: Routledge.

Benkler, Y. 2013. Commons and growth: The essential role of open commons in market economies. *University of Chicago Law Revue* 80(3): 1499–1555.

Bieler, A. 2017. Fighting for public water: The first successful European Citizens' Initiative, "Water and sanitation are a human right". *Interface: A Journal for and About Social Movements* 9(1): 300–326.

Birdsall, N., and Diofasi, A. 2015. *Global public goods for development: How much and what for.* CDG Notes. Washington, DC: Centre for Global Development. https://www.cgdev.org/publication/global-public-goods-development-how-much-and-what (accessed on July 15, 2018).

Bloemen, S., and Hammerstein, D. 2015. *The EU and the commons: A commons approach to European knowledge policy*. Commons Network in cooperation with Heinrich Böll Stiftung. Berlin/Brussels. http://commonsnetwork.eu/wp-content/uploads/2015/06/A-Commons-Approach-to-European-Knowledge-Policy.pdf (accessed on July 14, 2018).

Bollier, D. 2016. *State power and commoning: Transcending a problematic relationship.* A Report on a Deep Dive Workshop convened by the Commons Strategies Group in cooperation with the Heinrich Böll Foundation. http://commonsstrategies.org/state-power-commoning-transcending-problematic-relationship/ (accessed on July 15, 2018).

Bollier, D., and Helfrich, S. 2015. Overture. In Bollier, D., and Helfrich, S., eds. *Patterns of commoning*, 18–31. Amherst, MA: Commons Strategy Group and Off the Common Press.

Bowles, S., and Gintis, H. 2013. *A cooperative species: Human reciprocity and its evolution.* Princeton, NJ: Princeton University Press.
Brousseau, E., Dedeurwaerdere, T., and Siebenhuner, B. 2012. *Reflexive governance for global public goods.* Cambridge, MA: MIT press.
Brown, L. R. 2008. *Plan B 3.0: Mobilizing to save civilization (substantially revised).* New York: WW Norton & Company.
Brown, P. C. 2011. *Cultivating commons: Joint ownership of arable land in early modern Japan.* Honolulu: University of Hawaii Press.
Buchanan, J. 1965. An economic theory of clubs. *Economica* 32: 1–14.
Buck, S. J., 1998. *The global commons: An introduction.* Washington, DC: Island Press.
Caffentzis, G. 2012. A tale of two conferences: Globalization, the crisis of neoliberalism and question of the commons. *Borderlands* 11(2). http://www.borderlands.net.au/vol11no2_2012/caffentzis_globalization.pdf (accessed on July 15, 2018).
Capra, F., and Mattei, U. 2015. *The ecology of law: Toward a legal system in tune with nature and community.* Oakland, CA: Berrett-Koehler Publishers.
Clausen, L. T. 2016. Reinventing the commons: How action research can support the renewal of sustainable communities. In Hansen, H. P., Nielsen, B. S., Sriskandarajah, N., and Gunnarsson, E., eds. *Commons, sustainability, democratization: Action research and the basic renewal of society*, 29–52. New York and London: Routledge.
Crutzen, P. J. 2006. The "Anthropocene". In Ehlers, E., and Krafft, T., eds. *Earth system science in the Anthropocene*, 13–18. Berlin; Heidelberg: Springer.
Dardot, P., and Laval, C. 2014. *Commun, essai sur la révolution au XXI° siècle.* Paris: Le Découverte.
De Moor, T. 2011. From common pastures to global commons: A historical perspective on interdisciplinary approaches to commons. *Natures Sciences Sociétés* 19(4): 422–431.
De Schutter, O., and Pistor, K. 2015. Introduction: Towards voice and reflexivity. In Pistor, K., and De Schutter, O., eds. *Governing access to essential resources*, 3–15. New York: Columbia University Press.
De Waal, F. 2006. *Primates and philosophers: How morality evolved.* Princeton, NJ: Princeton University Press.
Dedeurwaerdere, T. 2014. *Sustainability science for strong sustainability.* Cheltenham, UK: Edward Elgar.
Diamond, J. 1997. *Guns, germs and steel: A short history of everybody for the last 13,000 years.* London: Vintage.
Díaz, S., Pascual, U., Stenseke, M., Martín-López, B., Watson, R. T., Molnár, Z., Hill, R., Chan, K. M., Baste, I. A., Brauman, K. A. and Polasky, S. 2018. Assessing nature's contributions to people. *Science* 359(6373): 270–272.
Engle-Merry, S. 1988. Legal pluralism. *Law & Society Review* 22(5): 869–896.
Ferrando, T. 2016. Il Sistema Cibo Bene Comune. In Quarta, A., and Spanò, M., eds. *Beni Comuni 2.0. Contro-egemonia e nuove istituzioni.* Milan: Mimesis.
Ferrando, T. 2017. Land rights at the time of global production: Multi-spatiality and "legal chokeholds". *Business and Human Rights Journal* 2(2): 275–295.
Ferrando, T., and Vivero-Pol, J. L. 2017. Commons and "commoning": A "new" old narrative to enrich the food sovereignty and right to food claims. *Right to Food and Nutrition Watch* 2017: 50–56. https://www.righttofoodandnutrition.org/files/02.rtfanw-2017_eng_17_12_article-5_web_rz.pdf (accessed on July 15, 2018).
Fraser, E. D. G., and Rimas, A. 2011. *Empires of food: Feast, famine and the rise and fall of civilizations.* London: Arrow Books.
Friedmann, H. 2015. Governing land and landscapes: Political ecology of enclosures and commons. *Canadian Food Studies* 2(2): 23–31
Fukuyama, F. 1989. The end of history? *The National Interest* (16): 3–18.
Gopal, L. 1961. Ownership of agricultural land in ancient India. *Journal of the Economic and Social History of the Orient* 4(3): 240–263.
Hampson, F. O., and Hay, J. B. 2004. Review essay: Viva Vox Populi quick check. *Global Governance* 10(2): 247–264.
Hardin, G. 1968. The tragedy of the commons. *Science* 162(3859): 1243–1248
Hardin, G. 1994. The tragedy of the unmanaged commons. *Trends in Ecology and Evolution* 9(5): 199.
Hardt, M., and Negri, A. 2009. *Commonwealth.* Cambridge, MA: Harvard University Press.
Harvey, D. 2007. Neoliberalism as creative destruction. *The Annals of the American Academy of Political and Social Science* 610(1): 21–44.
Hess, C., and Ostrom, E., eds. 2007. *Understanding knowledge as a commons: From theory to practice.* Cambridge, MA: MIT Press.

Hjorth Agerskov, A. 2005. *Global public goods and development: A guide for policy makers*. World Bank seminar series, Global Development Challenges Facing Humanity, May 12, Kobe and Hiroshima Universities. http://siteresources.worldbank.org/EXTABOUTUS/Resources/PublicGoods.pdf (accessed on July 15, 2018).
IPES-Food. 2016. *From uniformity to diversity: A paradigm shift from industrial agriculture to diversified agroecological systems*. Report 2. International Panel of Experts on Sustainable Food systems. http://www.ipes-food.org/images/Reports/UniformityToDiversity_FullReport.pdf (accessed on July 24, 2018).
IUC. 2009. *At the end of the end of history: Global legal standards: part of the solution or part of the problem*. The IUC Independent Policy Report, prepared by a group of lawyers at the International University. https://hal.archives-ouvertes.fr/hal-00404895/document (accessed on July 24, 2018).
Jeffrey, A., McFarlane, C., and Vasudevan, A. 2012. Rethinking enclosure: Space, subjectivity and the commons. *Antipode* 44(4): 1247–1267.
Jones, A. H. M. 1986. *The later Roman Empire, 284–602: A social, economic, and administrative survey*. Baltimore, MD: The Johns Hopkins University Press.
Kaul, I., Grunberg, O., and Stern, M. A. 1999. *Global public goods: International cooperation in the 21st century*. New York: Oxford University Press
Kaul, I., Conceição, P., Le Goulven, K., and Mendoza, R. U., eds. 2003. *Providing global public goods: Managing globalization*. New York: Oxford University Press.
Kent, G. 2015. Food systems, agriculture, society: How to end hunger. *World Nutrition* 6(3): 170–183.
Kropotkin, P. 1902. *Mutual aid: A factor of evolution*. Reprinted in 2009. London: Freedom Press.
Laerhoven, F., and Berge, E. 2011. The 20th anniversary of Elinor Ostrom's governing the commons. *International Journal of the Commons* 5(1): 1–8.
Leonard, T. C. 2009. Origins of the myth of social Darwinism: The ambiguous legacy of Richard Hofstadter's social Darwinism in American thought. *Journal of Economic Behavior and Organization* 71: 37–51.
Linebaugh, P. 2008. *The Magna Carta manifesto: Liberties and commons for all*. Oakland, CA: University of California Press.
Macpherson, C. B. 1971. *The political theory of possessive individualism (Hobbes to Locke)*. Oxford: Clarendon Press.
Madison, M. J., Frischmann, B. M., and Strandburg, K. J. 2010. Constructing commons in the cultural environment. *Cornell Law Review* 95(4): 657–709.
Mattei, U. 2012. First thoughts for a phenomenology of the commons. In Bollier, D., and Helfrich, S., eds. *The wealth of the commons: A world beyond market and state*. Amherst, MA: Levellers Press. http://wealthofthecommons.org/essay/first-thoughts-phenomenology-commons (accessed on July 15, 2018).
Mattei, U. 2013. *Bienes Comunes*. Madrid: Editorial Trotta.
Mattei, U., and Nader, L. 2008. *Plunder: When the rule of law is illegal*. London: John Wiley & Sons.
Mattei, U., and Quarta, A. 2018. *Punto di svolta. Ecologia, tecnologia e diritto privato. Dal capitale ai beni comuni*. Aboca Museum.
Mauss, M. 2002. *The gift: The form and reason for exchange in archaic societies*. London and New York: Routledge.
McCarthy J. 2005. Commons as counterhegemonic projects. *Capitalism, Nature, Socialism* 16(1): 9–24.
Montanori, M. 2006. *Food is culture: Arts and traditions on the table*. New York: Columbia University Press.
Moore, J. W. 2017. The Capitalocene, part I: On the nature and origins of our ecological crisis. *Journal of Peasant Studies* 44(3): 594–630
Musgrave, R. A. 1959. *The theory of public finance*. New York: McGraw-Hill.
Ostrom, E. 1990. *Governing the commons: The evolution of institutions for collective action*. New York: Cambridge University Press.
Ostrom, E. 2005. *Understanding institutional diversity*. Princeton, NJ: Princeton University Press.
Ostrom, E. 2009. *A polycentric approach to climate change*. Policy Research working paper WPS 5095. Washington, DC: World Bank.
Patel, R. 2007. *Stuffed and starved: Markets, power and the hidden battle for the world food system*. Melbourne: Black Inc.
Patel, R., and Moore, J. W. 2018. *A history of the world in seven cheap things: A guide to capitalism, nature, and the future of the planet*. Berkeley, CA: University of California Press.
Polanyi, K. [1944] 2001. *The great transformation: the political and economic origins of our time*. Boston, MA: Beacon Press.
Purdy, J. 2015. *After nature: A politics for the Anthropocene*. Cambridge, MA: Harvard University Press.
Quarta, A., and Ferrando, T. 2015. Italian property outlaws: From the theory of the commons to the praxis of occupation. *Global Jurist* 15(3): 261–290.

Quilligan, J. 2012. Why distinguish common goods from public goods? In Bollier, D., and Helfrich, S., eds. *The wealth of the commons: A world beyond market and state*. Amherst, MA: Levellers Press. http://wealthofthecommons.org/essay/why-distinguish-common-goods-public-goods (accessed on July 15, 2017).
Renger, J. M. 1995. Institutional, communal, and individual ownership or possession of arable land in Ancient Mesopotamia from the end of the fourth to the end of the first millennium BC *Chicago-Kent Law Review* 71(1): Article 11.
Sahlins, M. 1972. *Stone age economics*. Chicago and New York: Aldine-Atherton, Inc.
Samuelson, P. A. 1954. The pure theory of public expenditure. *The Review of Economics and Statistics* 36(4): 387–389.
Schelling, T. C. 1984. Self-command in practice, in policy, and in a theory of rational choice. *The American Economic Review* 74(2): 1–11.
Schuftan C. 2015. Climate, development. Food prices and food wars. *World Nutrition* 6(3): 210–211
Shiva, V. 2005. *Globalization's new wars: Seed, water and life forms*. New Delhi: Women Unlimited.
Unger, R., and Crawford, M. 1996. *Women and gender: A feminist psychology*. New York: McGraw-Hill, 2nd edition.
Ver Eecke, W. 1999. Public goods: An ideal concept. *Journal of Behavioral and Experimental Economics* 28(2): 139–156.
Verhaegen, E. 2015. La forge conceptuelle. Le "commun" comme réinterprétation de la propriété. *Recherches sociologiques et anthropologiques* 46(2): 111–131. http://rsa.revues.org/1547 (accessed on July 15, 2018).
Vila-Viñas, D., and Barandiaran, X. E. eds. 2015. *Flok society: Buen conocer*. Creative Commons BY-SA and GFL. http://book.floksociety.org/ec/ (accessed on July 8, 2018).
Vivero-Pol, J. L. 2017a. Food as commons or commodity? Exploring the links between normative valuations and agency in food transition. *Sustainability* 9(3): 442.
Vivero-Pol, J. L. 2017b. The idea of food as commons or commodity in academia: A systematic review of English scholarly texts. *Journal of Rural Studies* 53: 182–201.
Wall, D. 2014. *The commons in history: Culture, conflict, and ecology*. Cambridge, MA: MIT Press.
Wallerstein, I. 2016. The scholarly mainstream and reality: Are we at a turning point? In Wallerstein, I., ed. *Modern world-system in the longue durée*, 219–228. London: Routledge.
Walljasper, J. 2010. *All that we share: How to save the economy, the environment, the internet, democracy, our communities and everything else that belongs to all of us*. New York: The New Press.
Wolf, M. 2012. The world's hunger for public goods. *Financial Times*, January 24, 2012. https://www.ft.com/content/517e31c8-45bd-11e1-93f1-00144feabdc0 (accessed on July 15, 2017).
Workshop on Governing Knowledge Commons. 2014. *An introduction to knowledge commons*. http://knowledge-commons.net/downloads/Knowledge%20Commons%20Description.pdf (accessed on July 15, 2018).
Wright, E. O. 2013. Transforming capitalism through real utopias. 2011 Presidential Address. *American Sociologist Review* 78: 1–25.

PART I

Rebranding food and alternative narratives of transition

2

THE IDEA OF FOOD AS A COMMONS

Multiple understandings for multiple dimensions of food

Jose Luis Vivero-Pol

What is the dominant narrative of food?

Food has been defined as a commodity (Bush, 2010; Bahel et al., 2013), a commons (Dalla Costa, 2007; Karyotis and Alijani, 2016), a public good (Akram-Lodhi, 2013; Agyeman and McEntee, 2014), a private good (Samuelson, 1954) and a human right (Ziegler, 2001; De Schutter, 2014). Moreover, food has been recognized as an important cultural element (Van Esterik in this volume) and a power device (Frye and Bruner, 2012). Actually, food can be understood as a network of meanings (Szymanski, 2016), some of which may even be contradictory (e.g., how can food be a right and a commodity at the same time if rights are not tradeable?, or how can food be a basic need and a cultural determinant if human needs are basically equal and universal?).

The multiple meanings of food are phenomenological features that render every food item a sort of social agent. These meanings, situated in time and space, can be constructed and reconstructed depending on how various groups in society influence the policy arena, the social imagination and specific socio-technical practices.

At present, the complexity of visions and understandings seems to have been forgotten and abandoned. An obsession with cheap food that is financially accessible for the greatest number of customers, regardless of the environmental and social effects that this actually entails, has become the defining feature of food and the dominant economic system (Patel and Moore, 2017). This understanding of food, as an object that must be traded for as little money as possible, frames the functioning of the industrial food system, where the goal of commoditized food is maximizing profit out of its production, not adequately feeding people in a sustainable manner.

The idea and potential of the commons, which is vividly present in the history of the world (including of the Western world), appear to have been forgotten. More often than not, the commons have been appropriated by an economic understanding that classifies the goods based solely on two inner properties – rivalry and excludability – and then assigns the most suitable allocation mechanisms to each good. This approach is both theoretical and reductionist. It is evident that defining commons by just two features represents an impoverishment of highly diverse, place-based, time-dependent human constructions. The resources governing institutions, cultural trajectories, dominant narratives and moral principles that sustain the commons are all complex and fluent, as Elinor Ostrom (1990) taught us, so that distorting reductionism and simple models should be avoided (Frischmann, 2013). And yet, the economic meaning of commons is still dominant, reinforced by other normative constructs, such as the tragedy of the commons (Hardin, 1968), absolute proprietary regimes, private property as natural law, individualism, social Darwinism, competition over cooperation as a main driving motivation and the theory of rational choice. All of them have become the intellectual pillars that sustain the neoliberal socio-economic regime.

This chapter, being the first in a long list of texts addressing commons and food from different epistemic regards (or schools of thought, as described here), draws from the different interpretations of the term "commons" produced by academics and grassroots activists and commoners (see the introduction to this volume). It aims to pave the way for a non-stringent interpretation of the concept and its applicability to food and food systems. In that sense, this text may offer some useful insights to the following chapters that approach food, the food systems and the commons from multiple angles and different understandings of what a commons is and how food fits within that definition. Along those lines, I seek to respond to two questions, namely a) *How have the different schools of thought defined food within the private/public/commons typologies?* and b) *What is my rationale for defining food as a commons?*

What are the commons? Multiple understandings

Before delving into the details of the analysis of food as a commodity or a commons, I would like to present my understanding of "commodity", "commodification" and "commons". I follow Arjun Appadurai (1986), who defined "commodities" as anything intended for exchange and the "commodification" of a good as a situation in which its exchangeability for some other thing is its socially relevant dimension. Commodification has also been described as the symbolic, discoursive and institutional changes through which a good or service that was not previously intended for sale enters the sphere of money and market exchange (Gómez-Baggethun, 2015). The commodification of any good or service not only puts a price to it but also erodes its original values for society (Sandel, 2013), ultimately making them disappear. Typically, a commodity is a special kind of manufactured good or service associated with capitalist modes of production and embedded in the market society (Radin, 1996). As money-mediated commodity exchanges unfold, the symbolic ties and reciprocity logic that traditionally accompanied pre-capitalist transactions fade away (Mauss, 1970). Likewise, the absolute commodification of food, heralded by the industrialization and neo-liberalization of the food system, has brought the absolute dominance of the economic dimension of food and the undervaluation of those dimensions that cannot be valued in monetary terms, such as food as a human right, an essential resource for our survival or a cultural determinant.

The other concept that will appear extensively here (and in the entire book with multiple meanings) is "commons" (often written in plural). Commons are material and non-material goods that are jointly developed and maintained by a community and shared according to community-defined rules (Kostakis and Bauwens, 2014). They are goods that benefit all people and are fundamental to society's well-being and people's everyday lives, irrespective of their mode of governance (Bloemen and Hammerstein, 2015). For some political scholars, the practice of "commoning" is what creates the commons (Dardot and Laval, 2014). Whether material or non-material, natural or man-made, commons are often compounded of four elements: a) natural or cultural resources, b) the communities who share the resources, c) the commoning practices they use to share equitably and d) the purpose and moral narrative that motivates and sustain the commoning practices by the community.

More importantly, the notion of commons has been extensively and increasingly used as a paradigm of convergence of different struggles against neoliberal capitalism and the multiple enclosures it entails. In the last decade, the term has been deployed as a catch-all concept (Perilleux and Nyssens, 2017), with academic contributions evolving in parallel with practical developments by activists (Quarta and Ferrando, 2015). However, this widespread use of the commons terminology, often in a very uncritical way, has infused them with a mystic aura of social avant-garde and all the virtues of horizontal and fair governance (Verhaegen, 2015). In doing so, it risks becoming an empty slogan. As Rodotà already warned in 2013, if "everything is a commons, nothing is a commons" (Rodotà, 2013: 8). Therefore, the commons vocabulary should be better defined and their conceptual boundaries determined, so as to defend its uniqueness and prevent the void of its transformational power. Moreover, this boundary setting would be extremely helpful to inform the applicability of the "commons" term to food.

Epistemic regards on food in academia and grassroots activism

In order to navigate the private/public/commons distinction and apply it to food, it is essential to discuss the application of these concepts to food. In particular, it is interesting to assess how these three notions have been used and defined by economic, legal and political scholars along with grassroots activists. As imaginable, these groups use different epistemic toolboxes (i.e., values, knowledge, vocabularies and ideologies) to frame food differently. The dominant understanding in the industrial food system is that food is not a commons, but a private good and a commodity. Obviously, what one thinks food is depends upon how one perceives and judges it, and each of these different conceptions is connected to different beliefs and ways to know (Kaplan, 2012).

Throughout time and geographies, different realms of academic disciplines have addressed the commons and food by using different cognitive tools, accumulated knowledge and personal values, all of them forming particular epistemologies. As the multiple meanings of food and commons are relational, situated and context-specific, the valuation of food as a commodity or a commons must be understood as a social construct that can be modified by the instituting power of collectivities, be that alternative food movements, indigenous groups, food city councils, nation states, etc. In the final section of this chapter, I present my rationale to justify why valuing and governing food as a commons is the paradigm change the world needs to transition towards a different Anthropocene (Moore, 2017) where multiple food systems are designed to feed everybody adequately, sustainably within planetary boundaries and fairly to producers and consumers.

Revisiting the economic approach to food as a private good: social constructs can be modified

The economic typology of goods developed by economists during the second half of the 20th century (Samuelson, 1954; Musgrave, 1959) pivots around two features of goods, rivalry and excludability. These qualities are supposed to be inherent to the goods and not just socially constructed. Goods that are excludable and rival are termed private goods, and goods that are non-rival and non- (or difficult to be) excludable are termed public goods, to be better allocated by public means as they are considered as "market failures". So, the term "public goods" in economic terms is equivalent to my approach to the term "commons" in political terms. In this section only, both terms will be exchangeable.

In strict economic terms, food is rivalrous. If I eat a cherry, it is no longer available for others to eat. However, cherries are continuously produced by Nature (with or without the direct interaction of humans), so there is no limit to the number of cherries potentially available on Earth. As long as the replenishment rate outpaces the consumption rate, the resource is always available: given this characteristic, food is considered a renewable resource with a never-ending stock, such as air. So, the food I eat would not prevent others from eating food, although they could not eat the same piece I already ate. Actually, the world is producing far more food than is needed to adequately feed everybody and satisfy other non-consumption uses. However, one-third of the total food produced is wasted, and more than 40% of non-wasted food is used to feed livestock and produce biofuels, not to feed humans. A paradoxical disconnection thus exists between the potentially unlimited availability of food and its accessibility. This, I argue, is closely dependent on the assumption that food is a commodity that follows the laws of the market rather than those of Nature and Society.

Theoretically speaking, food is also excludable as we can prevent anyone from getting access to food, either by physical means, by pricing it at unaffordable costs or by making illegal to access food without paying a price. However, if the exclusion from food were to be enforced without reservations, the person trying to access food would die of starvation and it would thus eliminate the subject who tried to access the good. Regardless of this moral consideration, the ontological features of rivalry and excludability render food a private good that, according to the scholarly economists, is better allocated through market mechanisms, more specifically through capitalist market.

However, there is an ethical and political sense of the term "public good" (in economic terms) that needs to be distinguished here, invoking an important difference between "can" and "ought to". That is, a good is a public good any time individuals "ought not" to be excluded from its use. This is the condition the philosopher John O'Neill (2001, this volume) rightly called "a normative public good". The economic and normative meanings are logically distinct. A good from which individuals can be excluded is not necessarily one from which they ought to be excluded. Following that rationale, the case against the consideration of food as a commodity with absolute property and exclusionary rights should not be that food is not rival or excludable, but instead a case that it ought not to be excluded for many reasons, the most relevant being its essential nature as a vital resource for the human body. As the excludability feature is modulated by moral considerations, it is a normative property and not an ontological one.

Consequently, there is a moral notion that posits that food ought to be a public good (or a commons in the political vocabulary), in the sense of being available to all, as it is essential to all and the *sine qua non* pillar of human life. The goods that any community defines as normative commons or public goods, from which members should not be excluded, define the relationships of need, care and mutual obligations that are constitutive of that community (O'Neill, 2001, this volume), since social norms may deny exclusive property rights over certain common goods. Furthermore, the ideas of food commons or food as a public good, which are at odds with current mainstream thinking, can also be justified by applying nuanced approaches by Robert Musgrave (in Sturn, 2010: 304) and John Kenneth Galbraith (1958: 111). Musgrave accepted that public goods were a result of political decisions, although Samuelson rejected that notion (Demarais-Tremblay, 2014). Another neo-classical economist, John Kenneth Galbraith (1958: 111) stated that public goods "*must be provided for everyone* if they are to be provided for anyone, and they must be paid for collectively or they cannot be had at all" (italics are provided by author). Defining food as a good from which no member ought to be excluded is nothing but a political construct that helps define which type of community, society and nation we want to be members of.

The economic approach of defining food as a private good and a commodity, based on inner properties of goods, thus appears in contradiction to other approaches to goods based on phenomenological understandings that are context-specific, knowledge-based and place-situated. Other schools of thought see food differently.

Collective proprietary rights in food systems are recognized by legal scholars

Similarly to economists, legal scholars have recognized the possibility of collective regimes that own and govern the commons. Their approach is mainly driven by the question of titling and proprietary rights, which become the basis of defining the commons. The historical diversity mentioned in the introduction of this chapter is reflected in the current world's richness of proprietary schemes over natural resources (Schlager and Ostrom, 1992), where different bundles of rights can be identified and assigned to specific food resources. The legal proprietary regimes, which are nothing but social contracts situated in specific times and places, have varied between societies, civilizations and historic periods within civilizations.

While the economic definition of public–private goods only appeared in the 1950s, for millennia societies have used a mix of institutional settings – including commons, public goods and private goods – to govern resources and interact with Nature. Actually, the private arrangements that dominate industrial agriculture in the Global North are not equally prevalent in other areas of the world, where subsistence, traditional and agro-ecological types of agriculture are the norm and collective ownership of common resources are widespread. Speaking in numbers, two billion people in (mainly) rural parts of the world still depend on the commons (forest, fisheries, pasturelands, croplands and other natural resources) for their daily food (Weston and Bollier, 2013), with over 2.6 billion living in, and using, forests and drylands actively managed in commons or through common property arrangements (Meinzen-Dick et al., 2006). A great majority of small-scale, traditional farmers still have mixed proprietary arrangements for food resources (Bové and Dufour, 2001), such as the 500 million sub-Saharan Africans who still rely on communal lands as just one small example (Kugelman and Levenstein, 2013). The Food and Agriculture Organization (FAO) estimates that about 500 million hectares around the world are dedicated to agricultural heritage systems that still maintain their unique traditions and collective proprietary schemes (Altieri and Koohafkan, 2007). Moreover, 30% of all forests of the world are still owned or managed by communities under legally based collective proprietary schemes (Vira et al., 2015).

For sure, commons are not unfamiliar in the Western world. Despite the highly privatized and increasingly neoliberal character of Europe, and despite centuries of encroachments, misappropriations and legal privatizations, common lands harboring common resources (that can either be governed through collective arrangements or owned collectively) still amount more than 12.5 million ha (EUROSTAT with data from 2013[1]), or 5% of EU land. If we also include forest, mountainous and coastal areas, the figure can reach 9% of the total (Brown 2005 in Brown 2006a). Notwithstanding the current design of the EU Common Agricultural Policy (CAP) and the disincentive to the maintenance of collective institutions to manage the commons (Sutcliffe et al. 2013), common lands represent 9% of the surface of France (Vivier, 2002), 10% of Switzerland, 7.1% of Romania (Sutcliffe et al., 2013), 5.4% of Portugal (Serra et al., 2016: 172), 4.2% of Spain (Lana-Berasain and Iriarte-Goni, 2015) and 3.3% of the United Kingdom. These commons are widespread, rich in biodiversity (Brown, 2006b) and strongly linked to family farming (Sutcliffe et al., 2013). However, the relevance of the socio-economic importance of the food-producing commons in Europe is hardly noticed by general media and is hence neglected by the public authorities and mainstream scientific research. And yet, they survive by being meaningful to the Europeans living nearby.

The political regard of food as a commons (or public good)

If there is no doubt that commons are a reality in the contemporary world, the same cannot be said about the political consideration of food as a commons. For instance, little attention has been paid to the global commons related to agriculture and food security (FAO, 2002). Only recently have some authors started to elaborate rationales to justify why Food and Nutrition Security should also be treated as a global commons or global public goods (Page, 2013; Vivero-Pol, 2015). Food and nutrition security fits nicely with Kaul et al.'s (1999) definition of global public goods, namely "outcomes that tend towards universality in the sense that they benefit all countries, population groups, and generations".

This political disdain is, in my opinion, affected by a narrow understanding of food that does not take into consideration the daily practices that operationalize food as a commons. If we think of producing, transforming, cooking and sharing food, we realize that very often these are moments based on the multiplicity of food meanings, and that cannot be easily reduced to individualism, ownership and profit maximization. When food is collectively produced, jointly transformed, distributed on the basis of its value-in-use and eaten by people in a convivial manner, we have then a commons. When different resources are used (many of them also considered as commons, such as water, seeds, land, oceans, forests, traditional agricultural knowledge), considering the ecological and social limits to exploitation along with the duty to reproduce and respect Nature, we may have a commons. When we think about food not as energy for bodies but as the visible moment of a complex set of socio-ecological dynamics, we may think of food as a commons as well. This valuation of food as a commons is more evident at a local level, in place-restricted rural and urban communities. Although the food commons are better expressed at local level, the underlying narrative can be expanded from niche to niche until the place-restricted meaning covers a broader area and a much greater constituency.

Under a commons approach to food, communities and their members are placed at the center of the governing and stewardship process to co-produce, manage and benefit material and non-material resources essential for their livelihoods. This approach basically challenges the traditional idea of development, based on private property and wealth accumulation, and replaces it with reflexive democracy, participation, community-based development and rights-based approaches (e.g., to food, water, clean air, land and seeds as human rights and commons), indigenous rights, self-determination, the right to communal ownership and peasant rights. Additionally, food's essentiality for human survival adds another feature that supports its valuation as commons and forces us to challenge the premises and consequences of commodification: no food, no life. Therefore, food should never be denied to humans.

It does not matter where, how and when. It does not matter which gender, race, religion, political ideology or class people belong to. Everyone needs to eat. Moreover, as discussed in the last section of this chapter, food is connected with multiple aspects of life and ecology that inherently complexify its nature. A political, legal and economical redefinition of food as a commons (or public good) that subverts the accumulative and competitive paradigm of the commodity would provide a strong platform to achieve global goals like zero hunger and the protection of both land and marine resources.

Converging narratives of grassroots movements and local food innovations

As food systems are complex and combine a plurality of sub-systems, there is an urgent need to combine multiple and partial solutions into a viable transition with shared values and multiple but convergent praxis. Grassroot movements and local food (social) innovators seem to have engaged with the idea of commons and with its potential way before scholars did. Alternative civic food innovations, both customary and contemporary, present multiple narratives to confront the dominant near-monolithic discourse of the industrial food system. Both streams often converge on shared values about the valuation of food dimensions and walk-the-talk on building alternative niches (Ferrando and Vivero-Pol, 2017). These niches, as yet unconnected nodes of discontent and struggle, can now knit a crowd-sourced alternative to the industrial mainstream. Actually, many customary and contemporary food systems share the rejection of the normative valuation of food as a commodity. Other than their price in the market, the diverse civic food innovations have the valuation of the food dimensions in common, the convivial aspects of food production, collection, preparation and consumption and peer participation on an equal footing in designing, constructing and governing locally embedded food initiatives. Moreover, one person's concern for the wellbeing of others (either neighbor, community, city, region or planet) is a distinctive feature of well-nourished customary communities (Kent, 2015) as well as a standing value in urban alternative food networks.

Local transitions towards the organization of local, sustainable food production and consumption are taking place today across the world. Food is being produced, consumed and distributed through a multiplicity of open structures and peer-to-peer practices, aimed at co-producing and sharing food-related knowledge and items. Civic collective actions for food are generally undertaken, initially at local level, with the aim of preserving and regenerating the commons that are important for that community. Three examples will serve to highlight this movement: the Food Commons in California (USA), the Food Policy Council in Cork (Ireland) and the Walloon Network of Local Seeds (Belgium).

The Food Commons[2] is an initiative launched by a group of scholars, activists and organic agriculture entrepreneurs from California to re-connect food production and consumption. Their main goals are the preservation of common benefits throughout the value chain and to achieve a sustainable profitability. Institutionally speaking, the initiative encompasses a) a Food Trust (a non-profit entity to steward critical assets), b) a Food Hub (a locally owned cooperative that facilitates the food chain logistics) and c) a Food Fund (a community-owned financial institution). These three institutions are designed and governed by the people as members of the Food Commons initiative. Then, local food policy councils are growing in Western countries (Scherb et al., 2012; Carlson and Chappell, 2015), and the Cork Food Policy Council[3] in Ireland typifies their goals and motivations. The initiative enables people, eaters and producers, to regain control over the local food system and to foster community relations and shared values around local, organic and non-commodified food. Finally, civic collective actions for food have been mushrooming in Belgium over the last decade. They can largely be regarded as civic movements (e.g., AMAP – Associations to Support Peasants' Agriculture and GAS - Solidarity Purchasing Group) or social enterprises (e.g., GAC – Joint Purchasing Groups in Walloonia and Voedselteams in Flandre), considering the importance given to the social bonds between eaters and between eaters and food producers, and the importance given to food products and their accessibility. "Eating better", "improving local economies" and "transforming the food system" epitomize the three main attitudes behind the participation in these collective civic actions (Dedeurwaerdere et al., 2017).

Understandings of commons and food are multiple and phenomenological

This chapter has unfolded the elements of four schools of thought that have approached the normative nature of food (either as a private/public good or as a commodity/commons): three from within the academic domain (but whose narratives extend far beyond academia) and one encompassing the understanding of grassroots activists, practitioners of commons initiatives (dubbed "commoners") and some engaged scholars. For legal scholars, commons are usually place-restricted and determined by property entitlements. For economists, commons are determined by the inner properties of the resource. For activists and some political scholars, commons are a human-made praxis of collective governance and self-organized institutions. Therefore, the latter epistemology (to which I subscribe) posits that commons are neither types of resources with ontological properties, nor types of proprietary rights, but rather ways of acting collectively based on participation, self-regulation and self-negotiated principles and goals. Further, this praxis can be local, with clear physical boundaries, as studied by Elinor Ostrom, or embrace the whole human race with a global good. Additionally, activists posit that capitalism evolved to its current status by enclosing and privatizing everybody's commons, so the clash between both narratives is evident.

As one can appreciate in this book (e.g., in the Introduction or in numerous chapters), the vocabulary of the commons includes different interpretations of the same resource (be that water, knowledge or food), different meanings for the same term and tensions between different schools of thought with their own supporting values. The normative considerations (private, public and commons) are different and conflicting in some cases and certainly are diverse and evolving. That points to the commons-commodity categories as phenomenological and always situated in a particular place and historical period (Mattei, 2012; Szymanski, 2016).

The rationale to consider and govern food as a commons

After having presented the variety of epistemic regards of food, I would like to present my own approach to understanding, valuing and governing food as a commons. I support the social construction of any given material and non-material resource as a commons, if human societies so consider. Commons are forms of governance created collectively for resources that can be owned collectively, privately or publicly. So, the governance of the commons is different from the ownership of the commons. Governing a resource as a commons, however, is triggered by two important features: the essentiality of the resource to humans and the desire to institute a collective governance of that resource whereby every person affected by the resource has a role in its enjoyment and custody. According to those features, food is essential to every human, materially and spiritually, and it has been produced and distributed through non-market mechanisms for more than 2000 centuries (Shepperson, 2017; Vivero-Pol, 2017a). Both features qualify it as a commons.

My understanding of food as a commons contains a theoretical framework, an operational notion and a moral notion. The *theoretical framework* of food as a commons is based on the multiple dimensions of food relevant to humans which, for the sake of methodological appropriateness, have been reduced to six (see Vivero-Pol, 2017b). Those dimensions of food (see Figure 2.1) – as an essential resource, a cultural determinant, a human right, a public good, a natural resource and a tradeable good – cannot be adequately valued by the market through the value-in-exchange method, reducing the multiple meanings of food to a monetary valuation. Food is understood as a commodity when the economic tradeable dimension prevails over the non-economic; and as a commons when the different non-economic dimensions that are relevant to humans (value-in-use) are equally and properly valued. It is therefore the multidimensionality of food and its importance for every human being that endows this resource with the commons category. However, food can work as a commodity in particular circumstances, although not exclusively and not always. Therefore, governing food as a commons implies the human rights and public good dimensions become as relevant as, and sometimes even more than, the commodity dimension.

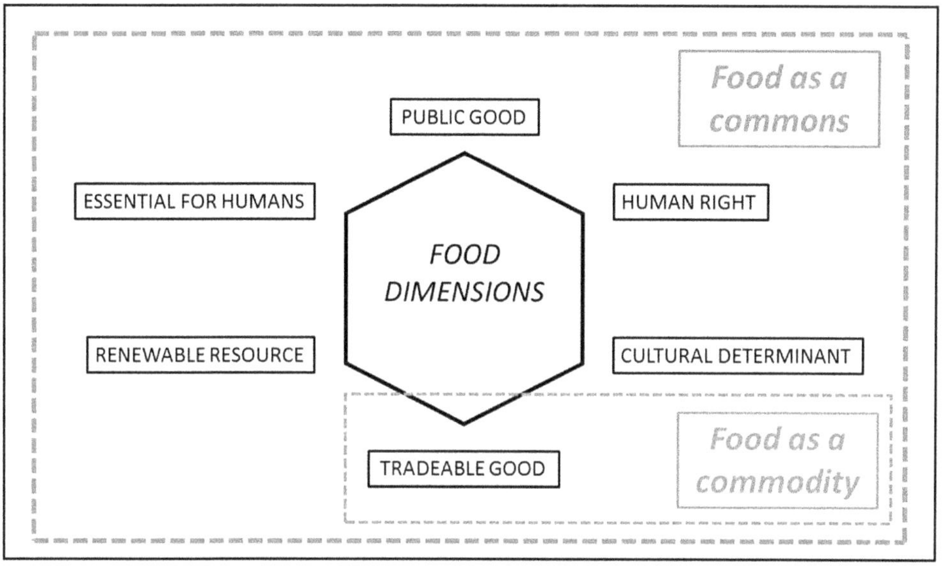

Figure 2.1 The multiple dimensions that render food a commons.

Source: Vivero-Pol (2017b).

Food as an essential resource for humans

Food is one of the three essential resources humans require to keep their vital functions, along with water and air. Food is a satisfier of the human need of subsistence (Maslow, 1943) that strongly impacts our capabilities and agency (Sen, 1999). Along those lines, previous influential scholars adopted a normative point of view (based on values) and considered the existence of "primary goods" from which the other goods are derived (Rawls, 1971); those primary goods deserve a special treatment in our society. A just society requires all humans to have the capability to live the lives they have reason to value (Nussbaum, 2011). To do it is unacceptable on the normative grounds that every human cannot satisfy his food needs when a) we already grow enough food to feed adequately the current and the expected population in 2050 (Holt-Giménez et al., 2012) and b) food scarcity has been artificially created through human-made enclosing mechanisms and political and institutional choices (De Schutter and Pistor, 2015).

Food as a cultural determinant

Food is often regarded as a defining pillar of cultures and civilizations. Everything having to do with food, such as its collection, capture, cultivation, preparation and consumption, represents a cultural act (Montanori, 2006). Different types of food have often been endowed with sacred beliefs (fish and bread in Christian beliefs; people made of corn among the Mayan peoples) and their production and distribution has been (and still is) governed by non-market rules (Fraser and Rimas, 2010). In many countries, social life revolves around meals. And not just society-wise – food is central to our identities as individuals and as members of a society (Fischler, 1988) and helps to shape the meanings of home (De Maret and Geyzen, 2015). Food also plays a key role in creating social bonds with relatives, friends and colleagues, since humans tend to eat together (commensality), thus reflecting the social relationships of individuals (Sobal and Nelson, 2003).

Food as a human right

Being that food is an absolute need (determined by our physiological needs) and the same in all cultures and historical periods (Max-Neef, 1992), access to food has been framed as a legitimate right to which citizens are entitled, and which society at large has an obligation to respect and provide for (Stavenhagen, 2003). So, food is formally considered a binding human right recognized under international law. The right to food protects the right of all human beings to feed themselves in dignity, either by producing their own food, purchasing it or receiving it from welfare systems (United Nations, 1948; 1999). In legal terms, the freedom from hunger dimension of the right to food falls under the category of "basic rights" (Shue, 1996), since it is necessary to have some basic needs met (socio-economic rights) before a wider set of liberties and moral rights (civil and political rights) can be enjoyed. Designating a good as a human right means that, under no jurisdiction and no circumstances, may that good be denied to anybody (Clapham, 2007: 120).

Food as natural resource

Natural ecosystems are an almost unlimited source of edible plants and animals, ranging from game and bush meat, fish and fowl, to vegetables, fungi or fruits (TEEB, 2015). In highly urbanized Europe, wild food is still consumed by more than 100 million people and is provided by more than 150 species (Schulp et al., 2014). Actually, wild foods are quite "à la mode" in the 21st century, having entered into the domain of haute-cuisine and healthy foods and remedies (Łuczaj et al., 2012). The marine species represent another interesting case. Fish stocks, especially those in international waters, are generally accepted as common goods (Bene et al., 2011). De Schutter and Pistor (2015) posited that some natural resources (e.g., water, air, food) are essential for the survival of every human being because they are neither context specific (applicable to all cultures and settings) nor relative (as caloric needs are determined by our physiological needs). Essential resources ought to have a special legal and political consideration. Additionally, the authors argue that the scarcity of those essential resources has been artificially created by human institutions and norms, first among them market mechanisms.

Food as a tradeable good

Food has also been a tradeable good since the start of agricultural societies. But its transformation into a capitalist commodity occurred in the 20th century with unbranded or undifferentiated foods from multiple producers (e.g., staple grain, beef meat or tomatoes) dispossessed of any kind of attribute other than the marketable features (safety, durability, beauty) (Lukács, 1968). Food as a commodity is the backbone of the food industry and is one of the biggest areas of economic activity worldwide, representing around 10% of the global gross domestic product (Forbes, 2007). As a commodity, international food trade, which only accounts for 23% of global food production (D'Odorico et al., 2014), is dominated by a few transnational companies (Clapp and Fuchs, 2009). And yet, the commodification process can be reversed (Lind and Barham, 2004), since food has become a commodity only for a very particular stream of the westernized neoliberal economy that so deeply pervades our lives.

Food as a public good

Food also has a public dimension that so far has not been properly valued. I subscribe to the notion that any given good is a result of "deliberate policy choices" made by societies (Kaul et al., 2003) according to what is perceived as a public need, rather than according to certain inherent characteristics of non-excludability and non-rivalry (Wuyts, 1992). Actually, public goods can be generated through collective choices (e.g., voting in a referendum in Slovenia to declare water a public good to be enshrined in the Constitution), funded by collective payments (e.g., taxes or public budgets) and owned through private, public and collective proprietary regimes (Capra and Mattei, 2015). Public goods, in the political sense, can be produced by governments if a society decides that all citizens should have access to them because their social or economic benefits are important or essential, regardless of the ability to pay. Food evidently qualifies as such.

Food as a commons

Food as a commons values the multidimensionality of food and does not assign a special primacy to the economic dimension as the current industrial food system does. Every eater should have a say in how food resources are managed (and is thus termed a food citizen by Gomez-Benito and Lozano, 2014), an idea that has been termed food democracy by De Schutter (2014). Actually, the development of "food citizenship", as opposed to "food consumption", requires moving beyond food as a commodity (Welsha and MacRaeb, 1998). Moreover, according to many, every eater should be guaranteed fair and sufficient access to this resource, regardless of his/her purchasing power.

Regarding the valuation of the six food dimensions, my assumptions are as follows:

a The recognition of these food dimensions is universal (although food as a human right is contested in some countries), but individuals differ in the weight and priority assigned to each dimension.
b Food dimensions shape our relationship to food and food-producing systems.
c Societies value food dimensions differently (i.e., they are context-specific).
d Valuation of food dimensions triggers human agency and the political stance vis à vis the food system, differentiating a food consumer from a food citizen.
e Food dimensions connect multiple elements and drivers that interplay in the food systems (e.g., biodiversity, climate change, gender or poverty).

On the other hand, the *operational conceptualization* of food as a commons puts emphasis on the social practices around food-producing systems (governance, institutions and customs). Commoning is the action of cultivating, processing, exchanging, selling, cooking and eating together. For example, when applied to food, commoning may be sharing the art of hunting (i.e., "monterias" – wild goat or boar hunting by elite people in Spain, or antelope hunting by Pygmy people in Bostwana); sharing traditional rice landraces in the Yunan province of Southern China to combat diseases and pests (Hanachi, et al., 2016); the social protocols of auctions of fish captures in Ireland; and all the traditions that express social belonging and solidarity in maize production, harvesting and religious rituals among the Guatemalan Mayas (Mariano, 2013).

Finally, *a moral notion* entails that food is a commons because it is undeniably fundamental to people's lives and is a cornerstone of human societies, regardless of how it is governed or who owns it. As something essential to people, the food commons belong to and shall be governed by people. In that sense, considering food as a commons carries a deeper, more subversive moral claim as to who owns Earth's food and food-producing resources (water, land, seeds and agricultural knowledge etc.), questioning John Locke's rationality to justify the appropriation of everybody's natural resources by individuals by simply putting your own work on those resources (see Timmermann in this volume).

A disruptive paradigm shift: the food commons system

The food commons are thus comprised of four elements: a) natural and non-material resources (foodstuff, cooking recipes, traditional agricultural knowledge); b) the communities who share the resources (which may be local, national or global, because we all eat); c) the commoning practices people use to produce, transform and eat food; and d) the moral narrative that sustains the main purpose of the food system (i.e., to produce food sustainably to feed the people adequately). All food commons benefit from a relational approach between the good, the purpose for using that good, the community that agrees on that purpose and the governing mechanisms to achieve that purpose.

The end-goal of a food commons system should not be profit maximization but rather increasing food access in ways that are fair to producers and consumers, build community and shorten the distance from field to table (Johnston, 2008), all while stewarding natural resources for future generations. This represents a worldview different from the dominant paradigm of the industrial food system; it is based on shared customary and contemporary models of social organization for food production and consumption, non-monetized allocation rules and sharing practices, principles of peer production based on commons (resources, knowledge, values), social economy and the importance of the commonwealth and happiness and the well-being of our communities. The commons dimension of food is about caring, collectiveness, equity, responsibility and stewardship. Embeddedness and direct democracy from local to global are also relevant features, linking the food commons with agro-ecology and alternative food systems. Consideration of the food commons invokes a radical paradigm shift from individual competitiveness and endless growth as the engines of progress towards collective cooperation and de-growth/frugality as drivers of happiness and the common good. This normative valuation may certainly sustain a transition pathway that, first, provides for sustainable nutrition for all and, second, provides meaning and not just utility, to food production, trading and consumption (Anderson, 2004). The food commons encompass ancient and recent history (customary valuations of food in different civilizations as well as modern and urban civic collective actions for food), a thriving alternative present (the myriad alternative food networks that share, barter and exchange food by means of non-monetized mechanisms) and an innovative, utopian and just vision for the future (Vivero-Pol, 2015).

To imagine and re-construct that aspirational and inspirational narrative shift, I propose to use the best epistemic tools of each school of thought so that food as a commons can be governed as a public good (by the governments) and as an essential resource that has to be traded under specific regulations (by the private actors). A different economic school could replace its ontological consideration of food as a commodity for a phenomenological understanding where value-in-use can become aligned with value-in-exchange. A critical legal school of thought could accept that multiple proprietary regimes and entitlements are valid and functional where food is at stake. Furthermore, a less conventional political school of thought could legitimize different governing mechanisms for an essential resource for human and societal survival, thereby devising particular arrangements in which civic food actions, public policies and moral economies can co-exist. Finally, the grassroots activist school of thought reminds us that the ultimate purpose of a successful food system is to feed us all within planetary boundaries, and without mortgaging the food-producing resources of future generations. Unfortunately, this is currently not the case with the dominant industrial food system.

The practice of commoning has the power to create new traditions and revitalize old ones, review laws and establish a different legal and political institutional framework (Charbonnier and Festa, 2016; Dardot and Laval, 2014; Capra and Mattei, 2015; Ruivenkamp and Hilton, 2017). This is what actually frightens the consolidated duopoly of the industrial food system (the state and the food market): self-organized communities and social food movements, based on different narratives and moral grounds (i.e., food as a commons), may design governing and allocation mechanisms and legal frameworks that are different from those that maintain the globalized free market of food. Re-commoning food systems may attack the legal and political scaffolding that sustains the hegemony of the market and state elites over every eater and more than two billion food producers. Convivial, relational and important for individuals and societies, food is a perfect agent of change (McMichael, 2000) and has transformative power (Vivero-Pol, 2017c). Re-commoning food may help us to re-create sustainable forms of food production (agroecology), new collective practices of governance (food democracies) and alternative policies (food sovereignty) to regain control over the food system by the most relevant actors (eaters and producers), from the current dominant actors (agri-food corporations and governments). Going even further, the re-construction of the entire food system as a commons is a revolutionary idea defended by some Italian scholars (Pettenati et al. in this volume).

History has taught us that for centuries, food has been valued and governed as a commons in different civilizations. Legal and political scholars demonstrate that this consideration is still alive in many customary food systems and is being reconstructed in innovative contemporary food initiatives. So, considering food as a commons is not a utopian idea, understood as a "no place" (incorrectly interpreted from Greek οὐ "not" and τόπος "place"), but a "good place" to aspire to (derived from Greek εὖ "good" or "well"), as Fonte and Cucco explore in detail in this volume. Food as a commons could be a transformative vision that leads to a different transition pathway, away from this unsustainable and unfair industrial food system and towards a better one in which everybody can eat well three times each day, because food is not just governed as a commodity. Food shall be seen differently, not only a resource whose main purpose is profit-maximization for private accumulation but also as a life-enabler, a source of pleasure and culture and a human right. As Marcel Proust once said: "The real voyage of discovery consists not in seeking new lands, but in seeing with new eyes". Let's travel together to rediscover food with new eyes. The paradigm shift from viewing food as a commodity to revaluing it as a commons will take many decades, and the opposing forces will be paramount. This book, to which this chapter contributes, is just a first step in the right direction.

Notes

1 With statistics only reflecting data from 13 EU members and referring only to available agricultural land, not including forest or coastal areas. https://ec.europa.eu/eurostat/statistics-explained/index.php?title=Farm_structure_survey_-_common_land#Statistics_on_common_land.
2 http://www.thefoodcommons.org/governance/ (Accessed June 12, 2018).
3 http://corkfoodpolicycouncil.com/ (Accessed June 12, 2018).

Bibliography

Agyeman, J., and J. McEntee. 2014. Moving the field of food justice forward through the lens of urban political ecology. *Geography Compass* 8(3): 211–220.

Akram-Lodhi, H. 2013. *Hungry for change: Farmers, food justice and the agrarian question*. Black Point, Nova Scotia: Fernwood.

Altieri, M. A., and P. Koohafkan. 2007. *Globally important ingenious agricultural heritage systems (GIAHS): extent, significance, and implications for development*. Rome: FAO. http://www.fao.org/docrep/015/ap021e/ap021e.pdf (Accessed on June 15, 2018).

Anderson, M.D. 2004. Grace at the table. *Earthlight* 14(1): 10–13.

Appadurai, A. 1986. Introduction: commodities and the politics of value. In *The social life of things: commodities in cultural perspective*, ed. A. Appadurai, 3–63. Cambridge: Cambridge University Press.

Bahel, E., W. Marrouch and G. Gaudet. 2013. The economics of oil, biofuel and food commodities. *Resource and Energy Economics* 35(4): 599–617.
Bene, C., M. Phillips and E. H. Allison. 2011. The forgotten service: food as an ecosystem service from estuarine and coastal zones. In *Ecological economics of estuaries and coasts. Reference module in earth systems and environmental sciences*, eds. M. van den Belt and R. Costanza, 147–180. Treatise on estuarine and coastal science, vol. 12. London: Academic Press.
Bloemen, S., and D. Hammerstein. 2015. The EU and the commons: A commons approach to European knowledge policy. Commons Network in cooperation with Heinrich Böll Stiftung Berlin/Brussels. http://commonsnetwork.eu/wp-content/uploads/2015/06/A-Commons-Approach-to-European-Knowledge-Policy.pdf (Accessed on August 14, 2018).
Bové, J., and F. Dufour. 2001. *The world is not for sale. Farmers against junk food.* London: Verso.
Brown, K. M. 2006a. New challenges for old commons: the role of historical common land in contemporary rural spaces. *Scottish Geographical Journal* 122(2): 109–129.
Brown, K. M. 2006b. *Common land in Western Europe: anachronism or opportunity for sustainable rural development?* IASCP Europe Regional Meeting: Building the European Commons: From Open Fields to Open Source. 23–25 March, Brescia. https://www.researchgate.net/publication/42761067_Common_Land_in_Western_Europe_Anachronism_or_Opportunity_for_Sustainable_Rural_Development (Accessed on June 15, 2018).
Bush, R. 2010. Food Riots: Poverty, power and protest. *Journal of Agrarian Change* 10(1): 119–129.
Capra, F., and U. Mattei. 2015. *The ecology of law: toward a legal system in tune with Nature and community.* Oakland, CA: Berrett-Koehler Publishers.
Carlson, J., and M. J. Chappell. 2015. *Deepening food democracy.* Minneapolis, MN: Institute for Agriculture and Trade Policy. https://www.iatp.org/documents/deepening-food-democracy (Accessed on June 15, 2018).
Charbonnier, P., and D. Festa. 2016. Bien communs, beni comuni. *Tracés. Revue de Sciences humaines* 16: 187–194. http://traces.revues.org/6622 (Accessed on June 15, 2018).
Clapham, A. 2007. *Human rights. a very short introduction.* Oxford University Press.
Clapp, J., and D. Fuchs, eds. 2009. *Corporate power in global agrifood governance.* Cambridge, MA: The MIT Press.
Dalla Costa, M.R. 2007. Food as common and community. *The commoner* 12: 129–137.
Dardot, P., and C. Laval. 2014. *Commun, essai sur la révolution au XXI° siècle.* Paris: Le Découverte.
de Maret, O., and A. Geyzen 2015. Tastes of homes: exploring food and place in twentieth-century Europe. *Food and Foodways* 23(1–2): 1–13, doi:10.1080/07409710.2015.1011980.
De Schutter, O. 2014. The transformative potential of the right to food. Report of the Special Rapporteur on the right to food to the UN Human Rights Council. A/HRC/25/57, 24 January 2014. http://www.srfood.org/images/stories/pdf/officialreports/20140310_finalreport_en.pdf
De Schutter, O., and K. Pistor. 2015. Introduction: toward voice and reflexivity. In *Governing access to essential resources*, eds. K. Pistor and O. De Schutter, 3–45. New York: Columbia University Press.
Dedeurwaerdre, T, O. De Schutter, M. Hudon, E. Mathijs et al. 2017. The governance features of social enterprise and social network activities of collective food buying groups. *Ecological Economics* 140: 123–135.
Desmarais-Tremblay, M. 2014. *On the definition of public goods. Assessing Richard A. Musgrave's contribution.* Documents de travail du Centre d'Economie de la Sorbonne 2014.04 https://halshs.archives-ouvertes.fr/halshs-00951577/document (Accessed on June 15, 2018).
D'Odorico, P., J. A. Carr F. LaioL. Ridolfi and S. Vandoni. 2014. Feeding humanity through global food trade. *Earth's Future* 2: 458–469. doi:10.1002/2014EF000250.
FAO (Organisation of United Nations of Food and Agriculture). 2002. *The state of food and agriculture.* Rome: FAO.
Federici, S. 2014. *Caliban and the witch. Women, the body and primitive accumulation.* New York: Autonomedia.
Ferrando, T. 2016. Il sistema cibo come bene comune. In *Beni communi 2.0. Contro-egemonia e nuove istituzioni*, eds. A. Quarta and M. Spanò, 99–109. Milan: Mimesis Edizioni.
Ferrando, T., and J. L. Vivero-Pol. 2017. Commons and 'commoning': a 'new' old narrative to enrich the food sovereignty and right to food claims. *Right to Food and Nutrition Watch* 2017: 50–56. https://www.righttofoodandnutrition.org/files/02.rtfanw-2017_eng_17_12_article-5_web_rz.pdf.
Fischler, C. 1988. Food, self and identity. *Social Science Information* 27: 275–292.
Forbes. 2007. The world's biggest industry. *Forbes*, 15 November 2007. https://www.forbes.com/2007/11/11/growth-agriculture-business-forbeslife-food07-cx_sm_1113bigfood.html#6c698f08373e (Accessed on June 17, 2018).

Fraser, E. D. G., and A. Rimas 2010. *Empires of food: feast, famine, and the rise and fall of civilizations.* Toronto: Free Press.
Frischmann, B. M. 2013. Two enduring lessons from Elinor Ostrom. *Journal of Institutional Economics* 9(4): 387–406.
Frye, J., and M.S. Bruner, eds. 2012. *The rhetoric of food: Discourse, materiality, and power.* New York and London: Routledge.
Fuys, A., E. Mwangi and S. Dohrn. 2008. *Securing common property regimes in a globalizing world: synthesis of 41 case studies on common property regimes from Asia, Africa, Europe and Latin America.* ILC 'Knowledge for Change' series, 3. Rome: CAPRI programme and International Land Coalition. http://www.land-coalition.org/sites/default/files/documents/resources/ilc_securing_common_property_regimes_e.pdf (Accessed on June 15, 2018).
Galbraith, J. K. 1958. *Affluent society.* Boston, MA: Houghton Mifflin.
Gómez-Baggethun, E. 2015. Commodification. In *Degrowth: a vocabulary for a new era*, eds. G. D'Alisa, F. Demaria and G. Kallis, 67–70. New York: Routledge.
Gomez-Benito, C., and C. Lozano. 2014. Constructing food citizenship: theoretical premises and social practices. *Italian Sociological Review* 4(2): 135–156.
Hanachi, M., T. Dedeurwaerdere and J. L. Vivero-Pol. 2016. *Overcoming the tragedy of the commons in crop disease management. The role of locally evolved institutional arrangements in the YuanYang Terraces traditional agro-ecological system.* Paper presented at the international conference "Du vivant au social: les semences en question" (Lovain-la-Neuve, October 6, 2016).
Hardin, G. 1968. The tragedy of the commons. *Science* 162(3859): 1243–1248.
Holt-Giménez, E., A. Shattuck, M. Altieri, H. Herren and S. Gliessman. 2012. We already grow enough food for 10 billion people … and still can't end hunger. *Journal of Sustainable Agriculture* 36(6): 595–598.
Johnston, J. 2008. Counterhegemony or bourgeois piggery? Food politics and the case of foodShare. In *The fight over food: Producers, consumers, and activists challenge the global food system*, eds. W. Wright, and G. Middendorf, 93–120. University Park, PA: Pennsylvania State University.
Kaplan, D. M. 2012. Introduction. In *The Philosophy of Food*, ed. D. M. Kaplan, 1–14. Berkeley, CA: University of California Press.
Karyotis, C., and S. Alijani. 2016. Soft commodities and the global financial crisis: Implications for the economy, resources and institutions. *Research in International Business and Finance* 37: 350–359.
Kaul, I, P. Conceição, K. Le Goulven and R. U. Mendoza, eds. 2003. *Providing global public goods: managing globalization.* New York: Oxford University Press.
Kaul, I., O. Grunberg and M. A. Stern. 1999. *Global public goods: international cooperation in the 21st century.* New York: Oxford University Press.
Kent, G. 2015. Food systems, agriculture, society. How to end hunger. *World Nutrition* 6(3): 170–183.
Kostakis, V., and M. Bauwens. 2014. *Network society and future scenarios for a collaborative economy.* London: Palgrave MacMillan.
Kugelman, M., and S. L. Levenstein. 2013. *The global farms race: land grabs, agricultural investment and the scramble for food security.* Washington, DC: Island Press.
Lana-Berasain, J. M., and I. Iriarte-Goni. 2015. Commons and the legacy of the past. Regulation and uses of common lands in twentieth-century Spain. *International Journal of the Commons* 9(2): 510–532.
Lind, D., and E. Barham. 2004. The social life of the tortilla: food, cultural politics, and contested commodification. *Agriculture and Human Values* 21(1): 47–60.
Łuczaj, L., A. Pieroni, J. Tardío, M. Pardo-de-Santayana, R. Sõukand, I. Svanberg and R. Kalle. 2012. Wild food plant use in 21st century Europe: the disappearance of old traditions and the search for new cuisines involving wild edible. *Acta Societatis Botanicorum Poloniae* 81(4): 359–370. doi:10.5586/asbp.2012.031.
Lukács, G. 1968. *History and class consciousness.* Cambridge, MA: The MIT Press.
Mariano, L. 2013. Hunger in the spaces of culture. Ch'orti maya indigenous visions. *AIBR: Revista de Antropología Iberoamericana* 8(2): 209–231.
Maslow, A. 1943. A theory of human motivation. *Psychological Review* 50(4): 370–396.
Mattei, U. 2012. First thoughts for a phenomenology of the commons. In *The wealth of the commons. A world beyond market and state*, eds. D. Bollier and S. Helfrich, n.p. Amherst, MA: Levellers Press. http://wealthofthecommons.org/essay/first-thoughts-phenomenology-commons (Accessed on June 15, 2018).
Mauss, M.. 1970. *The Gift: Forms and functions of exchange in archaic societies.* London: Cohen & West.
Max-Neef, M. 1992. Development and human needs. In *Real-life economics–understanding wealth creation*, eds. P. Ekins and M. Max-Neef, 197–213. London: Routledge.

McMichael, P. 2000. The power of food. *Agriculture and Human Values* 17: 21–33.
Meinzen-Dick, R., E. Mwangi and S. Dohrn. 2006. *Securing the commons: what are the commons and what are they good for?* CGIAR Systemwide Program on Collective Action and Property Rights. Policy Brief no. 4. http://www.ifpri.org/publication/securing-commons (Accessed on June 15, 2018).
Montanori, M. 2006. *Food is culture. Arts and traditions on the table*. New York: Columbia University Press.
Moore, J. W. 2017. *Capitalism in the web of life. Ecology and the accumulation of capital*. London: Verso Books.
Musgrave, R. A. 1959. *The theory of public finance*. New York: Macmillan.
Nussbaum, M. C. 2011. *Creating capabilities: the human development approach*. Cambridge, MA: Harvard University Press.
O'Neill, J. 2001. Property, care, and environment. *Environment and Planning C: Government and Policy* 19: 695–711.
Ostrom, E. 1990. *Governing the commons: the evolution of institutions for collective action*. New York: Cambridge University Press.
Page, H. 2013. *Global governance and food security as global public good*. New York: New York University. http://cic.nyu.edu/sites/default/files/page_global_governance_public_good.pdf (Accessed on June 15, 2018).
Patel, R., and J. W. Moore. 2017. *A history of the world in seven cheap things: a guide to capitalism, nature, and the future of the planet*. Oakland, CA: University of California Press.
Perilleux, A., and M. Nyssens. 2017. Understanding cooperative finance as a new common. *Annals of Public and Cooperative Economics* 88(2): 155–177.
Pettenati, G., and A. Toldo. 2016. Il sistema alimentare locale è un bene comune? In *Cibo, cittadini e spazi urbani. Verso un'amministrazione condivisa dell' Urban Food Policy di Torino*, eds. D. Ciaffi, F. De Filippi, G. Marra and E. Saporit, 15–17. Quaderno Labsus. Rome: Laboratorio per la Sussidiarietà. http://www.labsus.org/wp-content/uploads/2017/01/CIBO-CITTADINI-E-SPAZI-URBANI.pdf (Accessed on June 15, 2018).
Quarta, A., and T. Ferrando 2015. Italian property outlaws: from the theory of the commons to the praxis of occupation. *Global Jurist* 15(3): 261–290.
Radin, M. J. 1996. *Contested commodities*. Cambridge, MA: Harvard University Press.
Rawls, J. 1971. *A theory of justice*. Cambridge, MA: Belknap Press.
Rodotà S. 2013. Constituting the commons in the context of state, law and politics. In *Economics and the common(s): from seed form to core paradigm*. A report on an international conference on the future of the commons, 6–8. Berlin: Heinrich Böll Foundation.
Ruivenkamp, G., and A. Hilton. 2017. Introduction. In *Perspectives on commoning. autonomist principles and practices*, eds. G. Ruivenkamp and A. Hilton, 5–17. London: Zed Books.
Samuelson, P. A. 1954. The pure theory of public expenditure. *Review of Economics and Statistics* 36: 387–389.
Sandel, M. J. 2013. *What money can't buy: The moral limits of markets*. New York: Farrar, Straus and Giroux.
Scherb, A., A. Palmer, S. Frattaroli and K. Pollack. 2012. Exploring food system policy: a survey of food policy councils in the United States. *Journal of Agriculture, Food Systems and Community Development* 2(4): 3–14. doi:http://dx.doi.org/10.5304/jafscd.2012.024.007.
Schlager, E., and E. Ostrom. 1992. Property-rights regimes and natural resources: a conceptual analysis. *Nature* 413: 591–596.
Schulp, C. J. E., W. Thuiller and P. H. Verburg. 2014. Wild food in Europe: a synthesis of knowledge and data of terrestrial wild food as an ecosystem service. *Ecological Economic* 105: 292–305.
Sen, A. K. 1999. *Development as freedom*. Oxford: Oxford University Press.
Serra, R., P. Ferreira, I. Skulska, M. Alavez-Vargas, A. Salgado, J. Arriscado-Nunes and R. Garcia-Barrios. 2016. Education for sustainability in the context of community forestry. In *Biodiversity and education for sustainable development*, eds. P. Castro, U. M. Azeiteiro, P. Bacelar-Nicolau, W. Leal Filho and A. M. Azul, 169–183. Switzerland: Springer International Publishing.
Shepperson, M. 2017. How ancient lentils reveal the origins of social inequality. London: *The Guardian*, October 11. https://www.theguardian.com/science/2017/oct/11/lentils-origins-of-social-inequality?CMP=share_btn_tw (Accessed on June 15, 2018).
Shue, H. 1996. *Basic rights*. Princeton, NJ: Princeton University Press, 2nd ed.
Sobal J., and M. K. Nelson 2003. Commensal eating patterns: a community study. *Appetite* 41(2): 181–90.
Stavenhagen, R. 2003. *Needs, rights and social development*. Overarching Concerns Programme Paper Number 3. Geneva: UNRISD.
Sturn, R. 2010. Public goods before Samuelson: interwar Finanzwissenschaft and Musgrave's synthesis. *The European Journal of the History of Economic Thought* 17(2): 279–312.

Sutcliffe, L., I. Paulini, G. Jones, R. Marggraf and N. Page. 2013. Pastoral commons use in Romania and the role of the Common Agricultural Policy. *International Journal of the Commons* 7(1): 58–72.
Szymanski, I. F. 2016. What is food? Networks and not commodities. In *The Routledge Handbook of Food Ethics*, eds. M. Rawlinson, and M.C., Ward, 7–15. London and New York: Routledge.
TEEB. 2015. The economics of ecosystems and biodiversity for agriculture & food: an interim report. Geneva, Switzerland: United Nations Environment Programme. http://img.teebweb.org/wp-content/uploads/2015/12/TEEBAgFood_Interim_Report_2015_web.pdf.
UN 1948. *Universal declaration of human rights*. General Assembly Resolution 217 A (III). UN Doc. A/810, at 71 (1948).
UN 1999. *The right to adequate food* (Art. 11). General Comment of the Committee on Economic, Social and Cultural Rights, Geneva, 26 April–14 May 1999. E/C.12/1999/5.
Verhaegen, E. 2015. La forge conceptuelle. Le "commun" comme réinterprétation de la propriété. *Recherches sociologiques et anthropologiques* 46(2): 111–131. http://rsa.revues.org/1547 (Accessed on June 15, 2018).
Vira, B., C. Wildburger and S. Mansourian, eds. 2015. *Forests, trees and landscapes for food security and nutrition. A global assessment report.* IUFRO World Series 33. Vienna: IUFRO.
Vivero-Pol, J. L. 2015. Food is a public good. *World Nutrition* 6(4): 306–309.
Vivero-Pol, J. L. 2017a. The idea of food as commons or commodity in the academia. A systematic review of English scholarly texts. *Journal of Rural Studies* 53: 182–201.
Vivero-Pol, J. L. 2017b. Food as commons or commodity? Exploring the links between normative valuations and agency in food transition. *Sustainability* 9(3): 442. doi:10.3390/su9030442.
Vivero-Pol, J. L. 2017c. Transition towards a food commons regime: re-commoning food to crowd-feed the world. In *Perspectives on commoning: Autonomist principles and practices*, eds. G. Ruivenkamp and A. Hilton, 185–221. London: Zed Books.
Vivier, N. 2002. The management and use of the commons in France in the eighteenth and nineteenth centuries. In *The management of common land in North West Europe c. 1500–1850*, eds. M. De Moor, L. Shaw-Taylor and P. Warde, 143–171. Turnhout, Belgium: Brepols Publishers.
Welsha, J., and R. MacRaeb. 1998. Food citizenship and community food security: lessons from Toronto, Canada. *Canadian Journal of Development Studies / Revue canadienne d'études du développement* 19(4): 237–255.
Weston, B. H., and D. Bollier. 2013. *Green governance: ecological survival, human rights, and the law of the commons*. New York: Cambridge University Press.
Wolf, M. 2012. The world's hunger for public goods. *Financial Times*, January 24. https://www.ft.com/content/517e31c8-45bd-11e1-93f1-00144feabdc0 (Accessed on June 15, 2018).
Wuyts, M. 1992. Deprivation and public need. In *Development policy and public action*, eds. M. E. WuytsM. Mackintosh and T. Hewitt. Oxford: Oxford University Press/Open University.
Ziegler, J. 2001. The right to food. Report by the Special Rapporteur on the right to food to the UN Commission on Human Rights. E/CN.4/2001/53, para. 14.

3

THE FOOD SYSTEM AS A COMMONS

Giacomo Pettenati, Alessia Toldo and Tomaso Ferrando

Introduction

In the last years, academics and non-academics alike have increasingly been discussing the commons as an old–new framework to engage with the interactions between humans and nature. Despite the multiplicity of communal practices that characterize current food systems all over the world, it has only been recently that the idea of food as a commons has received attention behind the doors of universities and research centres. It took even longer for it to permeate the sphere of public and political debates (Dalla Costa, 2007; Rundgren, 2016; Sumner, 2011; Vivero-Pol 2017a, 2017b).

As demonstrated by several contributions to this edited volume, contemporary endeavours around food and the commons cover a wide range of approaches and visions, but they tend to share a strong critique of the commodification of food as an object of consumption and the desire to challenge the underlying assumptions of excludability, rivalry and exchange value. These authors affirm that food should be removed, like air and water, from market logics and constraints: it must be appreciated through the lens of utility and needs rather than price, profit and consumerism. When there was a plan to privatize water, people occupied streets in Bolivia (Assies, 2003), organized bottom-up movements and binding referendums in Italy (Fantini, 2012; Mattei, 2013) and manifested their dissent in India, the United States and several other places in the world. However, the number of those who question the fact that access to food is mainly dependent on economic transactions or, in fewer cases, on the consumers' possibility of accessing land and farming for self-consumption, is still limited. Luckily, the landscape seems to be changing both at the national and international level. This chapter aims to contribute to the strengthening of this raising tide.

The shift in paradigm is taking place through different forms and at different levels. In the legal framework, for example, Stefano Rodotà (2012) highlighted that the diffusion of a de-commodified vision had been increasingly consolidating through "the legal recognition of new words that are criss-crossing the world: free software, no copyright, free access to water, *food*, medicines, internet, and of the fact that these forms of access assume the form and strength of fundamental rights." In his words, this is proved by the fact that "[t]he United Nations General Assembly [...] recognized everyone's right to adequate food" (Rodotà, 2012, p. 316)[1] and therefore attributed international legal status to access to food as an intrinsic requirement of humanity.

In the areas of food studies, urban food planning and local food policies (Morgan, 2009; 2013), the holistic approach of the commons has been utilized to criticize a fragmented understanding of food and to support the idea of systemically engaging with all the dimensions of food, which, by its own nature cannot be reduced to its economic representation only. More generally, the commodification of food has been recognized by different authors as one of the causes of the structural fallibility of the corporate food system (Russi and Ferrando, 2015; Vivero-Pol, 2017a), the social injustices that are associated with transnational food chains (Pothukuchi and Kaufman 1999; Power 1999) and food insecurity (Friedmann 1993).

Following on from some of the experiences mentioned above, this chapter combines our different backgrounds as planners, geographers and lawyers to highlight the importance of a paradigmatic and political shift from commodity to commons and to go beyond the 'sole' redefinition of the understanding, practice and role of food. The real intelletual and practical step forward, we believe, is to use de-commodifying power of the commons to redefine (i.e. subvert) the entirety of the food system and thus help rewrite and reconsider each element that make food possible. In our eyes, food cannot be dissociated from the broader series of socio-economic interactions that lie behind it. Marx clearly affirmed that the wage hides all the unpaid work that goes into profit (Cox and Federici, 1975). Similarly, we believe that a critical de-commodification of food and its transition into a commons must approach food as the outcome and generator of contingent and dynamic socio-spatial relationships, most of which are hidden behind the final price and the act of consumption. In simpler terms, we think that it is not possible (or accurate) to aim 'only' to the vision of food as a commons, but that a radical and ecological transformation of our understanding of food must pass through thinking, performing and engaging with the whole food system as a commons. As recently discussed by Holt-Giménez (2017), we agree that what really matters is not food as the final moment of a complex system of capital, labor, land and water, but the relationships within the food system, and the way we govern them.

In order to present the alternative and formulate some suggestions, we have decided to adopt a horizontal and vertical approach to the topic, dividing the chapter as follows. The first part briefly introduces the concept of food systems and focuses on the different components that may be individually conceived as commons and the way in which de-commodification may be exercised. This is what Laura Nader (1980) would call a vertical slice, i.e. an engagement with the other commons that are mobilized (and often appropriated) underneath food as a commons. The second part of the chapter shifts from elements to phases, adopting a horizontal perspective that connects the different dots of a simplified food chain. However, rather than describing their functioning, we offer some preliminary thoughts on what it would mean to relate them to the paradigm of the commons. The third section engages with practices (Alternative Food Networks) and food policies (Urban Food Strategies) that in our opinion could represent privileged platforms to conceive, implement and consolidate the idea of food systems as a commons. In the Conclusions, along with an invitation to further dialog and debate, we step into the consumers' shoes and offer some reflections on the role that they could have in constructing and practicing the food system as a commons.

Before we continue, two caveats are needed, which we hope the reader will overcome through the combination of the other contributions contained in this volume. Firstly, we decided to be selective and to pick only a few of the elements that intervene in the definition of the complexity of the food system. We refer to the other authors in this volume for a broader and deeper analysis of some of them, such as seeds, agricultural knowledge, the cultural aspects of food, etc. Secondly, we are conscious that our attempt to think about the food system as a commons may lead to some questions about scalability and potential to satisfy the nutritional needs of the whole population. To this set of questions, we can only reply by mentioning that "the significance of smallholder agriculture is not limited to a subgroup of low-income countries" but, rather, "smallholder[s] play a role in the EU, OECD countries, and in developing countries, including Brazil, India, China that have reached 'middle income' status in the past 15–20 years" (HLPE, 2013: 28). If small-scale farmers and the food that they produce are the reason why there is food on the planet, communing and scaling up shall be easier than imagined.

Food system made of multiple commons

Pothukuchi and Kaufman (1999) define a food system as the combination of all the activities connected to production, transformation, distribution, consumption and post-consumption of food. In addition, their analysis incorporates all the institutions and the regulatory activities that link to these phases. Like food as a final output, which is composed of a multiplicity of different elements, the food system is the dynamic and heterogeneous ensemble of 'organic' and 'inorganic' elements, 'social' and 'natural' constructions (Murdoch et al., 2000), 'material' and 'immaterial' building blocks. Rather than a coherent and uniform space, the food system is a complex *assemblage* of people, resources, places, interactions, relationships, practices and politics. This is even more the case if we think that, despite production being local by definition, the distance between consumers and producers is progressively increasing, the food chain is exponentially the object of desire of transnational financial speculators, and few gatekeepers operate in positions of quasi-monopoly when it comes to connecting land to fork.

It is there, where the territory of food production meets the transnationalism of trade, 'consumption' and 'post-consumption', that we realize that the *commoning* of food at the end of the chain cannot be considered enough. There can be no food as a commons when the rest of the network is still organized and assembled around the commodification of Nature (e.g. land, seeds, soil and water), labour and all the other elements and relationships that make the food system possible. We thus believe in the importance of de-commodifying food, but we think that it would be essential not to be satisfied with this achievement and to go beyond what we could call the 'fetishism of the 'de-commodity''. Leaving aside – for now – the linear idea of phases of production, in this section we aim to make visible the complexity that constitutes the food system and engage with some of the elements (water, soil, labour and food infrastructures and landscape) that are essential to food production and therefore often the object of appropriation, enclosure and exploitation.

Water

Too often, discussions around food do not properly consider the intrinsic relationship between food and water. However, access to water and its use are increasingly contested, way beyond the traditional struggle between privatization of the network and bottling. For example, it has been increasingly proved that the production of food (livestock and fish in particular) is among the first responsible for water consumption in several parts of the world. Despite that, when droughts hit, as in California in 2015 or the Lazio Region (Italy) in 2017, local governments adopted measures aimed to reduced individual consumption of water (such as banning the irrigation of plants in the backyard) but did not take any initiative aimed at reducing the amount of water that was absorbed by farmers and livestock producers.

Similarly, the link between water and food should also be considered in light of the globalization of trade and construction of long-distance food chains. Research on the 'global virtual water network' produced by the University of Trento helps us to understand that transnational movements of goods, and in particular of food, are nothing more than a reallocation of natural resources from one part of the world to another, including of the water content of fruit, vegetables, meat, etc. (Antonelli and Sartori, 2014).[2] When supermarkets in the United Kingdom display red peppers that were produced in Morocco, what they are doing is generating profit out of the combination of water, soil, labour, etc. extracted in the Global South and whose value (both in terms of utility and exchange value) is appropriated in the Global North. This different understanding of food as water raises important questions in terms of social justice, environmental degradation, long-term sustainability of production, etc. Moreover, it forces a critical audience to go beyond the idea of 'who owns the water at the source' and to engage with the broader compatibility of transnational trade with the idea that water (even when transformed into food) should be considered and treated as a commons.

Soil

The reader is certainly aware of Hardin's *Tragedy of the Commons* (1968) and the way in which the depletion of soil by grazing sheep was utilized to prove the need for individual property rights. Leaving aside the validity of the theory, which has been widely debated and dismissed, this example shows that the fertility of soil occupies a central spot in the overall debate around the commons. Like Hardin, Ostrom (1990) and Marx 1969) clearly identified the restorative capacity of the soil as essential in guaranteeing the reproduction of a system, whether in terms of its ecological stability or in terms of the limits of capitalism as an extractivist form of organizing the relationships between society and nature (Moore, 2015; Mattei and Capra, 2015).

In more concrete terms, soil is increasingly under pressure both from within the agricultural realm (loss of biodiversity, use of fertilizers, intensification of production, expansion of the livestock frontier, etc.) and from outside it (urbanization, land conversion, climate change, etc.). For example, Italy loses 35 ha of agricultural land every day to give space to urban areas (ISPRA, 2016), a circumstance that not only affects the landscape but also has clear repercussions in terms of labour, water usage, food security and dependence on international food trading. The defence of land and soil, along with water, therefore represents a crucial step in the struggle against commodification and appropriation. As discussed in the chapter by Ferrando and Maughan (this volume), movements like Grow Heathrow, People4Soil, Campi Aperti and Mondeggi Bene Comune offer clear examples of the close interconnection between access to land, fertility of the soil, biodiversity, food security and the establishment of a food system that is regenerative rather than extractivist.

On the contrary, the green scheme implemented by the latest Common Agricultural Policy appears to go in the opposite direction, focusing on a superficial attempt to include environmental considerations in the framework of conventional food production without a proper assessment of the intrinsic link between soil, lives and agroecological practices. As a consequence, the green scheme lacks any reference to access to land as a prerequisite of a socially and environmentally sustainable agricultural policy (given the possibility of breaking down large-scale industrial production into a multiplicity of agroecological farms), and contains the provision of direct payments and subsidies that facilitate the abandonment of unsustainable activities but do not reward those who were already respecting the environment and the soil with their agricultural practices. Differently from the rhetoric proposed by the European Commission or digital agriculture, a true transition can only occur through the abandonment of the vision of land and soil as homogeneous and standardized spaces (Scott, 1998), and through their recognition as complex and living ecosystems that are constantly redefined by their interaction with human and non-human beings.

Labour and food infrastructures

Try to grow your own food, and you will become aware of the amount of time and labour that is required. Talk to a farmer, and this will be even more evident. However, one of the most striking features of the conventional food system is its ability of making labour invisible. For an increasing number of people, food is not grown or brought to the supermarkets: it appears. An exception is when labour is transformed into a commercial sale point – as in the case of the Colombian coffee farmer who, being portrayed as stereotypical and smiley cherry picker who rides a donkey and appears not to be struggling, allows coffee multinationals to provide consumers with the idea of immediacy and familiarity. In these cases, labour itself becomes a commodity and an opportunity to generate even more value than what is extracted through production. A well-portrayed smile makes consumers forget the thousands of miles that separate land from cup, hides the role of coffee traders and the hyper-concentrated market and convinces each drinker of the biggest lie, i.e. that because we know who grew our coffee, farmers must know to whom and for how much the product of their labour is sold.

Together with labour, the commodification of food has been very good at hiding the importance of capital and infrastructures in producing what is eaten. Along with people, the fetishization of food suppresses questions and concerns surrounding to the ownership of the means of production, the distribution of the value that is added through transformation, the power dynamics that characterize each relationship along the chain and the geographical distribution of production, transformation and consumption. Following on from the coffee example, although consumers are aware of the thousands of small-scale farmers who harvest the beans, they seem to be oblivious that those beans will be traded only by a bunch of transnational companies, or that few rosters all over the world control more than 50% of the transformation process. On the contrary, production can be collective and generate different socio-economic dynamics. This is the case of the communal bread ovens that still exist in several villages in the Italian Alps (Sasia, 2014) or communal land whose products are shared among the members of the farming community and the people in the surrounding areas. In these cases, the link between labour, capital, personal development, community cohesion and ecological limit appears evident and truly transformative. As discussed below, we thus believe that a truly commons-based food system could not avoid questions of ownership, dignity of labour and distribution of value and resources.

Food as landscape

As the reader would certainly recognize, food is much more than its material reality and is closely connected with the immaterial and cultural. Movements and organizations that aim to protect and safeguard food and are evidence of the intrinsic interconnections between food and a collective way of being, but also between food, history and identity. This is the case, for example, of processes of patrimonialization of food culture such as the inclusion of food cultural knowledge in the UNESCO list of Intangible Cultural Heritage (Brulotte and Di Giovine, 2014; Matta, 2016; UNESCO, 2003).

However, the intangible relevance of food goes beyond its cultural and 'traditional' aspects and concerns the way in which food production and consumption shape the landscape and vice versa. As discussed by Carolyn Steel (2006) in her account of how London was fed throughout centuries, or as more recently proposed by Jason W. Moore with regards to the mutually constitutive relationship between society and nature, there is little doubt that the world that we see around us today has been shaped and defined by food. From the transition from nomadism to sedentarism, from the occupation of new land to feed the growing Roman military, to the food routes that were connecting the four corners of the planet and that drastically changed what is grown and eaten and by whom (think of the way in which corn, potatoes and tomatoes have transformed entire regions of the world after being extracted from Latin America). Any attempt to engage with the food system as a commons would thus have to take into consideration the power that food has in transforming and being transformed by society and nature, and consider them as two elements of the same equation rather than as antagonistic or separated.

After this quick depiction of the vertical slice below food as a commons, we hope the reader may be convinced that engaging with food means engaging with water, soil, land, labour, culture and landscape, and much more. Food is the expression of a material and immaterial complexity that is often hidden behind the 'stuff' that people buy, transform, consume and throw away. This is even more the case at a time of mediatization of food, TV programs, food-based theme parks[3] and '*dancing mozzarelle*' (Bukowski, 2015). The food system disappears. Food is a show. Consumers – and not people – interact with an object aimed at giving individual physical or emotional pleasure, but not at creating connections and community. A view of food as a commons cannot ignore it and cannot be satisfied with fixing only some of the ecological and social inconsistencies of a commodified food system. In the next section, we thus take a detour from the individual manifestations of food as a commons and offer some rough thoughts on what it would mean to adopt a commons-based approach to the food system.

From elements to phases: a commons-based approach to the food system

We have just discussed the possibility of looking at the food system through the individual components that constitute it. From land to labour, it is possible to single out various elements of the food chain and discuss what changes would be needed to move away from a proprietary conception to a communal understanding and a collective form of governance. However, we are aware that a fragmented approach to water, land and soil, infrastructures, animals, biodiversity, etc., may weaken the strength of the commons as a holistic approach to nature and society, and therefore give way to partial and short-term spatial and temporal fixes (Harvey, 2003) rather than a systemic redefinition of the food system. For this reason, we propose to take the 'fetishism of the food commodity' – the idea that food is objectified and transformed into an object without history or complexity – seriously, to trace and map not only the individual elements but also the hidden combinations of socio-economic relationships that lie behind what humans and non-human beings eat. By paying attention to elements and phases, we propose a radical alternative to the conventional food regime (McMichael, 2009), an imaginary that is fully informed by the idea that food as a commons cannot be produced by practices that are non-ecological and socially unjust.

In the process of *commoning* the food system, we should engage with the ideological scaffolding that keeps the current regime alive. Firstly, the redefinitions of growth and production techniques (for both plants and animals) must be placed at the centre. However, this would also face the greatest opposition. At a time of changing dietary habits and mounting pressure to increase productivity, there is little doubt that conventional agriculture and industrial animal production represent two of the most serious threats to the achievement of an ecologically and socially just food system. Would we accept the provision of universal access to food in the name of the commons when this food is rooted in monoculture and loss of biodiversity, large-scale and capital intense farming, strong dependence on water and oil-based chemical inputs, exploitative labour practices, support of unhealthy diets, and based on an inefficient use of resources and a very heavy carbon footprint? These are all conditions that appear to be incompatible with the paradigm and principles that we are advocating. Yet, they increasingly lurk behind the dream of ending hunger and redistributing what is produced on the planet. Malthus is brought back to life and his vision used to defy the possibility of a commons-based food system. Similarly, would we accept technological innovations that offer remedies to some of the hardest environmental and economic problems of farmers but at the same time contribute to an increase in the objectification and control of nature because algorithms, genetic modifications and satellites are constructed in order to automatically respond to nature's variability and inherent unpredictability?

A second area of investigation should be that of food distribution, i.e. the actors and institutions that create and operate the logistics corridors that make (some) food available to (some) consumers. Often marginalized and understudied, after the food–fuel–finance crises of 2007–2009, the physical network that distributes food around the world has increasingly been understood as the missing element of any food policy (Sonnino and Faus, 2014). The way in which large-scale distribution produces and accumulates economic value defines how the food chain affects both producers (who often have to accept decreasing prices and join a global competition among farmers) and consumers (whose choices are often limited and determined by efficiency and logistical constraints rather than by their own needs and rights). A redefinition of the food chain under the paradigm of the commons should thus take into consideration the way in which distribution patterns and procedures affect the space where they take place, how they redefine the identity and lives of the people who are involved (Tsing, 2015) and how they impact the environment that is transformed into a logistic corridor and absorbs most of the externalities that are produced (Cowen, 2015).

Thirdly, a systemic analysis of food as a commons would require engaging with consumption as a practice that is varied and hard to map. However, no serious attempt at commoning the food system could avoid considering aspects such as the spaces where food is consumed; the cultural implications of food; the limits and effects of consumers' choices; the modalities, forms and time of food consumption; and the injustice behind the geographical and social diversity between the spaces where food is prepared and consumed. On the other hand, a commons-based food system could not ignore the centrality of the act of cooking as a social, cultural and democratic moment that should always connect the consumption of food with the act of preparing it, possibly in community kitchens where inequality and idiosyncratic behaviours can be overcome by a sharing approach and the horizontality of relationships. Even more than the two previous phases, the quality and dignity of work, and in particular of reproductive and care labour, cannot be ignored and should be put at the forefront of a systemic transformation (Federici, 2004; Alessandrini, 2011).

Finally, a commons-based analysis of the food system should pay attention to the ability of the existing setting to guarantee that no food is lost or wasted. However, this does not mean accepting that people living in food poverty are fed with the food that no one else has bought, as is the case for most of the contemporary interventions around food waste. Rather than supporting a common vision of the food system, the mere reallocation of the excesses legitimates the underlying assumption that food is a commodity and therefore can be produced in abundance and wasted with little or no economic cost for the responsible. On the contrary, actors and institutions should be organized in a way that production, transportation and consumption satisfy both the needs of the human population and deal with nature and resources in an ecological and regenerative way. Aesthetic standards, long-distance traveling, excesses in production and purchase, aggressive marketing strategies and any activity that leads to the generation of surplus and waste should not be covered up by the work of charities and food banks, but thoroughly analysed and exposed for their social and environmental unacceptability.

Of course, what we have just presented is a simplistic and pretty rough reconstruction of the areas that would be particularly affected by a redefinition of the food system according to the vision and objectives of the commons. However, we are convinced that this first step can help thinking systemically about food and the transformations that would need take place if we were to construct a new network based on the pillars of equality, justice, democracy, ecology and bottom-up participation. Rather than a methodological or analytical tool, the shift from food as a commons to food systems as a commons is a political drift that requires the recognition of the limits of fragmentation of the current approach to food (as if it were composed of a multiplicity of independent areas such as education, health, poverty, etc.).

All over the world, conceptualizations and experiences such as the *City-Region Food System*, the *local food system* (Hinrichs, 2003) and the *Systèmes Agro-Alimentaires Localisés – SIAL* (Cerdan et al., 2017) are trying to intensify the connections between the different phases and highlight the importance that each has with people, places and nature. In the section that follows, we look at Alternative Food Networks and Urban Food Strategies as two forms of engagement with the systemic nature of food and two opportunities to implement and integrate a commons-based approach to the way in which actors, institutions, places, rights, resources and nature are coordinated around food. What appears clear is the inherent and inalienable connection between food and territory, so that the latter becomes the starting point for our interpretation of the food system through the prism of the commons.

Food system as a commons: theories and practices

In the previous sections, we have discussed why the different components and relationships that constitute a food system should be considered as a commons and speculated on what social and ecological transformation would take place if this was to happen. Sceptical of the economic definition of the commons, which only focuses on rivalry and excludability, we have adopted a position that is closer to what may be considered a political approach to the commons (Mattei, 2013), i.e. we recognize the importance of taking the food system outside of the market–state dualism and rebuilding it around horizontal subsidiarity and the ecological limits of the planet. In this view, we believe in the adoption of a systemic vision of the food system that actively and responsibly involves all the actors, places, phase and dimensions of the food chain.

In our proposal, we find certain similarities with what Magnaghi (2012, p. 4) calls a 'return to the territory', i.e. a material and immaterial shift of society in opposition to the process of de-territorialization which characterizes our times. In his words:

> in contemporary society [we see that the] cure of the territorial commons has become increasingly feeble and distracted while processes of privatization of these same goods and their use are expanding and with them the de-territorialization of production and consumption. [...] in order to strengthen the concept of territory as a commons it is not enough to consider the territory in itself as a public good [...] it is necessary that it is considered, it sounds obvious, as a commons, that cannot be sold nor the object of *usucapio*, similarly to what happens to historical civic lands and that is characterized by an autonomy of use which makes it different from property (both public and private). It is here that we can start a search of new forms of management, based on participatory processes and active citizenship, which would bring back to life the meaning and the principles (although not necessarily the form) of civic uses.[4]

Like the process of re-territorialization, the idea of the food system as a commons requires the redefinition of the underlying institutions of the food chain and the attribution of new roles to all its actors. Apart from the dualism between private and public, commons-based food systems require producers, transformers, traders, consumers and all the other participants to be actively involved in a self-conscious and democratic process of co-production and co-determination of the food system. All the moments are connected and relevant, including that of consumption, which becomes an opportunity to critically and politically engage with the rest of the chain, although not the sole leverage. This would mean regaining control over the food system and aiming towards true and effective food democracy.

In this view, we are intrigued by the idea of *food citizenship* advanced by Wilkins (2005, p. 271) as the "practice of engaging in food-related behaviours that support, rather than threaten, the development of a democratic, socially and economically just, and environmentally sustainable food system." This new form of citizenship, which is at the same time aware, responsible and active, appears crucial to the vision of the food system as a commons, and is therefore welcomed. However, it will be important to avoid a simple linguistic shift from consumers to food citizens and co-producers, and promote instead a radical understanding of the socio-economic–ecological implications of the food system. Otherwise, this transition would simply reinforce ongoing attempts to 'green-wash' the existing system and to convince consumers that they can really make a difference, while they are simply scratching the surface of the iceberg (Ferrando, 2017).

Among the multiple experiences that may be considered expressions of a commons-based approach to food systems, we have decided to present and offer some thoughts on two of the axes that Wiskerke (2009) utilizes to define alternative food geographies: a) Alternative Food Networks and b) Urban Food Strategies. Although they are limited in space and scale, we believe that these examples provide important element for intellectual and political engagement. Avoiding any easy optimism and being conscious of the risk of legitimizing niche and gentrifying practices, we thus hope that they can be taken as inspiration for future and different interventions that may exclusively rely on members of the community or even combine a bottom-up push from the people with the facilitating and coordinating role of a renewed and truly democratic public authority.

Alternative Food Networks as co-production of food

Most of the time, Alternative Food Networks (AFNs) are defined by referring to what they are not. Rather than a lack of identity and character, this is a sign of the heterogeneity of the practices that fall into these categories: whether formal and informal, linked with production, distribution or consumption, AFNs propose and practice models that oppose the conventional food system, the centrality of agribusiness and the dominance of Large-Scale Retail Distribution. The alternative character does not depend, therefore, on one single element. On the contrary, it can be connected with the establishment of horizontal relationships among the members of the network, the shortening of the distance between production and consumption, the acknowledgment and redistribution of power throughout the chain, a focus on the ecological value of the chain and its interactions with the planet, the use of food as a means to struggle for social and spatial justice or other features that clash with the dominant corporate food system (Goodman et al., 2012).

Some help in the identification of common features, although not an exhaustive methodology, comes from Jarosz (2008), according to whom AFN experiences are often characterized by the presence of at least one of the following elements: a) a reduction of the distance between producers and consumers; b) the construction around small-scale farming and/or a low or positive ecological impact; c) purchase schemes that are not exclusively based on the exchange value of food but also consider the social interaction between producers, distributors and consumers and the socio-economic utility that it generates; or d) a clear identification of the political nature of the process and of the role that a different food system can play in achieving social, economic and environmental sustainability.

A point of interaction between AFNs and the commons relies, in our opinion, in the motivation that pushes an increasing number of producers, distributors and consumers in certain parts of the world (and certain parts of communities) to join these networks rather than conventional food spaces like supermarkets. Although we are aware that several cases of labour and environmental abuses have been reported in the case of practices that claim to be alternative, we believe that most AFNs offer a valid example of the way in which the food system can be constructed from the bottom up and participated in by its members, and have each of its phases constructed not as an end in themselves but as elements of a dynamic and continuous dialog between food, actors, institutions and the planet (Hendrickson and Haffernan, 2002; Sonnino and Marsden, 2006). Leaving aside those examples where participation is due to mere financial considerations or convenience (Tregear, 2011), there are several realities where AFNs express their members' desire to be critically and actively engaged in a food system that is better in terms of economic, social and environmental impact.

Out of the different examples, we consider Community Supported Agriculture (CSA) to best represent the possibilities of thinking, building, managing and implementing a food system that produces much more utility than the one deriving from food consumption and the impacts of which are perceived well beyond the circle of actors involved. As with AFNs, even CSAs cannot be easily defined. Like most AFNs, they refer to a variety of partnerships between farmers and consumers. What is unique, we believe, is the level of consumer involvement: on the one hand, they pay in advance for a share of the annual (or bi-monthly) production and obtain the right to a share of whatever the farm will produce; on the other hand, most of the schemes require them to pick and distributes shares every week, so as to strengthen the relationships between growers and consumers and also among consumers, and they can also contribute with hours of work in addition to or in replacement of the payment.

Thanks to the direct financial and personal involvement of consumers, CSAs challenge the idea of food systems as based on competition and food prices as based on market dynamics. "Farmers do their best to produce sufficient quantities, quality of food and variety to meet consumers' needs" (Junge et al., 1995), but price and food availability are not the sole determinant of the farmers' activity and their ability to continue doing their job. On the contrary, farmers and consumers establish a relationship that is based on solidarity and mutual support, where risks and benefits are shared between everyone so that ecological and socially sustainable food growing can continue (almost) independently from the volatility of prices and the direness of climate change.

Urban Food Strategies: institutional frameworks to experience a commons-based approach to food systems

An integrated approach to the multidimensionality of food and the complexity of the food system can also be institutionalized via the elaboration of transversal public policies that aim to coordinate and manage together all the areas of intervention that are related to food: education, health, agriculture, public procurement, urban planning, waste management, food aid, etc. In both the Global North and the Global South, local authorities have been engaged in the study, facilitation and implementation of policies and concrete measures that use their regulatory, administrative and financial power in order to oppose the worst effects of the hyper-commodification of food and of the food system.

In the North, the cities of Toronto (Blay-Palmer, 2009; Mah and Thang, 2013), New York (Morgan and Sonnino, 2010; Morgan, 2015) and San Francisco (Morgan and Sonnino, 2010) are considered among the pioneers. On the other side of the Atlantic, attempts to develop integrated urban food strategies have reached the United Kingdom (see Carey, 2013 for a discussion on Bristol; Partnership for Coventry, 2017), Northern Europe (Wiskerke, 2009; Cretella and

Buenger, 2016) and Greece (Skordili, 2013). Similarly, examples are visible in Australia (Caraher et al., 2013), China (Lang and Miao, 2013), Brazil (Rocha and Lessa, 2009) and several other parts of the Global South, where the debate around ecological and integrated food systems is closely intertwined with that around food sovereignty, access to land and the rights of the peasants (Borras et al., 2015; Ferrando and Vivero-Pol, 2017).[5]

Although different, all these experiences are characterized by the attempt to establish bottom-up processes of participation and dialog among all the actors who are more or less integrated into the food system (from the actual operators to academia, migrant workers and civic society), so that they can all contribute to the elaboration of a common vision and to the identification of the actions that are needed to achieve it. Often, urban food strategies capitalize on existing networks and connections, trying to move beyond their individual frameworks and to offer a coherent and coordinated opportunity to work together towards the identification of processes, practices and mechanisms that can guarantee (universal or targeted) access to healthy, local, socially just, ecological and culturally appropriate food (Sonnino, 2009). Each city and each authority has defined its priorities and its pattern. However, some choices appear to be recurrent and characterized by a high level of interconnection, including re-territorializing production and consumption, filling the gap between rural and urban, 're-moralizing' food systems (Morgan, 2010), improving food education and food literacy, establishing food hubs and community kitchens and peer-to-peer support in changing purchasing and consumption habits.

One of the strongest features of urban food strategies is their positive impact in terms of food security (Sonnino and Spayde, 2014), which is mainly due to the level of independence that local authorities and communities obtain vis-à-vis market dynamics and the abandonment of a reductionist approach to food security as a simple matter of feeding hungry people. As discussed by Lang (2010), the holistic and systemic approach embedded into urban food strategies provides other forms of support and integration beyond the division between providers and receivers of food. More importantly, they create new spaces for food (beyond the food banks and traditional places where food security is pivotal), which can use food (and food practices) to generate other forms of value that are not only economic but also spatial, social and cultural.

Despite our enthusiasm, the analysis of more than 60 sustainable food policies,[6] including the Milan Urban Food Policy Pact, has revealed a lack of the term 'commons' or any specific understanding of food policies as a way to construct a commons-based food system. However, we believe that these instruments represent an 'indirect narrative' of the food system as a commons: as a matter of fact, they offer different levels of recognition of the multiple dimensions of urban food systems; talk about participation and ecology; are based on bottom-up processes and on coordination with the public authority as a facilitator of collective processes; and understand the importance of avoiding a fragmented attitude towards the complexity and interdependency of food production, transformation, transportation, consumption and post-consumption. Whether institutionalized or not, the idea that food systems have to be thought of and governed as a commons is already embedded in several experiences around the world. Rather than starting from scratch, it is thus a matter of awareness, narrative, and consolidation.

Conclusions

The publication of an edited volume on *Food as a Commons* is evidence of the intellectual and practical desire for transitioning away from the idea of food as a commodity in order to find a long-term solution to the social, economic and environmental injustice that characterizes the corporate food regime (McMichael, 2009). In this regard, our contribution has proposed moving a step further and recognizes the necessity of combining the paradigm of the commons with a holistic approach to the food system. Rather than focusing on food as a commons, we have proposed engaging with the entire food system as a commons (or, in other words, adopting a commons-based approach to the food system), recognizing that food is the final expression of an intricate network of people, institutions, spaces, nature, power and several other elements. If we want to achieve the goal of democratic, just and ecological food for all, there is no alternative than constructing, organizing and practicing each of these moments and the whole food system as a commons.

Needless to say, we are aware that our proposal is not risk-free. First of all, we realize that if everything is a commons, nothing is a commons. For this reason, the possible links between food systems and the paradigm of the commons must be critically unpacked and assessed. Especially when it comes to constructing public narratives and claiming political interventions. A second, more specific, risk is that an excessive reliance on bottom-up mechanisms and processes for commoning the food system may represent an excuse for public actors to dismiss their legal and political responsibilities in terms of food security and food justice, socio-spatial equality, economic and environmental sustainability and control over anti-social and anti-ecological practices, in particular when it comes to regions of the world with a high level of inequality between areas and communities. Commoning the food system means changing the assumptions around each phase of the food chain, identifying and redistributing responsibilities, redefining practices and spaces for deliberation and identifying which actors shall be rewarded and which excluded. In no way shall it be leveraged to legitimize the neoliberal dream of the minimum state, or the creation of niches of sustainable and healthy food systems surrounded by a sea of conventional practices.

Notes

1 Translation by the authors. Italics added by the authors.

2 http://web.unitn.it/ssi/26959/the-global-virtual-water-network-social-economic-and-environmental-implications

3 We are making reference to the Fabbrica Italiana Contadina (FICO), the latest food project launched by Oscar Farinetti, the founder of Eataly, together with Eataly World, Coop Alleanza 3.0 and Coop Reno. FICO is a 100,000-square-meter theme park where food is at the same time an attraction and a commodity to buy.

4 Translation by the authors. In Italy, the term "usi civici" refers to the right of collective use that some communities can exercise on public or (less frequently) private goods, such as natural resources, pastures, woods and so on. The right to "usi civici" is still ruled by Italian national law (l. 1766/1927).

5 The authors have been recently involved in similar processes in the cities of Turin and Coventry. In both circumstances, the actors of the food system, members of the academia, city councils, civil society and some representatives of the private sector have gathered to discuss possible forms of integration and coordination between existing policies and projects, with the idea of combining a short-term intervention against food hunger with a long-term redefinition of the local food system. At the European level, in 2016 IPES-Food launched an interesting attempt to improve the coherence and coordination of European policies that affect the food system, with the aim of establishing a European Common Food Policy by 2019.

6 We have analysed all the food charters and food policies produced by the members of the UK Sustainable Food Cities (www.sustainablefoodcities.org) along with 20 documents released by cities and regions in the USA, Canada and Northern Europe.

References

Alessandrini, D. (2011), Immaterial Labour and Alternative Valorisation Processes in Italian Feminist Debates: (Re)exploring the 'Commons' of Re-production, *feminists@law*, 1(2), 1–28.

Antonelli, M. and Sartori, M. (2015), Unfolding the Potential of the Virtual Water Concept. What Is Still Under Debate?, *Environmental Science & Policy*, 50, 240–251.
Assies, W. (2003), David versus Goliath in Cochabamba: Water Rights, Neoliberalism, and the Revival of Social Protest in Bolivia, *Latin American Perspectives*, 30(3), 14–36.
Blay-Palmer, A. (2009), The Canadian Pioneer: The Genesis of Urban Food Policy in Toronto, *International Planning Studies*, 14, 401–416.
Borras, S., Franco, J. and Monsalve Suárez, S. (2015), Land and Food Sovereignty, *Third World Quarterly*, 36 (3), 600–617.
Brulotte, R. and Di Giovine, M. (2014), *Edible Identities: Food as Cultural Heritage*, London and New York, Routledge.
Bukowski, W. (2015), *La Danza delle Mozzarelle*, Rome, Alegre.
Caraher, M., Carey, R., Mac Conell, K. and Lawrence, M. (2013), Food Policy Development in the Australian State of Victoria: A Case Study of the Food Alliance, *International Planning Studies*, 18(1), 78–95.
Carey, J. (2013), Urban and Community Food Strategies. The Case of Bristol, *International Planning Studies*, 18(1), 111–128.
Cerdan, C., Boucher, F., Autier, D. and Fournier, S. (2017), Les systèmes agroalimentaires localisés, in P. Caron, E. Valette, T. Wassenaar, G. Coppens d'Eeckenbrugge, V. Papazian (eds), *Des territoires vivants pour transformer le monde*, Paris, Editions Quæ, 136–140.
Cowen, D. (2015), *The Deadly Life of Logistics*, Minneapolis, MN, University of Minnesota Press.
Cox, N. and Federici, S. (1975), *Counter-Planning from the Kitchen: Wages for Housework, A Perspective on Capital and the Left*, Brooklyn, NY, New York Wages for Housework Committee.
Cretella, A. and Buenger, M. S. (2016), Food as Creative City Politics in the City of Rotterdam, *Cities*, 5, 1–10.
Dalla Costa, M. R. (2007), Food as Common and Community, *Commoner*, 12, 129–137.
Fantini, E. (2012), Water as Human Right and Commons: Themes and Practices in the Italian Water Movement, *Pace diritti umani/Peace Human Rights*, 1, 15–40.
Federici, S. (2004), *Caliban and the Witch: Women, the Body and Primitive Accumulation*, New York, Automedia.
Ferrando, T. (2017), Certification Schemes and Labelling as Corporate Governance: The Value of Silence, in G. Baars, A. Spicer (eds), *The Corporation: A Critical, Multi-Disciplinary Handbook*, Cambridge, Cambridge University Press, 372–382.
Ferrando, T. and Vivero-Pol, J. L. (2017), Commons and 'Commoning': A 'New' Old Narrative to Enrich the Food Sovereignty and Right to Food Claims, *Right to Food and Nutrition Watch*, 10, 50–56.
Friedmann, H. (1993), After Midas' Feast: Alternative Food Regimes for the Future. In P. Allen (ed.), *Food for the Future: Conditions and Contradictions of Sustainability*, New York, Wiley and Sons, Inc., 213–233.
Goodman, D., Du Puis, M. and Goodman, M. (2012), *Alternative Food Networks*, London, Routledge.
Hardin, G. (1968), The Tragedy of the Commons, *Science*, 162(3859), 1242–1248.
Harvey, D. (2003), *The New Imperialism*, Oxford, Oxford University Press.
Hendrickson, M. K. and Heffernan, W. D. (2002), Opening Spaces Through Relocalization: Locating Potential Resistance in the Weaknesses of the Global Food System, *Sociologia Ruralis*, 42(4), 347–369.
Hinrichs, C. (2003), The Practice and Politics of Food System Localization, *Journal of Rural Studies*, 19(1), 33–45.
HLPE (2013), *Investing in Smallholder Agriculture for Food Security*, A Report by the High Level Panel of Experts on Food Security and Nutrition, Rome, Committee on Food Security.
Holt-Giménez, E. (2017), Introduction, in J. M. Williams, E. Holt-Giménez (eds), *Land Justice: Re-imagining Land, Food, and the Commons*, Oakland, CA, Food First, 1–14.
ISPRA – Istituto Superiore per la Protezione e la Ricerca Ambientale (2016), *Consumo di suolo, dinamiche territoriali e servizi ecosistemici*, Rome, ISPRA.
Jarosz, L. (2008), The City in the Country: Growing Alternative Food Networks in Metropolitan Areas, *Journal of Rural Studies*, 24, 231–244.
Junge, S. K., Ingram, R., Veerkamp, G. E. and Blake, B. (1995), *Community Supported Agriculture Making the Connection. A 1995 Handbook for Producers*, Davis, CA, University of California, Davis.
Lang, T. (2010), Crisis? What Crisis? The Normality of the Current Food Crisis, *Journal of Agrarian Change*, 10(1), 87–97.
Lang, G. and Miao, B. (2013), Food Security for China's Cities, *International Planning Studies*, 18(1), 5–20.
Magnaghi, A., ed. (2012), *Il territorio commons*, Florence, Firenze University Press.
Mah, C. L. and Thang, H. (2013), Cultivating Food Connections: The Toronto Food Strategy and Municipal Deliberation on Food, *International Planning Studies*, 18(1), 96–110.

Marx, K. (1969), *Theories of Surplus Value*, 3 vols, Moscow, FLPH.
Matta, R. (2016), Food Incursions into Global Heritage: Peruvian Cuisine's Slippery Road to UNESCO, *Social Anthropology*, 24(3), 338–352.
Mattei, U. (2013), Protecting the Commons: Water, Culture, and Nature: The Commons Movement in the Italian Struggle Against Neoliberal Governance, *South Atlantic Quarterly*, 112(2), 366–376.
Mattei, U. and Capra, F. (2015), *The Ecological of Law. Toward a Legal System in Tune with Nature and Community*, Oakland, CA, Berett Kohler.
McMichael, P. (2009), A Food Regime Analysis of the 'World Food Crisis.' *Agriculture and Human Values*, 26(4), 281.
Moore, J.W. (2015), *Capitalism in the Web of Life*, London and New York, Verso.
Morgan, K. (2009), Feeding the City: The Challenge of Urban Food Planning, *International Planning Studies*, 14, 341–348.
Morgan, K. (2010), Local and Green, Global and Fair: The Ethical Foodscape and the Politics of Care, *Environment and Planning A*, 42, 1852–1867.
Morgan, K. (2013), The Rise of Urban Food Planning, *International Planning Studies*, 18, 1–4.
Morgan, K. (2015), Nourishing the City: The Rise of the Urban Food Question in the Global North, *Urban Studies*, 52(8), 1379–1394.
Morgan, K. and Sonnino, R. (2010), The Urban Foodscape: World Cities and the New Food Equation, *Cambridge Journal of Regions, Economy and Society*, 3(2), 209–224.
Murdoch, J., Marsden, T. and Banks, J. (2000), Quality, Nature, and Embeddedness: Some Theoretical Considerations in the Context of the Food Sector, *Economic Geography*, 76(2), 107–125.
Nader, L. (1980), The Vertical Slice: Hierarchies and the Children, in G. M. Britan, R. Cohen (eds), *Hierarchy and Society: Anthropological Perspectives on Bureaucracy*, Philadelphia, PA, Institute for the Study of Human Issues, 31–43.
Ostrom, E. (1990), *Governing the Commons*, Cambridge, Cambridge University Press.
Partnership for Coventry (2017), *Feeding Coventry: Coventry Food Charter*, available from http://www.coventrypartnership.com/feeding-coventry/ [last accessed 23 November 2017].
Pettenati, G. and Toldo, A., (2016), Introduzione. Il Cibo è un bene comune?, *AA. VV, Commons/Comune: geografie, luoghi, spazi, Memorie Geografiche della Società di Studi Geografici*.
Pothukuchi, K. and Kaufman, J. (1999), Placing the Food System on the Urban Agenda: The Role of Municipal Institutions in Food Systems Planning, *Agriculture and Human Values*, 16, 213–224.
Power, E. (1999), Combining Social Justice and Sustainability for Food Security. In M. Koc, R. MacRae, L. J. A. Mougeot and J. Walsh (eds), *Hunger Proof Cities: Sustainable Urban Food Systems*, Ottawa, International Research Development Center, 30–37.
Rocha, C. and Lessa, I. (2009), Urban Governance for Food Security: The Alternative Food System in Belo Horizonte, Brazil, *International Planning Studies*, 14, 389–400.
Rodotà, S. (2012), Postfazione. Una strategia globale contro lo human divide, in M. R. Marella (ed.), *Oltre il pubblico e il privato. Per un diritto dei commons*, Verona, Ombre corte.
Rundgren, G. (2016), Food: From Commodity to Commons, *Journal of Agricultural and Environmental Ethics*, 29, 103–121.
Russi, L. and Ferrando, T. (2015), Capitalism A Nuh' Wi Frien'. The Formatting of Farming into an Asset, from Financial Speculation to International Aid, *Catalyst: A Social Justice Forum*, 6, 7.
Sasia, G. (2014), *Un commons nelle Alpi Occidentali. I forni comunitari a Rore*, Master's Thesis in Educational Sciences, Università di Torino.
Scott, J. C. (1998), *Seeing Like a State*, New Haven and London, Yale University Press.
Skordili, S. (2013), Economic Crisis as a Catalyst for Food Planning in Athens, *International Planning Studies*, 18(1), 129–141.
Sonnino, R. (2009), Feeding the City: Towards a New Research and Planning Agenda, *International Planning Studies*, 14, 425–435.
Sonnino, R. and Marsden, T. (2006), Beyond the Divide: Rethinking Relationships Between Alternative and Conventional Food Networks in Europe, *Journal of Economic Geography*, 6, 181–199.
Sonnino, R. and Faus, A. M. (2014), Sostenibilità e sicurezza alimentare: gli anelli mancanti, *Agriregionieuropa*, 10(36), www.agriregionieuropa.univpm.it.
Sonnino, R. and Spayde, J. (2014), The New Frontier? Urban Strategies for Food Security and Sustainability. In T. Marsden, A. Morley (eds), *Sustainable Food Systems: Building a New Paradigm*, London, Earthscan, 186–205.
Steel, C. (2006), *Hungry City. How Food Shapes Our Lives*, London, Vintage Books.

Sumner, J. (2011), Serving Social Justice: The Role of the Commons in Sustainable Food Systems. *Studies in Social Justice*, 5(1), 63.
Tregear, A. (2011), Progressing Knowledge in Alternative and Local Food Networks: Critical Reflections and a Research Agenda, *Journal of Rural Studies*, 27(4), 419–430.
Tsing, A. L. (2015), *The Mushroom at the End of the World: On the Possibility of Life in Capitalist Ruins*. Princeton, NJ, Princeton University Press.
UNESCO (2003), Convention for the Safeguarding of the Intangible Cultural Heritage.
United Nations (2014), Department of Economic and Social Affairs, Population Division. *World Urbanization Prospects: The 2014 Revision, Highlights (ST/ESA/SER.A/352)*, 7.
Vivero-Pol, J. L. (2017a), Food as Commons or Commodity? Exploring the Links between Normative Valuations and Agency in Food Transition, *Sustainability*, 9(3), 442.
Vivero-Pol, J. L. (2017b), Transition Towards a Food Commons Regime: Re-commoning Food to Crowd-Feed the World, in G. Ruivenkamp, A. Hilton (eds), *Perspectives on Commoning: Autonomist Principles and Practice (In Common)*, London, Zed Books, 185–221.
Wilkins, J. L. (2005), Eating Right Here: Moving from Consumer to Food Citizen, *Agriculture and Human Values*, 22(3), 269–273.
Wiskerke, J. S. C. (2009), *On Places Lost and Places Regained: Reflections on the Alternative Food Geography and Sustainable Regional Development, International Planning Studies*, 14, 369–387.

4
GROWING A CARE-BASED COMMONS FOOD REGIME

Marina Chang

Introduction

Since the global food crisis, a plethora of discourses from food movements has arisen. No matter which and how many different pathways and/or ideologies each one advocates, there is a broad-based desire for a structural change as systems shift from one regime to another (Friedmann, 2005, 2009; McMichael, 2005, 2009a,b). Recognising the fragmented quality of current food movements, there is an urgent need for strategic alliances (Holt-Giménez and Shattuck, 2011) and 'convergence in diversity' (Amin, 2011). As we search for answers, the recent surge of the commons seems to provide some promising insights. The chapter begins with an explanation of the food regime theory and its different variations, then presents the notion of the commons food regime as a new kind of food regime, and proposes a holistic, interconnected, and intersectional idea of care as the core to construct and grow a transformative new food regime. Growing a care-based commons food regime responds to a call for a new food regime in the 21st century.

Rethinking the food regime theory

The concept of food regimes was first developed in the 1980s, influenced by the French school of regulation theory as it applied to the non-agricultural economy (Aglietta, 1979; Atkin and Bowler, 2001). Friedmann and McMichael (1989:5) linked 'international relations of food production and consumption' to 'periods of capitalist accumulation'. As a historical concept, it defined specific points in time that oriented international food production and consumption (Friedmann, 1993). The first regime was marked by colonial forms of agrarian development, wherein commercial farming was specialised, and Europe imported wheat and meat in exchange for European manufactured commodities. The second regime was defined by the rise of industrialised, durable food production, such as grains and the emergence of a grain–livestock complex centred in the USA.

Although general periods of consolidation of power and movement behind one food regime often provide stability to the larger agri-food system, this does not mean a lack of change but rather stabilised tensions and a juggling of contradictory movements. Thus, when a crisis occurs there is a corresponding period of instability and a confluence of multiple trajectories that seek solutions and propose alternatives to be developed into a dominant model. As a matter of fact, multiple sets of different ideologies, institutions, regulations and policies can be revealed in each of the regimes. Similarly, the transition from the first to second regimes (via the Great Depression) consisted of a crisis in which almost every old aspect of the former was dramatically transformed. In a new and more intense moment of crisis, we should thus start from the assumption that 'systems can be changed' (Campbell and Dixon, 2009:264) and that revolutionary moments – even if they have been incapable of bringing justice to most of the world's population – are still capable of triggering theoretical and empirical engagement with the 'composition' and 'de-composition' of a new food regime.

So, we should ask, are we at the verge of a new food transition? For sure, although it is hard to identify linearity. While authors propose different reconstructions of the contemporary condition, little consensus can be found. Burch and Lawrence (2009) argue that we are now in a 'financialised food regime' that leaves little escape from the mechanisms of commodification and speculation. Whereas McMichael favours the notion of a 'corporate food regime' characterising the neoliberal world order, Friedmann is more cautious about identifying such a phenomenon at present, choosing instead to speak of an emerging 'corporate-environment' food regime (2005). Unlike other food regimes scholars, Dixon brings the human body to the fore. She traces the ideologies of nutrition along with different food regimes, which are central to state and class relations. Campbell (2009) applies a socio-ecological resilience theory to examine the emerging global scale of environmental governance for food auditing to understand cultural politics and social legitimacy in the two contending new food 'regimes': 'Food from nowhere regime' versus 'Food from somewhere regime', both regimes being mutually constitutive; the poor still suffer from the former and the privileged consumers demand the latter.

As Le Heron and Lewis (2009) suggest, food regimes scholars seem to connect their own participation in a variety of political movements to their academic work, which provides a new architecture of food regime theorisation. These direct engagements help them to explore the key challenges and contradictions in the contemporary food systems. Therefore, food regime theory is often seen as a more engaging type of intellectual endeavour, bringing theory to point to a new direction in the politics of that world. As Friedmann (2009:335) suggests, 'Thinking about food regime transition and/or emergence is partly empirical, partly definitional.' Wherever we are in the spectrum of food regimes, we consider essential to take Friedman's 'invitation' (ibid.) seriously and join the conversation on the present and future of food regime theory.

Another new kind of food regime: the commons food regime

For generations, the commons were assumed to be fading away. However, they never disappear from human history. After Hardin's (1968) influential article 'Tragedy of the Commons', theoretical work on the commons was developed in a rather rapid way. Indeed, in recent years we have observed 'the growth of the commons paradigm' (Bollier, 2007) or, to use Clippinger and Bollier's (2005) words, a 'renaissance of the commons'. This vast arena is now inhabited by diverse groups and disciplines ranging from different political interests and philosophies across many geographical regions, both inside and outside academia. People are increasingly finding the term 'commons' crucial in addressing issues of social dilemmas, degradation, and sustainability of a wide variety of shared resources. According to Hess (2008), a variety of approaches of the 'new commons' – different from the traditional natural commons and not necessarily applicable to Ostrom's design principles – evolve or come into being. We are now getting used to hearing about the need to protect the commons from enclosure, producing new commons through collaboration and networks based on voluntary reciprocity, using commons as a pedagogical device, creating new economic models, and rediscovering the commons.

In the context of an edited volume that rotates around the notion of the commons, here the more general term commons is chosen to describe the complexity and variability of different kinds of resources connected with food and to embrace a more inclusive term. It is close to Clippinger and Bollier's (2005, cited in Hess, 2008:35) definition of 'commons' as 'a social regime for managing shared resources and forging a community of shared values and purpose'. In their vision, the term 'commons regimes' identifies a self-organising governance system for shared resources, with a mixture of formal and informal institutions characterised by the application of rules, norms, and principles shaped by a full range of contextual factors.

The concept of a 'complex adaptive system' is useful to highlight the dynamic, non-linear, and emergent nature of governing the natural commons in the social–ecological systems. This invites us to consider what kind of governance system may encourage commons-based creative solutions, and how to manage potentially destructive elements that are inherent to complex systems. Due to its key characteristics of being robust, adaptable, and efficient, commons governance – i.e. self-organisation, self-governance, and collective actions (cooperation) – has been identified as one of the most important governance systems towards real world sustainability in dynamic landscapes such as our complex world (Ostrom, 1990, 2009; Olsson et el., 2004; Levin, 2006; Vincent, 2007; Floke, 2007).

However, from a social constructivist perspective, certain types of discourse would be interpreted and implemented in commons governance in any particular context. Drawing insights from political ecology, Armitage (2008:22) asserts that exploring multiple pathways and trajectories of change, as well as processes of self-organisation and interpretation of these processes, undoubtedly relies on human values and power relations. This reflection, however, points to a major ideological divide. As demonstrated by the other contributions in this volume, the increasing popularity of the commons discourse represents a warning call for critical engagement with its vocabulary and understanding. A deeper analysis of the socio-political nature of the revival of the commons seems to reveal that among the web of complex drivers, there are roots in the anti-capitalist paradigm. The next section draws out challenges from anti-capitalist perspectives, which argue for a radical transformation as the existing system is untenable.

Challenges from anti-capitalism perspectives

The first challenge relates to the historical roots and current relevance of the traditional paradigm of the commons. While acknowledging Ostrom and her colleagues' contributions to the study of commons, Caffentzis (2004, 2010) argues that their work hardly supports anti-capitalist and alter-globalisation movements and sees the commons as an ideal test case for social theory and management, and suggests that studying commons is like studying a firm. According to Caffentzis, Ostrom's problems relate therefore to an issue of governance and management demanding appropriate institutional designs and knowledge about how to achieve efficiency, equity, and sustainability of shared resources in order to avoid a 'tragedy of commons'. Anti-capitalists, in his view, look to the larger class context to understand what determines the dynamics of commons governance. The class clash occurs because while Ostrom and her colleagues focus on the situation to find out exogenous variables and the appropriate institutional arrangements that drive the tragedy of the commons or that prove its non-existence, anti-capitalists see these 'exogenous variables' as part of the violence of the history of commons and enclosures.

The second challenge derives from the understanding that modern capitalism cannot change its unsustainable mode of production, marked by its nature of exploitation in terms of its relationship to nature and to humans on the one hand, and its obsession with continuous economic growth in the pursuit of profit on the other. Reading capitalism through the lens of systems theory, Biel (2011) argues that the dominant mode of production not only depletes physical resources but also exhausts social and ecological repair. Capitalism destroys and alienates society a condition that appears incompatible with Ostrom and her colleagues' assertions that human beings are 'adaptive creatures' with an innate self-organising capacity for collective decision-making and problem-solving. Simply put, people at least have to be fed by healthy and nutritious food, and have access to all sorts of resources, both tangible (e.g. natural resources, shelters, and health care) and intangible (e.g. knowledge, culture, and security), as pre-conditions that allow them to adapt, develop, and unleash such potential. Biel (2011) asserts that, within capitalism, commons represents historical battles over 'commons regimes' as this is the mode of organisation that capitalism hates and fears, precisely because its effectiveness makes appropriation difficult. However, the usual hostility of capitalism towards commons does not exclude the possibility of allowing it to happen – a form of co-option in the promotion of the commons.

The third challenge derives from the incapacity of the traditional discourse around the commons to adopt the holistic and intersectional vision of rebuilding a society through the generation of commons as an alternative to capitalism. Brownhill et al. highlight (2012) the importance of ecosocialism, with its Marxist influence, as a political and intellectual paradigm that can challenge the rhetoric of production and power relations, and at the same time transform our relations both with each other and with nature. In the eco-socialist framework, 'de-alienation' challenges the concept 'de-growth' as being too slow and too late. With its focus on the issue of 'overconsumption', detached from social and historical understandings and struggles, it is argued, 'de-growth' may run the risk of falling into a kind of green capitalism. 'De-alienation' is a gendered interpretation of Marxist 'alienation' – historically, women have been separated from their means of survival and have been 'colonised' and 'enslaved' by men, religions, the state, and the region (Brownhill 2012; also see Federici 1998). Brownhill et al. (2012) emphasise women's strategic position in the battle of the global anti-capitalist movement, grounded in women's crucial and contended responsibility for and stewardship over aspects of fertility that are pre-conditions of capitalist accumulation and social control. It is clear that the challenge to the cross-cutting problems on a global scale requires agency on multiple fronts – across classes, genders, and social movements: the construction of a commons food regime and of a new society requires therefore the engagement of every individual within a complex interaction among different social agents – a unity and integration of people, life, society, and ecology, both materially and spiritually.

From an anti-capitalist perspective, governing the commons becomes, in this sense, a conscious project of rebuilding human capacity. While higher priority is given to ground-up, community-based initiatives, governing the commons is multi-levelled and multi-scalar, aiming at an entirely new planet in the 21st century. However, commons is not only an end but also a journey. According to De Angelis, instead of wishing for a mythical 'utopia' at the end of the tunnel, the act of commoning itself represents a set of values within a given community that is outside capital, grounded in everyday practices for the reproduction of livelihoods (2007:243). Unlike Hardt and Negris's position that there is nothing outside capitalism and the multitude (2000), De Angelis argues for an outside that is alive, lived, and crafted in spaces of sharing, conviviality, and communality through every single act from our body, connecting different struggles throughout the planet, and reclaiming 'the beginning of history'. He states, 'it is through production of commons that new values practices emerge' (2007:239).

If we consider each food regime as a seed of change, a production of commons through new values practices, what if we sow millions of such seeds and nurture them with care? Could we not arrive at a new forest? If this kind of thinking is allowed, then we need to explore the idea of care and its associated values in more detail.

Care is the core of growing a commons food regime

So far, this chapter has revealed that the concept of a commons food regime is in itself a rich combination of two areas of research, namely, a food regimes theory and commons regimes. With existing strands of thoughts, we can sketch out a myriad of pictures of what a commons food regime may look like or mean to different people. Such an approach (which is undertaken by Pettenati, Toldo, and Ferrando in their contribution to this volume) would run against the grain of this project which, at its heart, is concerned with a specific kind of new food regime. At issue here is the challenge of finding a core – 'a unity of diversity' and/or 'convergence in diversity' in order to create synergistic outcomes in a world held together by an array of disciplines, organisations, institutions, movements, and forms of discursive power, and at a multitude of sites across the social domain.

In this section, we propose the idea of care as the core of growing a commons food regime, as 'care' represents an 'attitude' and 'orientation' – a way of relating to other associated values (Tronto, 1994, 1999; Staeheli and Brown, 2003; Gleeson and Kearns, 2001; Haylett, 2003), a recognition of our intersubjectivity as 'the essential, primary and fundamental structure of subjectivity' (Levinas, 1985:95) and a worldview (*cosmovision*) that humankind, instead of being apart from nature, is an integral part of nature.

Geography and economy of care

In recent years, 'care' has been found to be a useful concept in many disciplines, addressing a wide range of dimensions and practices which are aimed not only at academic audiences but also at 'real world' readers (Popke, 2006). We start with what Morgan (2010) describes as 'politics of care'. Acknowledging the limitations of ethical consumerism as being enough to address the challenge of the threat from climate change, Morgan suggests we consider the potential to challenge the current unsustainable food system through a more progressive food policy, a new politics of care. This new politics of care exhibits two defining characteristics: first, they operate in the public instead of private realm and, second, they are practised both locally and globally, demonstrating a 'telescoping perspective' that aims to reduce the tension of geographical divides and the marginalisation of certain voices and communities. Its inclusive approach echoes Smith's assessment that anyone who promotes the idea of care 'need[s] to consider how to spin their web of relationships widely enough that some people are not beyond its reach' (Smith, 2000:97). Morgan also seeks to bring the complexity and multi-dimensional view of politics of care to the emergent global governance regimes, such as the vision of food sovereignty as well as other development projects, including the Kyoto 2 protocol in terms of reduction of green-gas emission and introduction of pro-poor trade regulation. Politics of care, in this sense, has implications that go all the way to how we view and perform traditional notions of democracy and citizenship. Through conscious and constant negotiation and deliberation, it can also be used as a vehicle to mobilise more political resources and collective obligations to counter the devastation caused by climate change in the world we are facing now.

In recent geographical scholarship, there has been a renewed interest in applying the concept of 'care' empirically through studying people's own perceptions, thinking about and understanding our food situations. For example, in Dowler et al. (2010)'s work, three defining themes were mapped in people's 'care-full' motivations: care for local economies, environment, and future generation; care for health and wholeness; and care about transparency and integrity in food systems, including matters of science and governance (ibid.:212). The study of this complex geography of sites of care brings up two notable points that demonstrate the intersectional and multi-dimensional nature of the relationship between care and food. First, care for food contributes to expression of identity and culture, a way of demonstrating people's 'ethical values' in their social practices at multi-levels and multi-scales. Second, both producers and consumers should thus be viewed as 'nuanced economic and social agents, who may know some things and who have an aptitude for learning more, and who are well able to articulate and practice the complexities of sourcing and using food' (ibid.:217). As Kaplan (2000) rightly points out, it is by exploring the social and symbolic perspectives of food that we can begin to grasp the multi-dimensional and complex dynamics of food and caring.

Since food is the foundation of all economies, another compelling argument for care comes precisely from an alternative economic system to the current capitalist economy – a caring economy, which deserves our better attention. A caring economy is not an end product, but a different orientation and a developmental process. Much of the work in this area is inspired by a series of feminist advancements in the development of economic theory and its integration with the prospects for caring. Himmelweit (2007), for example, shows that care is the development of a relationship, with an unequal distribution of caring responsibility and caring needs, as well as the impacts from social norms at different localities and societies. Through analysis of the evolution in labour markets and movements within and between the paid and unpaid economies, she argues that rather than leaving distributional decisions to market forces, public policy should play an important role in preventing an uncaring future, as policy intervention not only can and should reflect the social norms and practices of a society but can also facilitate changes in those norms and practices at the same time. Similarly, as early as in 1975, Federici and Cox's article, 'Counter-planning from the Kitchen', already made a point that demanding a wage for housework, not housewives, is not to addressed to husbands or even all men, but to the state, and is a revolutionary strategy because it changes the power relations and undermines the division of labour in a capitalist society. This perspective prompts us to introduce the concept of the caring economy that would be particularly important for a commons food regime.

Indeed, according to Mies and Bennholdt-Thomsen's 'subsistence perspective', 'there cannot only be no commons without community, but also no commons without economy … hence, reinventing the commons is linked to the re-invention of the communal or commons-linked economy' (2001:1021). The proposed alternative vision to global capitalist economies comes through the prioritisation of ordinary people's capacity for self-sufficiency and self-reliance, living quasi-premodern, as members of peasant-like, small-scaled, decentralised, and regionalised communities. They reject top-down decision-making, or the 'men's house politics' of Athenian 'democracy' – limited to a small group of elites – and applaud the collective, inclusive, and consensus-building form of politics, which they term 'taro field politics', with the first principle that politics is not separated from people's everyday practices and everyone is empowered through practising their politics (Mies and Bennholdt, 1999:207–212).

If the control of our food lies at the heart of the creation of an ecological, democratic, and just future, we must defend all the resources and means for food production from being enclosed and/or destroyed. This means that any resource essential to people's lives should be protected and controlled by the people of a village, a tribe, and a community, in both rural and urban areas. Resources include natural resources such as land, water, and biodiversity as well as their culture, language and knowledge. Strong connections are evident between the survival of commons and subsistence, as they are all linked within living communities. However, Mies and Bennholdt recognise that rights to control and governance the commons do not exist per se, even when they are communal. It requires huge efforts in terms of raising people's awareness through different kinds of social movements, mostly at the parish level, and concerted coordination and support between different groups to create synergy to defend their autonomy, values, and beliefs. Thus, they insist, 'no subsistence without resistance and no resistance without subsistence' (ibid.:213).

The insights of Levinas's philosophy

In the previous sections, we have discussed the concept of care predominantly from an autonomous perspective at various scales and levels. Yet, when we introduce care into the picture, it is hard to neglect Emmanuel Levinas, who is best known for his establishment of a heteronomous ethics, an ethics built not upon the self but the other (Simmons, 1999). There are three applications in Levinas's philosophy that might illuminate the significance of care to a commons food regime.

First and foremost, contrary to other commentaries challenging the proximity principle of care, at the core of Levinas's philosophy there is a clear understanding that the notion of relationship with the other must be conceived in terms of proximity, that is, before the presence of another 'face', a face-to-face encounter with another human being. We should be aware of the role of the body as an interface to experience the other. The face, Levinas says, 'is a living, naked presence; it is expression' (1969:66). A face before us is a living reminder calling for care of and respect for that person, to take care of another human person. The face has a unique authority over us, which provides a possibility to disobey our ordinary structures of being – seeking the self-preservation on which most Western philosophy has been based.

Second, in parallel to this proximity, there is the notion of the asymmetry of the ethical relationship, the non-reciprocal relation of responsibility – being called by another and responding to that other. As Colliere (1982, cited in Lavoie et al., 2006:231) puts it, 'care does make sense and has value only if it takes into account what is precious for people, what has meaning for them or contributes to give meaning again to their life.' Surely it is hard to accept such a strong assertion, a one-way connection, implying a rare moral obligation, especially in a society which normally promotes equality and egalitarianism as a virtue. In this asymmetrical relationship, Levinas encourages us to submit ourselves to learn from others, because 'the content of the other's instructions is ethical; it is a call to learn through action, through a response' (Diedrich et al., 2003). This response must be concrete, infinite, and asymmetrical generosity as we are called to respond to another incarnate being; it can be so extreme, according to Levinas, that the ego must be capable of 'giving the bread out of his mouth, or giving his skin' (1981:77). He insists on 'the significance of compassion at the very heart of care, this 'event of love', this typically human possibility that consists in 'suffering for' others, to share in the suffering of the other' (Lavoie et al., 2006:230).

Third, Levinas expands his care ethics from the face-to-face to the realm of politics, which is made up of many impersonal organisations and institutions. Levinas's care ethics (or care perspective) can be regarded as a radical response to the ancient Greek philosophy of justice, grounded on an autonomy-based ethics of justice (Diedrich et al., 2003). Simmons (1999) argues that far from being apolitical – as some have thought – Levinas's philosophy starts and ends in politics: it fundamentally changes the nature of politics, primarily on a metaphysical level, 'as prima philosophia' (Levinas and Kearney, 1986:29). Contrary to Hobbes's argument that politics could lead to ethics, Levinas and Rotzer claim that 'politics must be controlled by ethics: the other concerns me' (1995:59). For Levinas, '[p]olitics does not subsume ethics, but rather it serves ethics. Politics is necessary, but it must be continually checked by ethics' (1999:98). This political thought has a specific implication that cannot support a liberal freedom, an egoistic freedom (or freedom of the individual being), but only the freedom of ethics where civic duty and active cultivation of responsibility are the key constituencies (Diedrich et al., 2003). If Plato's model of the best 'king' is the one who is best in control of himself, Levinas envisions the one who is in an ethical relationship with the other (Simmons, 1999:99). In this regard, in order to keep politics checked by ethics, there is a need to call for a 'permanent revolution' (ibid.), trying to be the most just in whatever and wherever possible, 'a rebellion that begins where the other society is satisfied to leave off, a rebellion against injustice that begins once order begins' (Levinas, 1989:242).

All things considered, Levinas's radical perspective appears insightful to a commons food regime. First, his principle of proximity reminds us of being an ever watchful eye alert to our surroundings and responsible for both the people and events around us. Care for our local communities and food-related resources around us, though not exclusive, serves as a point of departure to grow a commons food regime. Second, the asymmetrical relationships in terms of moral obligations towards the other suggest that caring practices are not necessarily self-evident ways to behave but rather difficult tasks which require solid motivations and ongoing learning. Growing a commons food regime is a vision and process, more specifically a learning process to help us build our capacity, individually and collectively, to practise caring and associated values through commoning. Third, Levinas's metaphysical thoughts on the relationship between ethics bring us back to the centre of the commons: ordinary people living ordinary lives and coming together to cooperate in governing shared resources, whose very existence, views, and discoveries should be taken into account whenever we consider any commons food regime. For example, most activists tend to think they are in the right, that their own vision of what a world should be is correct, and forget to think of 'the other'. While Levinas refuses to propose laws or moral rules or unifying solutions, he does warn us that politics without ethics – a 'permanent revolution' and a 'rebellion' of our own thoughts – will result in violence and unwanted consequences.

A new worldview (cosmovision)

Despite these strengths, it must be added that Levinas structures his philosophy on the human being itself and ends up falling short on feminist grounds (e.g. Irigaray, 1991, 2005; Kristeva, 1988; Benso, 2000; Hirst, 2009) and – more broadly – on other forms of life. In order to address humans and non-humans, and an ethical and political philosophy, all-encompassing inclusivity towards others, also at a metaphysical level, we argue that a new worldview (*cosmovision*) is required. In the words of the veteran agrarian leader Faustino Torrez (cited in Rosset, 2013: 765), '[t]o transcend classical agrarian reform, we need a rebirth of the concept that goes beyond land as a means of production, incorporating notions of space, territory and cosmovision.' In pursuit of new narratives, alternative to modern Western culture and the capitalist world, many social movements identify with the concept of food sovereignty, for example in Africa, Latin America, and India. Similarly, it is evident that a number of indigenous perspectives are particularly pertinent in this context movement, for example, *buen vivir* in Latin America (Thomson 2011; Gudynas, 2011), *swaraj* in India (Shiva, 2010), and *Ubuntu* in Africa (Naicker, 2011). While they have different cultural and historical roots, these *cosmovisions* share common ideas of the caring about both nature and people. Humanity is a part of and interacts dynamically and reciprocally with all other elements (life systems and living beings) of nature.

Here, Navdanya's[1] journey of promoting seed as commons is a good example. Navdanya means 'nine seeds' or 'new gift' – symbolising protection of biological and cultural diversity. Underpinning Navdanya is the Gandhian ideal of *swaraj*, translated as capacity for self-rule, self-heal, and self-realisation – both individually and collectively. This is grown from within, from the strengths, perspectives, wisdom, and experiences of caring for people, communities, and nature (The Swaraj Foundation[2]). This self-organisation and self-healing characteristic of a living organism is particularly important for the ecosystem to adapt, to learn, and to rebuild itself. Through a process that could be considered similar to governing the commons, *swaraj* has stimulated active citizen participation, rejuvenated people's relationship with nature and the ecosystem, and fostered collaborative innovation. In this way, the movement bolsters the capacity of individuals and communities to shape changes to institutions, policies, and practices to transform the unsustainable food system. From an ancient Indian culture, all living things count, whether fish in the oceans, trees in the forest, or earthworms in the soil.

> The universe is the creation of the supreme power meant for the benefit of all creation; each individual life form must therefore learn to enjoy its benefits by forming a part of the system in close relationship with other species. Let not any one species encroach upon others' rights.
>
> *(Shiva, 2010:95)*

Most importantly, Vandanan Shiva asserts that we should explore ways to live at peace with nature. At the Earth University, a university of the seed, people come together to learn how to build a new kind of freedom for the whole of humanity with the help of 'living economies', 'which concern human creativity, and mimic nature's diversity, self-organisation, and complexity ... every community is its own centre. Connected to others in mutuality and support' (2005:72). Such a vision of 'living economies' cannot be achieved without a holistic understanding of the spirit of care and caring.

As we are aware of numerous definitions of 'care' that are available, some much less radical and inclusive than others, we have indicated what types of 'care' or 'caring practices' may lie at the basis and within a commons food regime. Care, as we see it, is a quality, both an innate characteristic in human beings and a difficult task that we need to learn and to develop, which is embedded in the core of any emergent commons food regime. Care is an attitude and a worldview shared across a wide range of social agents, 'forging a common platform based on some commons grounds' (Amin, 2011:xvii), to unite different discourses and movements. In this sense, care, as the core to a commons food regime, ensures that we are well-connected to both the world and history. In contrast to most theories in social sciences, a commons food regime is not only a concept for analysing what's happening in the world, but also a clear direction to help 'us commoners' to organise ourselves to collaborate, cooperate, and govern our food-related resources whenever and wherever is possible more effectively and sustainably.

Although there is an increasing attention to the significance of care, it is far from a mature state of affairs. As Tronto puts it, 'we are still too early in the evolution of care thinking to dismiss some avenues of thought as unproductive' (1999:116). Support for this assessment can be seen in Baier's (1997) discussion of 'moral reflections'. The word 'morality' – as it declares by its own root, *mores* – refers to our customs of co-existing, constituting morality's fundamental subject matter. However, this is not a given matter but demands our willingness, persistence, attributes, and wisdom. Moral reflection, she says,

> as a social capacity, is still in its infancy, with many rough corners still to be rubbed smooth. A commons of the mind is by no means assured, where morality and political morality is concerned. Yet we cannot renounce the project of trying to establish such a commons.
>
> *(Baier, 1997:63)*

With care, we remember our values and how we act in the world, are open-minded enough to be challenged from rather different perspectives, and are ready to embrace new ideas and ways of doing things.

Entering a new epoch of history

At the beginning of this chapter we introduced the evolution of food regimes theory since its first theorisations in the 1980s. Following a food regime scholar's attitude – a desire to look for the transition towards a contingent and contested development, politically open to multiple potential outcomes – we join Friedmann's 'invitation' to the conversation of the food regimes theory.

Given that the current landscape of food movements is complex, diverse, and, above all, fragmented, a coherent and strategic approach that can respond to the strong calls for 'convergence in diversity' across food movements to the question of how we can actually engage with the change is urgently needed. Following this, we explored the concepts of commons regimes in a complex world. The traditional and (so far) mainstream focus of commons theory has been predominately developed by Ostrom and her colleagues in response to Hardin's 'Tragedy of the Commons'. More recently, the rise of the commons has been evident and a myriad of new commons and new conceptions has emerged. However, of all the challenges to the mainstream approach to commons, the most severe and transformative perhaps arguably are those originating from an anti-capitalist perspective.

The notion of 'growing a commons food regime' aims to move from a passive conceptualisation to actively being engaged in the transformation of our current food regime. To deepen our understanding of a specific kind of commons food regime and find the core that connects and unites a myriad of discourses, disciplines, organisations, and movements working broadly in the politics of food, we propose the idea that care is the core of a commons food regime, as care represents an 'attitude', an 'orientation', and a 'worldview'. We endeavour to encompass a wide range of concepts and ideas on care.

While we share our vision and values of how we live in the world, we are also committed to being open-minded enough to be challenged from rather different perspectives and are ready to embrace new ideas and ways of doing things. Indeed, we human beings are both 'products' of our own systems and at the same time also the designers of a new system (a regime). As Boyle (cited in Ferguson, 2009:71) puts it, '[t]here is only one history of importance and it is the history of what you once believed in and the history of what you came to believe in.' As a way of concluding this chapter, we would like to assert that growing a care-based commons food regime is like entering a new epoch of history: the pattern is not written, but we make history by living, experiencing, generating, reproducing, and protecting the commons. We write new pages of history by putting care at the core of our actions. Only then, we will realise that we are growing together an ecological and just food systems in the 21st century.

Notes

1 See http://www.navdanya.org/.
2 See http://www.swaraj.org/.

References

Aglietta, M. (1979) *A theory of capitalist regulation: the US experience*. London: New Left Books.
Amin, S. (2011) 'Food sovereignty: a struggle for convergence in diversity', in E. Holt-Giménez (ed.) *Food movements unite!: strategies to transform our food system*. Oakland: Food First Books. pp. ix–xviii.

De Angelis, M. (2007) *The beginning of history: value struggles and global capital.* London: Pluto Press.
Armitage, D. (2008) Governance and the commons in a multi-level world. *International Journal of the Commons.* 2 (1), 7–32.
Atkins, P. and Bowler, I. (2001) *Food in society: economy, culture, geography.* London: Arnold.
Baier, A. (1997) *The commons of the mind.* The Paul Carus lecture series 19. Chicago: Open Court.
Benso, S. (2000) *The face of things: a different side of ethics.* New York: SUNY Press.
Biel, R. (2011) *The entropy of capitalism.* Leiden: Brill.
Bollier, D. (2007) 'The growth of the commons paradigm', in C. Hess and E. Ostrom (eds.) *Understanding knowledge as a commons: from theory to practice.* London: MIT Press. pp. 27–40.
Brownhill, L. et al. (2012) Degrowth? How about some 'de-alienation'? *Capitalism Nature Socialism.* 23 (1), 93–104.
Burch, D. and Lawrence, G. (2009) Towards a third food regime: behind the transformation. *Agriculture and Human Values.* 26 (4), 267–279.
Caffentzis, G. (2004) A tale of two conferences: globalization, the crisis of neoliberalism and question of the commons. A talk prepared for the Alter-Globalization Conference, August 9, 2004. San Miguel de Allende, Mexico.
Caffentzis, G. (2010) The future of 'the commons': neoliberalism's 'Plan B' or the original disaccumulation of capital? *New Formations.* 69 (1), 23–41.
Campbell, H. (2009) Breaking new ground in food regime theory: corporate environmentalism, ecological feedbacks and the 'food from somewhere' regime? *Agriculture and Human Values.* 26 (4), 309–319.
Campbell, H. and Dixon, J. (2009) Introduction to the special symposium: reflecting on twenty years of the food regimes approach in agri-food studies. *Agriculture and Human Values.* 26 (4), 261–265.
Clippinger, J. and Bollier, D. (2005) 'A renaissance of the commons: how the new sciences and Internet are framing a new global identity and order', in R. Ghosh (ed.) *CODE (Collaborative ownership and the digital economy).* Cambridge, MA: MIT Press. pp. 259– 286.
Diedrich, W. et al. (2003) Towards a Levinasian care ethic: a dialogue between the thoughts of Joan Tronto and Emmanuel Levinas. *Ethical Perspectives.* 13 (1), 33–61.
Dowler, E. et al. (2010) 'Doing food differently': reconnecting biological and social relationships through care for food. *The Sociological Review.* 57 (s2), 200–221.
Federici, S. (2004) *Caliban and the witch: women, the body and primitive accumulation.* New York: Autonomedia USA.
Federici, S. and Cox., N. (1975) Counter-planning from the kitchen. [online]. Available from: https://caringlabor.wordpress.com/2010/10/20/nicole-cox-and-silvia-federici-counter-planning-from-the-kitchen/ (Accessed on January 5, 2018).
Ferguson, M. (2009) *The Aquarian conspiracy: personal and social transformation in our time.* New York: Penguin Group USA.
Folke, C. (2007) Social–ecological systems and adaptive governance of the commons. *Ecological Research.* 22 (1), 14–15.
Friedmann, H. (1993) The political economy of food: a global crisis. *New Left Review.* 197, 29–57.
Friedmann, H. (2005) 'From colonialism to green capitalism: social movements and emergence of food regimes', in F. Buttel and P. McMichael (eds.) *New directions in the sociology of global development.* Research in rural sociology and development, vol. 11. Amsterdam: Elsevier. pp. 227–264.
Friedmann, H. (2009) Discussion: moving food regimes forward: reflections on symposium essays. *Agriculture and Human Values.* 26 (4), 335–344.
Friedmann, H. and McMichael, P. (1989) Agriculture and the state system: The rise and decline of national agricultures, 1870 to the present. *Sociologia ruralis.* 29 (2), 93–117.
Gleeson, B. and Kearns, R. (2001) Remoralising landscapes of care. *Environment and Planning D.* 19 (1), 61–80.
Gudynas E. (2011) Buen vivir: today's tomorrow. *Development.* 54 (4), 441–447.
Hardin, G. (1968) The tragedy of the commons. *Science.* [online] 162 (3859), 1243–1248.
Hardt, M. and Negri, A. (2000) *Empire.* Cambridge, MA: Harvard University Press.
Haylett, C. (2003) Class, care, and welfare reform: reading meanings, talking feelings. *Environment and Planning A.* 35 (5), 799–814.
Le Heron, R. and Lewis, N. (2009) Discussion. Theorising food regimes: intervention as politics. *Agriculture and Human Values.* 26 (4), 345–349.
Hess, C. (2008) *Mapping the new commons.* [online]. Available from: http://papers.ssrn.com/sol3/papers.cfm?abstract_id=1356835 (Accessed 8 April 2009).

Himmelweit, S. (2007) The prospects for caring: economic theory and policy analysis. *Cambridge Journal of Economics*. 31 (4), 581–599.

Hirst, A. (2009) *Eating the other: Levinas's ethical encounter*. [online]. Available from: http:// espace.library.uq.edu.au/view/UQ:188919 (Accessed 26 March 2012).

Holt-Giménez, E. and Shattuck, A. (2011) Food crises, food regimes and food movements: rumblings of reform or tides of transformation? *The Journal of Peasant Studies*. 38 (1), 109–144.

Irigaray, L. (1991) 'Questions to Emmanuel Levinas: on the divinity of love', in M. Whitford (ed.) *Luce Irigaray Reader*. Oxford: Taylor & Francis. pp. 178–189.

Irigaray, L. (2005) 'The fecundity of the caress: a reading of Levinas, totality and infinity, "Phenomenology of Eros"', in L. Irigaray (ed.) *An ethics of sexual difference*. Continuum International Publishing Group. pp. 154–183.

Kaplan, E. (2000) Using food as a metaphor for care middle-school kids talk about family, school, and class relationships. *Journal of Contemporary Ethnography*. 29 (4), 474–509.

Kristeva, J. (1988) 'Interview with Julia Kristeva', in E. Baruch and E. Hoffman (eds.) *Women analyze women: in France, England, and the United States*. New York: NYU Press.

Lavoie, M. et al. (2006) The nature of care in light of Emmanuel Levinas. *Nursing Philosophy*. 7 (4), 225–234.

Levin, S. (2006) Learning to live in a global commons: socioeconomic challenges for a sustainable environment. *Ecological Research*. 21 (3), 328–333.

Levinas, E. (1969) *Totality and infinity: an essay on exteriority*. Pittsburgh, PA: Duquesne University Press.

Levinas, E. (1981) *Otherwise than being or beyond essence*. London: Springer.

Levinas, E. (1985) *Ethics and infinity*. Pittsburgh, PA: Duquesne University Press.

Levinas, E. (1989) 'Ideology and idealism', in S. Hand (ed.) *The Levinas reader*. Oxford: Wiley-Blackwell. pp. 235–248.

Levinas, E. (1999) *Totality and infinity*. Pittsburgh, PA: Duquesne University Press.

Levinas, E. and Rotzer, F. (1995) *Conversations with French philosophers*. Atlantic Highlands: Humanity Books.

McMichael, P. (2005) Global development and the corporate food regime. *Research in Rural Sociology and Development*. 11, 265–299.

McMichael, P. (2009a) A food regime analysis of the 'world food crisis'. *Agriculture and Human Values*. 26 (4), 281–295.

McMichael, P. (2009b) A food regime genealogy. *The Journal of Peasant Studies*. 36 (1), 139–169.

Mies, M. and Bennholdt-Thomsen, V. (1999) *The subsistence perspective: beyond the globalised economy*. London: Zed Books.

Mies, M. and Bennholdt-Thomsen, V. (2001) Defending, reclaiming and reinventing the commons. *Canadian Journal of Development Studies/Revue canadienne d'études du développement*. 22 (4), 997–1023.

Morgan, K. (2010) Local and green, global and fair: the ethical foodscape and the politics of care. *Environment and Planning. A*. 42 (8), 1852–1867.

Naicker I. (2011) The search for universal responsibility: the cosmovision of Ubuntu and the humanism of Fanon. *Development*. 54 (4): 455–460.

Olsson, P. et al. (2004) Adaptive comanagement for building resilience in social–ecological systems. *Environmental Management*. 34 (1), 75–90.

Ostrom, E. (1990) *Governing the commons: the evolution of institutions for collective action*. The political economy of institutions and decisions. Cambridge: Cambridge University Press.

Ostrom, E. (2009) A general framework for analyzing sustainability of social-ecological systems. *Science*. 325 (5939), 419–422.

Popke, J. (2006) Geography and ethics: everyday mediations through care and consumption. *Progress in Human Geography*. 30 (4), 504–512.

Rosset, P. (2013) Re-thinking agrarian reform, land and territory in La Via Campesina. *The Journal of Peasant Studies*. 40(4): 721–775.

Shiva, V. (2005) *Earth democracy: justice, sustainability and peace*. London: Zed Books.

Shiva, V. (2010) Earth democracy: beyond dead democracy and killing economies. *Capitalism Nature Socialism*. 21 (1), 83–95.

Simmons, W. (1999) The third: Levinas' theoretical move from anarchical ethics to the realm of justice and politics. *Philosophy & Social Criticism*. 25 (6), 83–104.

Smith, D. M. (2000) *Moral geographies: ethics in a world of difference*. Edinburgh: Edinburgh University Press.

Staeheli, L. and Brown, M. (2003) Where has welfare gone? Introductory remarks on the geographies of care and welfare. *Environment and Planning A*. 35 (5), 771–777.

Thomson B. (2011) Pachakuti: Indigenous perspectives, buen vivir, sumaq kawsay and degrowth. *Development*. 54(4): 448-454.
Tronto, J. (1994) *Moral boundaries: a political argument for an ethic of care*. New York: Routledge.
Tronto, J. (1999) Care ethics: moving forward. *Hypatia*. 14 (1), 112–119.
Vincent, J. (2007) Spatial dynamics, social norms, and the opportunity of the commons. *Ecological Research*. 22 (1), 3–7.

5

NEW ROLES FOR CITIZENS, MARKETS AND THE STATE TOWARDS AN OPEN-SOURCE AGRICULTURAL REVOLUTION

Alex Pazaitis and Michel Bauwens

Introduction

It has become widely acknowledged that the current on-going crisis represents a turning point in the global economy. However, it is neither the first nor, most probably, the last of these moments in history. In fact, it has been shown that such decisive moments tend to appear every five to six decades, following a recurrence of cyclical progressions, which Kondratieff (1935/1926) statistically presented in his "long waves". Schumpeter (1982/1939), building on the analysis of the long waves, further discussed the cyclical behaviour of the capitalist economy, provoked by surges of technological innovation. Subsequently, departing from the Schumpeterian understanding of the economy, Perez (1983) postulates that those recursive patterns are not explicitly an economic phenomenon. They are rather explained as a result of a dynamic harmony and disharmony of the techno-economic sphere, on the one hand, and the socio-institutional, on the other.

The root cause of these patterns is conceived within the techno-economic domain, where technological revolutions cause discontinuities in the trajectory of technical change, leading to mismatches with the established institutional framework. This process eventually results in a shift of the 'techno-economic paradigm', i.e. the 'common sense' or the set of best practice principles that guide the engineering and economic behaviour of a certain time (Perez, 2002; 2004). Each techno-economic riddle has a socio-institutional solution, and once a match with the new paradigm is achieved, the potential for a period of prosperity and development is unleashed.

This process of 'creative destruction', as it is often delineated in the Schumpeterian tradition, exposits the powerful dynamic of technological innovation in re-shaping the world. Long periods of prosperity throughout the history of capitalism are characterised by and named after the core industries which had become the propellers of development of the time – from the Industrial Revolution to the Railway Era; and from the Age of Electricity to the Age of the Automobile (Perez, 2004). Likewise, the contemporary Information and Communications Technology (ICT) revolution has triggered an ever-growing discussion over the Information Age (Castells, 2010).

Nevertheless, it must be emphasised that technological revolutions, as in fact any type of revolution in the wider sense, do not necessarily lead to one inevitable social outcome. Much like social revolutions, they are organic and often destructive events that do not fall within the control of any particular social force. At the same time, the key role that technologies play in societal evolution has to be recognised. Technology is understood as a moving frontier which expands the sphere of the feasible, creating new possibilities for certain social groups that are able to deploy it.

Therefore, technology is itself a field of social struggle, as different social forces invest in the new opportunities to benefit from it (Feenberg, 2002). When social groups take control of a certain technology, then social, political and economic systems can effectively be transformed. In the neo-Schumpeterian tradition (Freeman, 1974, 1996; Perez, 2002), crises convey some of the basic functions of capitalism and are considered to be windows of opportunity for institutional change that rejuvenates the system. From a different perspective, Kostakis and Bauwens (2014) point out that crises, similar to the current crisis, can tentatively lead to something more than a socio-institutional regeneration of capitalism.

Foti (2017) adds an interesting characterisation of variations in economic cycles. He contrasts accumulation crises, i.e. crises in the supply of capital due to falling profit rates and capital-restrictive regulations, with regulation crises, where the issues are of demand due to an overaccumulation of capital, weak regulation and weak redistribution. The former typically lead to inflation crises and revolutionary waves of protest, as in the 1970s, whereas the latter are characterised by deflation and reformist processes, which is arguably the case today. This means that there is a conjuncture favourable for a broad reform of regulatory regimes and new forms of industrial and economic organisation.

The issue is further complicated with the deepening ecological crisis. There is a growing need for a shift from extractive production regimes, which exhaust natural resources and human capacities, to generative forms of production, which not only maintain their resource base but also enrich it. Today's 'jobsian' (Foti, 2017) mode of neo-liberal regulation overuses material resources and exhausts the soils, just as it exhausts the workers, creating ever-more precariousness. In contrast, a generative reform, for instance in the agri-food sector, would be oriented towards a just distribution of value among land-workers, along with practices that progressively regenerate the soil and the resources used, as in the cases of agroecology, permaculture and organic agriculture.

The role of a 'green' reformation is matched by the increasing importance of distributed ICT and manufacturing technologies and their capacity to mutualise productive resources. A potential synergy would combine mutualisation of knowledge, shared physical infrastructures, and (re) localisation of productive capacities. Commons-oriented communities can now emerge on global scale, based on mutualised technical and scientific knowledge in various fields of production or provisioning systems, while distributed manufacturing has the potential to radically shorten supply chains. Mutualisation of unused resources through a generative sharing economy has the potential to drastically diminish the thermodynamic load of the current production system.

Hence, new modes of social production and new models of value creation and distribution can emerge from radical socio-technical transformations, which, in the long term, have the potential to transcend the system as a whole. These aspects can bring about deep political and social change: a 'phase transition' in the main modality by which humanity allocates its resources.

A phase transition is made apparent within two moments in history, where there are significant fundamental differences in the dominant productive relations and processes. From the slave-based system of the Roman Empire to feudal serfdom, and from feudalism to capitalism, we can identify such profound alterations in the most vital aspects of human societies, including key raw materials and energy resources; technologies; types of territorial exploitation and financial systems. Above all, changes occur in global political dominance and the type of social contract and governance.

Bauwens and Niaros (2017), in the context of a Commons Transition Plan for the city of Ghent, Belgium, illustrate how such deep changes are already at work. There has been a tenfold increase in the number of commons-oriented civic initiatives since the crisis of 2008, roughly from 50 in 2006 to 500 in 2016, in a city of 300,000 inhabitants (Bauwens and Onzia, 2017). In Ghent, all provisioning systems are characterised by attempts that entail mutualised infrastructures. In Ghent, 93 of the 500 mapped initiatives are related to food, and there is no doubt that a sizeable generative food provisioning system already exists in the city and its bio-region. For instance, Gent en Garde[1] is a transition platform that endorses the demands of civil society for fair, organic and local food. It created, among other things, the Urban Agriculture workshop,[2] which is a working group of individuals and organisations whose mission is to create a more sustainable and healthy food ecosystem in Ghent. This development of the commons in the food sector is again partly connected to the public organisations in the city that are gradually building political support.

In the following sections, we attempt to explore a tentative path for such a transition, identifying the premises that would empower and legitimise a new political order. In the first section, the political economy of a prefigurative society emerging from a P2P phase transition is presented. The second section provides a theoretical approach of a new reconfiguration of the state, referred to as 'the partner state', which will support a potential transition by enabling and empowering social production. In the third section, we introduce a modest policy approach towards the partner state. A set of transformative policy proposals is presented, providing examples related to food and agriculture, which would potentially set up an open-source agricultural revolution. Finally, the last section attempts to synthesise the analysed aspects of the previous sections to formulate a proposition of an integrated model for a commons-based, sustainable agricultural system.

A P2P-driven phase transition[3]

On the verge of a phase transition, typically, a dominant system increasingly starts to show weaknesses in devising solutions to a series of systemic crises, while different classes of actors, including the ruling elites, as well as the subordinate productive communities, seek for solutions. Different patterns of response are being developed, which, at first, are used by the system and remain subsumed by the dominant paradigm, but at the same time form a new model, which will seek to gradually emerge and replace the old one. They form new social structures to enable and support the changes in the modalities of production and value creation. When these tensions are no longer absorbed by the dominant system, political and social turbulence eventually leads to revolution.

In the current phase, that is, the industrial society, the illusion of a natural abundance as the basis of the current socio-economic system and the ever-increasing negative market externalities create an unsustainable environment. On the other hand, artificial scarcity is imposed on naturally abundant resources, such as cognitive processes, including agriculture, and knowledge production. The commodification of the most vital means of human subsistence, including food and the basic means of production, has led to intimidating disparities and inequality. At the same time, the on-going economic crisis has amplified the existing inequalities, which further restrains the ability of the system to absorb the tensions.

In this picture, three types of patterned responses can be observed: (a) sustainable production, which introduces responsible approaches to account for the ecological limits and the cumulative effect of the negative environmental externalities; (b) cooperative forms of organisation and the social and solidarity economy, which strive to create egalitarian practices for value production and distribution and emphasise social justice; and (c) peer-to-peer (P2P) collaboration and commons-oriented production, associated with the sharing of resources, emerging from a recognition of the natural abundance of immaterial commons, such as technical and scientific knowledge, software and design.

Our approach puts the latter in the epicentre, emphasising its potential to compound the fabric of a new economy and society. P2P is simultaneously a relational dynamic and mode of exchange that has emerged from the radical diffusion of ICT and the internet. The development of GNU/ Linux and myriad Free/Open-Source Software (FOSS) projects, as well as the free encyclopaedia Wikipedia, exemplified a unique capacity for individuals to relate to each other and communicate in a permissionless fashion. This entailed a new mode of production, which Benkler (2006) called 'commons-based peer production' (CBPP): a new modality of value creation and distribution, where individuals self-organise and contribute to the creation of universally accessible digital goods. P2P is closely related to the practice of commoning, in the sense that it enhances the capacity for the creation and maintenance of shared resources. Even though it is primarily associated with the digital commons of information, including knowledge, software and design, these resources are closely bound to the generative capacity of any type of production, digital or physical. After all, information is 'the fundamental source of power and productivity' in the Information Age (Castells, 2010: 21).

P2P and the commons can thus prescribe the premises and a common vision for the necessary convergence of the three patterns described previously. For example, open forms of cooperativism and social and solidarity economic entities can operate in synergy with the commons to create livelihoods for the contributing communities. In convergence with the sustainability principles, new economic models such as an open-source circular economy can promote the regeneration of resources and the support of environmental stability for the current as well as the coming generations. The commons can function as the foundation for a pluralistic commonwealth, where multiple forms of value creation and distribution co-exist: a core institutional arrangement which will guide all other social forms towards achieving the maximal common good and individual freedom.

Such a convergence is necessary so that the contributing communities can be emancipated from the old decaying productive relations and reconstruct the necessary social and political institutions. In the present form of social and economic order, cooperation is subsumed under competition. Collaborative processes occur internally within hierarchically structured corporate entities, which compete with each other in markets. A P2P-driven phase transition would compel the reversal of this relation. In the P2P ecosystem, value is created in collaborative processes, while new types of ethical entrepreneurial coalitions co-create commons along with the productive communities. At the same time, a safety net of for-benefit associations supports the common infrastructure and protects and enriches these commons. Competition thus takes place within the sphere of collaboration. Value is created and distributed through the commons, while a new type of economy generates livelihoods for the contributors around this commons.

Cooperative, reciprocal organisation and economic democracy offer a prototype for a new political order, which moves beyond authority and hierarchy, promotes inclusion and participation and focuses on the service of the community above profit (Restakis, 2010, 2015): a political economy for a 'generative democracy' (Restakis et al., 2015), i.e. a type of democracy that is constantly re-created through distributed social production. A combination of CBPP for abundant resources and a reciprocity-based cooperative organisation for scarce material resources would further empower social reproduction and provide livelihoods for the contributors

(Bauwens and Kostakis, 2014). This type of synergy, referred to as 'open cooperativism' (Conaty and Bollier, 2015), could assist CBPP to move from a proto-mode of production to an autonomous and integrated mode of production, able to sustain itself and its contributors.

Nevertheless, as recent experience from radical political movements has shown, such a reconstruction of the productive relations cannot succeed within the old political order, where power is concentrated in a professional political elite operating in a market state form that is dominated by private financial interests. It is therefore necessary that the state itself and the dominant political structures are challenged and transformed as well. New hybrid forms of deliberative and participatory democratic governance have to be adopted in order to assign the political initiative to the civil society for setting the political agenda and executing the public services.

The P2P dynamics are already planting the seeds for the institutions of a new societal model. In this prefigurative society, citizens, markets and the state obtain new roles and importance, which can be summarised as follows:

a The citizens participate in common value creation and the civil society becomes productive. The commons gradually shift from the periphery to the very core of social and economic organisation and form the fundamental institutions.
b The market becomes 'ethical' and adopts generative as opposed to extractive economic practices. Cooperative and solidarity-based economic models determine the allocation of financial and physical resources.
c The state becomes the 'partner state', facilitating and enabling social production through participatory democratic governance. The political objectives shift away from the service of the political and economic elites and are directed towards the maximisation of personal and social autonomy.

What would be necessary for a successful transition to this model is the reconstruction of these prefigurative value-creating production systems. At the same time, the social and political power that relates to these social configurations would start to gain influence. The already-existing CBPP practices would need to be properly transformed so as to be able to assure their own self-reproduction. The new type of 'ethical economy' (Arvidsson and Peitersen, 2013) provides the context and vital space to encircle the commons and enable forms of non-commodified production and exchange (Bauwens and Kostakis, 2015). With the term 'ethical', we refer to an economic paradigm in which value creation serves the community before business, a model where the business logic has to accommodate to the social logic.

The initial steps have been made by the contributing communities of the FOSS and Wikipedia, as well as various open design initiatives around the globe. Ethical entrepreneurial coalitions have started to emerge, organising the productive communities in egalitarian and democratic economic entities and creating added value on top and along the commons. Cases like Enspiral (Pazaitis et al., 2017), the open enterprise Sensorica and Wikihouse (Bauwens et al., 2018) and numerous platform co-ops (Scholz and Schneider, 2017) have recently gained eminence, exemplifying hybrid modes of operation balancing between the commons and the capitalist marketplace. Specifically in the agricultural sector, projects like Farm Hack and the L'Attelier Paysan cooperative illustrate the potential of these models in food production (Giotitsas and Ramos, 2017). Support is provided by institutions like the Free Software Foundation, the P2P Foundation, Creative Commons and Wikimedia Foundation, which function as a safety net. They serve to protect and enrich the commons created by the communities through a variety of legal, technical and institutional tools and a global, interconnected community of knowledge and practice.

These emergent forms of social production have thus sparked an on-going transformation. However, a qualitative phase transition would require the reconstitution of powerful political and social movements. In order to transcend capitalism, a sustainable ecosystem would need to be created, to function as a counter-economy to the current model. At the same time, this ecosystem will function as the prototype for the transformation of the state. Elsewhere (see Kostakis and Bauwens, 2014; Bauwens and Kostakis, 2014, 2015; Bauwens et al., 2018) we have provided an extensive discussion of a tentative trajectory for this transitionary process. In this chapter, our focus is on the role of the state and targeted policies aiming to create a sustainable alternative to the dominant model. In the following section, we will further discuss the role of the partner state as enabler and facilitator of the political economy emerging from P2P relations.

The partner state: a theoretical approach

The partner state, first theorised by Italian political scientist Cosma Orsi (2005, 2009) and further developed by Kostakis and Bauwens (2014), is a state form that empowers the social creation of value by its citizens and enables autonomous social production. It protects the infrastructure of cooperation for the whole of society (Bauwens, 2012; Bauwens and Kostakis, 2015).

As a state form it is deducted from the micro-economic arrangements of CBPP and the social and solidarity economy with the new type of for-benefit associations that support them. CBPP relies on two premises for its reproduction: a commons of collectively managed resources and an infrastructure facilitating cooperation. These require adequate support and protection from enclosures, while the maintenance of the technological infrastructure comes with significant costs as well. Thus, the open-source communities have created for-benefit associations, a new form of social institution in service of the communities and everyone that contributes to the commons. In comparison with traditional non-profits and non-governmental organisations, for-benefit associations introduce an important social innovation. Stemming from the world of FOSS, and due to the non-rival – in fact anti-rival – nature of the digital commons, these institutions operate from a standpoint of abundance in relation to the commons. In this sense, their role does not focus on regulating scarce resources for a commanded community but rather consists of pro-actively enabling and empowering open cooperation based on shared resources.

Nevertheless, this type of sharing is not a 'free for all' situation. While the digital resources are abundant, the physical infrastructures involve both financial and ecological costs. Moreover, the 'for-benefit' character already suggests that the orientation of these associations is to provide for the common good of all the related participants, which in this case are the contributors to a certain open-source project. If we transpose this characteristic on a territorial scale, an institution that provides for the common good of a certain group of people associated with a specific type of social relationship is one of the main functions of the state. This has been the basis of the concept of the partner state: a set of institutions that protect the common good and enable citizens to create value. As such, the conceptualisation of the partner state can function on any territorial level: local, regional, national, transnational and global.

From a Marxist point of view (Miliband, 1965), the state is also an instrument of class rule and reflects the balance of forces in a particular social order. However, a transformation towards the partner state should not be viewed as a struggle of the commoners to oust the privileged classes and seize state power, only to get locked in to their own materialistic class interests (Troncoso in Bollier, 2016).[4] The partner state cannot be the instrument of a privileged rule alone; it needs to manage the common good. Just as the 'invisible hand' of the market is a myth, so too an invisible hand of the commons. Stemming from the immaterial world, P2P communities at most lack mechanisms for distribution of power, like authority, prices or democracy. Their governance is truly poly-archic and power is distributed according to meritocracy and on an ad-hoc basis.[5] Similarly, commoners tend to care about their commons and lack a general vision of the society as a whole. That specific provision for the whole requires its own specific set of institutions.

In this direction, Silke Helfrich (in Bollier, 2016)[6] attempted to redefine the way we think about the commons. The commons must be understood as 'an important form of transpersonal rationality and coordination – a new category that describes the individual-in-relation-with-others' (Bollier, 2016: 24). Helfrich proceeds to suggest that, even though there may be no commons without commoning, commoning is not necessarily the only kind of contribution to the commons. And here we find the role of the state in ensuring the rights of all citizens and supporting constructive relations on top of and along with the commons. In the Hegelian (1820) notion, the state is viewed in a broader sense, encapsulating the community as a whole, including its institutions. From this point of view, the state constitutes the sphere of full, actual and genuine freedom and is considered as the individual's utmost end. The partner state encapsulates this perception by enabling the individual to pursue ends larger than his/her own personal good.

In other words, state power may be a matter of social struggle. However, any post-capitalist aspiration of a state would ideally deem material struggle obsolete. In fact, Marx himself would probably be the first to celebrate this transformation. Therefore, the state should be reimagined as a Greek *polis*; a 'structured human living-together' (Drechsler, 2001: 6). In the *polis*, the citizen and the state are mutually dependent to achieve genuine happiness for both. It is therefore one of the main functions of the state to allocate power to those social structures that would better serve its paramount purpose of existence; as Aristotle has stated it: 'the good life' (*Politika*, I 1252b). The extent to which this is achieved, in turn, legitimises state power for the larger part of the society.

The neo-liberal state legitimises its dominance over its citizens and a general prioritisation of business over welfare, based on the assumptions of the quasi-democratic functions of free markets. A divine-like set of functions and 'laws' that are assumed to be in operation in markets are expected to appraise efficiency for commodity exchange, creating a positive-sum game for the majority of society. Furthermore, following Milton Friedman's (1953) 'positive' epistemology in economics, these assumptions have to a very large extent been left unchallenged until today, even though it is becoming ever more obvious that they are false and wildly unrealistic.

So, how can a P2P transition establish a sustainable political economy while avoiding similar fallacies? How can the partner state gain legitimacy in advancing its purpose to ensure a 'good life' for the whole society? CBPP may have the potential to redefine citizen involvement and democratic participation. Simultaneously, such processes create demand for effective state reforms that build upon the essence and the importance of abundance, distribution and intrinsic positive motivation (Kostakis, 2011). In the following section, we briefly present a cluster of certain policies that would aim to empower a critical mass of people to participate in CBPP and earn sustainable livelihoods. Furthermore, some examples are illustrated with relevance to agriculture and food in order to provide connection to the most important means of human subsistence that insure human well-being. As more people would be able to support themselves and improve their living conditions through this process, this type of political approach will gain in legitimacy and eventually be able to overthrow and replace the old political order.

Setting up an open-source agricultural revolution

In our understanding, the ideal pursuit of a revolutionary struggle for political power within the confines of the current dominant system would envision the partner state as its political outcome. A first step in this direction would be a cluster of policies with the foremost mission of empowering direct social value creation, protecting the commons and promoting democratic participation. Such a policy mix, which represents a fine balance between government regulation, private-market freedom and autonomous civil society projects, has been meticulously discussed elsewhere under the concept of 'the Partner State Approach' (Kostakis and Bauwens, 2014).

In the following subsections, we highlight some general directions, which aim to operate on different levels of the economy and society. Where applicable, references are provided from existing initiatives related to food and agriculture in order to illustrate some empirical cases that build up this alternative political economy. A potential transformation towards a resilient and sustainable food system would require transformations at multiple levels and diverse approaches of governance (Vivero-Pol, 2017). These experiments attempt to nest in the current system, simultaneously exploiting and challenging it and, eventually, transcending it.

It should be noted that the proposals derive from a composition of the analysis of the Partner State Approach (Kostakis and Bauwens, 2014; Bauwens and Kostakis, 2015), along with various policy-related approaches for the commons and the partner state (Restakis, 2015; Restakis et al., 2015; Pro-Comuns, 2016).

Ethical marketplace and cooperative organisation

A stream of policies would be directed to support and enhance the ethical economy and cooperative organisation. Coalitions of ethical economic agents should be created around the commons, engaging a diverse set of stakeholders in CBPP. Support should be provided to shared/reciprocal forms of ownership and open/commons-oriented business models. The creation of support structures for open commercialisation would maintain and enrich the commons and provide interconnection with global commons-oriented communities, such as open design communities. At the same time, institutional and legislative reform would be necessary in order to shape an appropriate framework for the operation of the ethical economy. Education and training institutes should disseminate the theory and practice of cooperativism and the values of reciprocity and service to the community, and support the development of the cooperative culture.

The mainstream commercial sector should be reformed to minimise the negative social and environmental externalities, and convergence with the social and solidarity economy should be incentivised. Hybrid economic forms, like fair trade and social entrepreneurship, should be empowered through targeted policy measures and financial support. Support infrastructures (e.g. technologies, facilities, etc.) should be developed and maintained, designed to promote the commons as well as openness and sustainability. Alternative financial instruments should be developed, including crowdfunding and seed funding schemes, as well as debt-free public financing and complementary currencies.

In the agricultural sector, a shift to sustainable, resilient and responsible production practices should be promoted, along with the establishment of complementary fair systems of distribution and consumption. Alternative approaches to this direction are movements like Community Supported Agriculture, including various networks and projects, such as FairShare, Urgenci and Growstuff, which support communities to adopt sustainable, localised farming and consumption practices. Ethical marketplaces like Fairmondo, the Food Assembly, Farmdrop and the Open Food Network have been developed to create integrated networks of local producers and consumers, promoting fair trade practices and responsible consumption patterns. Cooperative coalitions like the Land Workers' Alliance, Alemany Farm and the Dune Costiere community provide support for local producers, advocate for proper policies and provide education and awareness on sustainable farming practices and food security. Finally, community kitchens and urban pop-up restaurants promote a different approach to food in general based on communal relations and provide viable solutions to current societal crises, like the refugee crisis in Europe.

Technology and distributed production

Policies should promote open technologies and distributed localised production through the provision and support of the knowledge commons. The creation of open manufacturing spaces should be promoted, such as makerspaces, FabLabs and micro-factories. Through the convergence of local manufacturing technologies with sharing practices and community-based forms of governance, and supported by institutions dedicated to the expansion and diffusion of productive knowledge, such spaces may serve as vehicles for citizen-driven transformations (Niaros et al., 2017). Investments in science and technology should be aligned with the commons and the co-creation of productive knowledge, while all publicly funded research should be released under commons-based licenses.

Numerous significant projects and initiatives which emerged from the open-source movement have enabled significant aggregation of knowledge and practical tools related to agriculture. Projects like Farm Hack, L'Atelier Paysan cooperative, the Open Source Ecology, Open Source Beehives, Aker, MyFood, the P2P Food Lab and Open Land Labs offer a variety of social innovations and technological solutions for agriculture and farming, from open-source software applications and digital platforms to open hardware tools and technologies. The promotion of the open-source mind-set and the sharing of knowledge, practices and designs create an ecosystem that coalesces around a global digital commons for sustainable agriculture.

Democratic governance and the public sphere

Finally, a stream of policies should operate on a meta-level for the state itself, which should learn from the social economy and organically transform itself. The state should redefine its role through the empowerment and support of the civil society and the production of social value. Openness and transparency should be maximised and democratic governance should be promoted, through systematised participation, deliberation and real-time consultation with citizens, including online and offline facilitation, liquid voting and participatory budgeting. De-bureaucratisation should be forwarded through the commonification of public services and public-commons partnerships, while community-driven infrastructures and networks should be supported and enhanced. The public realm should be re-claimed by the commons through joint management and regeneration of workspaces and public spaces and the development of collaborative commons-oriented projects.

With the objectives to ensure food security and safety on a local/regional level, public authorities have put forward integrated agendas for the development of sustainable food systems. Prominent cases, like the Vancouver Food Strategy and the Canberra City Farm project, design strategies that include, inter alia, the promotion of fair and sustainable food production practices; the empowerment of grassroots community food initiatives; the promotion of learning; and the development of skills and social competences for socially, economically and environmentally responsible practices.

Vancouver and Canberra are only two of the 128 cities around the globe that have signed the Milan Urban Food Policy Pact,[7] developing an international framework of reference for the promotion of sustainable and just urban food systems. At the same time, on a grassroots level, numerous projects, like Prinzesinnengarten and the Urban Farming Guys, aim to revitalise unattractive and counter-productive urban space while providing safe, locally produced and fairly distributed quality foodstuffs.

These have been only a few indicative examples from the countless initiatives that are striving to create a new approach to agricultural and food production and consumption. Even though significant awareness has been gradually raised throughout the past decade, these practices remain to a large extent fragmented and unable to generate a strong counter-current to the dominant productive model. In the final section, we provide a proposition for the creation of the necessary conditions for such a convergence.

COFARMIN: a blueprint for a commons-based agricultural system

Our rough proposition towards a commons-oriented productive model in the agricultural sector aims to devise the techno-economic blueprints for innovation and sustainability. The main idea is based upon a potential conjunction of CBPP with the emerging capabilities of distributed manufacturing technologies. These include any type of technologies that enable customised local manufacturing of physical items, from desktop manufacturing equipment such as three-dimensional printers and computerised numerical control machines to more traditional benchtop tools like drills, cutters and screwdrivers.

This model, codified as 'design global, manufacture local' (DGML) (Kostakis et al., 2015) has sprouted from successful commons-oriented projects which focus on P2P technologies and open hardware, such as Farm Hack and L'Atelier Paysan cooperative. These cases demonstrate how a technological project can leverage a knowledge commons to engage a global community in its development (Dafermos, 2015). They furnish concrete examples of how commons-based technologies and practices along with distributed, localised manufacturing can enhance the autonomy of people and transform all sectors of production in the direction of economic and environmental sustainability. Just as networked computers have democratised the means of information production, the emergence of local distributed manufacturing is democratising the means of making.

In brief, DGML denotes a productive process where design is developed, shared and improved as a global digital commons, whereas the physical manufacturing occurs on-demand at the local level using shared infrastructures (Kostakis et al., 2016, 2017). The dynamics of this model lie on the distributed access to information, including knowledge and design, as well as on the physical means of making. Contrary to the mass-production industrial paradigm, which relies on economies of scale, DGML rests on commons-based economies of scope. While the advantages of scale rest on high-capital entry and cheap global transportation, the commons-based economies of scope share infrastructure costs in terms of intangible and tangible productive resources. Furthermore, distributed manufacturing technologies have been claimed to hold the potential to eventually revolutionise the manufacturing industry, by generating the premises to convey peer production, as it emerged from open-source software and the digital commons, into the physical realm.[8] Kostakis et al. (2015, 2016, 2017) have recently demonstrated the prospects of this model, emphasising its unique dynamics in terms of design-embedded sustainability, resilience, scale and strong collaboration impetus.

COFARMIN, standing for COoperative FARMing INfrastructures, is structured on three interlocking levels (Figure 5.1): (a) a digital techno-social platform of global knowledge commons, including code and designs; (b) a network of local makerspaces equipped with distributed manufacturing technologies; and (c) the local society and economy of productive communities. These digital and physical infrastructures connect the global commons-oriented communities of open design with the makerspaces and the local societies. The techno-social

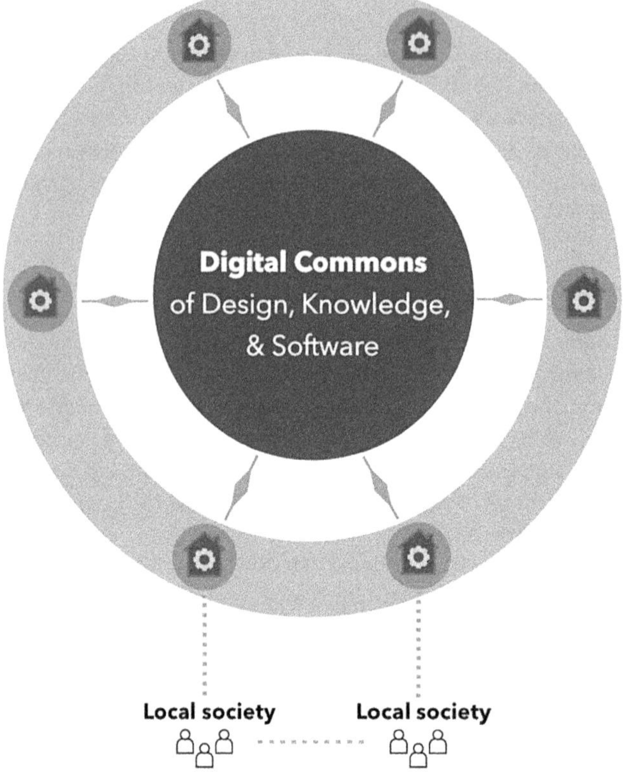

Figure 5.1 The three interlocking layers of the COFARMIN ecosystem.

Source: Nikos Exarchopouloos and Vasilis Kostakis.

platform provides a repository and readable library of already developed open-source and open hardware solutions related to agricultural production. Local farmers, producers and growers, as well as hackers, designers, engineers and activists, engage in creative interaction to exploit the global knowledge in order to develop tailor-made, innovative solutions to address local challenges. Local makerspaces will facilitate this interaction by providing technical means and expertise to customise and materialise the selected solutions in a cost-effective and sustainable manner. In this process, a circulation of open-source software and open hardware technologies is initiated within a collaborative networked environment, thus further contributing to the knowledge commons repository. The resulting collective intelligence empowers people, through participation and interaction, to adopt more sustainable productive patterns, as well as collaboratively develop innovative solutions to local societal challenges.

DGML envisions a bottom-up participatory paradigm which introduces new, decentralised and distributed systems of production and provisioning; inclusive governance; and commons-based value and open innovation. It may be argued that some of the least developed parts of the world need some of the most advanced technologies. ICT and distributed manufacturing may be the globally imagined tools that act locally in response to certain problems and needs. A diverse set of stakeholders, from small-scale producers, commons-oriented grassroots communities and individuals to micro-enterprises, as well as public institutions, would benefit from this (g)local process.

The emerging economic model enables a productive modality that is small-scale, on-demand, decentralised, resilient and locally controlled, yet simultaneously developed and designed on a global basis. The on-site local contributors are brought together in a networked environment and benefit from the socially produced use value, while enriching and expanding the commons sphere.

Moreover, a multi-stakeholder approach is developed, including open design communities and local producers as well as entrepreneurs, which is expected to result in generative synergies. Ethical entrepreneurial coalitions emerge to create added value on the commons, while open forms of cooperative organisation ensure the maintenance of the common infrastructures and provide livelihoods for the contributors. An ever-spreading virtuous spiral of collaborative creation and social innovation mobilises all the relevant stakeholders to move towards an integrated, sustainable commons-based system of agriculture.

Proposals for a partner state-enabled transition in food provisioning

The general logic of our proposals draws upon lessons from urban commons transitions in cities, such as the cases of Ghent, Barcelona and Bologna. These are considered convincing prefigurative forms of partner states, presenting important institutional innovations and processes that streamline cooperation between the city and commoners. Furthermore, such configurations may have surfaced on city level, but they provide an alternative transnational governance structure that complements and transcends the current state institutions. With the current global order and the inadequacy of nation-states in addressing contemporary challenges, the cases of urban commons reconfigurations provide useful lessons for a transition from today's market-centric form of state institutions to commons-centric ones.

Specifically with regards to food, the focus is on the promotion and support of sustainable alternative systems of provisioning, which may as well be applied in every field of human provisioning, including housing, mobility, energy and social care. A first step would require the emergence of a critical mass of commons-based seed forms of provisioning practices (as those presented in this volume in the chapters by Rosset and Val, Fonte and Cucco or Balazs). They first appear as viable solutions to urgent systemic problems the dominant system is unable to solve. As such, they possess a capacity to mobilise citizens, while they become stronger through interconnection with each other and integration with other related domains. For instance, local organic producers may connect with community-based kitchens to cover vital needs for disadvantaged members of society in direct confrontation with traditional food supply chains. Simultaneously, alliances can be built with initiatives from complementary systems, such as energy cooperatives or local makerspaces. Civic mobilisation around such alternatives can create crucial pressures for increasing social and, eventually, political power.

In response, a partner state would develop necessary regulatory and institutional frameworks to support these alternatives, to gradually transit them from the margin to the centre of the system. Different forms of direct and indirect support can be provided concerning regulations in food-related supply chains but also complementary systems, which may even create greater impact. For instance, policies such as feed-in tariffs that incentivise certain forms of energy production over others may deem alternative systems more appealing, along with their associated initiatives. Similarly, regulatory measures targeting profit-oriented renting platforms, like AirBnB or Uber, can support local commons-based alternatives.

Lastly, provided proper institutional support, generative ethical market forms can be developed around the commons-based alternatives. With adequate resources flowing from the dominant system to the commons-based one, those seed forms of provisioning could expand and become 'normalised', thus shaping the new logic in their respective systems and territories. Evidently, these processes are tightly interwoven and should take place concurrently. Nevertheless, it is crucial that a significant number of initiatives are operating before political action can be mobilised or appropriate institutions can be designed. More importantly, this approach is highly context-specific and dependent on the structural characteristics of the respective political economy but also on many cultural and subjective aspects, which vary in every context.

Acknowledgements

Alex Pazaitis acknowledges financial support from the Estonian Ministry of Education and Research [grant numbers: B52, IUT (19-13)].

Notes

1 https://wiki.commons.gent/wiki/Gent_en_Garde
2 https://stadslandbouwgent.wordpress.com/charter
3 This section is based on 'P2P Revolution and Commons Phase Transition', by Michel Bauwens, available at: http://commonsstrategies.org/p2p-revolution-and-commons-phase-transition. (Originally published in Spanda Journal VI, 1/2015: "Systemic Change", edited by Helene Finidori, The Hague: Spanda Foundation, available at: http://www.spanda.org/SpandaJounrnal_VI,1.pdf).
4 From the discussions in the context of the Deep Dive Workshop convened by the Commons Strategies Group in cooperation with the Heinrich Böll Foundation, held from 28 February to 1 March in Berlin, Germany. Full report by David Bollier, available at: http://cdn8.commonsstrategies.org/wp-content/uploads/2016/07/State-Power-and-Commoning.pdf (last accessed: 14 September 2017).
5 For a more extensive discussion on the post-democratic governance of P2P communities, see Bauwens, M. (2012) 'Blueprint for P2P Society: The Partner State & Ethical Economy'. In: Shareable, available at: http://www.shareable.net/blog/blueprint-for-p2p-society-the-partner-state-ethical-economy (last accessed: 14 Sep 2017).
6 See endnote 2.
7 The Milan Urban Food Policy Pact has been subscribed by Mayors on the occasion of a major event during Expo 2015, on 15 Oct 2015 in Milan. For more information see: http://www.milanurbanfoodpolicypact.org.
8 Here of course we have to acknowledge the limitations related to the materials and energy resources necessary for the equipment. Also, even though the digital commons may be of anti-rival nature, there are costs related to the regeneration of human effort and the physical infrastructures.

References

Arvidsson, A. and Pietersen, N. (2013) *The Ethical Economy: Rebuilding Value after the Crisis*. New York, NY: Columbia University Press.

Bauwens, M. and Kostakis, V. (2014) From the Communism of Capital to Capital for the Commons: Towards an Open Co-operativism. *TripleC: Communication, Capitalism & Critique* [Internet], 12, pp. 356–361. Available from: http://www.triple-c.at/index.php/tripleC/article/view/561 [Accessed 15 September 2016].

Bauwens, M. and Kostakis, V. (2015) Towards a New Reconfiguration among the State, Civil Society and the Market. *Journal of Peer Production* [Internet], 7, July. Available from: http://peerproduction.net/issues/issue-7-policies-for-the-commons/peer-reviewed-papers/towards-a-new-reconfiguration-among-the-state-civil-society-and-the-market [Accessed 15 September 2016].

Bauwens, M. and Niaros, V. (2017) Changing Societies through Urban Commons Transitions. Heinrich Böll Stiftung. Available from: http://commonstransition.org/wp-content/uploads/2017/12/Bauwens-Niaros-Changing_societies.pdf [Accessed 02 February 2018].

Bauwens, M., Kostakis, V. and Pazaitis, A. (2018, forthcoming) *Peer-to-Peer: The Commons Manifesto*. London: Westminster University Press.

Bauwens, M., and Onzia, Y. (2017) Commons Transitie Plan voor de stad Gent. In opdracht van de stad Gent. Available from: https://stad.gent/sites/default/files/article/documents/Commons%20Transitie%20Plan%20Gent.pdf [Accessed 02 February 2018].

Benkler, Y. (2006) *The Wealth of Networks: How Social Production Transforms Markets and Freedom*. New Haven, CT: Yale University Press.

Bollier, D. (2016) State Power and Commoning. A Report on a Deep Dive Workshop convened by the Commons Strategies Group in cooperation with the Heinrich Böll Foundation, 28 February–01 March 2016, Berlin. *Commons Strategies Group* [Internet]. Available from: http://cdn8.commonsstrategies.org/wp-content/uploads/2016/07/State-Power-and-Commoning.pdf [Accessed 15 September 2015].

Conaty, P. and Bollier, D. (2015) Toward an Open-Cooperativism: A New Social Economy Based on Open Platforms, Co-operative Models and the Commons. Commons Strategies Group Workshop, Berlin, 27–28 August 2014. *Commons Strategies Group* [Internet]. Available from: http://commonsstrategies.org/towards-an-open-co-operativism [Accessed 15 September 2015].

Castells, M. (2000) *The Rise of the Network Society*. Oxford: Blackwell.

Drechsler, W. (2001) Good and Bad Government: Ambrogio Lorenzetti's Frescoes in the Siena Town Hall as Mission Statement for Public Administration Today. Discussion Papers, No. 20. Local Government and Public Service Reform Initiative. Open Society Institute.

Feenberg, A. (2002) *Transforming Technology: A Critical Theory Revisited*. New York: Oxford.

Freeman, C. (1974) *The Economics of Industrial Innovation*. London: Penguin Books.

Freeman, C. (1996) *The Long Wave in the World Economy*. Aldershot, UK: Edward Elgar.

Friedman, M. (1953) *Essays in Positive Economics*. Chicago/ London: University of Chicago Press.

Foti, A. (2017) *General Theory of the Precariat: Great Recession, Revolution, Reaction*. Amsterdam: Institute of Network Clusters.

Giotitsas, C. and Ramos, J. (2017) A New Model of Production for a New Economy: Two Cases of Agricultural Communities. *New Economics Foundation*. Available from: http://thesourcenetwork.eu/wp-content/themes/showcase-pro/images/A New Model of Production for a New Economy - FINAL.pdf; http://thesourcenetwork.eu/wp-content/themes/showcase-pro/images/A%20New%20Model%20of%20Production%20for%20a%20New%20Economy%20-%20FINAL.pdf [Accessed: 16 October 2018].

Hegel, G.W.F. (1820) *Elements of the Philosophy of Right*. Translated by H. B. Nisbet, Edited by Allen W. Wood (1991). Cambridge University Press, Cambridge.

Kondratieff, N.D. (1935) The Long Waves in Economic Life. *The Review of Economic Statistics*, 17(6), pp. 105–115.

Kostakis, V. (2011) Commons-Based Peer Production and the Neo-Weberian State: Synergies and Inter Dependencies. *Administrative Culture*, 12(2), pp. 146–161.

Kostakis, V. and Bauwens, M. (2014) *Network Society and Future Scenarios for a Collaborative Economy*. Basingstoke: Palgrave Macmillan.

Kostakis, V., Niaros, V., Dafermos, G. and Bauwens, M. (2015) Design Global, Manufacture Local: Exploring the Contours of an Emerging Productive Model. *Futures*, 73, pp. 126–135.

Kostakis, V., Roos, A. and Bauwens, M. (2016) Towards a Political Ecology of the Digital Economy: Socio-environmental Implications of Two Competing Value Models. *Environmental Innovation and Societal Transitions*, 18, pp. 82–100.

Kostakis, V., Liarokapis, M., Latoufis, K. and Bauwens M. (2018) The Convergence of Digital Commons with Local Manufacturing from a Degrowth Perspective: Two Illustrative Cases. *Journal of Cleaner Production*, 197(2), pp. 1684–1693.

Miliband, R. (1965) Marx and the State. *The Socialist Register*, pp. 278–296.

Niaros, V., Kostakis, V. and Drechsler, W. (2017) Making (in) the Smart City: The Emergence of Makerspaces. *Telematics and Informatics*, 34(7), pp. 1143–1152.

Orsi, C. (2005) The Political Economy of Solidarity: Production. *Federico Caffè Centre Research Report* 5, Roskilde University.

Orsi, C. (2009) Knowledge-Based Society, Peer Production and the Common Good. *Capital & Class*, 33, pp. 31–51.

Pazaitis, A., Kostakis, V. and Bauwens, M. (2017) Digital Economy and the Rise of Open Cooperativism: The Case of the Enspiral Network, *Transfer: European Review of Labour and Research*, 23(2), pp. 177–192.

Perez, C. (1983) Structural Change and Assimilation of New Technologies in the Economic and Social Systems, *Futures*, 15, pp. 357–375.

Perez, C. (2002) *Technological Revolutions and Financial Capital: The Dynamics of Bubbles and Golden Ages*. Cheltenham: Edward Elgar Pub.

Perez, C. (2004) Technological Revolutions, Paradigm Shifts and Socio-institutional Change. In: Reinert, E.S. (ed.) *Globalization, Economic Development and Inequality: An Alternative Perspective*. Cheltenham: Edward Elgar Pub, pp. 217–242.

Procomuns (2016) Procomuns statement and policies for Commons Collaborative Economies at European level. Version 3.0, Barcelona, May 2016. Licensed as a collective work: Public Domain https://creativecommons.org/about/cc0/. Procomuns [Internet]. Available from: http://procomuns.net/en/policy [Accessed 15 September 2015].

Restakis, J. (2010) *Humanizing the Economy: Co-operatives in the Age of Capital*. New Society Publishers.

Restakis, J. (2015) Civil Power and the Partner State. Keynote address at the Good Economy Conference, Zagreb 19–21 May 2015, *Commons Transitions* [Internet]. Available from: http://commonstransition.org/civil-power-and-the-partner-state/ [Accessed 15 September 2015].

Restakis, J., Araya, D., Calderon, M.J. and Murray, R. (2015) ICT, Open Government and Civil Society. *Journal of Peer Production* [Internet], 7 July. Available from: http://peerproduction.net/issues/issue-7-policies-for-the-commons/peer-reviewed-papers/ict-open-government-and-civil-society/ [Accessed 15 September 2015].

Scholz, T., and Schneider, N. (2016) *Ours to Hack and to Own: The Rise of Platform Cooperativism, a New Vision for the Future of Work and a Fairer Internet*. New York, NY: OR Books.

Schumpeter, J. A. (1982/1939) *Business Cycles*. Philadelphia, PA: Porcupine Press.

Vivero-Pol, J.L. (2017) The Food Commons Transition: Collective Actions for Food and Nutrition Security. In: Ruivenkamp, G. and Hilton, A. (eds.), *Perspectives on Commoning: Autonomist Principles and Practices*. London: Zed Books, pp. 325–379.

6
FOOD SECURITY AS A GLOBAL PUBLIC GOOD

Cristian Timmermann

Introduction

In standard economic parlance, public goods are defined as goods that are non-excludable and non-rivalrous in consumption (Stiglitz, 1999). Since these goods can be enjoyed by all, but require resources, political will and effort to be produced and maintained, they are often in short supply or absent entirely (Anomaly, 2015). Well-known examples of public goods are world peace, a stable climate, and scientific knowledge. Communities collaborate to develop a public good by pooling resources together. Over recent decades we have observed a variety of attempts to provide public goods that are specifically designed to be enjoyed by everyone worldwide, the so-called global public goods. The characteristic of global public goods is that they are meant to make "humanity as a whole the *publicum*, or beneficiary" (Kaul et al., 1999: 3). The aim of this chapter is to highlight the advantages and shortcomings of applying to food security the concept of the global public good, as such an examination is crucial in drawing comparisons with the central concept of this handbook: food as a commons.

To recognize food security as a global public good, we need an understanding of the concept of the "public good" that goes well beyond economic reasoning and incorporates normative and societal goals. Food security, similar to world peace, produces benefits from which humanity as whole will profit, irrespective of the level of individual well-being. Food is a basic need and continuous access to food is crucial to the enjoyment of other interests. Therefore, every human being has a fundamental interest in food security (Page, 2013) or, at the very least, their own food security. Yet which understanding of the global public good should be supported when we talk about food security? And what are the advantages of using the concept of the public good as a policy instrument? Wide, interdisciplinary interest in public goods has brought a number of different understandings and usages of the concept. The aim of this chapter is to answer these questions by discussing five prominent characteristics attributed to public goods and their relation to food security. We begin with a short introduction of the concept of "food security" and then briefly introduce five different understandings of the concept of "public goods" before moving on to discuss them in detail.

The multiple interpretations of food security

The concept of food security is deeply rooted in the human rights discourse and has repeatedly changed its emphasis over recent decades due to both dominant political and economic pressures and new insights gained from international development experience. To understand how food security is being constructed through time, one must first briefly consider the different interpretations of the concept and recognize the fact that interest groups defending the idea that food is a tradable good, and activists and organizations claiming that food is a special, protected good, tend to promote very different understandings.

Perhaps the most elementary and oldest understanding of food security is to calculate the total amount of food that is available in a country in proportion to the population (Burchi and De Muro, 2016). A country or region is labelled as food secure if the total availability of foodstuffs (usually measured in calories) matches the currently official standard per capita caloric requirement. This view ignores how food is distributed within the region and recognizes neither variations due to special needs, such as child-bearing, breastfeeding or hard physical work, nor special circumstances, such as old age or ill health (cf. Gilson, 2015). This understanding has been partially motivated by political factors, reducing chronic hunger being seen as a central factor in improving the stability of a country or region. Moreover, food insecurity could make socialist movements more attractive (Perkins and Jamison, 2008). A major problem of this perspective is that it focuses on food availability without paying sufficient attention to actual access to food (Jarosz, 2014). Neoliberal agriculture and food policies, which are particularly keen to rely on market incentives and international trade, have embraced this understanding as it is compatible with a highly industrialized, export-oriented food sector.

In light of these shortcomings, an alternative view was proposed that identifies a household as a measurement unit (Burchi and De Muro, 2016). While this perspective has the added advantage of recognizing important differences in terms of food availability within a country or region, it fails to be sufficiently alert to food insecurity within the household, particularly that suffered by female family members. Such revisions have seen the improvement of human agency as an important goal. The focus is on not only the absence of hunger but also how adequate nutrition could improve human functioning (Sen, 1981). With this change in mind, it becomes important not only that people have sufficient access to food but also that a new concept of "food utilization" has been introduced to highlight the key point that food is actually used to improve nutrition (Burchi and De Muro, 2016). This understanding remains of interest to the food industry as it gives a significant advantage to those food producers which can produce and sell food cheaply. As an approach that particularly welcomes food utilization, it often leaves the adequacy of foodstuffs as a secondary consideration.

A newer worry is therefore the issue of food adequacy, which accommodates not only personal and circumstantial needs but also cultural and religious preferences. This understanding is presently one of the most widely supported definitions in food advocacy, the concept not only demanding access to a continuous and sufficient supply of food but also leaving some room for food choices:

> Food security exists when all people, at all times, have physical and economic access to sufficient safe and nutritious food that meets their dietary needs and food preferences for an active and healthy life.
>
> *(1996 Word Food Summit, cited in FAO, 2008)*

This view focuses not only on whether food is being utilized based on nutritional standards but also whether it is perceived as adequate in terms of social, cultural, and individual preferences and needs. Such an understanding is especially important when addressing malnutrition, since people who avoid some of the available foodstuffs will have a less diverse diet (Thompson, 2015).

It is important to underline that food security is not the same as the absence of hunger. In order to have food security, people need to have their dietary needs met to live an active life. Hunger is often measured as eating less than the minimum dietary energy requirement for a sedentary life, estimated by the Food and Agriculture Organization (FAO) as 1800 kcal/person/day (FAO et al., 2015). As is well known, active people have greater food needs than their sedentary peers. Most people need access to sufficient food to safely endure the strains of physically demanding work, such as growing and preparing food as well as engaging in work that will allow them to purchase food. Actually, performing work is the only way to have the resources to buy food or to produce the food they need (Elver, 2016; Pogge, 2016).

As a general rule, the different food security concepts focus mostly on people having access to adequate food; they do not place a particular value on how and where food is produced, as is the case with the principle of food sovereignty that is currently supported by a plethora of peasant organizations and agroecologists (Agarwal, 2014; Timmermann et al., 2018). The food sovereignty discourse emerged from farmers' organizations and food advocacy groups to counter neoliberal practices and the globalization of food and agriculture by demanding democratic control over food and agriculture policies and over the means of production (Beuchelt and Virchow, 2012; Jarosz, 2014). In our present analysis, it is vital that one bears these crucial differences in mind.

Multiple understandings of a public good

In this context, the aim of this admittedly narrow snapshot of the academic literature is to identify five different understandings of the concept of public good and apply it to the establishment of food security. First, there is the aforementioned, standard economic understanding of public goods, that is, goods from which it is not possible to cost-effectively exclude other people and which are not depleted when consumed. Second, some authors claim that public goods incorporate a normative call. Nobody ought to be excluded from the good in question. This approach to the concept is supported by the notion of human rights, in this context the human right to adequate food (UN Committee on Economic Social and Cultural Rights, 1999). A third understanding emphasizes the benevolent nature of the good for the public. A fourth attribute of public goods is having a public dimension – in the sense of being visible – in contrast with the private dimension of goods that are meant for individual consumption. Here we defend the idea that a failure to address food security, in particular the consequence of hunger, is at odds with attempts to recognize others as equals, as it provokes in our society a tendency to both ignore and block images of those suffering from hunger. Most people need to ignore images of hunger in order to enjoy day-to-day life, and in this way, they marginalize the needy. Fifth, as a public good, food security is a social product that requires "jointness of production," demanding global, coordinated participation. As such, eradicating hunger must be a continuous global commitment. Let us proceed by examining in detail these different understandings.

The economic understanding of public goods

Perhaps the most influential definition is the classic economic understanding of public goods, which encompasses goods that are non-excludable and non-rivalrous in consumption (Drahos, 2004). Here some clarification is required. A food item, for example a potato or carrot, is both excludable and of rivalrous consumption. I can hide a potato and eat it on my own. If we consider all food, in the sense of the global food supply, food remains excludable, as access can be hindered, prohibited, or sanctioned. Yet, as long as we can retain the level of current food production to population ratio, food is not typified by rivalrous consumption, as there is sufficient food available to satisfy everyone's hunger (Ausín, 2010). The world currently produces more than sufficient food to feed all human beings (Tittonell et al., 2016). As far as food security itself is concerned, it is an economic public good: a food-secure world produces numerous benefits that can be enjoyed simultaneously and from which no-one can be practically excluded. Therefore, establishing food security as a global public good addresses the three major problems that characterize public goods (Vivero-Pol, 2017a). First, there is the free-rider problem: people will be able to enjoy the benefits of living in a food-secure world irrespective of whether they contributed to the existence of this public good. Second, since most countries and people cannot recover their investment in food security, we will have an "undersupply" of food security, i.e. some people will remain food insecure as long as there is not a sufficiently large economic incentive to remedy this situation. Third, single groups retain the full advantage of withholding their contribution towards food security, while the price paid for such omissions is paid by society at large, leading to the deterioration of the good, i.e. food insecurity.

Since this understanding sets a special value on the non-excludable advantages of living in a food-secure world, it is important to spell them out. Here it needs to be noted that these arise directly from living in a society that is free from the suffering which food insecurity involves and indirectly from the improved social environment food security creates. These advantages can be grouped thus: (1) moral benefits, (2) public health gains, (3) market opportunities, and (4) higher social stability.

First, and most obviously, there is the moral dimension. According to the majority of moral codes, a world in which people do not suffer an avoidable harm – hunger – is simply a morally better world than one in which a seventh of the population (or a ninth or any other percentage) lacks access to sufficient food to live a flourishing life (Lappé et al., 2013). There are several ethical perspectives which strongly condemn current levels of hunger. Let us review the benefits of food security under the utilitarian, virtue ethics, and liberal perspectives.

Utilitarian ethics generally dictate reducing the number of people who suffer and the intensity of their suffering. Implementing this demand requires resources to be spent on those who can more effectively transform these resources into happiness. This makes the reduction of hunger, as an obvious form of suffering, imperative. People who act according to this ethical reasoning will have to expend their resources alleviating the worst forms of suffering instead of using them for resource-intensive banal pleasures. (Singer, 2004). Money spent on luxuries, such as, for example, trips to exotic islands or jewellery, should rather be used to support agricultural development programmes and urban farming.

Similarly, in the context of virtue ethics, while others are desperately in need of help, practising the virtue of beneficence is a higher priority than enjoying resource-intensive banal pleasures. Caring for others should not only be confined to the emotional level but should also demand specific action from everyone, including lifestyle changes. For those following utilitarian reasoning or virtue ethics, a hunger-free world not only yields the advantage of eliminating a severe form of suffering but also renders morally acceptable the spending of resources upon personal pleasures, or earthly delights, such as enjoying a trip to the countryside (Alemany, 2012).

The case is somewhat more complex for ethical theories that embrace a strong element of liberalism. Generally, liberalism protects the enjoyment of one's freedom as long as one does no harm to others. The extent to which this no-harm principle limits our freedom is highly contested, as there is no consensus on what counts as harming others (Holtug, 2002). Some hardline libertarians defend the idea that one's freedom to extend one's fist ends just before someone

else's nose begins. For this group, harm is usually understood as physical violence. A very different understanding of harm is defended by those such as Thomas Pogge (2008), who considers participating in and sustaining unjust institutions as causing harm to others. If we embrace an extensive understanding of harm, we are required to establish institutions that have the power to protect people from suffering the negative consequences of our actions. Depending on the understanding of harm adopted, it would morally require – for example – redrafting trade policies and mitigating climate change owing to their negative effects upon food security in many regions of the world (cf. De Schutter, 2009; McMichael, 2017). If we consider the prevention of harm as an ethical goal, establishing food security would amount to a guarantee that the direct and indirect harmful effects of our actions would not cause harm beyond a clear threshold or bottom line.

Second, food security leads to an overall improvement in public health (De Schutter, 2011). People who suffer chronic malnutrition are also more likely to transmit disease. Here it is important to underline the difficulty for food-vulnerable people to procure food on their own. People who do not have continuous access to adequate food are more prone to fall sick, and when they do fall sick, they are less capable of working, which reduces their income (or ability to grow food) and thus further diminishes their health (Friel and Ford, 2015). Improving food security reduces the propensity of diseases to propagate, something that benefits all human beings.

Third, there is the economic factor. People who have access to an adequate diet are less likely to be an economic burden upon society, or even better, people who have enough to eat become capable of cooperating in economic, cultural, and social endeavours to the extent of their potential capacities. A very large portion of the global population suffers from insecurity to cover basic needs, which might not allow them to engage in productive market interactions, making them more of a cost than an opportunity to industrial nations (Homann, 2007). From the point of view of large commercial entities and public budgets, there is thus an economically rational interest in the reduction (or even elimination) of the worst forms of hunger and malnutrition as a quintessential element of increasing future market opportunities.

Lastly, large-scale hunger or food insecurity has a destabilizing effect that may lead to food riots and major civil unrest (Page, 2013). Food riots erupt when economic and political injustice has reached a tipping point; people take collective action to object to the direction society has been taking by attempting to bring an end to "business as usual" (Patel and McMichael, 2009). Fear, desperation, and a strong commitment to social justice can mobilize large masses of people to force heads of government from office. Such revolts may have substantial short-term negative effects for both rich and poor citizens. Establishing food security has the advantage that widespread discontent is reduced, which is a stabilizing factor of benefit to everyone living in food-secure areas.

A special advantage to adopting the economic understanding of public goods is that it highlights the fact that we cannot cost-effectively exclude from the benefits of living in a food-secure world those members of our global society who do not contribute to food security. The impossibility of excluding non-contributing people or countries from the benefits of food security brings a series of problems that need be tackled to ensure that sufficient governments commit to significant cooperation. As with any public good, the establishment of food security suffers from the problem of free-riders. People generally dislike carrying others, having to do more than their fair share to achieve a common goal from which everyone benefits. Here it is crucial to note that not contributing another person's portion simply to avoid the injustice of providing more than what one believes to be one's fair share hardly serves an excuse for well-off people to fail to care for people in dire need (Stemplowska, 2016).

Relieving hunger-related suffering is a higher priority than ensuring fairness in the distribution of burdens among wealthy contributors. People have a moral duty to contribute towards food security on the basis of capacity, not equality.

This understanding clearly acknowledges the fact that the good cannot be adequately provided by the market, as food security demands that all people, irrespective of purchasing capacity, have the opportunity to cover their nutritional needs. Being aware of the positive externalities entailed by food security is of fundamental importance in justifying policies that seek to stimulate collective action to establish this public good. However, as a public-good understanding that strongly highlights the positive externalities of the provision of the good, its political support will be strongly dependent upon how these externalities are perceived. The population will have to seek out the necessary information if it is to gain an accurate perception of these externalities, something it may have little incentive to do (Anomaly, 2015). The different understandings of food security may lead to varying judgments on when food security has been accomplished, thus posing the risk that political support may be terminated under the assumption that the public good has already been delivered and does not require additional resources.

The normative call for public goods

A second understanding of public goods emphasizes the importance of these goods being accessible to all, thus embracing a certain ideal of solidarity and aiming to a world where nobody is involuntarily denied access to food (cf. Brody, 1996; Kallhoff, 2014). In a stronger form, this moral imperative entails that individuals ought not to be excluded from using the good (Ausín, 2010). This latter understanding is what John O'Neill (2001, this volume) has termed a normative public good. While an economic understanding of public goods refers to goods from which individuals *cannot* be excluded, public goods in a normative sense are goods from which people *ought not* to be excluded (Vivero-Pol, 2017b).

According to the understanding of food security as a normative public good, through act or omission nobody should deprive others of the right to access food. The idea of food being a good from which nobody ought to be involuntarily excluded is also well anchored in the human rights discourse.

There are, however, several problems with this understanding. In order for food to be available, it first needs to be produced. In its most basic form, this requires that food producers have some type of incentive to engage in food production and that they have the necessary resources to produce food. To stimulate food production, the most common incentive system is to give owners of production facilities the right to decide what happens to their produce. As a consequence, most people need to provide some resources – usually money or labour – as a token of exchange to access food. To prevent people from being unable to access food, some countries have established social security programmes for their citizens so they all may access food while food producers receive remuneration for their work (e.g. *Bolsa Familia* in Brazil). There are other approaches to arranging access to food, although in today's world they are mostly reserved for emergencies or exceptional circumstances, such as community food banks, direct delivery of food rations, or food aid.

Here we must ask ourselves if there are any valid reasons for making exceptions that would allow the withholding of food. Under a very strict understanding of normative public goods, food can never be withheld, even in the most severe cases of punishment. People may refuse food, if they so wish, but they should not be denied access to food. Also, as recognized in human rights law, hunger should never be used as weapon of war (Elver, 2016). In a strict moral sense, this would prohibit any retaliation policies, embargos, or sanctions which have the direct or indirect effect of making it more difficult for the civil population to access or produce food. This still leaves open the question of whether someone deserves more food in cases of severe shortage, when there is no option other than rationing food supplies. Again, strictly speaking, a normative understanding of a public good would make past conduct irrelevant. In cases of severe shortage, food would have to be distributed according to urgency of need, a form of triage or a lottery system, and not by applying any concept of deserving, since this understanding implies that nobody deserves to be food insecure.

The inclusion in policymaking of the principle of food being a normative public good would have to be linked with massive limitations on the market for food products. To establish this public good, there would need to be a balance between rewards that are large enough to stimulate food production but are small enough to avoid inhibiting the population's access to food. Moreover, governmental aid programs would have to be established on a global scale to ensure people do not end up excluded from this public good due to financial limitations, thus necessarily constraining the current liberty of food producers and commercial entities to seek profits from food, agricultural inputs, water, and land (Wilkinson, 2015). To realize food security as a normative public good, we would need to centralize the dimension of food as a human right and reduce its current role as tradable commodity (Vivero-Pol, 2017b).

In a similar vein, this principle strongly affects people's liberty to discard food – a power people often assume is unconditionally attached to one's property (Strahilevitz, 2005). The current everyday understanding of ownership can make us forget that ownership has, from a normative standpoint, always been conditional and subject to usage restrictions. Famously, the 17th-century English philosopher John Locke justified the property rights over an object with the amount of one's labour that is deployed to obtain it or improve it. However, he made such rights conditional upon three provisos. One's enclosure of an object should be subject to (1) leaving enough and as good for others, (2) covering subsistence needs (or charity), and (3) the claim leading to non-wastage (Widerquist, 2010). As is evident, this affects the liberty to use food in purely idiosyncratic ways. Until people demand access to food to cover nutritional needs, others may not grab food beyond subsistence needs or discard food products to serve economic or other interests. The strong interest in not permitting hunger trumps the food owners' interest in retaining the liberty to exclude and discard food.

In Locke's property theory, wastage is also theologically condemned: God gave humanity plenty for everyone to enjoy. Allowing uncollected fruit to spoil on the ground while people are hungry would be ingratitude for what we have received from God, as we as humanity would be not making the most of what has been bestowed (Locke, 1689/1960). It is unclear if this non-wastage condition should be interpreted so far as to command the most efficient use of resources (Hull, 2009; Timmermann, 2017). Adhering to a non-wastage principle would render unacceptable some practices common today in the commercial sector, such as discarding produce for cosmetic reasons or for not having received the expected remuneration (Gjerris and Gaiani, 2013) whenever this produce is claimed for hunger relief. The principle would also oblige a reduction of food spoilage by establishing incentives to make food distribution more efficient and to work closer with food banks and communities to ensure the prompt use of food close to its genuine use-by date (cf. Giorda, 2014; Kenny and Sage in this volume). A strict interpretation of the non-wastage principle would not only prohibit the wasting of the edible food we need, but also the wastage (or even the inefficient use) of the resources needed to produce food. This would render unacceptable certain land and water distribution and management arrangements that fail to leave enough resources of good quality as needed to cover the needs of the whole population.

By recalling that food is crucial to the satisfaction of human needs, another line of reasoning argues that thereby people can exert a legitimate claim for food out of necessity (cf. O'Neill, this volume). Such a right originating in necessity is currently being revived from its historical roots to claim access to resources vital to secure subsistence (Mancilla, 2016; Van Duffel and Yap, 2011). A right of necessity may trump the property rights of people who have more than they need to cover their subsistence, provided that the case of dire need is not caused by the claimant's own fault. For example, a poor farmer who has lost her means of subsistence due to drought could claim a right to access land, water, and food to restore her means of subsistence (Mancilla, 2012).

The normative understanding of a public good, when applied to food security, has the prime advantage of making it absolutely clear that for no reason does anybody deserve to be food insecure. Such a view is nowadays embraced by a number of initiatives that aim at eradicating hunger – by working towards zero hunger. The idea of zero hunger strongly embraces the normative goal of ensuring that nobody is excluded from accessing food (Paes-Sousa and Vaitsman, 2014). To achieve the zero-hunger target exemplified by Sustainable Development Goal 2, countries need both to improve the sustainability, efficiency, and reliability of food production and distribution systems and to implement programmes tailored to the local social and environmental circumstances to ensure that sufficient, adequate food reaches everyone without discrimination. The demand for zero hunger has become even more pressing as humanity continuously produces more food than is needed to cover basic needs. Ending world hunger is a feasible goal and therefore should be prioritized (Pogge, 2016). As long as there is sufficient public awareness of the importance of adequate nutrition, this framework can be used to justify not only freedom from hunger but also the more stringent requirements of the later understandings of food security.

Public goods as beneficent goods

A third understanding of a public good interprets a good both as a (tangible or intangible) commodity and as something of positive value to the general public (in the sense of being universally beneficent). From this perspective, it is important that the public at large perceives the good as a welcomed social improvement. In this sense, the public interest is often contrasted with interests which are solely private (Lever, 2013), echoing the distinction Rousseau (1762) drew between *volonté générale* and *volonté de tous*. This does not necessarily suggest a strict separation between acts intended for individual profit-maximization and those addressing public welfare. Actually, a number of philosophical traditions argue the opposite – that a prolonged and general commitment to the well-being of society will ultimately lead to an increase in individual welfare.

The principle that food security should be a beneficent public good also requires a critical approach to the way in which food security is being achieved. Not only does the public good itself have to improve social welfare, but the way in which it is achieved also needs to be perceived as socially acceptable. Short-sighted policies aiming at food security may have negative effects on social relations, nature, and future generations. Let us briefly review some of these dangers.

Out of the principles of humanity, the most obvious case is that there is nothing laudable in having reached food security because the hungry starved to death and do not count any more as food insecure. A second major issue of concern is how food production affects agricultural workers. Regional specialization in seasonal produce demands a massive migrant workforce that is often undocumented and rarely benefits from labour protection laws (Loo, 2014). Similarly, conventional food production may negatively affect communal values and jeopardize traditional ways of life. Paying too high a price in terms of the suffering of agricultural workers in order to achieve food security is incompatible with an understanding of a public good that sees this good as benevolent.

Moreover, if we want to assess the benevolent nature of a good we need to bear in mind that public goods are also enabling goods (Vivero-Pol, 2017a) and make a distinction between food as a public good and food security as a public good. While the availability of food generally has a positive effect upon society, some caution is necessary in a food-insecure world. Food needs to be sufficiently well-distributed. When we have people with an abundant supply of food co-existing side by side with food insecure people, it is likely that we will observe cases of extreme exploitation (Patel and McMichael, 2009). While food security makes exploitation more difficult, situations where food is available in an insufficient quantity or unevenly distributed – scenarios we encounter while progressing towards food security – allow exploitation.

Another issue of concern lies in the extent to which food production destroys nature. While some might accept the destruction of nature as an acceptable price to pay for avoiding hunger, it is by no means mandatory that such a choice be made between saving nature and eating. Similarly, conventional food producers cannot morally excuse their enormous environmental impact by claiming that this form of food production is the only way to feed the world (Thompson, 2015). We already possess the knowledge to significantly reduce the environmental footprint of food production, as agroecological farms throughout the world demonstrate (Geertsema et al., 2016; Tittonell et al., 2016), yet such methods often demand more labour and do not allow the overwhelming corporate control of food supply chains (Timmermann and Félix, 2015).

There is also much debate on how far the interests of future generations need be considered (Gosseries and Meyer, 2009). Some argue that the interests of future generations should be part of the general public interest. Others claim that rights can only be claimed by those who are already born. Sustainability principles and the human rights discourse demand that our current food production efforts should not jeopardize the ability of future generations to produce the food they will need (Godfray et al., 2010). Currently, our food production systems face the charge of being environmentally unsustainable by using large amounts of fossil fuels, contributing to genetic erosion, reducing the quality of soil, and contaminating aquifers, rivers, and oceans (Schipanski et al., 2016). This calls for urgent policy changes and investment in research into sustainable agriculture to reduce the environmental footprint of food production in order for current food security not to jeopardize future food security (Tittonell, 2013). Food security cannot be perceived as something that benefits the public at large if achieved by sacrificing the fundamental interests of future generations (Korthals, 2004b).

Understanding food security as a global public good that should be universally beneficent has the advantage of stimulating one to think about the price society pays for reaching this goal. The fact that food is a basic need does not mean that food procurement is not to be balanced against other important social goals, such as the protection of the environment, animal welfare, indigenous communities, agricultural workers, and long-term sustainability (Korthals, 2004a; Lawrence, 2017). The road to this goal may pass through a series of injustices. In particular, relying on a concept of food security that does not fully embrace principles of long-term social and environmental sustainability may lead to the ignoring of such types of injustices. This understanding may grant us a stronger sensitivity which is appropriate for identifying such injustices and enabling an ethical evaluation of such trade-offs.

A good that has been made public

A fourth conception of a public good stresses the importance of the good being visible, a publicity requirement so to say, compared to private goods, which can be subject to requests for privacy (cf. Rabotnikof, 2005). How does being public relate to food security?

First, the people suffering food insecurity need to gain visibility. Currently a very large number of people are practically invisible to the media. In many regions of the world, richer neighbourhoods have isolated themselves from the sight of poverty, often by building high physical walls. Hunger leads to suffering, and for people who are at least moderately caring, the images of hunger are something difficult to constantly ignore, especially when confronted with a sudden increase in images of poverty. There is a general tendency to paper over the images of suffering in order to live life as usual. It has even been claimed that there is a widespread refusal to be near poverty or even the sight of poor people, often manifested in hate and hostility towards the poor, a phenomenon Adela Cortina (2017) has termed "*aporophobia*". Throughout the world, nothing reveals poverty as strongly as signs of hunger. Irrespective of how one reacts to poverty, it is far easier to enjoy a meal when one is not directly in front of hungry people. While the ability to shrug off other people's suffering varies significantly among individuals, it is clear that it is difficult or impossible to build constructive social relationships while not visibly caring for the urgent needs of those with whom one engages.

Second, due to physiological constraints, especially chronically food-insecure people can rarely fully partake in the public sphere and thus can neither enjoy the benefits nor contribute as peers to the joint commitments of being active members in a world community. The more food-insecure people are, the likelier it is that they will concentrate their thoughts on solely one thing: food (Ziegler, 2011). Hunger and malnutrition hinder full bodily and mental functioning and development, often impeding people's participation as peers in social interactions that are not considered an immediate priority.

Lastly, if working towards food security is subject to the demands of publicity, it requires governments to be open about the well-being of their people and about how food security is being achieved. Citizens should be aware of the levels of hunger and insecurity that characterize their territories and have adequate access to information on the progressive realization of the right to food and on the methodologies used to measure such developments (cf. Lappé et al., 2013). Food security becomes a public issue for which governments are publicly accountable.

By following the understanding of a public good in terms of a transparency requirement, we discover that establishing food security requires increased openness and accountability towards food and agricultural policies. Food security has the potential to bring back into the public sphere people who now suffer hunger by facilitating their peer interaction and providing them with the strength to engage in cooperative projects. However, it is crucial that this understanding does not rely on a concept of food security that does not adequately assess nutritional needs. Today's international trade policies have made it particularly lucrative to sell processed food that satiates but lacks sufficient vitamins and nutrients, so-called empty calories, making people appear food secure, but at the cost of widespread obesity and other forms of malnutrition (Lawrence, 2017).

A good that requires "jointness of production"

A fifth perspective takes public goods as objects that by necessity have to be produced by a large group of people, that is, goods which require "jointness of production" (Waldron, 1987). The successful production – and in this case also maintenance – of the good requires collective and in most situations coordinated action. The fact that food security may require international cooperation is also recognized in the International Covenant on Economic, Social and Cultural Rights (1966, art. 11.2). This understanding of a public good highlights the fact that people need to come together to establish such a good.

Food security, as a good that demands continuous coordinated action, aptly epitomises the problem of jointness of production. Food security demands that food be accessible and available, requiring work for its production, transportation, and preparation. In today's highly populated world, it is necessary to have plant breeders and extension services at work so that food production increases in proportion to population growth while reducing its environmental impact (McIntyre et al., 2009). Effective policing is also required to ensure that food delivery and production are not blocked, for reasons other than legitimate food safety concerns, by criminal organizations or interest groups. Policy analysts and state officials have to be vigilant that lobbyists do not obstruct the public interest in food production policies. Lastly, it would be negligent to fail to effectively govern and coordinate at every level the effective use of resources and to recognize the different needs of the affected population. To this end, one needs to first understand that hunger is a multidimensional problem and therefore requires a complex set of policy measures (Paes-Sousa and Vaitsman, 2014; Page, 2013).

Maintaining food security as a public good also requires collective action. Furthermore, while some problems need be tackled at a local level, others need be dealt with globally. For example, to ensure that people are actually absorbing the nutrients they need from the food they eat, local undertakings will have to improve the sanitary infrastructure, water quality, and public health in rural and urban areas (Pinstrup-Andersen, 2009). Similarly, many problems can only be handled effectively at a global level, as international cooperation is required. Environmental and social factors increasingly determine the ability to achieve and maintain food security around the globe. In order to maintain harvest yields, we are in need of global efforts to mitigate climate change and to control the propagation of pathogens caused by international travel and trade. Even if we adhere to the idea that addressing food security is a national responsibility, the means to achieve this target demand international coordination and cooperation, which requires making compromises regarding varying national priorities in order to reach international consensus (Chen et al., 1999). The interdependence of food production and distribution chains makes establishing food security a major challenge which requires cooperation at the national and international level (Page, 2013).

The fact that a critical mass of people is required to provide a public good is not necessarily an unfortunate outcome. The undertaking of a joint project creates a series of positive externalities, such as building networks and trust, improving dialogue capacity, and establishing conflict resolution mechanisms and information exchange systems, among other community skills. Once these social capabilities are developed, these same capabilities can be used to address other urgent problems (Mormina, 2018), such as mitigating climate change, halting human trafficking, and addressing neglected diseases.

Understanding food security as a public good that requires jointness of production immediately highlights some of the collective actions initiated by the provision and maintenance of this public good (i.e. community supported agriculture, open seed schemes, urban food sharing initiatives). However, in terms of this understanding, establishing food security as a public good does not necessarily entail the use of democratic institutions. The requirement of jointness of production demands only that a number of people have to come together to develop and maintain the public good and does not presuppose a specific motivation or decision-making mechanism (i.e. it could be just a network of wealthy philantro-capitalists). Unfortunately, since most of the prominent understandings of food security place no particular value upon democratic decision-making, by understanding public goods solely as requiring jointness of production, we are left without a strong normative tool to tackle the extremely high consolidation of the food sector and to improve the resilience of food systems by diversifying providers and retailers (Schiff and Levkoe, 2014).

Concluding remarks

Let us go back to our two initial questions: which understanding of a global public good should be supported if we want to strengthen global food security? What are the benefits of using the global public good concept as a policy instrument? We have seen that all five understandings of public goods shine light on the different, sometimes slightly overlapping problems and opportunities which arise in establishing food security. An economic understanding of public goods creates awareness of the positive externalities of realizing food security for society as a whole and the necessity of looking for solutions that go beyond the market to attain the benefits of living in a food-secure world. The normative call of public goods demonstrates that depriving people of food security is a barrier to social cohesion and clashes with many deeply-anchored concepts of justice. Perceiving public goods as beneficent goods obliges us to carefully assess the costs we are currently paying to achieve food security, in terms of social relations, environmental degradation, and the interests of future generations. The transparency requirement of public goods stimulates us to think about how we deal with food insecurity and to demand more openness in efforts to establish food security and assess any progress towards this goal. Lastly, understanding a public good as requiring "jointness of production" creates greater awareness of the fact that securing food needs is a complex task that requires a multiple set of solutions and coordinated collective action.

All five understandings of a public good reveal some key problems inherent in establishing and maintaining food security, in addition to the many benefits of living in a food-secure world. These five understandings complement each other in the problem–benefit analysis of a public good. However, there is a central issue that is largely absent from the public good perspective, which is how and why public goods are provided and maintained. The role and design of a decision-making mechanism is largely absent, and it appears that the idea of a public good is generally indifferent to how decisions are made. It is at this particular point that we may see the normative advantage inherent in the concept of food as commons and their strong ties to managing communities (cf. Ferrando and Vivero-Pol, 2017; Vivero-Pol, 2017b). Valuing and governing food and food systems as a commons, with the particular features of active participation and shared responsibilities, may inform the political decision to consider food security as a global public good, and its future governance.

Acknowledgements

I owe special thanks to José Luis Vivero-Pol and Tomaso Ferrando for their many insightful comments and suggestions. This work is supported by a postdoctoral fellowship, FONDECYT/CONICYT No. 3170068.

References

Agarwal, B. (2014). Food sovereignty, food security and democratic choice: Critical contradictions, difficult conciliations. *Journal of Peasant Studies*, *41*(6), 1247–1268.

Alemany, M. (2012). Una nota sobre la responsabilidad moral individual frente a la calamidad del hambre. *Obets. Revista de Ciencias Sociales*, 7(1), 15–29.

Anomaly, J. (2015). Public goods and government action. *Politics, Philosophy & Economics*, *14*(2), 109–128.

Ausín, T. (2010). El derecho a comer: Los alimentos como bien público global. *Arbor*, *186*(745), 847–858.

Beuchelt, T. D., & Virchow, D. (2012). Food sovereignty or the human right to adequate food: which concept serves better as international development policy for global hunger and poverty reduction? *Agriculture and Human Values*, *29*(2), 259–273.

Brody, B. (1996). Public goods and fair prices: balancing technological innovation with social well-being. *The Hastings Center Report*, *26*(2), 5–11.
Burchi, F., & De Muro, P. (2016). From food availability to nutritional capabilities: advancing food security analysis. *Food Policy*, *60*, 10–19.
Chen, L. C., Evans, T. G., & Cash, R. A. (1999). Health as a global public good. In I. Kaul, I. Grunberg, & M. A. Stern (Eds.), *Global public goods: international cooperation in the 21st century* (pp. 284–304). New York & Oxford: Oxford University Press.
Cortina, A. (2017). *Aporofobia, el rechazo al pobre: un desafío para la democracia*. Barcelona: Ediciones Paidós.
De Schutter, O. (2009). *International trade in agriculture and the right to food*. Geneva: Friedrich-Ebert-Stiftung.
De Schutter, O. (2011). *The right to an adequate diet: The agriculture–food–health nexus*. Geneva: United Nations.
Drahos, P. (2004). The regulation of public goods. *Journal of International Economic Law*, 7(2), 321–339.
Elver, H. (2016). The challenges and developments of the right to food in the 21st century: reflections of the United Nations Special Rapporteur on the right to food. *UCLA Journal of International Law & Foreign Affairs*, *20*, 1–40.
FAO. (2008). An Introduction to the Basic Concepts of Food Security. Retrieved from http://www.fao.org/docrep/013/al936e/al936e00.pdf
FAO WFP and IFAD (2015). *The State of Food Insecurity in the World 2015*. Rome: FAO.
Ferrando, T., & Vivero-Pol, J.L. (2017). Commons and 'commoning': A 'new' old narrative to enrich the food sovereignty and right to food claims. *Right to Food and Nutrition Watch*, *10*, 50–56.
Friel, S., & Ford, L. (2015). Systems, food security and human health. *Food Security*, 7(2), 437–451.
Geertsema, W., Rossing, W. A., Landis, D. A., Bianchi, F. J., Rijn, P. C., Schaminée, J. H., ... Werf, W. (2016). Actionable knowledge for ecological intensification of agriculture. *Frontiers in Ecology and the Environment*, *14*(4), 209–216.
Gilson, E. C. (2015). Vulnerability, relationality, and dependency: feminist conceptual resources for food justice. *IJFAB: International Journal of Feminist Approaches to Bioethics*, *8*(2), 10–46.
Giorda, E. (2014). Food waste. *Encyclopedia of Food and Agricultural Ethics*, 993–1000.
Gjerris, M., & Gaiani, S. (2013). Household food waste in Nordic countries: estimations and ethical implications. *Etikk i praksis-Nordic Journal of Applied Ethics*, 7(1), 6–23.
Godfray, H. C. J., Beddington, J. R., Crute, I. R., Haddad, L., Lawrence, D., Muir, J. F., ... Toulmin, C. (2010). Food security: the challenge of feeding 9 billion people. *Science*, *327*(5967), 812–818.
Gosseries, A., & Meyer, L. H. (Eds.). (2009). *Intergenerational justice*. Oxford & New York: Oxford University Press.
Holtug, N. (2002). The harm principle. *Ethical Theory and Moral Practice*, *5*(4), 357–389.
Homann, K. (2007). Globalisation from a business ethics point of view. In K. Homann, P. Koslowski, & C. Luetge (Eds.), *Globalisation and business ethics* (pp. 3–9). Aldershot, UK: Ashgate.
Hull, G. (2009). Clearing the rubbish: Locke, the waste proviso, and the moral justification of intellectual property. *Public Affairs Quarterly*, *23*(1), 67–93.
Jarosz, L. (2014). Comparing food security and food sovereignty discourses. *Dialogues in Human Geography*, *4*(2), 168–181.
Kallhoff, A. (2014). Why societies need public goods. *Critical Review of International Social and Political Philosophy*, *17*(6), 635–651.
Kaul, I., Grunberg, I., & Stern, M. A. (1999). Defining global public goods. In I. Kaul, I. Grunberg, & M. A. Stern (Eds.), *Global public goods: international cooperation in the 21st century* (pp. 2–19). New York & Oxford: Oxford University Press.
Korthals, M. (2004a). *Before dinner: philosophy and ethics of food*. Dordrecht: Springer.
Korthals, M. (2004b). *Ethics of global public goods in food and agriculture*. Rome: FAO Discussion Papers.
Lappé, F. M., Clapp, J., Anderson, M., Broad, R., Messer, E., Pogge, T., & Wise, T. (2013). How we count hunger matters. *Ethics & International Affairs*, *27*(03), 251–259.
Lawrence, G. (2017). Re-evaluating food systems and food security: A global perspective. *Journal of Sociology*, *53*(4), 774–796.
Lever, A. (2013). *Democracy, public goods and intellectual property*. Paper presented at the Patents and Poverty: Themes from the Work of Thomas Pogge, St. Gallen.
Locke, J. (1689/1960). *Two treatises of government*. Cambridge: Cambridge University Press.
Loo, C. (2014). Towards a more participative definition of food justice. *Journal of Agricultural and Environmental Ethics*, *27*(5), 787–809.
Mancilla, A. (2012). Noncivil disobedience and the right of necessity. A point of convergence. *Krisis*, *3*, 3–16.

Mancilla, A. (2016). *The right of necessity: moral cosmopolitanism and global poverty*. London: Rowman & Littlefield.

McIntyre, B. D., Herren, H. R., Wakhungu, J., & Watson, R. T. (2009). *International assessment of agricultural knowledge, science and technology for development (IAASTD): synthesis report with executive summary: a synthesis of the global and sub-global IAASTD reports*. Washington, DC: Island Press.

McMichael, A. (2017). *Climate change and the health of nations: famines, fevers, and the fate of populations*. Oxford: Oxford University Press.

Mormina, M. (2018). Science, technology and innovation as social goods for development: rethinking research capacity building from Sen's capabilities approach. *Science and Engineering Ethics*, doi:10.1007/s11948-11018-10037-11941.

O'Neill, J. (2001). Property, care, and environment. *Environment and Planning C: Government and Policy, 19*, 695–711.

Paes-Sousa, R., & Vaitsman, J. (2014). The Zero Hunger and Brazil without Extreme Poverty programs: a step forward in Brazilian social protection policy. *Ciência & Saúde Coletiva, 19*(11), 4351–4360.

Page, H. (2013). *Global governance and food security as global public good*. New York: Center on International Cooperation.

Patel, R., & McMichael, P. (2009). A political economy of the food riot. *Review, 32*(1), 9–35.

Perkins, J. H., & Jamison, R. (2008). History, ethics, and intensification in agriculture. In P. B. Thompson (Ed.), *The ethics of intensification* (pp. 59–83). Dordrecht: Springer.

Pinstrup-Andersen, P. (2009). Food security: definition and measurement. *Food Security, 1*(1), 5–7.

Pogge, T. W. (2008). *World poverty and human rights: cosmopolitan responsibilities and reforms* (2nd ed.). Cambridge: Polity.

Pogge, T. W. (2016). The hunger games. *Food Ethics, 1*(1), 9–27.

Rabotnikof, N. (2005). *En Busca de Un Lugar Común. El Espacio Público En La Teoría Política Contemporánea*. México, DF: UNAM, Instituto de Investigaciones Filosóficas.

Rousseau, J.-J. (1762). *Du Contrat Social Ou Principes Du Droit Politique*. Amsterdam: Marc-Michel Rey.

Schiff, R., & Levkoe, C. Z. (2014). From disparate action to collective mobilization: collective action frames and the Canadian food movement. In L. Leonard & S. B. Kedzior (Eds.), *Occupy the earth: global environmental movements* (pp. 225–253). Bingley: Emerald.

Schipanski, M. E., MacDonald, G. K., Rosenzweig, S., Chappell, M. J., Bennett, E. M., Kerr, R. B., … Lundgren, J. G. (2016). Realizing resilient food systems. *BioScience, 66*(7), 600–610.

Sen, A. (1981). *Poverty and famines. An essay on entitlement and deprivation*. Oxford & New York: Oxford University Press.

Singer, P. (2004). *One world: the ethics of globalization*. (2nd ed.). New Haven, CT & London: Yale University Press.

Stemplowska, Z. (2016). Doing more than one's fair share. *Critical Review of International Social and Political Philosophy, 19*(5), 591–608.

Stiglitz, J. E. (1999). Knowledge as a global public good. In I. Kaul, I. Grunberg, & M. A. Stern (Eds.), *Global public goods: international cooperation in the 21st century* (pp. 308–325). New York & Oxford: Oxford University Press.

Strahilevitz, L. J. (2005). The right to destroy. *The Yale Law Journal, 114*, 781–854.

Thompson, P. B. (2015). *From field to fork: food ethics for everyone*. New York: Oxford University Press.

Timmermann, C. (2017). Harvesting the uncollected fruits of other people's intellectual labour. *Acta Bioethica, 23*(2), 259–269. doi:10.4067/s1726-569x2017000200259

Timmermann, C., & Félix, G. F. (2015). Agroecology as a vehicle for contributive justice. *Agriculture and Human Values, 32*(3), 523–538.

Timmermann, C., Félix, G. F., & Tittonell, P. (2018). Food sovereignty and consumer sovereignty: Two antagonistic goals? *Agroecology and Sustainable Food Systems, 42*(3), 274–298. doi:10.1080/21683565.2017.1359807

Tittonell, P. (2013). *Farming Systems Ecology: Towards Ecological Intensification of World Agriculture*. Wageningen: Wageningen Universiteit.

Tittonell, P., Klerkx, L., Baudron, F., Félix, G. F., Ruggia, A., van Apeldoorn, D., … Rossing, W. A. (2016). Ecological intensification: local innovation to address global challenges. *Sustainable Agriculture Reviews, 19*, 1–34.

UN Committee on Economic Social and Cultural Rights. (1999). *General comment 12. The right to adequate food (article 11) (E/C.12/1999/5)*. Geneva: United Nations Economic and Social Council.

Van Duffel, S., & Yap, D. (2011). Distributive justice before the eighteenth century: the right of necessity. *History of Political Thought, 32*(3), 449–464.
Vivero-Pol, J. L. (2017a). *How do people value food? Systematic, heuristic and normative approaches to narratives of transition in food systems.* PhD thesis, Université Catholique de Louvain, Louvain-la-Neuve. https://dial.uclouvain.be/pr/boreal/fr/object/boreal%3A191763
Vivero-Pol, J. L. (2017b). Food as commons or commodity? Exploring the links between normative valuations and agency in food transition. *Sustainability, 9*(3), 442.
Waldron, J. (1987). Can communal goods be human rights? *European Journal of Sociology, 28*(2), 296–322.
Widerquist, K. (2010). Lockean theories of property: justifications for unilateral appropriation. *Public Reason, 2*(1), 3–26.
Wilkinson, J. (2015). Food security and the global agrifood system: ethical issues in historical and sociological perspective. *Global Food Security*, 7, 9–14.
Ziegler, J. (2011). *Destruction massive: Géopolitique de la faim*. Paris: Seuil.

PART II

Exploring the multiple dimensions of food

7
FOOD, NEEDS AND COMMONS

John O'Neill

Introduction

In market economies food is a commodity that is distributed through market exchange. Indeed, Karl Polanyi takes the degree to which food, together with land and labour, are marketable commodities to be a measure of the degree to which an economy is fully developed as a market economy:

> The rise of the market to a ruling force in the economy can be traced by noting the extent to which land and food were mobilized through exchange, and labor was turned into a commodity free to be purchased on the market.
>
> *(Polanyi, 1957: 255)*

Food is also a vital human need. It is a necessity grounded in human biology and physiology. Failures of access to good food are a source of malnutrition and, where the lack is severe, death. The provision of food is also a good that is central to human culture and relationships. It is a central component of everyday social existence. The fact that food is both a commodity and a vital human need is a source of ethical and social conflict. As a commodity, food is distributed by ability to pay. As such, failures of access to food need not be a matter of there not being enough food available, but can rather be a matter of individuals lacking the means of exchange to buy the food they need (Sen, 1981). Where individuals have no access to land and the other means to provide their own food and hence must rely on their ability to buy food, where they lack the money to do so, they will go hungry. In the terms that Sen puts it, the person will be at risk of starvation when his 'exchange entitlements', that is 'the alternative bundles of commodities that he can acquire in exchange for what he owns', do not contain 'any feasible bundle including enough food' (Sen, 1981: 3).

Food as a commodity can conflict with food as a need and claims that individuals make on the basis of their needs. This conflict has been central to social movements around food since the development of market capitalist societies. As E. P. Thompson notes, food riots are not simply the spontaneous response to hunger (Thompson, 1971: 136). They represent a moral response, grounded in claims of need, to new market modes of provision. One way of presenting this conflict is as a conflict of a moral economy of need with an amoral or non-moral market sphere. However, this would miss the nature and varieties of conflicts at issue. On one hand, the new market order itself was premised on particular moral assumptions and relations that were articulated by defenders of commercial society, most notably by Adam Smith. On the other hand, the moral economy of need can take a variety of different forms – from pre-capitalist forms of paternalism that recognised rights of necessity through to more egalitarian forms of mutual aid and solidarity grounded in the recognition of common neediness and vulnerabilities. This chapter explores these conflicts of moral economy with the aim of articulating and defending the continued relevance of these egalitarian moral economies for thinking about food and its provision. The chapter is in five sections. The first section is analytical. It outlines the sense in which food should be understood as a need and as a public good. The second explores the conflicts in the moral economies of the new market order and the older traditions that recognised claims of necessity. These claims of necessity are themselves often grounded in an appeal to the various versions of the view that food belongs to a commons. Claims of necessity and response to those claims of necessity uncover changing senses in which food was considered as commons historically. This contrasts with the defenders of the new market order – in particular Smith – who denied claims of necessity. The final sections consider two strands of radical response to the defences of the moral economy of the market. The third section explores appeal to common neediness that underpins the egalitarian traditions of mutual aid and the complex relations these traditions had to modern welfare provision. The fourth section turns to Marx's discussion of the commons and its continuing influence on more recent defences of the commons. The fifth section offers some concluding thoughts.

Food as a need and food as a public good

That humans need food is a claim that has special moral and political force. To understand that force requires an analysis of the particular kind of needs-claim being asserted. An important initial distinction to be drawn is that between purely instrumental and absolute or categorical uses of the concept of need (Wiggins, 1998: 9–10). Some needs-claims specify needs that are *purely instrumental* in the sense that they are the necessary means to an end that is itself optional. A person says 'I need £10,000'. I ask 'why?' To the answer 'I need the money for a ticket to travel around the world', the obvious reply is 'do you really need to travel around the world?' The end itself is an optional one. However, many needs-claims specify needs that are not purely instrumental since the ends are not optional in this sense. Rather, they are the means to the realisation of a minimal level of human flourishing, so that a person can be said to be harmed if the needs are not satisfied. The needs are *absolute* or categorical. Wiggins offers the following characterisation of the concept:

> I need [absolutely] to have x
> if and only if
> I need [instrumentally] to have x if I am to avoid being harmed
> if and only if
> It is necessary, things being what they actually are, that if I avoid being harmed then
> I have x.
>
> (Wiggins, 1998: 10)

Wiggins usefully outlines a number of additional dimensions of needs claims which are key to understanding the special demands that a claim like 'human beings need food' make on moral agents and public policy. The *gravity* of needs is a question of how bad the harm would be if the need is not met. The *urgency* of needs concerns how rapidly action must be taken to meet the need. Needs are *entrenched* to the degree that they cannot be changed. An important class of entrenched needs are *basic needs* which are grounded in laws of nature and unchangeable facts about the human condition. A need for some good is *non-substitutable* if the need cannot be met by other goods in ways that will avoid the significant harm. *Vital needs* are absolute needs that are deeply entrenched, grave and scarcely substitutable. Food is a clear example of a vital need. A person will be badly harmed if the need for food is not met; they are grounded in facts about human biology; and the specific nutrients required for health are not substitutable. Those nutrients are required if the person is to avoid serious illness or death.

Another important conceptual distinction that needs to be marked at the outset is that between *dispositional* and *occurrent* needs (Reader, 2007: 71–72; Thompson 1987: 11–12). Consider again the claim 'humans need food'. As a statement about all human beings, it refers to dispositional needs, that is, conditions that human beings must meet if they are to live a minimally flourishing or decent life. However, the claim that 'all humans need food' is false if the term is used in the occurrent sense. An occurrent need refers to a particular lack of what is needed in the dispositional sense. In the occurrent use of the term, only the hungry or starving have a need for food. A distinction should also be drawn between needs and the specific objects that might meet that need – what are often called need satisfiers (Doyal and Gough, 1991: 69–75). While many needs, like the need for food, are universal since they are deeply entrenched in the biological constitution of human beings, the particular ways that need is met are clearly specific to local communities and cultures. Culinary variety is one of the more obvious examples of cultural diversity. At the same time, that same variety points to the way in which food is not simply a need grounded in the biological nature of human beings. Food is also a central good in meeting other vital human needs for affiliation and sociality.

The claim that food is a human need is uncontentious. The claim that it is a public good is in one sense of the term simply false. It is not a public good in the standard economic sense, that is, a good that is (i) 'non-rival' in consumption – the consumption of the good by one person does not decrease that of others – and (ii) 'non-excludable' – individuals cannot be excluded from the benefit of using the good. Indeed, food looks like a paradigm case of a good that lacks these characteristics. The consumption of food does deprive others of that food, and individuals are easily excluded from its consumption. The hungry and ill-nourished often lack access to food. Public goods in the economic sense are goods from which individuals cannot be excluded. The term public good, however, is often used in an ethical and political sense that is distinct from this economic sense. It is used to refer to goods from which individuals ought not to be excluded from their use. Elsewhere I have referred to these goods as normative public goods (O'Neill, 2001: 699–700, 2007: ch.3). In that sense, one way in which food might be understood as a commons is that it is a normative public good.

What are the grounds for believing that some goods should be normative public goods? One set of grounds is that where goods are necessary for the satisfaction of vital needs they should be treated as normative public goods within a community or across communities. Needs make claims to goods from which individuals should not be excluded. The universal provision of health care on the basis of need invokes such claims. Food as a vital human need looks like a strong candidate for membership of this category of normative public goods. The claim that food does have such a status has long been at the centre of the conflicts with the framing of food as a commodity. In the following section, I outline some of the practical and theoretical contours of this conflict. I trace some moments in the history of claims that need grounds common use. Doing so allows a distinction between the different senses in which food could be understood as a commons.

Needs, markets and the moral economies of subsistence

Hunger is an occasion for violence. Food riots have been a response to hunger for centuries. However, as E. P. Thompson notes, the 'abbreviated and "economistic" picture of the food riot, as a direct, spasmodic, irrational response to hunger' (Thompson, 1971: 136) needs to be rejected. The food riot expresses a moral response to a particular economy for the provisioning of food.

> It is ... true that riots were triggered off by soaring prices, by malpractices among dealers, or by hunger. But these grievances operated within a popular consensus as to what were legitimate and what were illegitimate practices in marketing, milling, baking, etc. This in its turn was grounded upon a consistent traditional view of social norms and obligations, of the proper economic functions of several parties within the community, which, taken together, can be said to constitute the moral economy of the poor. An outrage to these moral assumptions, quite as much as actual deprivation, was the usual occasion for direct action.
>
> *(Thompson, 1971: 78–79)*

The riots associated with food reveal conflicts between different moral attitudes and practices of economic life. Thompson himself presents the food riots in the late eighteenth century in England as 'a last desperate effort by the people to re-impose the older moral economy as against the economy of the free market' (Thompson, 1968: 73). The older moral economy was premised on 'the immorality of forcing up the price of provisions by profiteering upon the necessities of the people' (Thompson, 1968: 68). It contrasts with the economic order of the market in which older regulations premised on the rights of necessity were rejected. The system of liberty promised by the market offered, it was claimed, a more rational system. As Smith put it, the freedom of trade, both within and between countries, was 'both from reason and experience, not only the best palliative of a dearth, but the most effectual preventative of a famine' (Smith 1776: iv.v.b.39). From this perspective, the moral economy of the rioter becomes an expression of moral prejudice that stands in the way of a more reasonable economic order:

> The laws concerning corn may everywhere be compared to the laws concerning religion. The people feel themselves so much interested in what relates either to their subsistence in this life, or to their happiness in a life to come, that government must yield to their prejudices, and, in order to preserve the public tranquillity, establish that system which they approve of. It is upon this account, perhaps, that we so seldom find a reasonable system established with regard to either of those two capital objects.
>
> *(Smith 1776: iv.v.b.39)*

The new market order and older traditions of moral economy stand in conflict with each other. One way to present this conflict is as a conflict of a moral economy with an amoral or non-moral economic order. The notion of a moral economy as a pre-capitalist ethic of subsistence is one that is articulated elsewhere in the early literature on the moral economy. James Scott, for example, makes similar observations about the moral economy of the peasant. It represents an ethic of subsistence:

> The pre-capitalist community was ... organized around this problem of the minimum income – organized to minimize the risk to which its member were exposed by virtue to its limited techniques and the caprice of nature. Traditional forms of

> patron-client relationships, reciprocity, and redistributive mechanisms may be seen from this perspective.
>
> *(Scott, 1977: 9)*

However, to present the conflict as one between a moral and amoral economy is not, I think, the right way of understanding the conflict. The work of Smith itself illustrates the way that the market can itself be understood as an economy that is constituted by a particular set of practices, beliefs and norms that have their own ethical character. The liberal story is roughly this, that markets form an ethical sphere in which free agents enter into voluntary contractual relations with each other that involve mutual recognition and respect of each other as free, independent and autonomous beings. It is central to Smith's defence of commercial society as a sphere in which individuals develop and express the virtues of independence. It is found in Hegel's account of the contractual sphere of the market as one in which individuals 'recognise each other as persons and property owners' (Hegel 1967: paragraph 71R). It runs through the standard account in modern liberal theory of markets as spheres in which self-owners relate to each other through voluntary contract. The conflicts Thompson and Scott note represent a conflict not between a moral economy and an amoral economy, but a conflict between different moral economies. The point is important in understanding what is at stake in the conflict. In particular, it allows for a fuller understanding of the conflicts not just between the moral economies of the market and the pre-capitalist paternalistic moral economy, but the conflict between the moral economy of the market and the more egalitarian ethic of solidarity that emerged in early responses to market society.

The contours of the conflicts around the older moral economy of need and subsistence and the moral economy of the market have been well mapped. The older moral economies were premised on the assumption that need gave a person claims on others. The claims of necessity were of two kinds. The first is the claim to food in times of famine, a claim often articulated in terms of common goods. The second concerns justice-based constraints on prices.

The classical defence of the claims of need was articulated philosophically by the scholastics. Aquinas offers a standard account. In response to the problem of how private property is justifiable given that 'according to the natural law all things are common property', Aquinas follows Aristotle's defence of private property and common use (Aristotle, 1948: II.5). While Aristotle rejects Plato's communism, he holds that goods held privately should be in common use. In Aristotle's work, the combination of private ownership and common use is applied particularly to subsistence in food:

> On the one hand, property ought not to be owned in common, as some writers have maintained – though it ought to be use in common and as friends treat their belongings. On the other hand, none of the citizens should go in need of subsistence.
>
> *(Aristotle, 1948: VII.x.9)*

The institution of common meals is presented as what best meets the requirements of common use that meets the needs of all citizens:

> The institution of common tables is generally agreed to be for advantage of well-ordered states … The right of dining at the common tables should be equally open to every citizen; but poor men will always find it difficult to contribute their quota for cost from their own resources.
>
> *(Aristotle, 1948: VII.x.9)*

The need for all citizens to be included is the reason that the Cretan system of common meals funded from public resources is presented as preferable to the Spartan system, in which each had to provide a contribution (Aristotle, 1948: II.ix.29).[1]

Aquinas follows Aristotle. The 'power to procure and dispense' of goods should be private, since this best ensures care for goods and their procurement:

> [E]very man is more careful to procure what is for himself alone than that which is common to many or to all ... [H]uman affairs are conducted in more orderly fashion if each man is charged with taking care of some particular thing himself, whereas there would be confusion if everyone had to look after any one thing indeterminately.
>
> *(Aquinas, 1920: II.II 66.2)*

However, the use of goods remains common. With respect to the use of external goods, 'man ought to possess external things, not as his own, but as common, so that, to wit, he is ready to communicate them to others in their need' (Aquinas, 1920: II.II 66.2). Need gives rights to the use of goods. Those in need have claims on the property of those with a superfluity of goods. In response to the question 'Is theft justifiable in the case of necessity?', Aquinas writes:

> [I]f the need be so manifest and urgent, that it is evident that the present need must be remedied by whatever means be at hand (for instance when a person is in some imminent danger, and there is no other possible remedy), then it is lawful for a man to succor his own need by means of another's property, by taking it either openly or secretly: nor is this properly speaking theft or robbery.
>
> *(Aquinas, 1920: II.II 66.7)*

The reason it is not theft is that, with respect to need, goods are common property: 'In case of need all things are common property, so that there would seem to be no sin in taking another's property, for the need has made it common' (Aquinas, 1920: II.II 66.7). The procurement is not theft since need itself gives a title to property: 'that which he takes for the support of his life becomes his own property by reason of that need' (Aquinas, 1920: II.II 66.7).

Claims of necessity are also taken to constrain prices on goods. The argument is again Aristotelian in origin. A central argument against the justice of raising prices in conditions of necessity was that such transactions should not be understood as free and voluntary acts. Thus, as Aquinas puts its it: 'There is a partial coercion when necessity threatens, as is evident in the case of one who jettisons cargo into the sea to save a ship' (Aquinas, 2003 XIII: Article 4). Or consider again the following from Francisco de Vitoria:

> [F]or commutative justice to obtain in human exchange, it is not sufficient that it is simply voluntary from both sides; it is necessary that it has nothing involuntary mixed with it, as is evident from throwing merchandise into the sea ... [T]hat which is done because of need, albeit simply voluntary, has something of the involuntary mixed with it. Consequently, in the case of such exchange it is not sufficient that it is simply voluntary; it is required that there be no need to exchange some object, and in this there may yet be something of violence because of need.
>
> *(Francisco de Vitoria Commentarios I–II. 77. 1, cited in Langholm, 1998: 111–112)*

Exchanges where one party is compelled by necessity are not voluntary exchanges and hence are not just. The appeal in both texts is back to Aristotle's discussion of the conditions for voluntary action. In particular they appeal to a class of 'actions done because of fear of greater evils or because of something fine' (Aristotle, 1999: III.1), which Aristotle had described as a 'mixture' of the voluntary and involuntary. They are not clear cases of involuntary actions since the acts are not the result of external force or ignorance. However, they are not acts that have a reasonable alternative. The case of the captain who has to throw his cargo overboard to save a ship to which the scholastics appeal is the standard example. In one sense acts of this kind look voluntary, since the actor 'does it willingly … it is up to him to do them or not to do them' (Aristotle, 1999: III.1). In another sense they look to be involuntary: 'the actions without qualification are involuntary, since no one would choose any such action its own right' (Aristotle, 1999: III.1). It is this account of involuntary action to which the scholastics appeal. Exchange compelled by necessity is not fully voluntary. Hence, justice requires constraints on prices in times of need. Higher prices in conditions of scarcity are a form of injustice. Typical is Aquinas: 'to sell something to another in need for much more than the value of a thing … would be unjust' (Aquinas, 2003, XIII, Article 4 reply to the seventh argument).

Elements of the scholastic position survive with increasing qualifications into the work of Grotius, Pufendorf and Locke.[2] Thus, for example, Grotius follows the broad scholastic position on the rights of necessity:

> In direst need, the primitive right of user revives, as if community of ownership had remained, since in respect to all human laws – the law of ownership included – supreme necessity seems to have been excepted. … Even among the theologians, the principle has been accepted that, if a man under stress of such necessity takes from the property of another what is necessary to preserve his own life, he does not commit a theft.
>
> *(Grotius, 2013: 2.2.6)*

The right to buy goods at a just price similarly survives:

> We affirm, therefore, that all men have the right to buy such things at a fair price, unless they are needed by the person from whom they are sought; thus in times of extreme scarcity, the sale of grain is forbidden.
>
> *(Grotius, 2013: 2.2.19)*

However, where in Aquinas claims of need in goods are what define the permanent condition of common use in private property, in Grotius they are exceptions in the application of laws of property where private property has historically superseded the original condition of common property (Salter, 2005: 285).

In the hands of Locke, claims of need are weakened still further. In Locke, claims of necessity are understood in terms of duties of charity rather than of rights of necessity: 'Charity gives every Man a Title to so much out of another's Plenty, as will keep him from extream want when he has no means to subsist otherwise' (Locke, 1988: 2.5.4.6). Similarly constraints on prices on the basis of justice are rejected. Justice in prices consists simply in the 'market price at the place where he sells'. Hence, 'he that sells his corn in a town pressed with famine at the utmost rate he can get for it does no unjustice against the common rule of traffic' (Locke, 1695: 86). It rather becomes an offence against the 'common rule of charity' when the buyer is left without the means of subsistence.

The writings of Adam Smith defending the new political economy are marked by the rejection of claims of necessity as claims of justice that give a person rights in the property of another.

It is a mark of the virtuous agent that poverty and even death are to be preferred to the injustice involved in taking what is the property of another:

> The poor *man* must neither defraud nor steal from the rich, though the acquisition might be much more beneficial to the one than the loss could be hurtful to the other … *There* is no *commonly honest man* who does not more dread the inward disgrace of such an action, the indelible stain which it would forever stamp upon his own mind, than the greatest external calamity which, without any fault of his own, could possibly befal him; and who does not inwardly feel the truth of that great stoical maxim, that for one *man* to deprive another unjustly of any thing, or unjustly to promote his own advantage by the loss or disadvantage of another, is more contrary to nature, than death, than poverty, than pain, than all the misfortunes which can affect him, either in his body, or in his external circumstances.
>
> *(Smith, 1982a: III.3)*[3]

A virtuous agent is one who stands on his independence, for whom poverty and death is better than transgressing the rules of property on which 'depend the whole security and peace of human society' (Smith, 1982a: III.3).

Smith's arguments here form part of a more general case against the paternalist moral economy of subsistence in pre-capitalist society. The argument for the new moral economy of the market turned not just on the claim that the market economy would better deliver food, particularly through its incentives to production and distribution through price. It also turns on the claim that the market economy fostered the virtues of independence against the personal dependence that was embodied in the paternalism of pre-commercial society:

> [C]ommerce and manufactures gradually introduced order and good government, and with them, the liberty and security of individuals, among the inhabitants of the country, who had before lived almost in a continual state … of servile dependency upon their superiors. This, though it has been the least observed, is by far the most important of all their effects.
>
> *(Smith, 1981: III.IV.4)*

The ties of patronage of pre-commercial society render individuals dependent on the gift of the wealthy: '[The great proprietor] is at all times … surrounded with a multitude of retainers and dependants, who, having no equivalent to give in return for their maintenance, but being fed entirely by his bounty, must obey him' (Smith 1981: III.IV.5–6). Through the division of labour and exchange relationships, these ties of dependence are broken. The worker's income is not tied to any particular individual: 'Though [the wealthy person] contributes, therefore, to the maintenance of them all, they are all more or less independent of him, because generally they can all be maintained without him' (Smith 1981: III.IV.11). While Smith acknowledged that the wage worker and servant do not have the independence of the artisan (Smith 1981: I.viii.48), and that the wage worker lays down a portion of 'his liberty' (Smith 1981: I.v.7), both are still free of the forms of personal independence of pre-commercial society. It is this independence that is one of the most important achievements of commercial society: 'Nothing tends so much to corrupt and enervate and debase the mind as dependency, and nothing gives such noble and generous notions of probity as freedom and independency. Commerce is one great preventative of this custom' (Smith 1982b: vi.6).

Smith's defence of the new moral economy of the markets was premised then not just on claims that commercial society better delivered human subsistence, but that it did so without the forms of personal dependence that were a feature of the pre-commercial moral economy of subsistence. However, this contrast between the moral economy of the market and the paternalist moral economy of subsistence does not define the field of the debate around need, food and destitution. Both need to be contrasted with more egalitarian appeals to an economy founded upon the recognition of claims of need. Two strands of egalitarian response can be analytically distinguished: forms of social solidarity and mutual aid that were a feature of early working-class responses to the facts of necessity; and the rejection of private property and defence of common ownership articulated most influentially by Marx. The next two sections examine these responses in more detail.

The moral economies of solidarity

The contrast Thompson draws between the moral economy of the crowd and the new political economy of the market turns on the place of necessity in human life and the social relations required to meet the claims of necessity. One appeal may have been, as Thompson notes, to paternalist traditions of support to those in need. However, what are also articulated in this period are appeals to a moral economy of mutual aid. One response that found expression in the period in which the new moral economy of the market emerged was an egalitarian moral economy realised in the forms of solidarity and mutual aid found in the early friendly societies and worker's associations. It is premised on the existence not simply of needs that require responses, but the fact of mutual neediness – the recognition that we are all needy and dependent creatures. It stands in opposition to the moral economy of the market in a radical sense. The moral economy of the market is premised on a picture of agents as independent beings who recognise each other as such, recognition embodied in the contractual relations entered into with each other. The contrast between this moral economy and that of the paternalist tradition lies in part in the ideal of independence it embodies – independence given radical expression in Smith's moral philosophy. The moral economy of solidarity that contrasts with the moral economy of the market does not deny the importance of personal independence as a virtue. However, it acknowledges, in the way that the moral economy of the market does not, that the achievement of independence is only made possible against the background of our mutual dependence as creatures of need. The assumption that pervades the moral economy of the market – that we are independent contracting agents – is itself premised on the unacknowledged ties of dependence and facts of neediness. The work of Smith is particularly important here in that it both notes the existence of dependence and need and the ways that modern commercial society moves forward by rendering this fact invisible.[4]

Smith does not deny the fact of human neediness: 'All the members of human society stand in need of each others assistance, and are likewise exposed to mutual injuries' (Smith 1982a: II.ii.3.1). Human neediness and dependence on others are grounded in our biological existence. Human beings come into the world in complete dependence on others (Smith 1982a:VI.ii.1.3). When illness and old age visit us, these facts of human dependence force themselves on us again.

> In the languor of disease and the weariness of old age, the pleasures of the vain and empty distinctions of greatness disappear ... Power and riches appear then to be, what they are, enormous and operose machines contrived to produce a few trifling conveniences to the body, consisting of springs the most nice and delicate, which must be kept

> in order with the most anxious attention ... They keep off the summer shower, not the winter storm, but leave him always as much, and sometimes more exposed than before, to anxiety, to fear, and to sorrow; to diseases, to danger, and to death.
>
> *(Smith 1982a: IV.1.8)*

However, while Smith acknowledges these facts of human existence, he argues that the progress of economic life requires that we forget that this is the case:

> But though this splenetic philosophy, which in time of sickness or low spirits is familiar to every man, thus entirely depreciates those great objects of human desire, when in better health and better humour, we never fail to regard them under a more agreeable aspect.
>
> *(Smith 1982a: IV.1.9)*

Social forgetfulness and self-deception about our neediness are a condition for the flourishing of commercial society: 'It is this deception which rouses and keeps in continual motion the industry of mankind' (Smith 1982a: IV.1.10).

It is in this context that the invisible hand makes its appearance in *The Theory of Moral Sentiments* as a link between 'natural selfishness and rapacity' of the rich and a 'distribution of the necessaries of life, which would have been made, had the earth been divided into equal portions among all its inhabitants' (Smith 1982a: IV.1.10). The invisibility of neediness and the social forgetting of the facts of human dependence they entail are a condition of market societies progressing.

Smith's defence of the moral economy of the market based on the art of social forgetting contrasts with the very different response to the facts of human neediness and dependence among those whose conditions made them more conscious of the ubiquity of human vulnerability. A feature of mutualism of working-class communities is that they make explicit the facts of mutual neediness and dependence in conditions of a rough equality of vulnerabilities. It receives expression in a solidaristic moral economy. That moral economy finds one articulation in the statements of the early friendly societies. Consider for example the following excerpt from Rules and Orders of the Honourable Society of Workington written on February 2, 1792:

> When we look upon mankind as being subject to an innumerable train of evils and calamities, resulting either from pain or sickness, or the infirmities of old age, which render them unable to procure even a scanty subsistence, when at the same time they are made capable of the noblest friendship, common prudence induces us so to form ourselves into society, that the insupportable condition of the individual may, by the mutual assistance and support of the whole, become tolerable.
>
> *(cited in Gray, 2001)*

Statements like this express a particular picture of the social relations demanded by the claims of need. The statement is premised on the recognition of common neediness. The members form a community premised on the awareness that they share conditions in which any, 'though innumerable train of evils and calamities', could suffer the loss of the very means of subsistence. Mutual aid is founded on the recognition of neediness. This moral economy expressed in the early friendly societies contrasts with the new moral economy of the market articulated by Smith. The moral economy of mutual aid starts not from social amnesia about the facts of neediness and dependence, but rather their acknowledgement. Personal independence is a virtue, but its achievement is only made possible against the background of mutual dependence grounded in common neediness.

The moral economy of solidarity and mutual aid has an ambivalent relationship to more recent forms of welfare provision. Consider another contrast with the picture of social solidarity articulated in the statements of the friendly societies noted above. Consider the contrast with Michael Ignatieff's defence of the bureaucratic provision of goods presented as an alternative to solidarity. Ignatieff, in his book *The Needs of Strangers*, offers the following observation on an example of an old man meeting his needs for food:

> I came upon one old man once doing his shopping alone, weighed down in a queue at a potato stall and nearly fainting from tiredness. I made him sit down in a pub while I did the rest of his shopping. But if he needed help, he certainly didn't want it. He was clinging on to his life, grasping for breath, but he stared straight ahead when we talked and his fingers could not be pried from his burdens.
>
> *(Ignatieff, 1990: 9)*

Ignatieff's own observations on the episode turn him to reflections on the way provisioning according to need must develop in modern society. Claims of need are not best addressed by acts of personal benevolence which threaten the independence of the recipient. They are best addressed rather through the transformation of needs into entitlements which are met through the bureaucratic procedures of the welfare state.

> [The old] have needs, and because they live within the welfare state, these needs confer entitlements – rights – to the resources of people like me ... The mediated quality of the relationship seems necessary to both of us. They are dependent on the state, not upon me, and we are both glad of it. Yet I am also aware of how this mediation walls us off from each other.
>
> *(Ignatieff, 1990: 9–10)*

The mediation of the meeting of needs through the state preserves the personal independence: 'The bureaucratic transfer of income among strangers has freed each of us for the enslavement of gift relations' (Ignatieff, 1990: 18). However, it comes with the cost of the loss of social solidarity: 'Yet if the welfare state does serve the needs of freedom, it does not serve the needs of solidarity. We remain a society of strangers.' (Ignatieff, 1990: 18) The social solidarity promised by the earlier roots of friendly societies is replaced by provision by bureaucratic means.

What are we to make of Ignatieff's argument? One point to notice about the argument is the assumptions he smuggles in about his own relation to neediness and those in need, an assumption which contrasts strongly with the earlier statements of the friendly societies such as that quoted above. Ignatieff in his argument presents the problem as if he himself stood above the fray of neediness. He considers the problem from outside the domain of need, as someone who is well resourced and able to intervene to support those who in contrast are within the domain of neediness. The book expresses, as Kate Soper notes, a detached and solipsistic stance that fails 'to communicate any identification with the "strangers" whose calamities it describes' (Soper, 1985: 111). The question the passages address is whether the 'resources of people like me' should be transferred to those in need through personal gift or through provision through the state. What is absent from the passages is any thought that motivates the statement of the friendly society – that through 'an innumerable train of evils and calamities' it is possible that he himself will stand in the place of the old man he assists, that he is not above the world of need, but a common citizen of that world. He is a modern example of the state of forgetfulness recommended by Adam Smith.

Ignatieff's perspective is not an accidental one. Inequality and market relations create a social world in which common neediness is often invisible. All humans share certain needs in the dispositional sense. It is in the occurrent sense that only some have needs. All humans need food in the dispositional sense. However, only the hungry have needs for food in the occurrent sense. The well fed do not. The rich and powerful have few occurrent needs. However, they have few occurrent needs only because their dispositional needs are met in ways that go unnoticed. As such, they appear to themselves and to others quite misleadingly as beings who are self-sufficient, standing above the fray of need. Neediness is apparent only among those with pressing occurrent needs, who are often the very people whose labour provides the resources to meet basic dispositional needs of those who appear above common neediness. In conditions of markets and inequality, the dependence of the wealthy on the labour of others is invisible, at the same time as they trouble themselves with the question of the transfer of resources to those with occurrent needs. The point is made nicely by Soper in a comment on Ignatieff's perspective on the pensioners he takes himself to be assisting from his wealth:

> [When] he claims that we can only bear the reality of our actual reliance upon each other by complexly mediating it through the 'numberless capillaries of the State', Ignatieff seems to imply that the underlying relations needing to be masked are those of charity. 'They are dependent upon the State', he writes, 'not upon me and we are both glad of it.' But pensioners are not merely beneficiaries but erstwhile toilers and benefactors themselves; and in speaking of a 'transference of some tiny portion' of his income into the pockets of those less fortunate than himself, Ignatieff overlooks the transference of some rather less than tiny portion of the revenue of their labours that in turn enables the fortunate to remain fortunate.
>
> *(Soper, 1985: 112)*

The dependence of the wealthy on the contributions of the 'needy' are rendered invisible by the working of wage labour and market relationships.

Ignatieff's account of the welfare state is one that contrasts bureaucratic provision with social solidarity. That account contrasts with a different tradition about welfare provision that is expressed in the work of Titmuss, in particular in his defence of an 'institutional–redistributive model of social welfare':

> It sees social welfare as a basic integrated institution in society providing both universal and selective services outside the market on the principle of need. Universal services, available without distinction of class, colour, sex or religion, can perform functions which foster and promote attitudes and behaviour directed towards the values of social solidarity, altruism, toleration and accountability.
>
> *(Titmuss, 1970: 263)*

Titmuss's work on the welfare provision traces its roots back to the moral economies of solidarity and mutual aid: '[T]he major impulse came from below – from the working-man's ethic of solidarity and mutual aid. It found expression and grew spontaneously from working-class traditions and institutions to counter the adversities of industrialism' (Titmuss, 1987: 122). It is in the 'great network of friendly societies, medical clubs, chapel societies, brotherhoods, co-operatives, trade-unions, and savings' that the origins of welfare provision can be found. In contrast to Ignatieff's picture of welfare provision as a process through which the fortunate pass down their gifts to those in need by bureaucratic means, welfare provision on Titmuss's account starts from the fact of general vulnerability to states of dependency:

> '[S]tates of dependency' arise for the vast majority of the population whenever they are not in a position to 'earn life' for themselves and their families; they are dependent people. In industrial societies there are many causes of dependency; they may be 'natural dependencies' as in childhood, extreme old age, and child-bearing. They maybe causes by physical and psychological ill-health and incapacity, in part these are culturally determined. Or they may be wholly or predominantly determined by social and cultural factors. These may be said to be the 'man-made' dependencies.
>
> *(Titmuss 1987: 46)*

One central source of such 'man-made' dependencies lies in the market economy itself which breaks down relations of care between strangers:[5] 'one of the functions of atomistic private market systems is to free men from any sense of obligation to or for other men' (Titmuss, 1970: 239). In this account, rather than standing in contrast to forms of social solidarity, universal welfare provision is seen as a condition of sustaining solidarity. The central question for Titmuss was how social solidarity among strangers on the basis of the recognition of common neediness is to be fostered. His better-known work on the gift relationship in blood needs to read against this background, that is, as an institution that fosters solidarity between strangers who recognise their vulnerability to injury. It is also against this background that he defends universality in welfare against means-tested benefits which make the provision of goods dependent on a person giving evidence of her poverty. To introduce a means test is to fail 'to understand the indignities of expecting the poor to identify themselves as poor people and to declare, in effect, that: I am an unequal person' (Titmuss 1987: 203). In conditions of inequality and the invisibility of mutual dependence, announcing neediness and dependence carries a stigma. The point remains an important one, in particular in the provision of food. Consider, for example, the following comments on the experience of receiving food from food banks in the UK, where need and stigma are both apparent:

> All the interviewees suggested that they had hesitated before coming to the food bank and most had felt a sense of embarrassment … For those with families, it was their children's needs that led them to overcome their embarrassment. One interviewee commented: 'It throws your pride out of the window … I am doing it for my kids, I am not going to make my kids suffer just because of my pride' (female, aged 34). A father of two children commented on how uncomfortable he felt: 'I was nervous coming here, I thought I had done something wrong … having to ask for food your ego takes a battering' (male, aged 40).
>
> *(Purdham et al., 2015: 8)*

The stigma associated with the provision of food through schemes such as food banks is widely commented upon. Where food as commodity is no longer available, if food takes the form of handouts directed at the needy, the fact of neediness itself becomes stigmatised. Social independence is compromised.

Common property, freedom and need

The discussion of the relationship between claims of necessity and food as commons thus far in this chapter has focused on the consumption of food. A weakness in any such focus is that it leaves out of the picture the background of production and, more specifically, relationships of power and ownership over the production of food against which many social vulnerabilities to the loss of subsistence need to be understood. The point is central to Marx's account of the commons.

An initial point to note is that Marx is in partial agreement with Smith on the achievements of commercial society in the development of personal independences. A feature of Marx's account of capitalism is the recognition of the freedom from personal dependence that commercial society achieved: 'In … the developed system of exchange … the ties of personal dependence, of distinctions of blood, education, etc., are in fact exploded, ripped up' (Marx, 1973: 165). However, this personal independence was marked by two features. Firstly, personal independence is replaced by new forms of objective dependence, the subordination of individuals to the impersonal workings of market forces:

> These *objective* dependency relations also appear, in antithesis to those of *personal* dependence (the objective dependency relation is nothing more than social relations which have become independent and now enter into opposition to the seemingly independent individuals; i.e. the reciprocal relations of production separated from and autonomous of individuals) in such a way that individuals are now ruled by *abstractions*, whereas earlier they depended on one another.
>
> *(Marx, 1973: 164)*

Full independence requires that objective dependence of this form is overcome through social relations being brought under the 'communal control' of individuals (Marx, 1973: 162). Secondly, the new independence of the worker forms one part of the worker's double freedom, the freedom to sell his or her labour power as owner of that commodity. The other component of that double freedom is that the worker is 'free from, unencumbered by, any means of production of their own' (Marx, 1887: 705). The process of primitive accumulation is 'the historical process of divorcing the producer from the means of production' (Marx, 1887: 705). One central component of that process is the loss of usufruct rights in common land through enclosure of the commons (Marx, 1887: ch.27). It is the loss of independent access to the means of subsistence that marks the specific forms of vulnerability characteristic of the wage worker. It underpins the point made by Sen noted earlier – that in absence of access to the means of subsistence, the worker relies upon 'exchange entitlements' and, where these do not contain 'any feasible bundle including enough food' (Sen, 1981: 3), the worker is at risk of starvation.

For Marx, it is the absence of commons in production, rather than just in consumption and use, that is the source of social vulnerability of the worker. This forms the background to Marx's defence of the principle 'each according to his need' in the *Critique of the Gotha Programme* (Marx, 1875: 87). The defence takes place in the context of the criticism of principles of 'fair distribution' (Marx, 1875: 83ff) for being concerned simply with the distribution of the 'means of consumption' (Marx, 1875: 85) rather than with the underlying patterns of ownership of the means of production that define different modes of production. Marx's own discussion of the principle of distribution according to needs presupposes common ownership. Marx rejects one version of the concept of a just price, that of the 'fair wage', on similar grounds. The appeal to 'a fair day's wage for a fair day's work' fails to address the underlying source of wage workers' social vulnerability in their separation from the means of production; hence the alternative slogan – 'the abolition of the wages system' (Marx 1865: 149). Marx's discussion in this regard breaks from the older Aristotelian tradition in which claims of necessity ground common property in the consumption of the necessities of subsistence, but not common property in their production.

This Marxian tradition has had its own influence on recent work on food as commons. A recent move by some scholars has been to reject the traditional view that the separation of the worker from the commons as a means of production is a process that is specific only to the transition from pre-capitalist to capitalist society, creating the conditions for capitalism. Rather, it is an ongoing mode of capitalist accumulation that continues beyond that period. A notable expression of this position is that articulated by Harvey:

> All the features of primitive accumulation that Marx mentions have remained powerfully present within capitalism's historical geography up until now. Displacement of peasant populations and the formation of a landless proletariat has accelerated in countries such as Mexico and India in the last three decades, many formally common property resources, such as water, have been privatized ... and brought within the capitalist logic of accumulation.
>
> *(Harvey, 2003: 145–146)*

These processes constitute a continuing form of accumulation by dispossession (Harvey, 2003, ch.4). Resistance to this separation is articulated as a defence of the commons. Access to the sources of provisioning of food, from the appropriation of unused land by landless peasant movements and seed exchange systems outside the market to urban gardens form part of a defence of spheres of commoning independent of both state and market (Akbulut, 2017: 400).

Conclusion

Food is a vital human need. As such, it is a good from which no person ought to be excluded. It is what I have called a normative public good. In our modern market society, food is a commodity. As such, where individuals and groups lack the means to provide their own food and the ability to pay for food through market exchanges, food is a good from which individuals can be and are actually excluded. This chapter has explored the conflict between food as a vital human need and food as a commodity. It has explored a variety of moral economies that were developed in response to the new moral economy of the market in which food is a commodity. All appealed in different ways to the concept of need and the claims of necessity. They also appealed in different ways to the idea of the commons. One older appeal to need as a basis for commons is that articulated in the scholastic view that claims of need give all common use rights in property. These claims were weakened and transformed into exceptions in the work of Grotius, and weakened still further in Locke, where rights of necessity were transformed into duties of charity. In the work of Smith, rights of necessity disappear. The economy is understood as a sphere in which independent agents relate to each other through commercial exchanges. As such the market breaks the paternalist relations of dependence that mark pre-commercial society. This picture of commercial society, however, depends, as Smith himself recognised, on the invisibility of dependence and vulnerability as a fact about the human condition. The picture of the market as a sphere of independent contracting agents that forms the basis of the moral economy of the market is premised on the unacknowledged ties of dependence and facts of neediness. Unacknowledged dependence still informs accounts of the welfare state, such as that developed by Ignatieff, that conceive of welfare as the transfer of wealth from the fortunate who stand above the sphere of neediness and dependence to the unfortunate who do not.

In contrast to the forms of unacknowledged dependence and neediness that inform the moral economy of the market, egalitarian forms of mutual aid that were developed in response to the market economy were grounded in the acknowledgement of dependence and common neediness. This tradition of moral economy has its own long history in forms of everyday mutual aid in working-class communities. It continues to be articulated in arguments for universal provision of basic welfare goods, such as that of Titmuss, which start from the practices of mutual aid and their basis in the acknowledgement of the fact of vulnerability to states of dependency.

The new forms of mutual aid of working-class communities had their roots, in part, in a common social vulnerability that was a result of the separation of workers from the means of production. While Marx, like Smith, notes the new forms of personal independence occasioned by commercial society, this personal independence was one part of the double freedom of the worker. The worker is also freed from any of their own means of production. The processes by which the workers are separated from the means of production in primitive accumulation include the loss of usufruct rights in common land. The claim that primitive accumulation is not simply an historical event but part of a continuing process of accumulation by dispossession informs recent works on the commons and commoning. Resistance to dispossession is articulated in the language of the commons. Food as commons on this account requires not just a commons in use but also a commons in production. What this political economic response still draws from the moral economy of working-class communities is an understanding of commons that is founded on recognition of mutual neediness.[6]

Notes

1 For a discussion, see Kraut (2002: 220–224).
2 For a discussion of Grotius and Pufendorf's views and their relationship to the earlier scholastic tradition, see Salter (2005).
3 The passage is an echo of Cicero: 'for one man to take from another and to increase his own advantage at the cost of another's disadvantage is more contrary to nature than death, than poverty, than pain and than anything else that may happen to his body or external circumstances' (Cicero 1991: III.21). On the status of claims of necessity in *The Wealth of Nations*, see Hont and Ignatieff (1983). For an excellent defence of the view that for Smith the needs of the poor do not give them rights based in justice, see Salter (2012), who responds to the revisionary account of those such as Fleischacker (2004, ch.10) who argue for the opposing view.
4 I discuss different aspects of Smith's position and the contrast with the mutualist tradition further in O'Neill (2006, 2011, 2015).
5 Cf. Kent (2018) and Chang (2018).
6 I would like to thank Bengi Akbulut, Richard Christian, Ben Fine, Ian Gough, John Salter, Julia Steinberger and Jose Luis Vivero-Pol for their comments on earlier versions of this chapter.

Bibliography

Akbulut, B. 2017. 'Commons', in C. Spash, ed. *Routledge Handbook of Ecological Economics*. London: Routledge.
Aquinas, T. 1920. *The Summa Theologica*. Second and revised edition, translated by Fathers of the English Dominican Province. Online edition: http://www.newadvent.org/summa/

Aquinas, T. 2003. *De Malo*. New York: Oxford University Press.
Aristotle. 1948. *Politics*. Translated by E. Barker. Oxford: Clarendon.
Aristotle. 1999. *Nicomachean Ethics*. Second edition, translated by T. Irwin. Indianapolis, IN: Hackett.
Chang, M. 2018. 'Growing a Care-Based Commons Food Regime', in J. L. Vivero-Pol , T. Ferrando , O. De Schutter and U. Mattei, eds. *Routledge Handbook of Food as a Commons*. London: Routledge.
Cicero. 1991. *On Duties*. Edited by M. T. Griffin and E. M. Atkins. Cambridge: Cambridge University Press.
Doyal, L. and Gough, I. *A Theory of Human Need*. Basingstoke, UK: MacMillan.
Fleischacker, S. 2004. *On Adam Smith's The Wealth of Nations: A Philosophical Companion*. Princeton, NJ: Princeton University Press.
Gray, P. 2001. *A Brief History of Friendly Societies* [Online]. Available at http://web.archive.org/web/20011225144332/http://www.afs.org.uk/research/researchpgrayhistorypage.htm [accessed: 8 September 2014].
Grotius, H. 2013. *On the Law of War and Peace*. Cambridge: Cambridge University Press.
Harvey, D. 2003. *The New Imperialism*. Oxford: Oxford University Press.
Hegel, G. 1967. *Philosophy of Right*. Translated by T. Knox. Oxford: Oxford University Press.
Hont, I. and Ignatieff, M. 1983. 'Needs and Justice in *The Wealth of Nations*', in I. Hont and M. Ignatieff, eds. *Wealth and Virtue: The Shaping of Political Economy in the Scottish Enlightenment*, pp. 1–44. Cambridge: Cambridge University Press.
Ignatieff M. 1990. *The Needs of Strangers*. London: Hogarth Press.
Kent, G. 2018. 'Community-Based Commons and Rights Systems', in J. L. Vivero-Pol, T. Ferrando, O. De Schutter and U. Mattei, eds. *Routledge Handbook of Food as a Commons*. London: Routledge.
Kraut, R. 2002. *Aristotle: Political Philosophy*. Oxford: Oxford University Press.
Langholm, O. 1998. *The Legacy of Scholasticism in Economic Thought*. Cambridge: Cambridge University Press.
Locke, J. 1695. 'Venditio', in Dunn, J. 1968. 'Justice and the Interpretation of Locke's Political Theory', *Political Studies*, 16: 84–87.
Locke J. 1988. *Two Treatises of Government*. Edited by P. Laslett. Cambridge: Cambridge University Press.
Marx, K. 1865. 'Value, Price and Profit', in *Collected Works*, 20, pp. 101–149. London: Lawrence and Wishart, 2010.
Marx, K. 1875. 'Critique of the Gotha Programme', in *Collected Works*, 24, pp. 75–99. London: Lawrence and Wishart, 2010.
Marx, K. 1887. *Capital Volume 1, Collected Works*, 35. London: Lawrence and Wishart, 2010.
Marx, K. 1973. *Grundrisse*. Harmondsworth: London.
O'Neill, J. 2001. 'Property, Care, and Environment', *Environment and Planning C: Government and Policy*, 19: 695–711.
O'Neill, J. 2006. 'Need, Humiliation and Independence', in S. Reader, ed. *The Philosophy of Need: Royal Institute of Philosophy Supplement*, 57. Cambridge: Cambridge University Press.
O'Neill, J. 2007. *Markets, Deliberation and Environment*. London: Routledge.
O'Neill, J. 2011. 'The Political Economy of Recognition', *The Adam Smith Review*, 6: 129–151.
O'Neill, J. 2015. 'Equality, Vulnerability and Independence', in A. Bielskis and K. Knight, eds. *Virtue and Economy: Essays on Morality and Markets*. Farnham, England: Ashgate.
Polanyi, K. 1957. 'The Economy as Instituted Process', in K. Polanyi, C. Arensberg and H. Pearson, eds. *Trade and Market in the Early Empires*. Glencoe, IL: The Free Press.
Purdam, K., Garratt, E. and Esmail, A. 2015. 'Hungry? Food Insecurity, Social Stigma and Embarrassment in the UK', *Sociology*, 50: 1–17.
Reader, S 2007. *Needs and Moral Necessity*. London: Routledge.
Salter, J. 2005. 'Grotius and Pufendorf on the Right of Necessity', *History of Political Thought*, 26: 284–302.
Salter, J. 2012. 'Adam Smith on Justice and the Needs of the Poor', *Journal of the History of Economic Thought*, 34: 559–575.
Sen A. 1981. *Poverty and Famines: An Essay on Entitlement and Deprivation*. Oxford: Clarendon.
Scott, J. 1977. *The Moral Economy of the Peasant*. New Haven: Yale University Press.
Smith, A. 1981. *An Inquiry into the Nature and Causes of the Wealth of Nations*. Indianapolis, IN: Liberty Press.
Smith, A. 1982a. *The Theory of Moral Sentiments*. Indianapolis, IN: Liberty Press.
Smith, A. 1982b. *Lectures on Jurisprudence*. Indianapolis, IN: Liberty Press.
Soper, K. 1985. 'The Differences of Need (Book Review of M., Ignatieff *The Needs of Strangers*)', *New Left Review*, I/152: 109–119.
Thompson, E. P. 1968. *The Making of the English Working Class*. London: Penguin.

Thompson, E. P. 1971. 'The Moral Economy of the English Crowd in the Eighteenth Century', *Past and Present*, 50: 76–136.
Thompson G. 1987. *Needs* London: Routledge and Kegan Paul.
Titmuss, R. 1987. *The Philosophy of Welfare: Selected Writings of Richard M. Titmuss*. Edited by B. Abel-Smith and K. Titmuss. London: Allen and Unwin.
Titmuss, R. 1970. *The Gift Relationship.* London: Allen and Unwin.
Wiggins, D. 1998. 'The Claims of Need', in *Needs, Values, Truth*. 3rd edition. Oxford: Clarendon Press, pp. 1–58.

8

COMMUNITY-BASED COMMONS AND RIGHTS SYSTEMS

George Kent

The goal and the pathway

Imagine that a group of like-minded people wanted to establish a new community at a specific site. How should they design it? The planners would have to address a broad range of issues, including physical facilities and economic and social arrangements. The design work could begin with the drafting of a charter for the community that would set out the major guiding principles for its operation. How should they design the community's food system? More specifically, how could the food system be designed to fulfill the vision of food-as-a-commons? What sort of community charter would be proposed by advocates of the food-as-a-commons idea, such as the contributors to this volume? A consensus statement on that would make our collective understanding of the concept readily visible and debatable. However, building that consensus could take months, maybe years. This chapter can be viewed as the initial concept paper, to be discussed while preparing the first draft of the community's charter.

In this chapter, the analysis of food-as-a-commons focuses on its construction at community level. This approach has pedagogical value because thinking about how things should work at the local level is easier than thinking on a much larger scale. This exploratory exercise should make it easier to address the challenges in creating commons-based systems at any level. This localized approach is strategically important because an all-at-once transformation of all the world's food systems and subsystems is not feasible. It is sensible to think in terms of an incremental approach. If commons ideas can be implemented successfully in many different local communities, they could serve as inspiring models. Others could adapt the principles to fit their own circumstances. The first efforts would serve as pilot studies, experiments. This approach has the advantage of accommodating diversity, while any singular global approach would be oppressive and trigger resistance.

This opening section sets out this chapter's goal and a proposed pathway to its achievement. Section 2 then explains the distinction between viewing food-as-a-commodity and food-as-a-commons. Section 3 discusses the historical shift toward food-as-a-commodity but shows that many places maintain the food-as-a-commons as their primary approach. The treatment of food as a commodity has led to huge increases in the wealth of owners of large-scale industrial food operations, whether on farms, factories, or chains of fast food restaurants. This has been accompanied by the increasing marginalization of workers in these businesses, as shown in Section 4. Historically, the food-as-a-commons perspective emerged much earlier than the modern food-as-a commodity perspective. It emerged naturally, as a way of meeting communities' needs. Section 5 explores the potential for deliberate design of local food systems based on viewing food-as-a-commons, and as a way of pushing back against the pressures of those who promote the commodity-centered view of food.

Drawing on developments in global human rights, Section 6 examines ways in which the basic principles of rights could be used as the basis for governance in local communities. Section 7 then discusses how these tools could be applied specifically in relation to local food systems. Section 8 argues that good governance at the local level is not about delivering charity efficiently but rather it is about making charity unnecessary. Section 9 points out the importance of considering not only the technology of food production but also the social and environmental context in which it is embedded. Agroecology, for example, could be a good basis for community-based food systems, but that will happen only if explicit attention is given to the social and environmental dimensions of the practice. The section highlights the importance of conviviality in living together well. The conclusion in Section 10 argues that by building a world of many strong communities in which people live well together, we would create a diverse world that works well for everyone. It would grow from the bottom up, mimicking the way in which natural systems emerge, meeting the needs of all their parts.

Food viewed as commons or as commodity

The distinction between food-as-a-commodity and food-as-a-commons can be made in several ways. In this chapter, it is associated with the historical shift in motivations for producing food. Pre-modern agriculture was undertaken to maintain the community's health. Modern food production, whether on farms or in factories, is driven by the pursuit of private wealth. In producing food-as-a-commons, the motivation is to benefit not only yourself but also your community and your environment. Production is likely to be organized in cooperatives, owned and managed by groups, with little distance between bosses and workers, and little distance between producers and consumers. In contrast, in self-interested treatment of food as a commodity, there is likely to be a single owner–manager with many subordinate workers. The product is viewed as a commodity, and its nutritive and other values are important mainly to the extent they affect sales and thus profits for the owners.

The deep motivation for food-as-a-commons is caring about the well-being of all people and the environment, now and in the future. The deep motivation of food-as-a-commodity is narrow self-interest through the pursuit of wealth for owners, and indifference toward and exploitation of people and the environment (Kent, 2016).

The idea of food-as-a-commons means food systems should benefit individuals, households, communities, nations, and the world, all at once. There are systems within systems, each looking after itself but also looking after the larger whole, like organs in an organism. Where tradeoffs and compromises become necessary, the systems should be biased in favor of the weakest among them, through processes that systematically reduce inequities. If individuals thrive, the group thrives. There is deep interest in the survival and thriving of the whole, including its environment, and not simply the survival of separate components or subsystems. The idea that one part might thrive at the expense of others is viewed as foolish and unacceptable.

The historic shift in motivations

The shift in motivations for food production has been part of the broader trend toward globalization, especially in trade. Centuries ago, people depended almost exclusively on local production, whether for food, clothing, or anything else. International trade centered on exotic luxury products such as spices and silk. The transformation was based on the growing trade in inexpensive bulk commodities. The shift is well illustrated in the history of islands (Chirico and Farley, 2015). In pre-contact Hawai'i, food was abundant and people were healthy. Taro, the people's basic staple food, was produced to meet the needs of the community. Then, in the 1860s, businessmen who came along with Protestant missionaries from the United States began to produce and export rice for profit. There was a large-scale shift from taro to rice production in Hawai'i. The close link between production and consumption was broken, and the distinction between farming for nutrition and farming for money became clear. The people whose taro supply was threatened were not the ones who benefited from rice exports. The export orientation in agriculture followed on a massive scale with the production of sugar and pineapples. By the end of the 20th century, rice, pineapple, and sugar production had all collapsed, and Hawai'i now imports about 85% of its food (Kent, 2015).

When the motivation for food production shifts to the desire to make money, there is no satiation. More is always better. Considerations such as limiting pollution and depletion of the environment are overridden. This shift of the motivation from producing community health to producing private wealth has been global in scope (Kaufman, 2012; Lindgren, 2013; Rosenthal, 2013; Tudge, 2013; UNCTAD, 2013; Vivero-Pol, 2017).

Historically, local non-industrial food systems had tight links between primary producers and final consumers, and real human relations between them. These systems still function in much of the world where farming is not tied to modern markets. It has been estimated that only 30% of the world's food supply is produced on industrial farms while half of the world's cultivated food is produced by peasants (Courtens, 2012, based on ETC Group, 2009). One research group says, "Smallholders are responsible for over 90% of all investment in agriculture and for up to 80% of all the food produced and consumed in the world" (Transnational Institute, 2015).[1] Pre-modern forms of agriculture are alive and doing well in many parts of the world, but they get little attention. Their effectiveness in providing good food supplies has been well documented (Inter Pares, 2004; Kuhnlein et al., 2009). These time-tested modes of food production are losing ground, often displaced by industrial food production.

In modern industrial food systems, producers and consumers are separated not only by distance but also by layers of wholesalers, processors, and investors, each with its own interests in the food system. They ship the products to the most lucrative markets, as illustrated by the global fish trade and the fruit and vegetable trade. In the industrial food system, food is directed to people with money, anywhere in the world, not to neighbors in need.

The growth of modern agriculture is mainly driven by increasing concentration of control and wealth, not increasing productivity, efficiency, or sustainability. The priority given to wealth rather than health is well illustrated by the inattention to health and environmental impacts by the promoters of baby foods (Baby Milk Action, 2016; Kent, 2017). Many people express dismay at the amount of food that is wasted in industrial food systems. Once it is recognized that the primary objective of such systems is to maximize cash value, not nutritive value, that waste of food is not surprising. We need to get beyond the naïve idea that the food production industry is driven by the desire to achieve good health for all.

Many large farms receive subsidies from government. Some subsidies are explicit, in the form of cash, and some are hidden, in the form of government services such as road-building, marketing assistance, and research. The subsidies provide incentives that lead to overproduction of certain commodities and result in distortions in the human diet and in the economic system. The overproduction of corn in the United States is well known (Pollan, 2007), and the global overproduction of rice has become more visible (Lobello, 2013). Little of the excess food goes to those who need it. It is more likely to be turned into a cheap ingredient for processed food or go to animal feed or used to produce biofuel.

It is not only farms that are helped by governments. The United States government provides huge services to infant formula manufacturers. Through a complex and opaque process, the result is that half the formula used in the country is distributed by the government at no direct cost to the families, a huge benefit for the manufacturers of formula (Kent, 2017). Subsidies to help the needy or protect the environment can make sense. There is no good reason to subsidize wealthy, large-scale food producers. If the public policy objective is to improve human nutrition, far more would be achieved by subsidizing well-managed small farms that produce basic fresh foods (Wiggins and Keats, 2013) and by providing better support for breastfeeding (Kent, 2017).

People who think the hunger problem can be solved simply by increasing food production, described by their critics as "productivists," generally ignore the question of how people with little money are to access the new increments in food supplies. They focus on agriculture, ignoring the fact that in high income countries consumer expenditures for food go mainly to processors, not farmers. The technological improvements they advocate are likely to benefit some farmers while driving others off the land. Frances Moore Lappé calls on us to "see through the productivist fixation that inexorably concentrates power, generating scarcity for some, no matter how much we produce" (Lappé, 2011).

Marginalization of food workers

Industrialized food production disconnects producers from consumers, and it also tends to exploit workers, the environment, and customers. The maldevelopment of modern agriculture is well illustrated in Guatemala, which has one of the world's highest rates of land concentration. The white elite occupies 65% of the arable land. The wealthy plantation owners focus on international exports such as coffee, sugar cane, and African palm oil, and use cheap, mostly indigenous labor (Tran, 2013).

Structurally, the situation in the United States is not very different. Big farms eat up smaller farms. The number of people employed in farming is small because they are displaced by machines, especially in the big industrial farming operations. The workers get low pay, while the owners prosper. Many are subsidized by the government. Advocates of large-scale modern agriculture often justify it by claiming economies of scale and efficiency in production. However, the key advantage of large farms is that they have one owner profiting from the work of machines and many poorly paid laborers (Food Chain Workers Alliance and Solidarity Research Cooperative, 2016).

Farm workers are among the poorest paid in the United States. The median annual wage for food workers in New York State is around $20,000, lower than for all other categories of workers (New York State Department of Labor, 2016). The income levels of many farm and restaurant workers are so low they remain food insecure and eligible for government food assistance programs (Sustainable Food Trust, 2016). Many work under degrading conditions (Oxfam America, 2016). Just as workers in manufacturing are steadily marginalized (Swanson, 2016), so too are food workers. In economic systems in which many people are discarded, there is widespread poverty and hunger alongside great wealth. Government programs such as the Supplemental Nutrition Assistance Program (SNAP) and the Special Supplemental Nutrition Program for Women, Infants, and Children (WIC) enable more poor people to work for low wages. They make it possible for employers to hire people for less money than they otherwise would have to pay. Thus, these subsidies to the poor also have the effect of subsidizing the rich.

In modern agriculture, the selection of what to produce and how to produce it depends on the ability to expropriate others' labor (Guo, 2016). The idea that there can be market-based solutions to poverty and food insecurity is doubtful because market-based approaches regularly widen the gaps between rich owners and poor workers (GRAIN, 2016). The income gap in California has been widening rapidly, due largely to the low incomes of farm workers in the state (Bohn and Danielson, 2016). Referring to SNAP and WIC, Bohn and Danielson said, "Policies that help bring low-income Californians more fully into the labor force and increase earnings hold promise for reducing inequality because families at all income levels rely most on their earnings from work." That is not true. At the high end, most income comes from what you own, not from what you do. The rate of income growth to individuals at the high end is consistently higher than it is for those at the low end, so income gaps widen over time.

There is widespread malnutrition among food workers. Some agencies have responded by devising interventions such as bio-fortified crop varieties in the hope of improving these food producers' nutrition status (Fiorella et al., 2016). In many cases, it would be better to get these people out of food production altogether. The problem is not the nutritional quality of their products but rather the fact that these workers' incomes are extremely low. They need better options. Instead of trapping poor farmers with subsidies to produce low-value crops such as wheat and rice, they could be supported in switching to crops that are more valuable in both economic and nutritional terms, such as vegetables, pulses, and fruits (Jones and Ejeta, 2016). People who are poor should be subsidized, but in ways that empower rather than trap them. On both the producers' and the consumers' sides, the dominant economic system benefits the rich far more than the poor, steadily widening the gap between them. The system does not care much for people without money. And, the evidence is clear, it does not care much for people who are hungry (Kent, 2016).

Designing community-based commons

Those who benefit from the dominant industrialized food system are not receptive to suggestions for radical changes in their operations. It may be possible to devise new approaches in spaces the industrialists do not occupy, in small local communities. The design of the community's food system would depend in part on the nature of the local physical environment and the surrounding support services and market potential. Primary food production could come from a mix of household gardens and community farms. There could be a variety of food processing operations and a variety of marketing arrangements, both inside and outside the community. There could be a variety of food markets and restaurants designed to enrich the life of the communities in many different ways, in contrast to the big-name franchise operations that are designed mainly to send profits to the company's distant headquarters. The community could simply prohibit those extractive operations.

The food system would have to be complemented by suitable housing arrangements and good infrastructure to deal with energy and transportation needs and waste disposal. Residents could find employment outside the boundaries of the communities, but many employment opportunities would be created within the community. The design of the arrangement could be guided by the diverse experience of many intentional (deliberately designed) communities around the world.[2]

The idea of food-as-a-commons is not about characteristics inherent in food items, but about attitudes toward food, people, and the environment, and the forms of governance they imply. Some supporters of food-as-a-commons feel that all food should be regarded that way, and that treating food as a commodity is just wrong. However, the view adopted here is that we should be open to a judicious mix of food-as-a-commons and food-as-a-commodity. Local food systems could be partly based on the management of community food enterprises of various kinds operating alongside a variety of conventional private operations. Anyone who wants to make and sells pies should be welcome to do that. People could have privately owned gardens at their homes and also participate in jointly operated community gardens. Just as in cohousing arrangements, there should be private spaces as well as public spaces, with people free to move among them as they wish.[3] Communities should be free to export and import food, so long as they consider the impacts on people and environments, now and in the future, both inside and outside the community.

The following sections discuss how the community's food system could be designed, in terms of basic considerations and guiding principles. The designers could work out the specific language for their charters in their situations.

Rights systems

The community's charter could articulate clear rights and responsibilities of the residents, as an explicit social contract. Before getting into that, it is important to clarify the meaning of rights in general. Some people think of rights as a body of accepted norms or standards. However, rights-based social systems involve more than that. There are institutional arrangements that must be put into place to make sure that rights are in fact realized.

Rights are *enforceable claims*. Thus, rights should be understood not simply as standards for behavior but as a component of rights-based social systems. This is a generic form. The global *human* rights system is one concrete example. It is based on a series of international treaties, built on the foundation established by Universal Declaration of Human Rights of 1948. *Human* rights are understood to be universal in scope and focused on the protection and promotion of human dignity. There are also local rights-based social systems, some related to the global human rights system and some that stand alone. Some rights are collective, such as the rights of indigenous peoples.

In any well-developed system of rights there are three major roles to be fulfilled: the *rights holders*, the *duty bearers*, and the *agents of accountability*. The task of the agents of accountability is to make sure that those who have the duty carry out their obligations to those who have the rights. Thus, to describe or design a rights system, we want to know:

A. The nature of the *rights holders* and their rights;
B. The nature of the *duty bearers* and their obligations corresponding to the rights of the rights holders; and
C. The nature of the *agents of accountability*, and the procedures through which they assure that the duty bearers meet their obligations to the rights holders. The accountability mechanisms include, in particular, the remedies available to the rights holders themselves.

These are the core components, the "ABCs" of any legal system of rights (Kent, 2011, pp. 65–73). They apply to the global *human* rights system, and also to the systems of rights established within nations or other administrative units. There can be systems of rights within prison or hospitals, for example. The prisoners or patients could have clearly articulated rights, the correlative obligations of staff members and others could be clearly spelled out, and there could be institutional means for holding the duty bearers accountable, including effective recourse procedures available to the prisoners or patients themselves. If they have no such recourse, it is not a properly functioning rights system.

In well-functioning rights systems, rights and duties must be clearly articulated and well known, and there must be systematic means of accountability to ensure that those who have the obligations do in fact carry them out. The most fundamental of these recourse mechanisms is effective institutional arrangements through which the rights holders themselves can insist that their rights are realized. Since young children and many disabled people cannot speak for themselves, there should be designated surrogates who can speak on their behalf.

Right to food

The human right to adequate food is a component of that global human rights system. The most authoritative interpretation of that right was spelled out in a United Nations document called General Comment 12 (UNECOSOC, 1999).[4] If it is used well, the right to food has the potential to bring about great improvements in the quality of life for many people (De Schutter, 2014).

It is important to protect and promote the right to food locally. For example, a right to food could be established in hospitals, prisons, school, and communities. It might be related to the global version, the *human* right to adequate food, or it could be entirely local. The right to food in India, for example, is local in the sense that it was established based on a line in the national constitution affirming the right to life. It did not grow out of the global human rights system. Of course, over time, that global thinking has influenced law and policy in India. In another example, I have shown how a school meals program could be based on locally formulated rights (Kent, 2010a).

In formulating rights relating to the food system in a community, one could draw ideas from General Comment 12. For example, the community could make a commitment to ensure the community members have the right to food as defined in General Comment 12:

> The right to adequate food is realized when every man, woman and child, alone or in community with others, has the physical and economic access at all times to adequate food or means for its procurement.
>
> *(UNECOSOC, 1999, para 6)*

Some of the provisions in that document could be simplified and adapted to fit the local situation. For example, paragraph 15 could be modified by saying:

"The Community Council has four types of obligations regarding the food and nutrition situation in the community:

- The Council has the obligation to ***respect*** existing access to adequate food, and must not take any measures that prevent such access.
- The Council has the obligation ***protect***, meaning the Council must not do anything that deprives individuals of their access to adequate food.
- The Council has the obligation to ***facilitate***, meaning it must strengthen people's access to resources and means to ensure their livelihood, thus ensuring their food security.
- The Council has the obligation to ***provide*** adequate food for individuals who cannot provide for themselves for reasons beyond their control."

Basic principles of this sort could be modified through democratic processes in the community. They could be explained in further detail in by-laws that accompany the charter. As a practical matter, rights do not work well if those in power are weak or do not care enough about the well-being of the rights holders. But if they do care and they have the capacity, well-designed rights systems can be useful for ensuring the delivery of important services.

Rights and responsibilities can be set out explicitly for any aspect of the food system, possibly in the form of contracts. For example, a community garden could have the roles of all participants set out in contracts. If the will is there, communities could establish special arrangements to ensure that infants and young children are well nourished. School feeding programs could be strengthened by establishing rights systems for them. Support services could be provided to pregnant women and new mothers, such as paid maternity leave. Hospitals and other birthing sites could be required to follow the requirements of the Baby Friendly Hospital Initiative, with regular checks to ensure their compliance (UNICEF, 2016). The community could adhere to the International Code of Marketing of Breast-milk Substitutes (World Health Organization, 1981), even if the national government does not require that. Communities could resist their governments' offers of free or highly subsidized infant formula (Kent, 2017). Some rights could be established at the local level even if they have not been established at any higher level. As mentioned earlier, school feeding programs could be based on locally formulated rights (Kent, 2010a).

Setting out explicit rights might be too bureaucratic for a small community, but it could be done at higher levels where governance arrangements are more formal. For small communities that wish to take control over their own circumstances, it might make more sense to write a simple community charter that sets out basic principles for how the community should function. That statement could be elaborated over time, as needed.

Making charity unnecessary

Many national and sub-national governments provide welfare programs for needy people. The poor are also served by many nongovernmental agencies, including churches and temples. Many set up soup kitchens and food banks. Some charitable programs are very creative. For example, in Spain there are Robin Hood restaurants that use earnings from the breakfast and lunch services to provide free dinners for needy people (Minder, 2016). The nongovernmental organization Heifer International promotes sharing systematically through Passing on the Gift, a program in which low-income people who receive donated animals "share the offspring of their animals – along with their knowledge, resources, and skills – an expanding network of hope, dignity and self-reliance."[5] Sharing can be facilitated by having people set up tables at farmers' markets to accept people's excess fruits and vegetables and give them to people who need them.[6]

The discussion in this section is about charity, understood as unidirectional, from the haves to the have nots. This is very different from the multidirectional sharing we commonly do with one another, discussed in the following section.

There are ways to do charity well. Caution is needed, however, because in many settings, sustained programs for the needy, whether governmental or nongovernmental, can silence the needy and allow unjust social systems to continue (Poppendieck, 1998; Riches and Silvasti, 2014). Badly designed charities can disempower those who are supposed to benefit, trapping them in their poverty and subservience.

We should be sensitive to the sharp point made by Nigerian novelist Chinua Achebe in his novel, *Anthills of the Savannah*: "While we do our good works let us not forget that the real solution lies in a world in which charity will have become unnecessary."[7] Achebe's story is about "how the fierce pursuit of self-interest comes at tremendous cost to the community as a whole."[8] This pattern is evident in many countries, both rich and poor.

Social and environmental dimensions

There are many ways in which local food systems can help to strengthen local communities. One basic example is the community garden, devoted to producing good food for the local community, rather than producing commodities for sale to maximize money income. Community gardens can be organized as cooperatives and managed democratically by the participating members. Many decisions would have to be made: What to plant? Who does what work? Who gets what produce? Who manages the financial aspects of the operation? The same types of questions would have to be raised in relation to other food-related activities such as processing and selling or organizing group meals. Who starts up the activity? Who owns it? Who does what? Who gets what? Saying that the people in the community would do things together is not enough. There is a need for clarity. The community's charter could provide guidance, in terms of principles, with the concrete details worked out at the level of individual enterprises. There are organizations that can advise on how to set up cooperatives in harmony with local cultural practices and traditions.[9] Northern Italy demonstrates the benefits of having entire regions organize their businesses as cooperatives.[10] The decisions about how to manage cooperative food enterprises could be made and communicated informally, but in many settings it would be advantageous to spell it all out. An informal Memorandum of Agreement might be enough, or it might be important to work out detailed, signed contracts with all who are involved.

Agroecology and community-based food systems

Many people view agroecology as a promising approach to developing community-based food systems (IPES Food, 2016). While some think of it as a new approach, it has deep roots in pre-modern non-industrial agriculture. A discussion of the potential of agroecology to feed Africa identifies major agroecological principles, listed here in shortened form:

- Enhance the recycling of biomass.
- Strengthen the "immune system" of agricultural systems through the enhancement of functional biodiversity.
- Provide the most favorable soil conditions for plant growth.
- Minimize losses of energy, water, nutrients, and genetic resources.
- Diversify species and genetic resources in the agroecosystem.
- Enhance beneficial biological interactions and synergies among the components of agro-biodiversity (AFSA, 2017, p. 8).

These are good recommendations, but they do not speak about the motivations of the people who operate agroecological farms. There is nothing that would prevent a greedy landowner from adopting them while at the same time exploiting the workers and harming the environment. There is a need to go beyond the technical elements and also discuss what would turn agroecology into *community-based* agroecology.

Concern for the well-being of others will not appear just because certain technical practices are followed. Virtuous methods of food production such as agroecology and permaculture can be used in questionable ways for questionable purposes (Holt-Giménez and Altieri, 2012, 2016). It is important to examine not only their productivity but also their social, environmental, and economic impacts. Managed well, they could sustainably produce not only good food but also stronger communities. This is recognized in what has been called social permaculture:

> The key insight of social permaculture is that, while changing individuals is indeed difficult, we can design social structures that favor beneficial patterns of human behavior. Just as, in a garden, we might mulch to discourage weeds and favor beneficial soil bacteria, in social systems we can attempt to create conditions that favor nurturing, empowering relationships.
>
> *(Starhawk, 2016)*

Attention has been given to agroecology on the global governance agenda (IPES Food, 2016), but there has been little discussion of governance within and around agroecological farms in the social context in which they are embedded. The basic principles for agroecology and other elements of food systems should include guidelines on social organization. Several major points would apply for any community-based food system.

Community-supported agriculture (CSA) is based on direct sales from farmers to consumers. It provides a good example of how the organizing principles for one component of local food systems can be based on principles set out in an explicit charter. In a press release, a leading proponent says,

> CSA is a tremendously flexible concept for consumer–farmer connections. It's an alternative system of distribution based on community values. The economics of direct sales make this a win–win solution for farmers and farm members. The farmer gets a decent price and the member pays less, since there is no middleman.[11]

The press release provides a link to the charter, which covers social and economic relationships as well as farming technology. It can be adapted by local communities as they wish.

Democratic management

Ownership and management of local food systems should be shared among those involved, and decision-making should be democratic. This applies to farms and factories, and also to the community as a whole. Local food sovereignty would be lodged in and exercised by the local governing body. The enterprise or other space that is co-managed would be the shared commons. The reference to food-as-a-commons is best understood as an approach to management of food systems, not as a property inherent in food itself. Thus, when resources of the ocean have been described as the common heritage of all mankind, the meaning was that they should be jointly managed. The Latin expression *res communis* was understood to mean that certain resources of the earth belong to all of us, now and in the future. As the "common heritage of all mankind," we must look after those resources as trustees, on behalf of future generations (Christy and Scott, 1965; Kent, 1978). Those resources belong to everyone and should be managed on behalf of everyone.

The *res communis* concept contrasts with *res nullius*, which means that certain resources belong to no one and are there for the for the taking. Wild foods in the forest and fish in the ocean generally are viewed as *res nullius*. Whoever gets to them and takes them first then rightfully owns them. The *res nullius* concept invites plunder because it does not call for any high-level management body mandated to look after the resources on behalf of future generations.

Self-reliance, not self-sufficiency

A clear distinction should be made between pursuing self-sufficiency, which means producing what you consume, and pursuing self-reliance, which means making your own collective decisions, rather than being the subject of decisions by others outside your community (Kent, 2010b). Self-reliance is lodged in local self-governance. There is nothing wrong with exporting and importing food so long as decisions about it are made with full regard for the well-being of people and environments both inside and outside the community.

Sharing

Most food is distributed through conventional marketing methods, but people in caring communities also share their gardens' produce with their neighbors, or share jams, breads, and cakes (see Balazs, this volume). Sharing can be carried to surprising extremes. In the British town of Todmorden, for example, people raise fruits and vegetables and invite others to harvest them even without asking.[12] In South Central Los Angeles, vegetable gardens are being placed in abandoned lots and traffic medians.[13] Small local seed libraries can be set up to help people start their own home gardens.[14] If it is well managed, the sharing of human milk through direct contacts or through human milk banks can be of great benefit not only for health but also for building the sense of community.[15] The sharing of mothers' milk is now supported in systematic ways (HMBNA, 2017; Kent, 2017; PATH, 2013). Food sharing is commonplace, especially in low-income communities (Morton et al., 2008). It can be enhanced through community festivals and pot-luck meals, perhaps based on a regular schedule. Soup kitchens of various forms could be established.[16] The Shareable website offers ideas for creative sharing, including many centered on food (Shareable, 2016). There are creative ways to facilitate meal sharing.[17] In some places, groups of friends take turns hosting meals at their homes in a regular cycle. Some cohousing communities organize frequent common meals.[18] Block parties and fiestas also help to build the sense of community.

In addition to sharing food, there are opportunities for sharing information, advice, and ideas. Dietary advice can be shared.[19] Many organizations encourage gardening, including some, such as Gardens for Health International, that promote gardening specifically to prevent malnutrition.[20]

Sharing has been studied as part of the *gift economy*, in contrast with the conventional *exchange economy* (Kropotkin, 1902; Titmuss, 1997). Many pre-modern food systems use non-market modes of exchange. They are beyond the comprehension of modern neo-classical economics, but they can work well (Sahlins, 1972). Most gifts are not about charity, but about connecting with other people, based on concern about their collective well-being.

Cultivating caring

In a study on hunger in Scotland, one of the core messages was "Food is about community and not just consumption." The authors called for recognition that "Projects which aim to build community around food often help to create the feeling of a place where people choose to go, rather than have to" (Independent Working Group on Food Poverty, 2016, pp. 13–14). One project that aims to help the poor

> uses music and surplus food to bring people together to catalyse the energy of communities across Ayrshire. People from across each area, of all ages and backgrounds, come together to sing, to cook, to play, to share advice or to just talk.
>
> *(Independent Working Group on Food Poverty, 2016, p. 34)*

The program helps needy people in a way that fully respects their dignity. This approach goes beyond conventional charity because it is designed to transform human relationships. Where there is healthy interaction, people are likely to shift away from exploitation and indifference and instead move toward more caring about one another's well-being. This is not about some sort of mawkish compassion to the downtrodden, but about being nice to each other routinely, in ordinary ways.

Caring about one's neighbors can be cultivated by encouraging community members to spend more time working and playing together. That would reduce the likelihood that anyone in the community goes hungry, even in the absence of a recognizable program for achieving that outcome (Kent, 2016). The interaction does not have to be about food, but hunger would disappear more quickly if the joint activity was about food. The sense of community can lead to many different food projects, and those projects in turn would help to build the sense of community. They would show the presence of caring and help strengthen it. If we care about each other regularly, in small increments, there will be fewer calls for large-scale rescue missions (Gawande, 2017).

There are many options. People could garden together, cook together, and eat together in many different settings. Food-related skills could be strengthened through the sharing of knowledge and hands-on experience. People who are facing difficulties could be offered food packages or meals, and could also be given support in learning how to grow food, shop better, and cook for themselves (Pascual and Powers, 2012). Community farms could raise food products mainly to meet their own needs, but at the same time produce a few specialty products for sale outside the community (Sanz-Cañada, 2016). People should not be treated simply as tools of industrial systems arranged by others for the benefit of others. Staying human means having an active social life as well. Food can play an important part in building conviviality in the community. Being together with friends over shared food and drink is an important part of life. We should not live just to work or work just to eat. Play is important, too.

One observer speaks of "the central role of conviviality: the pleasure of sharing food with others, of celebrating communal culinary traditions and life at large." He is saddened by the fact that "we rush from one task to the next and eating becomes just another chore to be slotted into our busy schedules" (Middelmann-Whitney, 2010; also see Gopnick, 2012). An observer with Slow Food USA makes the point that convivial meals contribute to the nourishment of bodies and also the nourishment of communities:

> I believe the best way to begin is through building meaningful human relationships, through linking people and communities together around a sense of common purpose. Groups of people become communities by sharing work, sharing struggles, and sharing food. This leads to real, personal relationships; a sense of co-dependence and co-commitment. Once you've shared a meal with someone, or worked on a project together, you view each other differently. You're more likely to take care of each other, and, I believe, you're more likely to stand together and work for change together.
>
> *(Viertel, 2012)*

For isolated individuals and families, having someone to eat with may be as important as having something to eat (Independent Working Group on Food Poverty, 2016, p. 35). For instance, Brazil's dietary guidelines now recommend, "Eat in company whenever possible" (Tsai, 2016). Guiding principles like these relating to local food systems should be articulated and continuously elaborated under the leadership of the local community. Communities can learn from each other but still retain their diversity as they attend to their local concerns.

Conclusion

Food operations and entire food systems should not be treated as if they were purely technical projects inside a terrarium, with us humans looking in from the outside. We and our environment are inside, actively functioning parts of local, national, and global food systems. Community-based food systems need to be designed so they work well for all of us. Advocates of agroecology or any other aspect of food systems should give more attention to the social and environmental dimensions of their practices, taking account of how they are affected by interactions between people, and between people and their multi-layered environments.

We cannot require conviviality in the law, but we can organize communities in ways that facilitate positive social interaction, minimize exploitation and indifference, and encourage caring. By setting up community-level food projects, treating food-as-a-commons, food systems can facilitate people's working and playing together and, in that way, support their caring about one another's well-being.

In strong, well-functioning communities, there is not much need for charity or for laws requiring that governments respect, protect, facilitate, and provide for their people. Charity and legal systems of right become less important. They may be maintained as safety nets for dealing with emergencies. When people live well together, formalized charity and human rights fade into the background, on standby, like the local fire department.

When people live well together in their local communities, they are less likely to threaten other communities, even those very different from their own. In a world made up of strong local communities, based on strong local food systems, with food functioning mainly as a commons, it would be possible to grow a global food system that works well for everyone, in diverse ways.

Notes

1 The Food and Agriculture Organization of the United Nations provides comparable information through its Family Farming Knowledge Platform, at http://www.fao.org/family-farming/background/en/.

2 Guidance on the design of intentional communities can be found on the Internet on sites such as that of the Fellowship of International Community: www.ic.org. The organization's magazine, called *Communities: Life in Cooperative Culture*, published a special issue on Food and Community in Summer 2015, as Issue 167. The magazine's website is communities.ic.org.

3 Cohousing Association. (2017) "What is cohousing?" *Cohousing Association of the United States*. October 4. http://www.cohousing.org/what_is_cohousing.

4 The Food and Agriculture Organization of the United Nations maintains a website and newsletter that discuss recent developments relating to the right to food at http://www.fao.org/righttofood/right-to-food-home/en/.

5 Heifer International. (2013) *Passing on the Gift*. Heifer International. http://www.heifer.org/ourwork/approach/passing-on-the-gift.

6 Cave, J. (2013) "6 people making a difference in Honolulu," *Honolulu Magazine* (November). http://www.honolulumagazine.com/Honolulu-Magazine/November-2013/6-People-Making-a-Difference-in-Hawaii/index.php?cparticle=1&siarticle=0#artanc.

7 Achebe, C. (1987) *Anthills of the Savannah*. Heinemann, p. 93. http://ieas.unideb.hu/admin/file_6618.pdf.

8 BookRags. (2007) *Anthills of the Savannah Premium Study Guide*. BookRags, Inc., p. 1. https://asrevision.wikispaces.com/file/view/anthillssavannah+studyguide.pdf.

9 Kohala Center. (2011) "Laulima center offers assistance for cooperatives," *Hawaii 24/7* (February 8). http://www.hawaii247.com/2011/02/08/laulima-center-offers-assistance-for-cooperatives/.

10 Luna, M. (2013) "Region in Italy reaches 30% coop economy," *Shareable* (July 25). http://www.shareable.net/blog/illustrious-region-in-italy-reaches-30-coop-economy?utm_content=kent%40hawaii.edu&utm_source=VerticalResponse&utm_medium=Email&utm_term=Read%20more&utm_campaign=Shareable%3A%20Region%20in%20Italy%20Reaches%2030%25%20Coop%20Economycontent.

11 Press Release, February 14, 2017. *Launch Date Approaches for Historic CS Farm Charter in the US and Canada.* http://csaday.info/wp-content/uploads/2017/02/Press-release-for-CSA-Charter.docx.

12 See Graff, V. (2011) "Eccentric town, Todmorden, growing ALL its own veg," *Mail Online.* http://www.dailymail.co.uk/femail/article-2072383/Eccentric-town-Todmorden-growing-ALL-veg.html; Incredible Edible website at http://www.incredible-edible-todmorden.co.uk; Warhurst, P. (2012) *How We Can Eat Our Landscapes.* YouTube. http://www.ted.com/talks/pam_warhurst_how_we_can_eat_our_landscapes.html?utm_source=newsletter_weekly_2012-08-10&utm_campaign=newsletter_weekly&utm_medium=email.

13 *Ron Finley: A Guerrilla Gardener in South Central LA. Ted Conversation.* http://www.ted.com/talks/ron_finley_a_guerilla_gardener_in_south_central_la.html.

14 Steinberger, J. L. and Peterson, S. (2016) *The Seed Saving Controversy.* Edible Feast. http://www.ediblefeast.com/eddyawards/vote/9358/seed-saving-controversy.

15 Davies, L. (2016) *Opinion: I Was Blown Away by the Bond Between Strangers Who Share Breast Milk.* TVNZ News. https://www.tvnz.co.nz/one-news/new-zealand/opinion-blown-away-bond-between-strangers-share-breast-milk.

16 Bayne, M. (2013) 'A crock-pot of soup as community organizer'. *Utne Reader* (March/April). http://www.utne.com/mind-body/community-organizer-zm0z13mazwil.aspx?newsletter=1&utm_content=03.08.13+Mind+and+Body&utm_campaign=2013+ENEWS&utm_source=iPost&utm_medium=email.

17 Johnson, E. M. (2013) "Survival of the … nicest? check out the other theory of evolution," *Yes! Magazine* (May 3). http://www.yesmagazine.org/issues/how-cooperatives-are-driving-the-new-economy/survival-of-the-nicest-the-other-theory-of-evolution?utm_source=ytw20130503&utm_medium=email.

18 Blank, J. (2014) *Common Meals in Cohousing Communities.* Cohousing: The Cohousing Association of the United States. February 1. http://www.cohousing.org/meals-2001; Villines, S. (2014) "Cohousing meal programs and leadership," *Sociocracy: A Deeper Democracy* (August 24). http://www.sociocracy.info/cohousing-meal-programs-and-leadership/.

19 To illustrate, the dietary guidelines issued by the Brazilian government in 2014 could be adapted to fit other contexts. See Ministry of Health of Brazil. (2014) *Dietary Guidelines for the Brazilian Population*, Second Edition. Brasilia. http://www.foodpolitics.com/wp-content/uploads/Brazilian-Dietary-Guidelines-2014.pdf and Monteiro et al. (2015).

20 The American Community Gardening Association is devoted to the idea of building community through gardening. See its website at http://www.communitygarden.org/about-acga/. Also see the website of Gardens for Health International at http://www.gardensforhealth.org/. Recommendations for policies to support community gardening are offered in Schukoske, J. E. (1999). "Community development through gardening: State and local policies transforming urban open space," *Legislation and Public Policy* Vol. 3, No. 35, pp. 351–92. https://communitygarden.org/wp-content/uploads/2013/12/schukoske.pdf.

References

AFSA (2017) *Agroecology: The Bold Future of Farming in Africa.* Alliance for Food Security in Africa. http://afsafrica.org/agroecology-the-bold-future-of-farming-in-africa/

Baby Milk Action. (2016) *Trade vs Health—WHO Opens the Door to Big Business While Trying to Protect Babies.* Baby Milk Action: Cambridge, UK. http://www.babymilkaction.org/archives/9786

Bohn, S. and Danielson, C. (2016) *Income Inequality and the Safety Net in California.* Public Policy Institute of California: San Francisco. http://www.ppic.org/main/publication.asp?i=1190

Chirico, J. and Farley, G.S., eds. (2015) *Thinking Like an Island: Navigating a Sustainable Future for Hawai'i.* University of Hawai'i Press: Honolulu.

Christy, F.T. and Scott, A. (1965) *The Common Wealth in Ocean Fisheries: Some Problems of Growth and Economic Allocation.* Johns Hopkins Press: Baltimore, MD.

Courtens, J.P. (2012) 'Blame industrialized agriculture, not organic farmers', *Letters to the Editor* (September 13). http://www.letterstotheeditor.com/blame-industrialized-agriculture-not-organic-farmers/

De Schutter, O. (2014) The *Transformative Potential of the Right to Food. Report of the Special Rapporteur on the Right to Food to the UN Human Rights Council.* A/HRC/25/57. http://www.srfood.org/images/stories/pdf/officialreports/20140310_finalreport_en.pdf

ETC Group (2009) *Who Will Feed Us? Questions for the Food and Climate Crises* (November). Action Group on Erosion, Technology and Concentration. http://www.etcgroup.org/content/who-will-feed-us

Fiorella, K.J., Chen R.L., Milner, E.M. and Fernald, L.C.H. (2016) 'Agricultural Interventions for Improved Nutrition: A Review of Likelihood and Environmental Dimensions', *Global Food Security* 8: 9–47.

Food Chain Workers Alliance and Solidarity Research Cooperative (2016) *No Piece of the Pie: U.S. Food Workers in 2016*. Los Angeles, CA: Food Chain Workers Alliance. http://foodchainworkers.org/wp-content/uploads/2011/05/FCWA_NoPieceOfThePie_P.pdf

Gawande, A. (2017) 'The Heroism of Incremental Care'. *The New Yorker*. January 23. http://www.newyorker.com/magazine/2017/01/23/the-heroism-of-incremental-care

Gopnick, A. (2012) *The Table Comes First: Family, France, and the Meaning of Food*. Vintage Books, New York.

GRAIN (2016) *Grow-ing Disaster: The Fortune 500 Goes Farming*. GRAIN. December. https://www.grain.org/article/entries/5622-grow-ing-disaster-the-fortune-500-goes-farming

Guo, J. (2016) 'The Sinister, Secret History of a Food that Everybody Loves', *Washington Post*. April 25. https://www.washingtonpost.com/news/wonk/wp/2016/04/25/the-secret-ancient-history-of-the-potato-that-could-change-the-story-of-civilization/?wpmm=1&wpisrc=nl_p1wemost-partner-1

Holt-Giménez, E. and Altieri, M. (2012) 'Agroecology, Food Sovereignty, and the New Green Revolution', *Agroecology and Sustainable Food Systems* 37(1): 90–102.

Holt-Giménez, E. and Altieri, M. (2016) *Agroecology 'Lite': Cooptation and Resistance in the Global North*. Food First: Oakland, CA. https://foodfirst.org/agroecology-lite-c

Independent Working Group on Food Poverty (2016). *Dignity: Ending Hunger Together in Scotland*. The Scottish Government. http://www.gov.scot/Publications/2016/06/8020/downloads#res502395

Inter Pares (2004) *Community-Based Food Security Systems: Local Solutions for Ending Chronic Hunger and Promoting Rural Development*. Inter Pares: Ottawa, Canada. https://interpares.ca/resource/community-based-food-security-systems-local-solutions-ending-chronic-hunger-and-promoting

IPES Food (2016) *From Uniformity to Diversity: A Paradigm Shift from Industrial Agriculture to Diversified Agroecological Systems*. International Panel of Experts on Sustainable Food Systems. http://www.ipes-food.org/images/Reports/UniformityToDiversity_FullReport.pdf

Jones, A. D. and Ejeta, G. (2016) 'A new global agenda for nutrition and health: the importance of agriculture and food systems'. *Bulletin of the World Health Organization* 94(3). http://www.who.int/bulletin/volumes/94/3/15-164509/en/

Kaufman, F. (2012) *Bet the Farm: How Food Stopped Being Food*. New York: Wiley.

Kent, G. (1978) 'Fisheries and the law of the sea: a common heritage approach'. *Ocean Management* 4: 1–20. http://www2.hawaii.edu/~kent/FisheriesandLOS.pdf

Kent, G. (2010a) 'Designing rights-based school feeding programs', *Societies Without Borders* 5(2): 152–173. http://scholarlycommons.law.case.edu/cgi/viewcontent.cgi?article=1116&context=swb

Kent, G. (2010b) 'Swaraj Against Hunger', *Gandhi Marg* 32(2:July–September): 149–168. http://www2.hawaii.edu/~kent/SwarajAgainstHunger.pdf

Kent, G. (2011) *Ending Hunger Worldwide*. Paradigm Publishers: Boulder, CO.

Kent, G. (2015) 'Food Security in Hawai'i'. In Jennifer Chirico and Gregory S. Farley, eds., *Thinking Like an Island: Navigating a Sustainable Future in Hawai'i*. University of Hawai'i Press: Honolulu.

Kent, G. (2016) *Caring About Hunger*. Irene Publishing: Sparsnäs, Sweden.

Kent, G. (2017) *Governments Push Infant Formula*. Irene Publishing: Sparsnäs, Sweden.

Kropotkin, P. (1902) *Mutual Aid: A Factor of Evolution*. https://www.marxists.org/reference/archive/kropotkin-peter/1902/mutual-aid/

Kuhnlein, H. V., Erasmus, B. and Spigelski, D., eds. (2009) *Indigenous Peoples' Food Systems: The Many Dimensions of Culture, Diversity and Environment for Nutrition and Health*. Food and Agriculture Organization of the United Nations and McGill University, Canada: Centre for Indigenous Peoples' Nutrition and Environment: Rome. http://www.fao.org/docrep/012/i0370e/i0370e00.htm

Lappé, F. (2011) 'The Food Movement: Its Power and Possibilities'. *The Nation*. September 14. https://www.thenation.com/article/food-movement-its-power-and-possibilities/

Lindgren, S. (2013) 'Bet the Farm: Spinning Wheat into Gold'. *UTNE Reader*. January/February. http://www.utne.com/politics/bet-the-farm-zm0z13jfzlin.aspx?newsletter=1&utm_content=01.02.13+Environment&utm_campaign=2013+ENEWS&utm_source=iPost&utm_medium=email

Lobello, C. (2013) 'Why Asia Is Letting Millions of Tons of Extra Rice to Waste'. *This Week*. July 31. http://theweek.com/article/index/247611/why-asian-countries-are-letting-millions-of-tons-of-extra-rice-go-to-waste

Middelmann-Whitney, C. (2010) 'Conviviality Now! Family Feasts for Body and Soul'. *Psychology Today*. May 1. https://www.psychologytoday.com/blog/nourish/201005/conviviality-now

Minder, R. (2016) 'In Spain, New Restaurants Nourish the Needy, and the Soul'. *New York Times*. December 19. http://www.nytimes.com/2016/12/19/world/europe/in-spain-new-restaurants-nourish-the-needy-and-the-soul.html?emc=edit_nn_20161220&nl=morning-briefing&nlid=2155033&te=1

Monteiro, C. A. et al. (2015) 'Dietary guidelines to nourish humanity and the planet in the twenty-first century. A blueprint for Brazil', *Public Health Nutrition* 18(13): 2311–22.

Morton, L.W., Bitto, E.A., Oakland, M.J., and M.S. (2008) 'Accessing food resources: Rural and urban patterns of giving and getting good', *Agriculture and Human Values* 25: 107–19. http://link.springer.com/article/10.1007%2Fs10460-007-9095-8#/page-1

New York State Department of Labor. (2016) *Occupational Wages*. Website. https://labor.ny.gov/stats/lswage2.asp#45-0000

Oxfam America (2016) *No Relief: Denial of Bathroom Breaks in the Poultry Industry*. Oxfam America. https://www.oxfamamerica.org/static/media/files/No_Relief_Embargo.pdf

Pascual, T. and Powers, J. (2012) *Cooking up Community: Nutrition Education in Emergency Food Programs*. WHY Hunger and National Hunger Clearinghouse. http://www.whyhunger.org/uploads/fileAssets/CUCFINAL1.pdf)

PATH (2013) *Strengthening Human Milk Banking: A Global Implementation Framework, Version 1.1*. Bill & Melinda Gates Foundation Grand Challenges Initiative: Seattle, WA. http://www.path.org/publications/detail.php?i=2433 http://www.path.org/publications/files/MCHN_strengthen_hmb_frame_April2015.pdf

Pollan, M. (2007) *The Omnivore's Dilemma: A Natural History of Four Meals*. Penguin: New York.

Poppendieck, J. (1998) *Sweet Charity: Emergency Food and the End of Entitlement*. Penguin Books.

Riches, G. and Salvasti, T. (2014) *First World Hunger Revisited: Food Charity or the Right to Food?* Second edition. Palgrave Macmillan: New York.

Rosenthal, E. (2013) 'As Biofuel Demand Grows, So Do Guatemala's Hunger Pangs'. *New York Times*. January 5. http://www.nytimes.com/2013/01/06/science/earth/in-fields-and-markets-guatemalans-feel-squeeze-of-biofuel-demand.html?_r=0

Sahlins, M. (1972) *Stone Age Economics*. Aldine-Ahterton: Chicago, IL. https://libcom.org/files/Sahlins%20-%20Stone%20Age%20Economics.pdf

Sanz-Cañada, J. (2016) 'Local agro-food systems in America and Europe. Territorial anchorage and local Governance of identity-based foods', *Culture & History Digital Journal* 5(1): e001.

Shareable (2016) *How Is Technology Changing the Way We Share Food? The SHARECITY Research Team to Find Out*. Shareable.

Starhawk (2016) 'Social permaculture—what is it?', *Communities*. Fellowship for Intentional Community. http://www.ic.org/social-permaculture-what-is-it/

Sustainable Food Trust (2016) *Panel Discussion: Food Justice*. YouTube video. https://www.youtube.com/watch?v=ZW_i4GBBvc0

Swanson, A. (2016) 'A Single Chart Everybody Needs to Look at Before Trump's Big Fight over Bringing Back American Jobs'. *Washington Post*. November 28. https://www.washingtonpost.com/news/wonk/wp/2016/11/28/theres-a-big-reason-trump-might-not-be-able-to-keep-his-promise-on-jobs/

The Economist (2016) 'Biotechnology: Seedy business'. *The Economist*. May 21.

Titmuss, R. (1997). *The Gift Relationship: From Human Blood to Social Policy*. The New Press: New York.

Tran, M. (2013) "Guatemala remembers Conflict Victims as New Battles Ignite Over Resources'. *The Guardian*. October 24. http://www.theguardian.com/global-development/2013/oct/24/guatemala-battle-resources

Transnational Institute (2015) *Investing for Development?* TNI. September 17. https://www.tni.org/en/publication/investing-for-development

Tsai, M. (2016) *Brazil's Dietary Guidelines: Eat Real Food, Together*. FoodTank. http://foodtank.com/news/2016/07/brazils-dietary-guidelines-eat-real-food-together

Tudge, C. (2013) 'World agriculture: Living well off the land', [Commentary] *World Nutrition* 4(7): 514–48. http://wphna.org/wp-content/uploads/2015/02/WN-2013-04-06-361-390-Colin-Tudge-Living-well-off-the-land-1.pdf

UNCTAD (2013) *Wake Up Before It Is Too Late: Make Agriculture Truly Sustainable Now for Food Security in a Changing Climate. Trade and Environment Review*. United Nations Conference on Trade and Development, Geneva, Switzerland, pp. 19–21. http://unctad.org/en/PublicationsLibrary/ditcted2012d3_en.pdf

UNECOSOC (1999) *Substantive Issues Arising in the Implementation of the International Covenant on Economic, Social* and *Cultural Rights: General Comment 12 (Twentieth Session, 1999) The Right to Adequate Food (art. 11)*. United Nation. Economic and Social Council. E/C.12/1999/ Geneva, Switzerland http://www.refworld.org/docid/4538838c11.html
UNICEF (2016) *The Baby-Friendly Hospital Initiative*. United Nations Children's Fund: New York. http://www.unicef.org/programme/breastfeeding/baby.htm
Viertel, J. (2012) 'Beyond voting with your fork: From enlightened eating to movement building', *Food First Backgrounder* 18(1). http://foodfirst.org/wp-content/uploads/2013/12/BK18_1-2012_Spring_Backgrounder_-_Beyond_Voting_with_your_Fork-_From_Enlightened_Eating_to_Movement_Building.pdf
Vivero-Pol, J.L. (2017) 'Food as commons or commodity? Exploring the links between normative valuations and agency in food transition', *Sustainability* 9.
Walljasper, J. (2016) 'How One Farm Is Reinventing Agriculture for Better Food and a Brighter Future'. *Shareable*. November 28. http://www.shareable.net/blog/how-one-farm-is-reinventing-agriculture-for-better-food-and-a-brighter-future
Wiggins, S. and Keats, S. (2013) *Smallholder Agriculture's Contribution to Better Nutrition*. Overseas Development Institute: London. http://www.ajfand.net/Volume13/No3/Reprint-ODI%20Smallholder%20agriculture%E2%80%99s%20contribution%20to%20Nutrition%202013.pdf
World Health Organization (1981). *International Code of Marketing of Breast-milk Substitutes*. WHO: Geneva. http://www.who.int/nutrition/publications/code_english.pdf

9

FOOD AS CULTURAL CORE

Human milk, cultural commons and commodification

Penny Van Esterik

What is culture?

Food is at the emotional heart of social life, but it is never an abstraction for us, because food becomes us. Culture and our bodily constitution help to explain how this transformation happens. Culture bridges the gastronomical, culinary and nutritional aspects of food by observing activities taking place in fields, factories and kitchens in different parts of the world. Nutritionists know the biochemical details about how our bodies absorb and use nutrients from our food. Anthropologists wrestle with the complexities of culture and how culture makes us human. In this chapter, I review culture as a conceptual tool, and then develop and apply it to understand food cultures. I then examine gender as one dimension of culture that often disappears from economic analysis of the global food system. In order to bring culture more directly into the commons, I use the example of human milk and the commoditization of both industrial substitutes and human milk itself to demonstrate the relation between culture and the commons.

Foods serve both nutritional and medicinal functions in addition to social and relational functions (Fischler, 1988). Culture accomplishes much more than defining the symbolic meaning of individual food items. Cultural knowledge helps solve the omnivore's paradox or dilemma about what to eat; traditional food rules and cultural cuisines provide guidelines to augment the testing and tasting practices of individuals. If the item is sweet, it is probably safe; if it is bitter, it might be poison, or at least medicinal. Today the paradox for Euro-American eaters might be stated as what to eat in the face of excessive industrial food choices, few of which are healthy options. Nevertheless, even industrial food choices are still shaped by culture.

Culture is a contested concept within anthropology. Post-modern critiques of the concept of culture have resulted in efforts to avoid claiming ethnographic authority by presenting a totalizing vision of the exotic other. Media and popular culture use a more essentialized and reified folk version of culture. That is, they tend to emphasize how "our" culture differs from "their" culture, without much attention to the fact that there are no discrete bounded cultures. Preoccupation with culture sometimes results in the exoticizing of other groups and overstressing their differences, or conversely, ignoring cultural differences and taking an ethnocentric perspective. More often, culture is not defined but rather used as a taken-for-granted concept with a taken-for-granted meaning. Thus, the cultural dimensions of food may mean different things to different people. For interdisciplinary work, it is useful to have a general working definition that can be shared widely to facilitate cooperation. For our purposes, let us consider culture as the cumulative knowledge and skills that people use to adapt to or cope with their surroundings, or the way of life characteristic of a group of people in a particular time and place.

To anthropologists, the definition of culture is shaped by the theoretical perspective guiding their research, much like the concept of the commons. The authors in this book share a working definition of the commons to refer to shared resources that can be accessed and used by the community that manages those resources (cf. Vivero-Pol, 2017). Many books have been written about the definition of culture; in fact, at least one book contains nothing but definitions of culture (Kroeber and Kluckholn, 1952). There is substantial overlap of meanings of the concept, with the result that some elements harken back to past meanings not currently in use.

These elements of past usage persevere in the domain of food studies. Interesting for our discussion, one of the Latin root meanings of culture links directly to food production: to cultivate or tend crops or animals. This meaning is retained in phrases such as wet rice culture or viniculture, referring both to the cultivation of wet rice (or grapes for wine-making) and to the cultural elements that usually accompany growing that crop. Other, more popular meanings evoke the idea of a cultured person whose manners and tastes mark class distinctions. For example, often artistic activities are regulated through Ministries of Culture. The free culture movement refers to artistic products such as images, music and movies shared through Creative Commons licenses (Siefkes, 2012). The idea of a cultured or refined person has resonance in some of the current interest in gastronomy. Taste as a marker of class and ethnicity partakes of some of these understandings of culture, as Bourdieu's work in *Distinction* (1986) demonstrates.

Early uses of the term culture in anthropology emphasized cultural classification, often with evolutionary implications. One of the earliest definitions of culture, by Sir Edward Tylor (1971), covered broad topics such as custom, belief, art, law and morals. Although out of date and out of sync with contemporary anthropological theory, some elements of Tylor's definition – particularly customs and material culture – remain relevant for the study of food culture. Observations on how people produce and consume food gave rise to the definition of food cultures, in the plural rather than the singular, stressing the wide variations in the way people feed themselves. Ideally, food cultures are defined through long-term ethnographic fieldwork based on participant observation.

Generations of North American anthropology students were introduced to the concept of culture through textbooks that identified the characteristics of culture as adaptive, learned or socially transmitted, shared, symbolic, valued, integrated and constantly changing (cf. Haviland, 2013). These features could also apply to cuisine as a key component of all food cultures (cf. Crother, 2013). Cultural cuisines develop over centuries of traditional food practices that are primarily transmitted person to person. Traditional cuisines provide ways to combine food items into recipes, and recipes into meals; they encourage a sense of balance to meals, and discourage under- and over-eating (Wilson, 2015:xxvii). But cuisines constantly change in response to new ingredients (or their loss) and new technologies. Cuisines are broadly shared in a community, and highly valued, although people recognize personal differences in the interpretation of cuisines and in levels of culinary skills. Even the symbolic meanings of key food items change over time. Think of the meanings Euro-Americans attribute to white bread vs. brown bread (Bobrown-Strain, 2012), or the religious significance of bread itself. The cultural meanings of certain unique foods like consecrated bread prevent them from being sold in the market.

Food is a vehicle for the practice of social and cultural life. But different theoretical approaches to culture present different opportunities for understanding the role of food in shaping social life. The shift to meaning in anthropology oriented the concept of culture more to the cognitive, drawing attention to webs of signification (Geertz, 1973) and away from material conditions. A focus on meaning, texts, discourse and cultural beliefs about food helps with our understanding of how people construct their food worlds. To speak of food or food systems as *only* culturally constructed is to miss the material nature of food and the vital role of nutrients for human survival. Marxist approaches to food stresses material conditions, including labour. Food that is produced and eaten is a material substance that must be acted upon to turn food items into culturally acceptable meals through the application of learned practical skills. There are material, developmental and physiological worlds that constrain culturally constructed choices around food.

People who work to reduce hunger and malnutrition in the world may be better served by applying theories of culture based on political economy than semiotics, while acknowledging that people's decisions about food may well be based on the symbolic properties of food. Those working to try to change people's minds about food through nutrition education or media messages need to understand the semiotics of food, while still acknowledging the material and social structures that constrain choices.

In this chapter, I present a middle ground between the material and the symbolic, between the uniqueness of all culturally constructed food systems and the universality of nutritional requirements. I consider food as a cultural complex incorporating inherited structures and adapting to current conditions. The food complex includes systems of production and consumption and is linked to all other aspects of culture. To avoid an otherwise impossibly complicated task, I consider the food complex as if it could stand alone as an autonomous system with an integrity of its own. This allows us to appreciate, for example, the differences between French and American food cultures (cf. Rozin et al., 2011), or British and Thai food cultures, without considering, for instance, how each food system is linked to religion and politics. Each food system incorporates distinct customs and established habits that must constantly negotiate current conditions; sometimes these conditions appear to have nothing to do with food per se – for example, cars, taxes or telephones.

Approaching food as commons requires conceptualizing culture broadly enough to include individual agency and experiences of taste and pleasure, community level meanings such as ritual uses and markets and levels most often addressed in food policy such as price and food safety. The next section explores how an understanding of food cultures may be of use to food activists and policy makers interested in food commons.

Food cultures

Whatever the definition you prefer to use, what is usually understood by the term culture is critically important to food studies and to the paradigm shift to food as commons. Although there are no discrete bounded cultures in today's world, food remains an effective culture-bearer. What does culture accomplish in the study of food? First, it draws attention to the diversity of customs and practices associated with obtaining and transforming food items into culturally appropriate meals. It is easy to forget that. WEIRD people (Western, educated, industrial, rich and democratic) should not be the only reference point for policy making. The numerous qualitative ethnographies, particularly of small-scale societies, provide constant reminders of the range of food items that are treasured. Food cultures define what is considered edible in local communities. Insects are an excellent source of protein, but only if they are considered edible. Recent efforts to process insects such as crickets and integrate them into industrial recipes represent one approach to dealing with the ecological constraints of the planet (Birgit et al., 2013). Fermented foods evoke disgust reactions in some parts of the world, while milk products evoke the same reaction in other areas. Clearly, the content of food cultures is not hard wired, in spite of experiments designed to show that infants, given a wide range of food items, will select a nutritious diet (cf. Davis, 1939); food cultures are flexible to accommodate personal taste preferences and local changes in food items.

Food insecurity often stretches the boundaries of edibility. Items that are usually considered inedible, such as insects, grass or human bodies, may be re-categorized as edible under extreme conditions of famine or war. Some would starve rather than consume pork or beef, items that are considered luxuries in other food communities. Food cultures include ways of dealing with excess and scarcity of food through techniques such as fermentation, pickling and drying.

Even food waste is culturally constructed. Leftover food might be considered the next meal, snacks for children, food charity, polluted garbage or sacred food from the gods. While processed food as commodity has a "best before" date to encourage new purchases, defining a short shelf life for a food item even when the products remain edible creates more waste (Lang, 2010). Genetically modified foods have set new aesthetic standards for the appearance of fresh food. In Canada, fruit and vegetables that are not genetically modified and are occasionally misshapen are sold at a discounted price as "naturally imperfect."

Reciprocity and intergenerational relationships are vitally important for food security. Sharing food across generations, for example, is a form of food security. Older generations may be repositories of important knowledge about the location and processing of edible wild plants in countries such as Lao PDR. In other countries, elders who live alone may be excluded from community food sharing practices, and may lose interest in eating. In the latter situation, new community initiatives such as "meals on wheels" have replaced kin-based cultural food sharing practices.

Eating together is the moral core of human society. Feeding and eating teaches us that we are all connected through food. Commensality – the act of eating together and sharing food – is a special kind of consumption. Despite that, economists and nutritionists seldom explore the complexity of commensality. However, this is one of the most important ways to understand relatedness and intimacy. Commensality also entails reciprocity. Meals shared insure that invitations to share future meals will follow. Reciprocity may include calculation, as with dinner parties where households decide who to invite and how elaborate the meal should be; sharing the cost of rounds of drinks or food; and, of course, the business lunch. These commensal processes can become competitive or exclusionary. For example, competitive feasts such as potlatches among First Nations on Canada's west coast display and redistribute vast quantities of food to claim and socially mark a change in status (Crowther, 2013). Patterns of commensality rule some people in the commensal circle and exclude others. The intricate cultural rules guiding social reciprocity are deeply embedded in food cultures.

Language is a key component of culture; language and culture are always intertwined. Jurafsky (2014) shows how the language of food operates in crafting menus. But food is also part of a broader system of communication, as Barthes (1961), Levi-Strauss (1966) and Douglas (1999) have shown. Food may be used to signify gender, class, ethnicity, nationalism and other dimensions of identity – significations often reiterated through stories and other media. Films such as *Like Water for Chocolate* (1992), *Babette's Feast* (1987), *Chocolat* (2000), *Chef* (2014) and *Julie and Julia* (2009) capture both the pleasure and the cultural context of eating. In addition to these Euro-American favourites, Chinese films such as *Eat, Drink, Man, Woman* (1994), *The God of Cookery* (1996), *Kung Fu Chefs* (2009) and *Cook Up a Storm* (2017) draw attention to the competitive context of Asian cooking. These films are reminders that eating is a vibrant cultural activity with emotional, symbolic and performative elements that are easily ignored when considering food as commodity.

Culture shapes how people interact with their environment, which is the ultimate source of food. Techniques for managing shared resources such as pasture land, forests and rivers are also important parts of food cultures and are often linked to kinship. When systems of kinship are ignored, it is more difficult to understand how common resources are managed. For example, lineages and clans may claim hereditary rights to certain food resources through matrilineages, patrilineages or both. Rights to land may be inherited through oldest or youngest sons or divided equally among all children.

Research confirms that some attributes of taste are hard wired in humans; the preference for sweet and dislike of bitter is part of our primate heritage. But flavour perception is also shaped by culturally informed experiences. Both culture and language shape how we learn to recognize and value taste (Shepherd, 2012).

Food cultures define normal and festive meal formats and meal cycles. Consider the Italian meal format of a light sweet breakfast and a heavy three course lunch compared with a Lao meal format of sticky rice with varied side dishes. Food events such as wedding dinners, church picnics, backyard barbeques and state banquets structure commensality and are culturally scripted. They often mark seasonal and work rhythms and are occasions where local communities share food experiences, from preparation to celebratory consumption.

Food culture helps to accomplish the goals of institutions like prisons, schools, hospitals, nursing homes and factories. School lunches socialize children into their food cultures (Allison, 1991) and prisons impose additional discipline on inmates through the kind of meals that are served. Ascetic practices and fasting for religious purposes differ from the extremes of fasting seen in eating disorders such as anorexia nervosa. But there are parallels in the ascetic logic and food discipline of pious monks and teenage anorexics (O'Connor and Van Esterik, 2015).

Ethnic and national identity is closely entwined with food. Even insulting stereotypes are based on assumptions about what others eat (frogs, limeys). We maintain distance from the other by not eating like the other. What is considered good to eat and what is not creates social boundaries of inclusion and exclusion which change over time. In South Asia, caste and subcaste distinctions are maintained through different food practices and prohibitions against eating with lower caste groups. Through food practices, customary hierarchies are created and reproduced. The food heritage of a community or a nation may be highly valued, such as Italian lardo (Leitch, 2003); but items can also be commoditized for commercial purposes through culinary tourism (Long, 2004) or by marketing specific ethnic foods such as unpasteurized cheeses or regional specialties (Van Esterik, 2006).

Sharing food is the basis of community and ensures individual and group survival. But it can also reflect relations between nations. State banquets and other kinds of feasts reveal a great deal about power relations between individuals and even nation states. Food cultures also incorporate practices like feeding supernatural entities such as gods, spirits, ghosts and ancestors (Pollan, 2013). From Santa Claus to communion, food exchanges are critical for creating and maintaining social relations between humans and other beings. Sacred food such as bread or corn may be eaten ritually; other foods, such as beef or pork, may be avoided as polluting. Foods that are imbued with supernatural power often play a role in herbal medicine and form the basis for pre- and proscriptions (or food taboos) for the purpose of healing or protecting vulnerable people, like pregnant women (Van Esterik, 1988).

Finally, anthropologists such as Laura Nader (1972) remind us to "study up." For people who are interested in changing the industrial food system, it is important to understand the logic, values and practices of the corporate food culture. Susan George (1976) understood this, as she called for researchers to study the rich in order to understand poverty. It would be foolish to challenge the hegemony of food industries without understanding the Davos culture that guides the operation of food industries. The bureaucratic regulations surrounding food safety and intellectual property rights are part of the constraints and opportunities the food industry must negotiate (and manipulate) to maximize their profits. These, too, are important parts of the food system. We cannot treat trade and economic dimensions of food as culture-free processes. There are too few studies of corporate food cultures; perhaps the commons movement can provide more of these.

Attending to the cultural dimensions of food requires working with the contingencies and unpredictability of human life, and these are difficult to quantify. Often analysis of the cultural dimensions of food gets caught up in the dualisms of Cartesian thinking. The distinctions between public and private, individual and society, and body and mind distort rather than clarify the analysis of cultural food processes. For example, street vendors of food in Southeast Asia often start making quantities of everyday food dishes within the home, move food into the public street for sale and feed families with the leftovers that do not sell. Our research models should reflect this flexibility and not become trapped in exclusive binary categories like public or private.

Food cultures are never static. They are always adapting to new conditions; they are creative systems, always substituting items, adding recipes, changing formats. As Goody (1982) reminds us, it is easy to criticize industrial food, but it did a great deal to improve food security for the poor. Culturally defined meal and taste preferences are notoriously hard to change from the top down through regulations or nutrition education. The real tragedy is when culturally valued comfort foods are replaced by industrial processed food products promoted by advertising, such as dried cereal.

Consider the cultural and class dimensions shaping how British breakfasts have changed over time from the days when porridge, tea and bread and jam were staples to the fry-up breakfast. The precursor to the British "fry up" breakfast of sausage, bacon, eggs, beans, mushrooms and fried bread was attributed to rural gentry in the 1300s, where meat as a status symbol displayed the wealth of the family. The meal was standardized in the Victorian era, and canonized in Mrs Beeton's *Book of Household Management* (1861). By the 1950s, about half the British population consumed some variation of a fry-up breakfast. I had my first British fry-up breakfast in Penang, Malaysia and my second in Nairobi, Kenya.

While this meal came to signify British cuisine, particularly in the colonies, at home it was replaced by American packaged commercial cereals. Now, the British consume 6.7 kg of highly processed extruded cereals full of salt, fat and sugar per person per year (Lawrence, 2010). Cornflakes entered the country in 1924, and a decade later, Kellogg opened its first manufacturing plant in Manchester. Intensive advertising helped this highly processed American product replace traditional British breakfasts. Highly processed foods have replaced unprocessed foods in many parts of the world and have significant negative health consequences, as explored by other chapters in this volume.

Globalization and commoditization of food diminishes the diversity of food cultures by reducing access to key valued food items and altering traditional patterns of food sharing. The spread of industrial food products, starting with commercial baby foods, has already begun to set the industrial palate from birth with the standardized taste of infant formula and commercial baby foods. Meanwhile, the market develops long-lasting fruit and vegetables that can survive long-distance travel. These processes contribute to the homogenization of taste, as generations of children acquire an industrial palate and learn to crave sugar, fat and the artificial vanillin found in commercial cookies, for example (Wilson, 2015:xv; Bentley, 2014).

In cash-strapped households, purchased processed food is less likely to be shared. But cultural assumptions about food sharing are persistent. Lao refugees in North America provide processed foods such as cookies and wrapped cakes for Lao monks, rather than the traditional sticky rice and fermented fish, an adaptation to their new social, cultural and economic food environments.

Food as commodity denies culture by undermining the moral economy of food sharing and commensality and undercutting the holistic conception of food as something more than fuel for individual consumption. Devaluing local knowledge of the food system has hastened the loss of knowledge about processing and preparing local foods. Often, the introduction of new technology meant to reduce the work of food preparation had unanticipated consequences. For example, the Mexican tortilla maker and the Japanese rice cooker had an impact on the workload of women, but it also altered the taste of the foods produced. With the adoption of rice cookers in Thailand, rice water was no longer available for medicinal purposes or for making baby food. Often new technologies result in the loss of cultural food processing practices. This deskilling and loss of knowledge is most visible when we take a closer examination of gender within food studies.

Gender and food praxis

Women have a unique relationship with food; thus, food studies needs a feminist lens. "Feeding and eating are profoundly meaningful in all cultures and are deeply entwined with gender relations" (Counihan, 1999:2). Nearly 20 years later, the Canadian Journal of Food Studies (2018) put out a special issue on feminist food studies, reviewing past contributions and demonstrating how far we still had to go to reduce "gendered food injustice." Allen and Sachs (2012:23) point out that

> Women perform the majority of food-related work, but they control few resources and hold little decision-making power in the food industry and food policy. And, although women bear responsibility for nourishing others, they often do not adequately nourish themselves.

Any cultural analysis of food and food systems must take note of the gendered division of labour in food production and preparation, and of the ideologies underlying them. Even food itself can be coded as masculine or feminine, both in small-scale societies and in Euro-American popular culture's obsession with the manliness of meat and backyard barbeques.

Research in feminist food studies identifies male bias in past research and policy around food and is acutely sensitive to power imbalances. It pays particular attention to women and girls and generally aims for achieving gender equality. Counihan (1999:11) reminds us that "power relations around food mirror the power of the sexes in general." While gender does not always have to be central in food studies, it does have to be examined as part of the equation. The gender bias of economics and law in the past went unheeded until work in feminist economics (Waring, 1988) and jurisprudence (Charlesworth et al., 1991) unpacked the importance of gender analysis. While Waring pointed out how women's work was not counted by economists, Charlesworth and colleagues exposed the gender bias in supposedly neutral laws. Food provides insights into the human condition; but these insights are often viewed as ephemeral by economists or lawyers – the professionals closest to the development of food policy under the current food regime.

The importance of gender for food studies cannot be overemphasized, but it exists in relation to many other differences such as ethnicity, sexual orientation, race and class. Feminist theory now incorporates elements of queer theory, including the recognition that sex and gender are culturally constructed, and in most parts of the world, a simple male–female binary based on sexual differences is not adequate for understanding culturally relevant gender categories.

Consider the case of Ambrosia as an illustration of intersectionality in food studies. Ambrosia, an American company, paid poor Cambodian women around $1 US for providing one ounce of breastmilk. Ambrosia then processed the milk and sold it in North America for $5 US per ounce, making a substantial profit. The company promoted the initiative as a way to provide income opportunities for poor Cambodian women and a way to encourage them to continue breastfeeding their own children for more than six months. In March 2017, Cambodia banned the company from exporting human milk on the grounds that it exploited poor mothers; now Cambodia treats breastmilk as a human tissue like blood that should not be commercialized (Cheang, 2017). Any analysis of this case must consider the intersection of gender, ethnicity, race and class to understand how a product like human milk, critical for human life, can be commoditized and given a monetary market value.

Food can be both empowering and disempowering to women, who are often considered as nourishers to their families and communities. Domestic cooks may be highly praised, but they are seldom recognized as professional chefs. Around the world, women are overwhelmingly responsible for feeding the family, with particular responsibility for the young, old and sick family members (DeVault, 1991). A materialist focus on the political economy of food directs attention to the division of labor in the production, processing and preparation of food – food work as women's work. Intra-family food distribution usually favours men and boys over women and girls, who eat least and last (Van Esterik, 1984). The gender bias in feeding men and boys is most obvious in South Asian communities, and results in differing rates of stunting and malnutrition. Such biases are easily dismissed as a part of traditional culture and thus unchangeable, rather than identified as the result of practices based on gender oppression. This is a reminder that attending to the cultural dimensions of food does not always simplify the analysis of food or gender inequality. Rather, the concept of culture can be used to maintain the status quo, justify oppression and even protect current food regimes.

We often see the coercive power of food and feeding in families, where women attempt to control mates and children through food, as Weismantel (1991) showed in Peru. Thompson et al. (2016) explore how parents respond to their children's food preferences. Mothers often cater to the food preferences of their husbands and children at the expense of their own preferences, using food as a way to express love and devotion to family. Picky eating and fussy children, long considered a food-secure North American problem, now appears well entrenched in countries such as China (Chee, 2000). Other eating disorders are increasingly appearing throughout the world, often defined in very different terms (Becker, 2007).

Customs encode women's knowledge of food. Recipes shared across generations are usually transmitted orally. These recipes become part of the collective commons of a community. In North America, community cookbooks produced by churches and schools capture this traditional culinary knowledge. In recent years, recipes that once belonged to the collective as part of the knowledge commons have become copyrighted and privatized in the form of expensive, professionally published cookbooks (Heldke, 2003). It is then difficult to attribute these recipes to the woman who originally shared them. The knowledge commons, or constructed cultural commons (Madison et al., 2012), might provide a way to share culinary knowledge, formally or informally.

Feminist food scholars draw particular attention to the bodies that consume food. This corporeal focus incorporates women's physical and emotional connections to food and eating. Food and gender are often linked around material bodies. For example, food and sex are metaphorically connected and carefully regulated in many societies; consider the linkages between the words for sex, candy and dessert in English (Jurafsky, 2014:103). The corporeal focus is also important in the examination of body image and eating disorders. Feminist food scholarship has drawn attention to the Euro-American idealization of the slender female body and the disparagement of body shapes that do not conform to strict but ethnocentric standards of beauty. At the same time, the industrial food regimes provide "empty and cheap calories that increase obesity" (Vivero-Pol, 2017:3).

The corporeal realities of eating disorders (O'Connor and Van Esterik, 2015), obesity and reproductive processes like pregnancy, postpartum care and breastfeeding all involve food and nurturing practices that centre around care (see Chang's chapter, this volume). The next section draws attention to human milk, a gendered food product that is individually consumed by infants but supported by the cultural knowledge and practices of women. This critical human food is currently threatened by both the loss of postpartum customs that support new mothers and the promotion of inferior industrial substitutes for mother's milk and for local complementary foods.

Human milk: private good, common heritage or marketable commodity?

Why examine infant feeding as part of the discussion of food and the commons? Infant food is not as sensitive to the historical transformations of food regimes as adult food, since newborns survive on a single food, milk – human, animal or synthetic. It is more difficult to exoticize infant foods such as milk and gruel. If attention to the cultural attributes of food tends to stress exotic differences in items and cuisines, infant feeding brings up commonalities based on the universal needs of all human newborns for milk.

Lactation as a mammalian trait has remained unchanged for millions of years. Unlike food for adults, changes in human milk are constant micro-changes related to the season, time of day and maternal diet, not the macro-changes that can be seen in other parts of the food system. Until recently, the only suitable food for a newborn was produced not agriculturally or industrially but rather within the maternal body. Changes in infant feeding practices have revolved around wet nursing, shared breastfeeding, use of animal milks and the development of industrial substitutes based on animal milk or soy milk. The production of these new infant foods involves some of the same technological adaptations seen in industrial food for adults, such as fortification and the introduction of genetically modified ingredients.

As food regimes changed over time, most women continued to breastfeed, untouched by policy shifts around trade agreements, food surpluses and food subsidies, for example. But food regimes did shape the products used to substitute for human milk. While the nutritional quality of infant formula has improved over the last century, particularly during the mercantile–industrial regime (1947–1973), Friedmann (2005) argues that after 1973, food regimes stabilized around corporate power. When the emerging health problems of processed foods were recognized, few considered the risks from using infant formula. Decisions about breastfeeding were deemed personal and framed as part of a lifestyle choice. But the costs of not breastfeeding have been estimated at $302 billion dollars annually. Improving breastfeeding could prevent 823,000 child deaths (under 5) and 20,000 annual deaths of women from breast cancer (Lancet, 2016).

Breastfeeding rates declined throughout the 1900s, particularly in urban populations of the developing world. In the United States, only 22% of mothers even initiated breastfeeding in 1972; by 1984, nearly 60% of women initiated breastfeeding (Martucci, 2015:155). By the eighties, UN agencies like the WHO and UNICEF confirmed the importance of human milk and began to promote breastfeeding and regulation of the infant formula industry. Today, breastfeeding rates are no longer declining; however, most countries fall short of meeting the WHO guidelines of exclusive breastfeeding for six months and continuing to nurse well beyond the first year. While the resurgence in breastfeeding is reassuring, at the same time global processes have made substitutes for human milk more accessible and acceptable around the world.

Breastfeeding as a process and human milk as a product are both unique and incommensurate. To say that human milk is a living food is more than a metaphor, since the product is actually composed of living cells that pass from the mother's body to the infant's body during breastfeeding, something that no industrial substitute can claim. Human milk adjusts to the contingencies of mother and infant, including the time spent apart, health status, time of day or night and mother's diet. Milk is thus personalized and adapted to individual differences.

Human milk is a place-based food linked to locality and taste preferences through the transfer of flavours to the infant from the mother's amniotic fluid and later through her milk. In this way, the flavour principles that guide household food patterns shape the taste socialization of infants (Mennella et al., 2009). Breastfeeding also teaches self-regulation of appetite. You cannot overfeed or force a breastfed baby to consume more milk; when infants are full, they stop working. This is not the case for other products fed through a feeding bottle, including complementary foods that can be force fed. As an example of food sovereignty, a breastfed infant consumes local food and has control over the food supply rather than being a passive consumer of an industrial product. But maternal milk is unique in that it is constantly replenished as long as someone needs it. The delicate balance between supply and demand is an example of food sovereignty at its most elemental.

Breastfeeding has always been deeply embedded in the moral economy and imbued with complex cultural meanings. When shared, it was usually exchanged by reciprocity and redistribution through real or fictive kinship links. While wet-nursing was often a paid market transaction, it too had a relational component, as in the creation of milk siblingship through wet nursing among Muslim women (Altorki, 1980). Cultural practices ensure that infants of women who produce less milk, or who may be unable or unwilling to breastfeed, survive. The use of the internet to match women who need milk with women who have extra milk is rapidly replacing wet-nursing and other cultural customs that guided milk sharing (Thorley, 2009).

More recently, the development of community-based milk-sharing networks have expanded and become more public since matches between women who have extra milk and women who need donor milk can now be made over the internet. Milk sharing internet groups operate in over 50 countries, with over 50,000 members (Palmquist and Doehler, 2014:141). Milk sharing may be an under-the-radar practice in some parts of the world. Community-based milk sharing undercuts efforts to commoditize human milk and sell it in the market. Breast milk exchanges among women carry some elements of non-market relations. Researchers are beginning to reveal the extent to which women try to learn about the women whose milk they are feeding to their infants (Falls, 2017; Palmquist and Doehler, 2014).

Donor milk banks discourage informal milk sharing because they consider it a riskier practice, and informal milk sharing means that fewer women would donate their excess milk to milk banks (Palmquist and Doehler, 2014:141). Sharing networks such as Eats on Feets (eof.org) and Human Milk for Human Babies (hm4hb) stress non-commercial altruistic milk sharing and exchange of a product that is priceless.

Can human milk be considered a common pool resource when it is shared in this way? Unlike pure public goods, common pool resources face problems of over use and have the potential to be over-exploited. Women's bodies are finite; can they be over-exploited? Think of the poor Cambodian women who sold their excess milk to Ambrosia. Did they deny milk to their own children in order to maximize the income they received for donating their excess milk?

There are elements of breastfeeding that reflect the collective logic of the commons. First, milk is a product of millions of years of mammalian evolution, and its development reflects adaptations of generations of mother–infant pairs. A woman who breastfeeds in Tanzania or Toronto is not starting from scratch but building on the experiences of her mother, grandmother and earlier generations; her milk reflects the nutritional status of these past generations of women. Similarly, the process of breastfeeding is sustained by the collective knowledge and customs of generations of women. The support mothers need to produce this critically important food is encoded in the cultural customs of postpartum care. The erosion of culturally defined postpartum customs such as resting for a month after birth and eating special foods can leave new mothers with no support and no confidence in their milk supply.

Without the postpartum knowledge and practices that support new mothers, lactation can fail. In North America, a lost generation of women who were not breastfed and whose mothers did not breastfeed depended on groups such as La Leche League to reintroduce the knowledge and skills necessary to solve breastfeeding problems and to provide a supportive environment for breastfeeding mothers. Critics have pointed out the limits of this environment, identifying the founding members as Catholic, middle-class white women from Chicago. But over time, La Leche League groups and other mother-to-mother support groups adapted to meet the needs of working mothers, students, lesbian mothers, immigrant groups and transgendered parents (Mac Donald et al., 2016). Using phrases such as nursing rather than breastfeeding (in order to include chestfeeding) and parent-to-parent support rather than mother-to-mother support, these groups have become more inclusive. The knowledge and practices made visible and exchanged in these groups could be considered part of the collective heritage of the commons.

The process of commoditization is well advanced in the infant feeding market where even human milk is treated as a commodity; at the same time, the market for inferior substitutes has expanded. The greatest demand for these products is in the emerging markets in Asia (Baker et al., 2016). The baby food industry is projected to reach $63 billion US by 2017 (Euromonitor International, 2001). As a comparison:

> The value of all arms transfer *agreements* with developing nations by the United States and foreign countries was over 71.5 billion USD. This was a substantial increase from 32.7 billion USD in 2010. In 2011, the value of all arms *deliveries* to developing nations was 28 billion USD, the highest total in these deliveries values since 2004.
>
> *(Grimmet and Kerr, 2012)*

Baby food is a very profitable commodity.

> Breastfeeding is a global food production system with unsurpassed capacity to promote children's food security and maternal and child health, but it is side-lined by trade negotiators who seek instead to expand world markets for cow's milk-based formula.
>
> *(Smith, 2015:1)*

Considering the immense profit made by infant formula companies, is it possible to view infant feeding outside the profit-driven market mechanism? Infant formulas are heavily promoted as equivalent or even superior to human milk. The use of these substitute milks is one of the most dangerous in vivo experiments of the modern world; the risks are rarely publicized, in spite of the many industrial accidents that are discovered after cases of infant morbidity and mortality are traced back to specific factories and brands. In December 2017, millions of packages of infant formula contaminated with salmonella made by the French dairy company, Lactalis, were recalled from at least 83 countries. The production of infant formula is also adding substantially to the pollution of the environment (cf. Formula for Disaster, 2015).

Human milk is difficult to market, as the makers of Ambrosia discovered. It has few tradable features; it is only moderately durable, does not have a beautiful appearance and is not standardized (cf. Vivero-Pol, 2017:3). The marketing of human milk and human milk products illustrates the danger of corporatizing the commons. Over 2000 patents have been registered for components of human milk by companies that recognize the potential of genetically engineered lactoferrin and human milk oligosaccharides for possible use in the treatment and prevention of diseases such as HIV/AIDS and cancer. Should for-profit companies be able to patent these products that originate in women's bodies?

American companies like Prolacta and Medolac extract human milk proteins, sugars and fats obtained from human milk from donors (who were sometimes unaware that their donations were used to make products for sale) in order to make special formulas and human milk fortifiers for use in hospitals for sick and premature infants. Prolacta estimates that the cost of feeding an infant the fortifier ranges from $5600 to $10,000 per hospital stay (Falls, 2017:132).

Medolac produces human milk products from a donor milk co-operative, which then processes the milk so that it can be stored safely at room temperature. Medolac also sells frozen human milk that does not need to be thawed before use. It charges about $5.90 US for a four-ounce pouch of milk. Donor mothers are paid $1 US per ounce for their milk. At least ten ounces of donor milk are needed to make one ounce of fortifier (Bamesberger, 2014).

Most human milk is purchased by parents. But not all; artists, body builders, cosmetics companies, ice cream and cheese manufacturers and milk fetishists represent niche markets that also buy human milk. (Gourmet chefs in China were strongly criticized for reportedly serving abalone in breastmilk.) Many entrepreneurs have purchased human milk through the market, mediated by the internet. Should the artisanal use of human milk be treated as just another on-line food purchase?

Without changes in the way we think about infant feeding, could human milk become so labor intensive or "expensive" that it is not worth the effort for women to produce it? Human milk is a renewable resource, like seed; but it is a food produced within women's bodies and produced only through the act of consumption by another individual. How does this change the way we think about renewable resources?

Food is the ultimate individual commodity according to the system developed by economist Elinor Ostrum (1990), in that people must individually consume food containing enough calories to maintain individual body function. Food is destroyed by ingestion and cannot be recirculated. In that scheme, human milk appears to be the epitome of an individualized private good since it is produced in an individual woman's body, consumed by another individual and no one else can enjoy that same milk again.

Human milk, like water, can be considered in many different ways, depending on how it is classified – common, public, private or reserved for a privileged group. Just as water should be considered as part of our shared human heritage that must be protected for the common good, not appropriated for private gain (Barlow, 2012:163), so too should human milk.

What are the policy implications of seeing human milk as part of the commons? If human milk were reframed as part of the collective commons and supported as a global public good, then the cultural knowledge and practices that support breastfeeding would be highly valued and protected from the misinformation and promotional practices of infant formula companies. Human milk would be recognized as a renewable resource co-produced by women, nature, human culture, biology and the environment. All humanity benefits from healthier infants who become the next generation of healthy adults. Commercial infant foods, when needed, would be considered an essential resource like water and made exempt from the marketing practices considered acceptable for other food commodities. Subsidized and carefully regulated, the best possible products would be available to people who needed them. Collective management of the commons would include protecting postpartum customs and services that support new parents. Knowledge and skills of breastfeeding management would be treated as a public good. The example of infant feeding and human milk expands the narrative of the commons and hints towards new ways to integrate culture and the commons.

Culture and the commons

Around the world, commons are culturally constructed based on vastly different ideologies and histories that underlie the collective use of resources. Like culture, there are many different ways to understand the commons. Common resources, much like cultural resources, are both material and non-material; the commons include both the goods shared as well as the social practices for managing those resources – what Meretz (2012:28) calls commoning. Common goods are shared according to cultural rules, often phrased as customs and traditions. Managing the commons requires understanding several relevant dimensions of culture, including the sexual division of labour, gender ideologies, hierarchies of power and systems of kinship. Without knowledge of the cultural rules underlying these processes, it would be impossible to manage the commons in a locally appropriate manner. Although customs constantly change, traditions should not be used to justify unfair practices.

The discourse of food as commodity contributes to the invisibility of culture in the global food system. Market-based policies focus on economics alone. Yet cultural traces emerge even in economic narratives in phrases such as local custom, invisible rules and intangible resources. The food industry understands the importance of cultural context well, as companies use cultural information to adapt their marketing practices to target and promote their food products.

It is easier to envision air, water, forests and irrigation systems as part of the commons, even if they are difficult to manage. It is harder to conceptualize culturally complex processes such as infant feeding, postpartum customs, cuisines, special food processing techniques and recipes as part of the collective commons. It appears easier to manage fisheries and forests as common pool resources than food banks, human milk banks or community kitchens, for example. But a holistic approach to food commons requires that the cultural dimensions of food and eating be considered.

Culture and the commons reinforce each other in several important ways. First, small-scale societies where non-market relations dominate provide examples of shared access to food resources through joint production and consumption. Culturally constructed nurturing practices shape how food resources are shared in a community and are highly valued. Community gardens, potluck dinners, organic buying clubs, communal kitchens, farmer's markets and CSAs in Euro-American communities replicate some of these collective models for sharing food resources.

Second, the knowledge commons provides a way to think about the shared common heritage of food knowledge in the form of recipes, cooking methods, food processing techniques, the ritual use of foods, local technology and indigenous wild foods, for example. Third, if spaces and places are culturally constructed, then building the commons as a spatial project and creating shared public space where people encounter one another will be an important part of the commons movement (Lutz, 2012:236). Eating and breastfeeding in public spaces makes affective labour visible, reinforces our interdependence and mutual neediness (O'Neill in this volume) and challenges patriarchal notions of the separation of public and private spaces (Doonan, 2018). But culture shapes what is considered a public space, the public activities suitable for using that space and who has access to public space. How will the commons accommodate differences in how public space is used? Will it monitor what women wear, or whether a woman can breastfeed or eat in public, for example?

Fourth, consider the commons as a way to accommodate cultural differences. Ruth Benedict, a founding mother of American anthropology, once said that the role of anthropology was to make the world safe for differences. Culture and the commons seek to retain as many solutions as possible to the problem of making sure everyone is adequately fed. Food as a commons draws attention to the lack of diversity in industrial food. The appearance of diversity is accomplished through packaging and branding. Industrial food contributes to the homogenization of taste, another product of the reduction in biodiversity. How would food as a commons protect biodiversity, including genetic resources, when more than half of all calories consumed by humans come from three crops: rice, wheat and maize?

The commons as a set of values about food is visible in global responses to short term emergency food shortages; common food stores can keep people alive during emergencies, but it is not sustainable and does nothing to transform the food system. Supporting breastfeeding mothers in emergencies, rather than distributing donated infant formula, is a sustainable practice and requires primarily skilled lactation support and countering assumptions that women refugees under stress cannot breastfeed. Recent initiatives such as IYCF-E (infant and young child feeding in emergencies) and SafelyFed provide this assistance in emergency situations.

A market focus on food as commodity erases culture by altering taste preferences of cultural cuisines and devalues nurturing practices such as postpartum care, home cooking, regional food preservation techniques, gardening, food sharing through feasting and commensality. The cultural loss is particularly obvious in North America, where cooking skills are not being passed to the younger generation. A market focus speeds up the tendency to replace unprocessed and locally processed foods with highly processed foods, many with the addictive properties of salt, fat and sugar. Industrial food encourages the development of fake food that replaces real food, much as margarine replaced butter and infant formula replaced human milk. How do these highly processed foods fit into local meal formats? Are they shared in the same way? In defence of fast food, recall how effective Chinese McDonald's was at easing the burden of commensality, making it possible for people to eat alone or to pay for their own meals (Watson, 1997). The market has succeeded in substituting the industrial palette for the Greek or Thai or Navajo palette. We see the resistance to these changes in communities that have shifted back to local foods and in initiatives like the Slow Food movement.

Marina Chang (in this volume) argues that care for food contributes to the expression of identity and culture and is a way of demonstrating people's "ethical values." Although the state may define policies to guarantee fair access to food, culture defines what is fair. For example, intra-family food distribution systems that favour males are considered fair based on culturally defined gender ideologies. Cultural resources for the caring commons will not be easy to integrate into commons work. Cultural attributes are hard to isolate and measure. They do not lend themselves to quantitative measures that often guide global food policy. Even food items themselves are more likely to be defined as calories or nutrients, in a practice known as nutritionism. Attention to cultural dimensions of food and eating reminds us that there can be no one-size-fits-all method for developing the food commons. Yet there remain dangers of ethnocentrism in food policy.

Finally, if this chapter has successfully demonstrated the importance of examining the cultural dimensions of food systems, then the next logical step would be to produce ethnographies of community-based commons, demonstrating how the commons work in different cultural settings.

Bibliography

Allen, P. and C. Sachs. (2012) Women and Food Chains: The Gendered Politics of Food. In *Taking Food Public*, edited by P. William-Forson and C. Counihan. New York: Routledge. pp. 23–40.
Allison, A. (2013) Japanese Mothers and Obentos: The Lunch-Box as Ideological State Apparatus. In *Food and Culture: A Reader.* Third edition, edited by C. Counihan and P. Van Esterik. New York: Routledge. pp. 154–172.
Altorki, S. (1980) Milk-Kinship in Arab Society: An Unexplored Problem in the Ethnography of Marriage. *Ethnology* 19(2), 233–244.
Baker, P., J. Smith, and L. Salmon. (2016) Global Trends in Commercial Milk-Based Formula Consumption: Is an Unprecedented Infant and Young Child Feeding Transition Underway? *Public Health Nutrition* 19 (14), 2540–2550.
Barlow, M. (2012) Water as a Commons: Only Fundamental Change Can Save Us. In *The Wealth of the Commons: A World Beyond Market and State*, edited by David Bollier and Silke Helfrich. Amherst, MA: Levellers Press. pp. 161–165.
Barthes, R. (1961) Toward a Psychosociology of Contemporary Food Consumption. Reprinted in *Food and Culture: A Reader.* Third edition (2013), edited by Carole Counihan and Penny Van Esterik. New York: Routledge. pp. 23–30.
Becker, A. (2007) Culture and Eating Disorders Classification. *International Journal of Eating Disorders* 40, S111–S116.
Beeton, I. (1861) *Beeton's Book of Household Management.* London: Chancellor Press.
Bentley, A. (2014) *Inventing Baby Food.* Berkeley, CA: University of California Press.
Birgit, A. et al. (2013) Potential and Challenges of Insects as an Innovative Source for Food and Feed Products. *Innovative Food Science and Emerging Technologies* 17, 1–11.
Bobrown-Strain, A. (2012) *White Bread: A Social History of the Store-Bought Loaf.* Boston: Beacon Press.
Bollier, D. and S. Helfrich, eds. (2012) *The Wealth of the Commons: A World Beyond Market and State.* Amherst, MA: The Commons Strategies Group, Levellers Press.
Bourdieu, P. (1986) *Distinction: A Social Critique of the Judgement of Taste.* Trans. R. Nice. London: Routledge, Kegan and Paul.
Chang, M. (2018) Growing a Care-Based Commons Food Regime. In this volume.
Charlesworth, H., Chinkin, C. and Wright, S. (1991) Feminist Approaches to International Law. *The American Journal of International Law* 85(4), 613–645.
Cheang, S. (2017) Cambodia suspends export of human breastmilk. *Toronto Star*, March 21.
Crowther, G. (2013) *Eating Culture.* Toronto: University of Toronto Press.
Counihan, C. (1999) *The Anthropology of Food and Body: Gender, Meaning and Power.* New York: Routledge.
Davis, C.M. (1939) Results of the Self-Selection of Diets by Young Children. *Canadian Medical Association Journal* 41(3), 257–261.
De Vault, M. (1991) *Feeding the Family.* Chicago, IL: University of Chicago Press.
Doonan, N. (2018) Voir le jour: Breastfeeding and the Commons. *Canadian Journal of Food Studies* 5(1): http://dx.doi.org/10.15353/cfs-rcea.v5i1.
Douglas, M. (1999) The Abominations of Leviticus. Reprinted in *Food and Culture: A Reader.* Third edition (2013), edited by Carole Counihan and Penny Van Esterik. Routledge. pp. 48–58.
Euromonitor International. (2001) Euromonitor Passport Information database for baby food. Euromonitor International Ltd, London. Retrieved from http://euromonitor.com/baby-food.
Falls, S. (2017) *White Gold: Stories of Breast Milk Sharing.* Lincoln, NE: University of Nebraska Press.
Federici, S. (2012) Feminism and the Politics of the Commons. In *The Wealth of the Commons: A World Beyond Market and State*, edited by David Bollier and Silke Helfrich. Amherst, MA: Levellers Press. pp. 45–54.
Fischler, C. (1988) Food, Self and Identity. *Social Science Information* 27, 275–292.
Formula for Disaster. (2015) Retrieved from http://ibfan.org/docs/climate-change-2015-english.pdf.
Friedmann, H., (2005) From Colonialism to Green Capitalism: Social Movements and Emergence of Food Regimes. In *New Directions in the Sociology of Global Development.* Emerald Group Publishing Limited. pp. 227–264.
George, S. (1976) *How the Other Half Dies.* New York: Pelican.
Geertz, C. (1973) *The Interpretation of Cultures.* New York: Basic Books.
Goody, J. (1982) *Cooking, Cuisine and Class.* Cambridge: Cambridge University Press.
Grimmet, R. and P. Kerr. (2012) Conventional arms transfers to developing nations 2004–2011. US Congressional Research Service Report to Congress, August 24. Retrieved from fas.org/sgp/crs/weapons.

Haviland, W. (2013) *Anthropology: The Human Challenge*. 14th edition. Cengage Learning.
Heldke, L. (2003) *Exotic Appetites: Ruminations of a Food Adventurer*. New York: Routledge.
Jurafsky, D. (2014) *The Language of Food*. New York: W.W. Norton.
Kroeber, A. and C. Kluckholn. (1952) *Culture: A Critical Review of Concepts and Definitions*. Papers of the Peabody Museum of Archaeology and Ethnology, vol.47, Harvard University.
The Lancet. (2016) *Editorial: Breastfeeding: Achieving the New Normal*. Vol. 387 (10017).
Lang, T. (2010) Crisis? What Crisis? The Normality of the Current Food Crisis. *Journal of Agrarian Change* 10(1), 87–97.
Lawrence, F. (2010) Drop that spoon! The truth about breakfast cereals. *The Guardian*, November 23.
Leitch, A. (2003) Slow Food and the Politics of Pork Fat: Italian Food and European Identity. *Ethnos* 68(4), 437–462.
Levi-Strauss, C. (1966) The Culinary Triangle. Reprinted in *Food and Culture: A Reader*. Third edition (2013), edited by Carole Counihan and Penny Van Esterik. New York: Routledge. pp. 40–47.
Long, L. (2004) *Culinary Tourism*. Lexington: University of Kentucky Press.
Lutz, S. (2012) Shared Space: A Space Shared Is a Space Doubled. In *The Wealth of the Commons: A World Beyond Market and State*, edited by David Bollier and Silke Helfrich. Amherst, MA: Leveller Press. pp. 236–238.
MacDonald, T. et al. (2016) Transmasculine Individuals' Experiences with Lactation, Chestfeeding and Gender Identity: A Qualitative Study. *BMC Pregnancy and Childbirth* 16, 106.
Madison, M. et al. (2012) Constructing Commons in the Cultural Environment. In *The Wealth of the Commons: A World Beyond Market and State*, edited by David Bollier and Silke Helfrich. Amherst, MA: Leveller Press. pp. 369–374.
Martucci, J. (2015) *Back to the Breast: Natural Motherhood and Breastfeeding in America*. Chicago, IL: University of Chicago Press.
Mennella, J. et al. (2009) Early Milk Feeding Influences Taste Acceptance and Liking During Infancy. *American Journal of Clinical Nutrition* 90, 780S–88S.
Meretz, S. (2012) The Structural Communality of the Commons. In *The Wealth of the Commons: A World Beyond Market and State*, edited by David Bollier and Silke Helfrich. Amherst, MA: Leveller Press. pp. 28–34.
Nader, L. (1972). Up the Anthropologist: Perspectives Gained from Studying Up. In *Reinventing Anthropology*, edited by Dell Hymes. New York: Pantheon Books. pp. 284–311.
O'Connor, R.A. and P. Van Esterik. (2012) Breastfeeding as Custom Not Culture: Cutting meaning down to Size. *Anthropology Today* 28(5), 24–25.
O'Connor, R.A. and P. Van Esterik. (2015) *From Virtue to Vice: Negotiating Anorexia*. New York: Berghahn Books.
Ostrom, E. (1990) *Governing the Commons: The Evolution of Institutions for Collective Action*. Cambridge: Cambridge University Press.
Palmquist, A. and K. Doehler. (2014) Contextualizing Online Human Milk Sharing. *Social Science and Medicine* 122, 140–147.
Pollan, M. (2013) *Cooked: A Natural History of Transformation*. New York. Penguin Press.
Rozin, P., A. Remick, and C. Fischler. (2011) Broad Themes of Difference Between French and Americans in Attitudes to Food and Other Life Domains: Personal Versus Communal Values, Quantity Versus Quality, and Comforts Versus Joys. *Frontiers in Psychology* 2, 177.
Shepherd, G. (2012) *Neurogastronomy*. New York: Columbia University Press.
Siefkes, C. (2012) The Boom of Commons-Based Peer Production. In *The Wealth of the Commons: A World Beyond Market and State*, edited by David Bollier and Silke Helfrich. Amherst, MA: Leveller Press. pp. 289–294.
Smith, J. (2015) Markets, Breastfeeding and Trade in Mothers' Milk. *International Breastfeeding Journal* 10(9), 1–7. doi: 10.1186/s13006-015-0034-9
Thompson, C., S. Cummins, T. Brown and R. Kyle. (2016) Contrasting Approaches to Doing Family Meals: A Qualitative Study of How Parents Frame Children's Food Preferences. *Critical Public Health* 26(3), 322–332.
Thorley, V. (2009) Mothers' Experiences of Sharing Breastfeeding or Breastmilk: Co-feeding in Australia, 1978–2008. *Breastfeeding Review* 17(1), 9–18.
Tylor, E. (1871) *Primitive Culture*. London: John Murray.
Van Esterik, P. (1984) *Intra-family Food Distribution: Its Relevance for Maternal and Child Health*. Ithaca, NY, Cornell Nutrition Surveillance Program.

Van Esterik, P. (1986) Feeding Their Faith: Recipe Knowledge Among Thai Buddhist Women. *Food and Foodways* 1 (1), 198–215.

Van Esterik, P. (1988) To Strengthen and Refresh: Herbal Therapy in Southeast Asia. *Social Science and Medicine* 27 (8), 751–761.

Van Esterik, P. (1999) Right to Food, Right to Feed, Right to be Fed: The Intersection of Women's Rights and the Right to Food. *Agriculture and Human Values* 16, 225–232.

Van Esterik, P. (2006) From Hunger Foods to Heritage Foods: Challenges to Food Localization in Lao PDR. In *Fast Food/Slow Food: The Cultural Economy of the Global Food System*, edited by Richard Wilk. Lanham, MD: Altimira Press.

Van Esterik, P. and R. O'Connor. (2017) *The Dance of Nurture: Negotiating Infant Feeding.* New York: Berghahn Books.

Vivero-Pol, J.L. (2017) Food as Commons or Commodity? Exploring the Links between Normative Valuations and Agency in Food Transition. *Sustainability* 9, 442.

Watson, J., ed. (1997) *Golden Arches East: McDonalds's in East Asia.* Stanford, CA: Stanford University Press.

Waring, M. (1988) *If Women Counted: A New Feminist Economics.* San Francisco, CA: Harper.

Weismantel, M. (1991) *Food, Gender and Poverty in the Ecuadorian Andes.* Philadelphia, PA: University of Pennsylvania Press.

Wilson, B. (2015) *First Bite.* New York: Basic Books.

10
FOOD AS A COMMODITY

Noah Zerbe

Introduction

The post–World War II era witnessed a fundamental transformation of the global food system. The international food regime established in the later part of the nineteenth century gave way to a system of food production heavily subsidized and shielded from international competition by the state in Western societies and their colonies. Food production was largely a national affair, rooted in the specific relations of production that were largely defined at the local or regional level. As agricultural productivity increased with the application of new technologies and improvements in seed varieties and cultivars, the scale and nature of production changed dramatically. By the late 1970s, this system began to break down, and a new food regime rooted in neoliberalism, freer international trade, and, perhaps most importantly, in the commodification of food, began to emerge.

In this chapter, I trace the commodification of food in political and economic terms from the end of World War II through the contemporary era. I argue that the rise and global expansion of neoliberal capitalism has fundamentally transformed the global food regime, resulting in the near absolute commodification of food and its transformation from a vital component of life into an instrument for speculative investment and profit at any cost, which do not benefit the producer or the consumer.

This process is, of course, highly uneven. Smallholder farmers, particularly those in the global south, are often disconnected from global markets even while they—and their livelihoods—are directly impacted by them.[1] For such producers, the marketization of food and food security can be felt primarily in two key areas. First, insofar as such farmers market a portion of their annual production, they are directly impacted by broader changes in global markets, particularly with respect to the challenges posed by price volatility (discussed below) and concentrated corporate control. Second, such farmers also face increasing competition from producers in the global north, who often benefit from state subsidies, price supports, extension services, and other forms of state intervention, affecting food and commodity prices in the global south.[2] Consumers along the global food chain also suffer, with cheap and unsustainably produced commodity food contributing to both under-nutrition (in the form of hunger and micronutrient deficiencies) and over-nutrition (in the form of obesity and the so-called diseases of prosperity) so that corporate actors can profit (Patel, 2012; De Schutter, 2014).

Food as a commodity

As Karl Polanyi (1944) observed, the development of the market society necessitated the treatment of all factors of production—including land (nature), labor (people), and money—as commodities,[3] with prices for such commodities set by the market. This commodity fiction is a necessary organizing principle for a market society and requires that the market is disembedded from broader political structures and societal habits. So, while pre-market societies may have subjected market transactions to non-market principles such as reciprocity, redistribution, or social obligation, in the market society, the market itself becomes the sole organizing principle for society, disembodied from broader social obligations (Wood, 2002; Sahlins, 1972; O'Neill in this volume).

To be clear, this distinction indicates neither that non-market–based systems were fundamentally equal, nor that trade did not take place. From the earliest days of agriculture, control over food surplus has been a major source of political power. Indeed, the introduction of settled farming systems, which originated about 10,500 years ago, likely increased the intensity of work, as farming proved to be more demanding (in terms of hours worked) than hunter–gathering. Indeed, some studies claim that the quality of life and health (at least in terms of nutritional quality) initially declined in the first permanent settlements for agricultural production (Sahlins, 1972; Ulijaszek et al., 1991; Latham, 2013). But the system of farming produced a surplus, which could be appropriated, creating inequality and permitting specialization (Bender, 1978).

The commodification of food over time, as discussed below, gave the market the central role in determining food production and consumption, as the dominant allocation mechanism. According to Wood (2002: 96–97), the market has an "unprecedented role in capitalist societies, as not only a simple mechanism of exchange or distribution, but the principle determinant and regulator of social reproduction." From this perspective, the neoliberal restructuring of the global food system that began in the late 1970s represents only the most recent development in a long and uneven history towards the commodification of food that began under colonialism and expanded under the first food regime, only to briefly retreat and then re-emerge under the second.

Echoing Polanyi (1944), Radin (1996: 104) observes that commodification or non-commodification must be "seen as largely hypothetical endpoints on a continuum of possible meanings and corresponding policy choices." In the context of the commodification of food, Radin's comments highlight two important points. First, commodification is always an incomplete and contested process. In some contexts, food is treated as a commodity. In others, it may not be. The most obvious division here is between subsistence producers, who largely grow food for their own household consumption and thus operate largely outside the commodified global food markets, and food consumers who must rely on global markets for their substance. But other examples are relevant as well. The mother who breastfeeds her infant child operates in a decommodified environment; the mother who purchases formula from Nestle does not (Van Esterik, this volume). The forager who gathers berries, mushrooms, or other food items operates in a decommodified environment. The consumer who purchases the same goods from a local supermarket does not. It is, in other words, not the item itself that determines its relative commodified or decommodified status, but rather the process of production and exchange. The prioritization of exchange value to foster accumulation rather than use value to feed people is the central goal of the commodification of food.

Second, this chapter starts from the assumption that the food system is structurally defined by the way in which its actors (farmers, agri-firm CEOs, politicians, eaters, etc.) think about food, nutrition, and food security. Conceptualizing food security (as opposed to food sovereignty) as access to sufficient nutrition, rather than as control over food choices, places the market in the center of issues of food production and food policy, effectively commodifying those policy frameworks (Timmermann, this volume).

In this framework, it is important to focus on the commodification of food as a theoretical and material process that was accelerated by two recent developments: the expansion of intellectual property rights and the dramatic acceleration of financialization of food and agricultural markets.

As the commodification of seeds is extensively addressed in another chapter (Frison and Coolsaet in this volume), just a brief mention is included here. Plant varieties have long been subject to intellectual property protection. But the Trade Related Intellectual Property Rights (TRIPs) Agreement, concluded as part of the establishment of the World Trade Organization in 1995, dramatically expanded protections afforded. Prior to the TRIPs Agreement, plant genetic resources were generally excluded from patent protection, although other forms of intellectual property protection may have been granted at the discretion of national governments. Indeed, the rights afforded plant breeders were relatively narrow, and plant genetic resources were governed under the principle of *res communis*, as the common heritage of humanity. Individual plants could be privately owned, but the knowledge embedded in the plants' genetic code was both difficult to protect practically and philosophically less clearly private property (Zerbe, 2007, 2009). The implementation of the TRIPs Agreement mandated much stronger protections for plant varieties, often including patents on gene lines that transcended the individual plant. This undermined traditional agricultural practices like seed saving and exchange between farmers, which were particularly common in the developing world. It also clearly demarcated the distinction between seed as agricultural input and seed as commodity observed by Kloppenburg (1988).

The commodification of food should not be seen as the natural status of food itself, but as the result of a long and contested historical—and political—process. That process began with European enlightenment and the transition to capitalism, accelerated under British imperialism and the colonial project, and reached its zenith in the contemporary era with the financialization of food itself (Capra and Mattei, 2015; Moore, 2015; Vivero-Pol, 2017a).

The post-war food regime

In the waning days of World War II, a new global food regime[4] began to emerge. Structured around the global supremacy of the United States, the system privileged US production, mirroring the embedded liberalism of the Bretton Woods political and legal architecture. The previous food regime, characterized by settler colonialism and British support for global free trade in food (Friedmann, 1987; Friedmann and McMichael, 1989), began to give way to a system that favored the United States and corresponded to its increasingly central role in the global economy (Winders and Scott, 2009; Winders, 2009).

Prior to that period, the global–colonial food system was marked by a preference for free trade and a rejection of mercantilist policies, a preference signaled most dramatically by the British repeal of the Corn Laws in 1849. The Corn Laws had protected British farmers by imposing high tariffs on imported grains, protecting both land owners and small farmers from competition with lower priced producers in the United States, Canada, and elsewhere. The repeal of the Corn Laws in Britain marked the symbolic beginning of a first global food regime, variously described as the colonial food regime (McMichael, 2009), the settler–colonial food regime (Friedman, 2005), or the British food regime (Winders, 2016). That regime, lasting from roughly 1860 until the outbreak of World War I, was built on a foundation of cheap, commodified food imports to Europe (primarily wheat and meat imports from the settler colonies in the Americas and Australasia) and of specialized tropical export crops from colonies in Asia, Africa, and parts of Latin America. Philosophically, the colonial food regime was characterized on the one hand by a broadly free-market–based approach to food trade, including low trade barriers and tariffs, and on the other hand by a general flow of trade from the settler colonies in the United States, Canada, Australia, and elsewhere to Great Britain and the rest of Europe. Such trade was essential to the ongoing industrialization of Great Britain, and indeed was central to British hegemony during that period (Friedmann, 1987: 129), while it simultaneously provoked a demand for greater social protections and economic nationalism on the part of both British workers and farmers.

Seen from this perspective, the settler–colonial food regime must be understood as a system that restructured food production, beginning the transition from systems of production for local consumption to production for global consumption. In doing so, it began to undermine local food systems, redirecting food production to global markets, thereby beginning the process of dismantling traditional, community-based networks of reciprocity and social obligation in favor of a system of food production maximizing value in exchange.

Of course, global trade in agricultural goods predated the establishment of the first food regime. Trade in spices began long before the European colonial project, and control of those commercial routes—and of access to other tropical crops—represented a driving force for early colonial endeavors. But as Wood (1998: 17) observes,

> nowhere, neither in the great trading centers of Europe nor in the vast commercial networks of the Islamic world or Asia, was economic activity, and production in particular, driven by the imperatives of competition and accumulation. The dominant principle of international trade everywhere was "profit on alienation."

The first food regime differed from earlier systems of food trade in two important ways. First, it marked the imposition of capitalist imperatives of price competition, efficiency, and, perhaps most importantly, market dependence into the realm of food. In this sense, it is not the absence of the state but rather the central organizing role of the market that defined the era.[5] Second, and relatedly, it saw the development of "the first price-governed international market in an essential means of life," namely, food (Friedmann, 2004: 125). It was, in other words, an early but important example of the commodification of food, driven by the state in pursuit of particular policy objectives.

The British food regime finally collapsed with the outbreak of World War I and was replaced after World War II by a new food regime rooted in the post-war Keynesian compromise and reflecting the global influence of the United States. Under the new food regime, variously described as the surplus food regime (Friedmann, 1993), the mercantile–industrial food regime (Friedmann, 2004), the US food regime (Winders, 2016), or the US-centered intensive food regime (McMichael, 2013), extensive state intervention in agricultural production was permitted at the domestic level, while at the global level, free trade was encouraged. Food surpluses—where they developed in the global north (and in particular in the United States)—were leveraged as foreign aid in support of American Cold War objectives. Cheap American commodity grain became the de facto basis of international agricultural trade, with subsidized American corn and wheat—often sold at prices below the cost of production thanks to generous support by the US government—displaced locally produced food, disrupting traditional diets and displacing local food systems.

The clearest expression of this new American food regime could be found in the General Agreement on Trade and Tariffs (GATT). The rules of the GATT system must be understood in their specific historical context. The agreement was negotiated in the immediate aftermath of World War II by 23 colonial powers and their former dependencies, notably excluding nearly all of Latin America, Asia, and Africa.[6] The creation of the GATT followed a rejection by the United States and the United Kingdom of both the proposal for an International Trade Organization and a World Food Board, an institutional framework intended to stabilize food commodity prices at a global level, similar to more narrowly focused international commodity agreements for coffee and a handful of other crops. Such an agreement, if implemented at a global level, likely would have set quantitative restrictions on exports and imports while simultaneously limiting price fluctuations for crops governed by the agreement.

The rejection of the World Food Board and the subsequent implementation of the GATT in 1947 moved to free international trade by implementing requirements for non-discrimination and national treatment of import and export.[7] As a result, the tariffs imposed on manufactured goods fell sharply under the regime. At the same time, GATT enabled the Keynesian compromise by permitting domestic protections for workers and citizens, shielding them from the most dramatic impacts of the free market (Przeworski, 1985).

The GATT reflected the specific foreign policy interests of the world's dominant power, the United States. By exempting agriculture from GATT's free trade provisions, the agreement permitted the United States (and soon thereafter Europe) to maintain extensive market interventions. In agriculture, domestic price supports, production controls, export subsidies, and other mechanisms to shield domestic food producers from global competition were permitted. Such market interventions at the domestic level provided strong protection for farmers ensured continued food surpluses in the United States. There, government programs helped maintain strong production, providing stable prices and managing surpluses by purchasing commodities from American farmers at guaranteed prices, thereby protecting American farmers and consumers from the worst effects of commodification.

GATT's permissive rules on agricultural support also permitted surplus disposal through donation or sale of agricultural commodities at subsidized prices internationally. Indeed, according to Friedmann (2017: 121), "This became a generally organizing principle of the regime that arouse implicitly as the US entered into trade and investment relations first with the devastated economies of Europe and Japan through Marshall Plan Assistance, and then with the colonies gaining independence during the next two decades" through food aid programming and concessionary and non-concessionary sales.

As US allies in Western Europe recovered from the devastation of World War II, the need for American grain and economic assistance under the Marshall Plan waned. Still facing massive grain stockpiles domestically, the US government turned to food aid for the developing world as a disposal mechanism for surplus American production. As Winders (2009: 339) observes, food aid permitted the United States

> to dispose of surplus agricultural commodities that accumulated through price support programs and [CCC] purchases. Thus the US could continue supply management policy while averting crises of excess surplus stocks … In contrast to the Marshall Plan [for Europe], PL 480 [food aid for the developing world] did not emphasize building up agriculture in other nations, but instead centered almost entirely on supplying cheap agricultural imports. The consequence was essentially to destroy much local agriculture in the periphery because local producers could not compete with imports subsisted by the US. The result was the elimination of competition for many agricultural world markets.

While the British food regime saw grain flows from the relatively less industrialized settler colonies in North America and Australasia to Europe, the American food regime saw grain flow from the now relatively developed economies of North America and Australasia to the relatively less developed economies of postcolonial Asia and Africa. Such trade, as Winders (2009) notes, served to undermine prospects for rural development in the global south.

By the late 1970s, the second food regime faced a number of challenges. The surplus production in the United States—and its disposal as food aid in the global south—began to break down in the early 1970s, as the cost of supply management programs became politically less palatable, and growing competition from producers in Europe, Canada, Australia, and elsewhere made disposing of surpluses more difficult. Declining American grain stockpiles also undermined the need to find places to dispose of surpluses. Domestically, competing interests within the agricultural sector in the United States, most notably between wheat producers who favored continue state support and corn producers who benefited less from state protections and favored liberalization of agricultural markets, also emerged.[8]

About the same time, a series of economic crises (a combination of high rates of unemployment and inflation, dubbed stagflation) and fiscal crises (signaled most dramatically by collapse of tax revenues and sharp increases in social expenditures necessitating emergency relief from the International Monetary Fund) rocked the global north. Consequently, the social protections that defined embedded liberalism increasingly began to give way to a global system of neoliberalism premised on the weakening and ultimate dismantling of the Keynesian welfare state (Ruggie, 1982; Harvey, 1995).

In the agricultural sector, the neoliberal policies being enacted similarly focused on reducing state support, particularly in the global south. Systems of government support that had been implemented in the agricultural sector all over the world, and which ranged from direct producer payments and price supports to extension services and agricultural research—often constructed using state funds and international loans—began to be dismantled. The International Monetary Fund, originally created to address market failures, including the failure of private firms to invest in research, development, and production of public goods, rolled back its support for such initiatives, and austerity was the rule of the day. The state was recast as the source of economic inefficiency, and the understanding of the need for and nature of state intervention to ensure the provision of public goods was dramatically curtailed. Plant breeding initiatives across the global south were shuttered. Support to farmers was restructured, and particularly in the developing world, subsidies and state support gave way to the imperatives of free market competition. Public nutrition programs were sharply curtailed, and food security was conceptualized primarily in terms of market access. Food, in short, was increasingly conceptualized in purer commodity form, stripped of its social and cultural meaning.

Perhaps nowhere is this clearer than in the Agreement on Agriculture (AoA). Concluded in 1994 as part of the Uruguay Round of GATT negotiations that resulted in the creation of the World Trade Organization (WTO), the Agreement on Agriculture reflected a move away from the second food regime. As described by Clapp (2017: 107), the Agreement "was a curious mix that demanded market-opening measures in the developing countries while at the same time allowing industrialized countries to continue to protect their own markets through a range of domestic support measures."

This was not—at least according to its proponents—how the Agreement on Agriculture was intended to come out. Indeed, the AoA was supposed to liberalize agricultural trade by reducing subsidies and other trade barriers in the global north while simultaneously expanding access to agricultural commodities around the world. The outcome, however, was very different. US and European Union protections were shifted into permitted categories, while market-intervention programs in the global south were subject to increase scrutiny.

Efforts to rekindle WTO talks following the failed Seattle Round in 1999 have largely stalled, with Ministerial Conferences in Doha (2001), Cancun (2003), Hong Kong (2005), Geneva (2009 and 2011), Bali (2013), Nairobi (2015), and Buenos Ares (2017) all failing to produce agreements. Disputes over agricultural trade have played a central role in the breakdown of the talks, with disagreements centering on two main tensions. On the one hand, the Cairns Group (a group of 20 agricultural exporting countries) has continued to pressure the United States, the European Union, and Japan to reduce their agricultural subsidies and liberalize trade. On the other hand, the G33 (a group of developing countries) has demanded the right to increase protections and market interventions in developing countries, particularly for the goal of maintaining food security, citing the principle of special and differential treatment.[9]

Towards a neoliberal food regime

What emerged in place of the second food regime was a system of food production that returned in some ways to the British food regime of the late nineteenth century, albeit in a more intensified and expansive form. The breakdown of both the Fordist model and the Keynesian compromise of the post-war era meant that programs intended to protect workers, farmers, and citizens from the worst excesses of the private paradigm were dismantled in favor of a neoliberal ideology of free market supremacism. The role of the state, particularly in the developing world, was sharply curtailed. Under the premise of structural adjustment, farmer support and food security programs in the global south were sharply curtailed. Extension services were reduced or eliminated, public seed breeding and distribution programs were privatized, and grain marketing boards were dismantled—all with devastating effects on smallholders. Ironically, many of these very programs had been built using loans from the World Bank and with technical assistance from the United States and other developed countries.

Meanwhile, in parts of the global north (most notably Australia, New Zealand, and Canada), farmer-support programs, which had been part of the implicit political compromise between labor and capital and were central to the development project more broadly under the Keynesian compromise, were reduced or eliminated. Price supports, direct producer payments, extension services to farmers—as well as labor protections more broadly—were dismantled.

On the contrary, the United States and the European Union continued to maintain farm support programs, albeit in shifting forms. In the United States, the power of labor unions was dramatically curtailed. Membership in the United Farm Workers, which had been a central site for political and economic struggle for farm laborer, declined.

In place of the Keynesian compromise, neoliberalism took hold.[10] This system signaled a sharp increase in the power of financial capital over public policy around the world. Philosophically, neoliberalism advanced the proposition that human well-being is best secured through liberation of individual entrepreneurialism and skills in a political and social environment defined by protection of strong private property rights, free markets, and free trade. The role of the state was not to interfere in market outcomes but rather to establish and preserve a framework appropriate for market determination of social outcomes (Harvey, 2006). Efficient outcomes determined through market relations thus became the primary (and in some cases, sole) determinant of value and worth.

The twin processes of globalization and neoliberalization also played out in the agricultural sector. Agriculture, of course, has long been global, with germplasm flowing from farmer to farmer and spices and other food items being traded. Indeed, the spread of genetic resources and plant cultivars from one region to another—most notably as part of the European colonial project after 1492, but more broadly as well—created the foundation for farming itself (McCann, 2005; Crosby, 2003). However, such trade was largely limited to the exchange of plant genetic material (via seed or plant clippings) or to high-value, low volume commodities that did not easily spoil. The development of new technologies, including mass transportation and refrigerated shipping, during the first food regime permitted the lengthening of food chains, reducing seasonality in food consumption (Hobsbawm, 1975), encouraging the massive growth of cities, and providing the foundation for industrialization. Such social, political, economic, and technological changes also provided the foundation for the commodification of food itself.

Such technologies also encouraged the graduation concentration of food processing. The late nineteenth and early twentieth century witnessed both the birth of the feedlot and the slaughterhouse (Schlosser, 2001) and the expansion of canning (Marx de Salcedo, 2015). Sites of food production and consumption were gradually separated and distanced from one another (Goodman at al., 1987). In the global north, the nature of farming itself began to change, as small farmers producing primarily for local consumption gave way to large, industrialized farming operations producing commodity crops for sale.

Patterns of farming in the global south also changed. Where domestic land tenure was weak or ill-defined and integration into global commodity chains permitted, farms were similarly consolidated and production was gradually retooled for export rather than domestic consumption. Where land rights—especially those that protected communal landholding—were stronger, smallholder producers were more often able to retain their customary ownership and use rights and continue to earn a living and support their families and communities through farming.

The compression of both time and space (seasonality and distance)[11] in the food production process continued throughout the twentieth century, albeit unevenly across geographic regions.[12] This process, particularly in the global north, enabled capital to overcome the traditional limits imposed by nature on the process of appropriation and accumulation in the farming sector, especially seasonality. Specific foods could be made available to regions far away from the site of production and outside their normal harvest time. Strawberries, apples, and other seasonal produce could be made available year-round by shifting the site of production from region to region and transporting commodities to the end consumer.

The distancing of food production and consumption chains paralleled the development and introduction of new technologies in the household. Refrigerators and freezers, and later microwave ovens, effectively expanded seasonality and fueled changes in the nature of household labor, carrying particular gender dynamics. Ensuring the reproduction of the household has, with few exceptions, been seen primarily as the responsibility of women. The gendered division of labor usually reserved the "public" and paid sphere of work for men and limited women to either unpaid housework or low-wage jobs similar to those in the household, such as "laundry girls," maids, or nannies (Waring, 1999; Cohen, 2004). As the number of women in the formal workforce in the global north increased, the gendered division of household reproduction largely remain unchanged. Women continued to perform the vast majority of the labor of household reproduction (Parker and Wang, 2013). New food technologies—along with other labor-saving devices in the household, from washing machines to vacuum cleaners—were frequently presented as emancipatory, freeing women from the drudgery of housework. In reality, however, such technologies did little to alter the underlying gendered division of household reproduction (Schwartz-Cohen, 1983).

At the same time, some reproductive work was moved outside the household and commodified in particular ways. As Marx de Salcedo (2012: 107) observes,

> even in the most technologically developed countries, housework has not been significantly reduced. Instead, it has been marketized, redistributed, mostly on the shoulders of immigrant women from the South and former socialist countries [...] As women's increased participation in waged work has immensely increased, especially in the North, large quotas of housework have been taken out of the home and reorganized on the market basis through the virtual boom of the service industry, which now constitutes the dominant economic sector from the viewpoint of wage employment. This means that more meals are eaten out of the home, more clothes are washed in laundromats or by dry cleaners, and more food is bought already prepared for consumption.

In other words, as women were increasingly employed outside the household, particularly in the global north, demand for labor-saving technologies in the house increased, and patterns of food consumption gradually shifted. People (read: women) made fewer but larger shopping trips, and consumption of pre-prepared meals increased. Consumption of food prepared by others (e.g., in restaurants) became far more common. And families began to eat apart in greater numbers.

The development of more integrated markets at the national (and later the global) level led to a decline in regional foods, as regional diets yielded to more uniform tastes, signaled most clearly by the displacement of local restaurants by regional, national, and increasingly international franchise chains offering uniform food around the world (Matthews, 2000; Ritzer, 2008). Food was disembedded from locality, resulting in a host of perverse outcomes from ultra-processed food devoid of cultural significance and greater pollution and environmental destruction from excess packaging, long transportation distances, and farming methods (like Confined Animal Feeding Operations, CAFOs) that externalized environmental consequences of production.

More recently, global food production has been marked by both differentiation and integration (Goodman at al., 1987; Watts, 1994). Agricultural input production and food processing and distribution networks are increasingly concentrated among a smaller number of transnational firms. Oligopolies developed both within the United States and globally, with a small number of firms controlling large portions of production in meat processing, soybean crushing, coffee, and countless other food processing areas, as well as in agrochemical and seed markets (IPES-Food, 2017; FAO, 2015; Saitone and Sexton, 2017; Hendrickson and Heffernan, 2007).

The result of such concentration in agricultural production chains means that transnational markets are often dominated by oligopolies, including when it comes to the production of agricultural inputs (such as seed, chemical inputs, and farm machinery), and the handling of outputs by grain processors and traders. Under such a system, traditional assumptions about the functioning of free markets break down. Indeed, as noted by the UN Food and Agriculture Organization (2015: 9),

> the organization of the modern food system into complex [Global Value Chains] raises questions about the assumption of competitive markets. In the agriculture sector, there is a high degree of concentration among firms both within countries and internationally, pointing to a lack of competition.

In this context, farmers have relatively fewer options to source agricultural inputs or sell their outputs. They became price takers.

In the process, food was increasingly constructed socially in a purer commodity form, devoid of its broader social, cultural, and humanitarian content and meaning. The production and trade in food was driven solely by concerns of profitability and accumulation rather than the needs of people and communities for nutrition and identity. Conceptualizing food as a commodity necessarily places the market in the central, mediating role for ensuring food security. Access to the global market becomes synonymous with ensuring food security. Indeed, as Vivero-Pol (2017b: 31) observes,

> this valuation of food [as a commodity] conditions the set of policies, economic mechanisms and legal frameworks that are put in place, privileging those that are aligned with the commoditized valuation and discarding or downsizing those that support other narratives of food.

Parallel to the increasing concentration taking place in both input and processing markets, the global food system has also witnessed a dramatic expansion in production and consumption of specialty (niche) products. Paralleling the broader trend towards flexible (just-in-time) production under Post-Fordism, agricultural commodities have become increasingly specialized and bulk commodities have become increasingly segregated. To take but one example: coffee had historically been divided into two broad categories in global markets (Arabica and Robusta), but it began to be differentiated and marketed along multiple, specialty lines, differentiated not just by variety but by quality, geographic origin, and specific conditions of production (e.g., organic, shade grown, bird friendly, fair trade, etc.).[13]

At the same time, new food items were marketed for sale in the developed world, and in many cases consumer identity became tied to the consumption of alternative foods. New markets were born, as foods traditionally not consumed in the West, like quinoa, amaranth, and millets, became increasingly popular, often to the detriment of indigenous communities for which those were staple crops (Romero and Shahriari, 2011).

The fact that food was considered and treated as any other global commodity facilitated the transformation of tastes, the ultra-processing of food products, the expansion of global production and consumption patterns, and the obliteration of seasonality, which, on their part, favored the commodification of social relations across larger and larger portions of the globe (Kloppenburg, 1988; Mann, 1990). The changes in agricultural production and consumption that emerged as part of the neoliberal food regime resulted in the transformation of local systems. Farmers who had traditionally produced for local (primarily household) consumption now produced for consumers in distant markets. Traditional crops grown in developing countries gave way to crops more suited to the tastes of consumers in the developed world. And, perhaps most importantly, the nature of food itself was transformed. Relations between producer and consumer came to be mediated solely by the price mechanism of the free market. The value of food became reduced to its market price.

The financialization of food markets

Just as important, the dismantling of regulation separating financial speculation from more traditional use of food commodity futures for price stability and risk mitigation following the passage of the Commodity Futures Modernization Act in 2000 has been associated with the "financialization of the food system" (Clapp, 2014, 2016). Indeed, this process can be seen as the ultimate form of commodification of food, with food transformed into a financial instrument completely devoid of any social, cultural, or political context or meaning.

The Commodity Futures Modernization Act overturned strict rules which limited financial speculation in agricultural futures markets that had been in place since the Great Depression. These rules were intended to ensure that futures markets were used only by those with a direct interest in farming and food production in order to create greater stability prices and ensure liquidity. The relaxing of such restrictions opened the door for billions of dollars to flow into the agricultural commodity markets, as hedge funds, pension funds, and investment banks looked for new financial instruments to maintain ever greater levels of profitability. Food became just another commodity to be bought and sold in order to generate returns to capital, independent of individual or community needs for nutrition and sustenance. Food futures also represented a vehicle to protect against inflation or declines in the value of the US dollar.

The massive infusion of capital into agricultural commodity markets had a number of consequences, including both sharp increases and dramatic fluctuations in food prices. As De Schutter (cited in Livingston, 2012) observed,

> What we are seeing now is that these financial markets have developed massively with the arrival of these new financial investors, who are purely interested in the short-term monetary gain and are not really interested in the physical thing – they never actually buy the ton of wheat or maize; they only buy a promise to buy or to sell. The result of this financialisation of the commodities market is that the prices of the products respond increasingly to a purely speculative logic. This explains why in very short periods of time we see prices spiking or bubbles exploding, because prices are less and less determined by the real match between supply and demand.

The capital flows are massive. Between 2007 and 2011, food commodity investment by banks and hedge funds alone increased from $65 billion to $126 billion annually (Worthy, 2011: 13). According to at least one source, purely financial players—those with no direct connection to the production, processing, distribution, marketing, or sale of food or agricultural commodities—account for the majority of investment in wheat futures (Livingston, 2012), while another source notes more generally that "Speculators [in 2008] have about 70 percent of the open interests in commodity markets. Ten years ago, they controlled roughly 30 percent of the market" (Magdoff, 2012). Total investment in food commodity futures peaked in April 2011, at a total value of $448 billion (Clapp, 2016: 151). By 2013, facing strong opposition from movements in civil society linking financial speculation in food commodity futures with global hunger, several high-profile banks announced they would no longer invest in such instruments. By 2015, total investment in food commodity derivatives stood at $276 billion, with an additional $156 billion held in indexed funds (Clapp, 2016: 131).

The massive inflow (and outflow) of speculative investment in agriculture has dramatically affected futures markets, leading to significantly greater price volatility[14] (Russi, 2013; McMahon, 2014). The dismantling of international commodity agreements, many of which were created in the aftermath of World War II to foster price stability, no doubt contributed as well.[15] But, following the liberalization of rules limiting speculative investment in agriculture and the elimination of commodity agreements, price volatility increased dramatically. Swings in corn prices, for example, increased from an average year-on-year fluctuation of 6.7 percent between 1960 and 1970 to an average of 19.6 percent per year between 1996 and 2010. Other commodities exhibited a similar pattern, with fluctuations in soybean prices increasing from 7.5 to 15.5 percent per year, wheat prices from 5.5 to 17.9 percent per year, Arabica coffee from 8.6 to 21.4 percent per year, and Robusta coffee increasing from 9.2 to 22.9 percent per year over the same period. Only cocoa seemed to buck this trend, remaining relatively volatile through the entire period (Winders et al., 2016: 82).

As a result, the traditional purpose of futures markets—to protect against dramatic fluctuations in prices and to ensure market liquidity—has given way to pure speculation completely disconnected from production. Food came to be seen only as a vehicle for profit and capital accumulation, its social and cultural context obliterated, and its necessity for sustaining human life ignored. In the process, consumers, particularly poor consumers in the global south, were driven out of the market by sharply higher prices, and farmers became increasingly subject to dramatic price fluctuations.

Conclusion

For Polanyi, the market society represented a theoretical ideal type. In reality, markets are never fully embedded or disembedded from broader social relations, and commodification is always an incomplete process (Radin, 1996). So, food products can work as commodities under specific circumstances and as commons under different ones. It is the relative degree of embeddedness or disembeddedness (or commodification and decommodification) that is critical. Thus, the defining distinction between market and non-market societies is the relative degree to which the economy is subject to limits and control by non-market factors. Such factors may range from community-held rights (for example, over land use or fishing resources) to regulatory limits on the idiosyncrasy of property or the accumulation of capital. Over the past 200 years, food has increasingly been subject to the logic of the market, represented in its purest commodity form. The gradual shift in global food regimes, from the open British regime to the protections afforded by the post-war American food regime, have increasingly given way to a global food system characterized by strong and absolute private property rights, high levels of financial speculation, and global networks of production and consumption with longer food chains that enable middlemen to accumulate more profit.

In the process, food itself has been redefined, treated as any other commodity, something to be bought and sold, not as a fundamental necessity for human survival or a natural product that could benefit all of us. The production and exchange of food, as with any other commodity, came to be driven by imperatives of profit and accumulation. Access to food increasingly was defined in terms of access to the global market, and those without the ability to purchase food were condemned to marginalization and starvation. The environmental, legal, and social consequences of the commodification have been largely ignored up to recent times. The broader cultural and social meanings of food, explored elsewhere in this volume, are obviated by such treatment. The result of the commodified food system is a perverse combination of obesity and malnutrition. Even setting aside the negative environmental externalities associated with industrialized agricultural production for just a moment, the current global food system is one that simultaneously produces millions of tons of food that waste and rot every year while people around the world go hungry due to their inability to afford the same food. Hunger, in other words, is the primary output of the global, commodified food system.

Notes

1 According to Herrero et al. (2017), small farms (defined as those no larger than 20 hectares) account for more than three-quarters of food production in Sub-Saharan Africa, South and Southeast Asia, and China. Very small farms (defined as those no larger than two hectares), produce about 30 percent of food production in Sub-Saharan Africa and South and Southeast Asia.

2 Perhaps the most well-known example of this is the situation faced by Mexican maize growers following the implementation of the North American Free Trade Agreement (NAFTA). See Nadal and Wise (2004) for a more detailed discussion.
3 Appadurai defines a commodity as "any thing intended for exchange" (2005: 35) or as an "object of economic value" (1986: 3). According to Radin (1996), commodification involves interconnected processes: objectification (alienation from personhood or community), fungibility (exchangeability and interchangeability), commersability (comparable in relative terms to other commodities), and monetization.
4 The concept of food regimes was first articulated by Friedmann and McMichael (1989) to highlight the historical and global evolution of the world food system in the context of broader political and economic changes. According to Friedmann (2004: 125), a food regime can be understood as "a relatively bounded historical period in which complementary expectations govern the behavior of all social actions, such as farmers, firms, and workers engaged in all aspects of food growing, manufacturing, distribution and sales, as well as governmental agencies, citizens, and consumers."
5 Indeed, in this respect, it is important to remember that "free trade" and other "market-based" systems of exchange are heavily dependent on extensive state intervention (cf. Wood, 2002a; Wood, 2003; Tabb, 2004).
6 The only founding members of GATT from the global south were: Brazil, Burma (Myanmar), Ceylon (Sri Lanka), Chile, China, Cuba, India, Lebanon, Pakistan, Southern Rhodesia (Zimbabwe), Syria, and South Africa. Of these, both Southern Rhodesia and South Africa were still governed by white settler governments.
7 The twin policies of national treatment and non-discrimination represent the core principles of both the GATT treaty and, later, the World Trade Organization (WTO). At the most basic level, national treatment, provided for under Article III of the GATT treaty, requires the treatment of goods produced in other signatories to the GATT in the same way as goods manufactured domestically. As outlined in Article I of the GATT treaty, non-discrimination (referred to as "General Most Favored Nation Treatment" in the document) requires the same tariff levels and non-tariff barriers be applied to all signatories of the GATT, effectively prohibiting preferential treatment or market access to a subset of GATT members. Together, national treatment and non-discrimination are the foundation of free trade in the current global economy.
8 Winders (2009) and Winders and Scott (2009) argue the competing (and conflicting) interests of Southern (cotton), Corn Belt, and Wheat Belt producers help to explain the development and evolution of both US farm and food aid policies from the Civil War onward. Specifically, they contend that a coalition of cotton and wheat farmers suffered from chronic surpluses and were more vocal and successful in their demands for government-backed supply management programs than corn farmers who, through the US Farm Bureau, called for more market-oriented agricultural policies. Importantly, it should also be noted that US corn producers benefited from greater demand for their crop, initially as livestock feed, then from high fructose corn syrup, and later from corn ethanol production, while simultaneously facing less international competition from foreign producers than wheat and cotton producers.
9 The principle of special and differential treatment is spelled out in Article XVIII of the GATT treaty. Such treatment, under the rules of the World Trade Organization, can range from longer implementation times to exemptions from specific requirements altogether.
10 For Harvey, neoliberalism is a project for the restoration of class power (1995: 13) established through the financializing of everything and the intensification of capital accumulation through financial instruments at the expense of other factions of capital (2006: 24).
11 The compression of space (in the sense of connecting distant sites of food production and consumption) should not be read as conflicting with Clapp's (2014) notion of distancing, which she asserts is a central component of the financialization of food, as discussed below. Indeed, in a sense, the compression of space is directly connected to the idea of distancing.
12 Time–space compression, driven by technological innovation and the pressure to intensify capital accumulation, was described by Harvey (1989) as the defining feature of postmodernity.
13 The exact size of the specialty coffee market is hard to determine for two reasons. First, the definition of "specialty coffee" (and associated terms like "bird friendly") is contested. Second, some specialty processors engage in direct negotiations with producer groups rather than purchasing coffee on commodity exchanges. Starbucks, for example, purchases most of its coffee through private contracts rather than via commodity markets. Such mechanisms frequently exclude small farmers who are unable to

meet the stringent certification requirements, even if they are capable of producing according to those standards (Jaffee, 2007). Nevertheless, according to statistics from the International Trade Center (2011), trade in specialty coffee represented 10–15 percent of global coffee exports.

14 Using just one example, Clapp (2016) concludes that the underlying dynamics of supply and demand, combined with traditional explanations like weather, could only explain about 20 percent of the food price rise in the lead up to the 2007–08 global food crisis.

15 Examples of such agreements include the International Wheat Agreement (established in 1949), the International Coffee Agreement (established in 1962), and the International Cocoa Agreement (established in 1972). Similar agreements—some of which were perhaps less formal—were created for sugar, cotton, and wool. For a more detailed discussion, see Winders et al. (2016) and Fridell (2007).

Bibliography

Appadurai, A. (1986). "Introduction: Commodities and the Politics of Value." In A. Appadurai, ed. *The Social Life of Things: Commodities in a Cultural Perspective.* New York: Cambridge University Press.

Appadurai, A. (2005). "Definitions: Commodity and Commodification." In M. Ertman and J. Williams, eds. *Rethinking Commodification: Cases and Readings in Law and Culture.* New York: New York University Press, p. 35.

Bender, B. (1978). Gatherer–Hunter to Farmer: A Social Perspective. *Archaeology and Religion*, 10 (2), 204–22.

Capra, F. and U. Mattei (2015). *The Ecology of Law: Toward a Legal System in Tune with Nature and Community.* Oakland, CA: Berrett-Koehler.

Clapp, J. (2014). "Financialization, Distance, and Global Food Politics." *The Journal of Peasant Studies*, 41 (5), 797–814.

Clapp, J. (2016). *Food.* London: Polity.

Clapp, J. (2017). "Food Security and Contested Agricultural Trade Norms." *Journal of International Law and International Relations*, 11 (2): 104–15.

Coffee Research Institute (2006). *Coffee Trade: New York Coffee Exchange.* Retrieved March 1, 2010, from http://www.coffeeresearch.org/market/coffeemarket.htm.

Cohen, P. (2004). "The Gendered Division of Labor: 'Keeping House' and Occupational Segregation in the United States." *Gender and Society*, 18 (2), 239–52.

Crosby, A. (1972 [2003]). *The Columbian Exchange: Biological and Cultural Consequences of 1492.* Westport: Praeger.

De Schutter, O. (2014). *The Transformative Potential of the Right to Food.* Report submitted to the United Nations Human Rights Council, 24 January, UN Doc. A/HRC/25/57 http://www.srfood.org/images/stories/pdf/officialreports/20140310_finalreport_en.pdf (accessed April 7, 2018).

Food and Agriculture Organization (2015). *The State of Agricultural Commodity Markets 2015-16.* Retrieved June 17, 2017, from http://www.fao.org/3/a-i5090e.pdf.

Fridell, G. (2007). *Fair Trade Coffee: The Prospects and Pitfalls of Market-Driven Social Justice.* Toronto: University of Toronto Press.

Fridell, G. (2014). *Coffee.* London: Wiley.

Friedmann, H. (1987). "Family Farms and International Food Regimes." In T. Shanin, ed. *Peasants and Peasant Societies* (pp. 258–76). Oxford: Basil Blackwell.

Friedmann, H. (1993). "The Political Economy of Food: A Global Crisis." *New Left Review*, (197), 29–57.

Friedmann, H. (2004). "Feeding the Empire: The Pathologies of Globalized Agriculture." *The Socialist Register, 2005.* New York: Monthly Review Press, pp. 124–143.

Friedmann, H. (2017). "Agriculture and the Social State: Subsidies or Commons?" *Journal of International Law and International Relations*, 11 (2), 116–130.

Friedmann, H. and McMichael, P. (1989). "Agriculture and the State System: The Rise and Decline of National Agricultures, 1870 to the Present." *Sociologia Ruralis*, 29 (2), 93–117.

Goodman, D., Sorj, B. & Wilkinson, J. (1987). *From Farming to Biotechnology: A Theory of Agro-Industrial Development.* New York: Blackwell.

Harvey, D. (1989). *The Condition of Postmodernity: An Enquiry into the Origins of Cultural Change.* Cambridge: Blackwell.

Harvey, D. (1995). *A Brief History of Neoliberalism.* Oxford: Oxford University Press.

Harvey, D. (2006). *Spaces of Global Capitalism: Towards a Theory of Uneven Geographic Development.* London: Verso.

Hendrickson, M. & Heffernan, W. (2007). *Concentration of Agricultural Markets*. Retrieved March 1, 2010, from National Farmers' Union: http://www.nfu.org/wp-content/2007-heffernanreport.pdf.
Herrero, M., P.K. Thornton, B. Powers, J.R. Bogard, R. Remains, S. Fritz, J.S. Gerber, G. Nelson, L. See, K. Waha, R.A. Watson, P.C. West, L.H. Samberg, J. Steeg, E. Stephenson, M. van Wijk, and P. Havlik (2017). "Farming and the Geography of Nutrient Production for Human Use: A Transdisciplinary Analysis." *The Lancet Planetary Health*, 1 (1), e33–e42.
Hobsbawm, E. (1975). *The Age of Capital, 1848–1875*. London: Weidenfeld and Nicolson.
International Panel of Exports on Sustainable Food Systems (IPES-Food), (2017). *Too Big to Feed: Exploring the Impacts of Mega-Mergers, Consolidation, and Concentration of Power in the Agri-Food Sector*. Retrieved 28 February 2018 from www.ipes-food.org.
International Trade Center (ITA) (2011). *The Coffee Exporters Guide*, 3rd edition. Geneva: ITA. Retrieved 28 February 2018 from http://www.intracen.org/.
Jaffee, D. (2007). *Brewing Justice: Fair Trade Coffee, Sustainability, and Survival*. Berkeley: University of California Press.
Kloppenburg, J. (1988). *First the Seed: The Political Economy of Plant Biotechnology, 1492–2000*. Madison, WI: University of Wisconsin Press.
Latham, K. (2013). "Human Health and the Neolithic Revolution: An Overview of Impacts of the Agricultural Transition on Oral Health, Epidemiology, and the Human Body." *Nebraska Anthropologist*, 28, 95–102.
Livingston, A. (2012). "The Real Hunger Games: How Banks Gamble on Food Prices and the Poor Lose Out." *The Independent*. April 1.
Magdoff, F. (2012). "Food as Commodity." *Monthly Review*, 63 (8), 15–22.
Mann, S. (1990). *Agrarian Capitalism in Theory and Practice*. Chapel Hill, NC: University of North Carolina Press.
Marx de Salcedo, F. (2012). *Revolution at Point Zero: Housework, Reproduction, and Feminist Struggle*. Oakland: PM Press.
Marx de Salcedo, F. (2015). *Combat-Ready Kitchen: How the US Military Shapes the Way You Eat*. New York: Current.
Mathews, G. (2000). *Global Culture/Individual Identity: Searching for Home in the Global Supermarket*. Oxford: Taylor and Francis.
McCann, J. (2005). *Maize and Grace: Africa's Encounter with a New World Crop, 1500–2000*. Cambridge, MA: Harvard University Press.
McMahon, P. (2014). *Feeding Frenzy: Land Grabs, Price Spikes, and the World Food Crisis*. London: Profile.
McMichael, P. (2009). "A Food Regime Genealogy." *Journal of Peasant Studies*, 36 (1), 139–69.
McMichael, P. (2013). *Food Regimes and Agrarian Questions*. Halifax: Fernwood.
Mintz, S. (1996). *Tasting Food, Tasting Freedom: Excursions into Eating, Power, and the Past*. Boston, MA: Beacon Press.
Moore, J (2015). *Capitalism and the Web of Life: Ecology and the Accumulation of Capital*. London: Verso.
Nadal, A. and T. Wise (2004). "The Environmental Costs of Agriculture Trade Liberalization: Mexico-US Maize Trade Under NAFTA." Working Group on Development and Environment in the Americas Discussion Paper No. 4 (June). Retrieved 27 February 2018 from ase.tufts.edu/gdae.
Parker, K. and W. Wang (2013). *Modern Parenthood: Roles of Moms and Dads Converge as They Balance Work and Family*. Washington, DC: Pew Research Center. Retrieved 26 February 2018 from assets. pewresearch.org.
Patel, R. (2012). *Stuffed and Starved: The Hidden Battle for the World Food System*. Brooklyn, NY: Melville House.
Polanyi, K. (1944). *The Great Transformation: The Political and Economic Origins of Our Times*. Boston, MA: Beacon.
Przeworski, A. (1985). *Capitalism and Social Democracy*. New York: Cambridge University Press.
Radin, M.J. (1996). *Contested Commodities: The Trouble with Trade in Sex, Children, Body Parts, and Other Things*. Cambridge, MA: Harvard University Press.
Ritzer, G. (2008). *The McDonaldization of Society*. Los Angeles, CA: Pine Forge Press.
Romero, S. and Shahriari, S. (2011). Quinoa's Global Success Creates Quandary at Home. *The New York Times*. March 20, A6.
Rossi, L. (2013). *Hungry Capital: The Financialization of Food*. Alresford: Zero Books.
Ruggie, J.G. (1982). "International Regimes, Transactions, and Change: Embedded Liberalism in the Postwar Economic Order." *International Organization*, 36 (2), 379–415.

Sahlins, M. (1972). *Stone Age Economics*. Hawthorne, NY: Aldien de Gruyter.
Saitone, T.L. and R.J. Sexton (2017). "Concentration and Consolidation in the U.S. Food Supply Chain: The Latest Evidence and Implications for Consumers, Farmers and Policymakers." *Economic Review*. Retrieved 28 February 2018 from www.kansascityfed.org/.
Schlosser, E. (2001). *Fast Food Nation: The Dark Side of the All-American Meal*. New York: Houghton Mifflin.
Schwartz Cohen, R. (1983). *More Work for Mother: The Ironies of Household Technologies from the Open Hearth to the Microwave*. New York: Basic Books.
Tabb, W. (2004). *Economic Governance in an Age of Globalization*. New York: Columbia University Press.
Ulijaszek, S., G. Hillman, J.L. Boldsen and C.J. Henry (1991). "Human Dietary Change [and Discussion]." *Philosophical Transactions: Biological Sciences*, 334 (1270), 271–79.
United Nations Conference on Trade and Development (2006). *Tracking the Trend Towards Market Concentration: The Case of the Agricultural Input Industry*. New York: UNCTAD.
Vivero-Pol, J.L. (2017a). "Transition Towards a Food Commons Regime: Re-Commoning Food to Crowd-Feed the World." In G. Ruivenkamp and A. Hilton, eds. *Perspectives on Commoning: Autonomist Principles and Practices* (pp. 325–379). London: Zed.
Vivero-Pol, J.L. (2017b). *How do people value food? Systematic, heuristic and normative approaches to narratives of transition in food systems*. PhD Thesis, Université Catholique de Louvain. https://dial.uclouvain.be/pr/boreal/fr/object/boreal%3A191763
Waring, M. (1999). *Counting for Nothing: What Men Value and What Women are Worth*. Toronto: University of Toronto Press.
Watts, M. (1994). "What Difference Does Difference Make?" *Review of International Political Economy*, 1 (3), 563–70.
Winders, B. (2009). "The Vanishing Free Market: The Formation and Spread of the British and U.S. Food Regimes." *Journal of Agrarian Change*, 9 (3), 315–44.
Winders, B. and Scott, J. (2009). *The Politics of Food Supply: U.S. Agricultural Policy in the World Economy*. New Haven, CT: Yale University Press.
Winders, B., Heslin, A., Ross, G., Wekster, H., and Berry, S. (2016). "Life After the Regime: Market Instability with the Fall of the US Food Regime." *Agriculture and Human Values*, 33, 73–88.
Worthy, M. (2011). *Broken Markets: How Financial Market Regulation Can Help Prevent Another Global Food Crisis*. World Development Movement.
Wood, E.M. (1998). "The Agrarian Origins of Capitalism." *Monthly Review*, 50 (3), 14–31.
Wood, E.M. (2002). *The Origins of Capitalism: A Longer View*. London: Verso.
Zerbe, N. (2007). Contesting Privatization: NGOs and Farmers' Rights in the African Model Law. *Global Environmental Politics*, 7 (1), 97–119.
Zerbe, N. (2009). "Setting the Global Dinner Table: Exploring the Limits of the Marketization of Food Security." In J. Clapp and M Cohen, eds. *The Global Food Crisis: Governance Challenges and Opportunities* (pp. 161–78). Waterloo: Wilfred Laurier University Press.

PART III

Food-related elements considered as commons

11

TRADITIONAL AGRICULTURAL KNOWLEDGE AS A COMMONS

Victoria Reyes-García, Petra Benyei and Laura Calvet-Mir

Introduction

Traditional[1] knowledge systems consist of the information, beliefs, traditions, practices, institutions, and worldviews developed and sustained by Indigenous and rural communities. Traditional knowledge systems are often seen as an adaptive strategy to the environment in which communities live and include information, practices, and institutions related to many spheres of community life, from medicinal to astronomical or agricultural information (Berkes et al. 2000; Reyes-García et al. 2016). As an adaptive strategy, these knowledge systems provide place-based communities with information to deal with everyday life issues, while considering the long-time maintenance of the ecological system in which they are embedded. Traditional agricultural knowledge (TAK) refers to the locally adapted systems developed by farmers and other place-based communities around the world through their interaction with nature in order to support food production in a way that is environmentally and culturally adapted to the specific characteristics of each context (Malezieux 2012; Gilles et al. 2013; Vandermeer and Perfecto 2013). TAK systems encompass information about how to recognize and efficiently manage agricultural landscapes and agroecosystem elements (i.e., knowledge on storage and culinary characteristics of crop landraces) (Riu-Bosoms et al. 2014; Calvet-Mir et al. 2010).

Over the last decades, components of traditional knowledge systems have been incorporated in the development of commercial products, and thus commodified. Some well-known examples of these processes include the patenting of traditional medicinal knowledge by pharmaceutic companies (Soejarto et al. 2005; Heinrich 2015) and the appropriation of TAK with commercial purposes (Whitt 1998; Brush 2004). This highly contested move has raised concerns about the need to develop benefit-sharing agreements between knowledge holders and commercial companies (Siebenhuner et al. 2005; Engels et al. 2011), but – more importantly – it has led to a harsh debate on how to protect traditional knowledge systems from misappropriation (Shiva 1997; Macilwain 1998; Moran et al. 2001). Although at a practical level the debate seems to revolve around the relation between local people and commercial companies, at a more theoretical level the debate ultimately concerns the governance of traditional knowledge systems. Some authors have argued that the contribution of traditional knowledge holders should be protected through the application of some sort of Intellectual Property Rights (Brush 2004). Other authors, however, oppose the privatization of knowledge already in the public domain, suggesting that this knowledge should be governed as a public good (Smale et al. 2004; Shiva 2004). Departing from these two positions, in this chapter we explore the potential of a third approach to the governance of traditional knowledge: the commons framework, or the idea that traditional knowledge systems could also be peer-governed by knowledge users.

The idea that knowledge can be governed under the commons framework is not new *per se*. For several years, scholars working on the governance of digital knowledge have argued that knowledge – defined as intelligible ideas, information, and data regardless of the form in which it is expressed or obtained – should be considered under the commons framework (Boyle 2003; Hess and Ostrom 2007; Bollier and Helfrich 2014). Here we follow this line of thought and explore two examples in which TAK is conceived and managed under the commons framework. In the next section, we discuss the theoretical framework supporting the idea that knowledge can be governed as a commons, i.e., as a resource used by a group of people who have self-developed a set of rules to manage the social dilemmas derived from its collective use. We then illustrate the governance of TAK under the commons framework presenting two case studies. The first case illustrates local governance of TAK as commons by a close community with tight social bonds; the second case provides an example of how digitalized TAK could be governed by a peer-to-peer governance system. Given the contrast between the two cases, we then discuss how the commons framework could also contribute to articulate the sharing of TAK at different scales. We conclude by exploring the degree to which a commons-based governance can be considered a contestation to commodification and enclosure movements that threaten the maintenance of TAK systems, and thus people's ability to sustain environmentally and culturally adapted food systems (Boyle 2003; Lakshmi Poorna et al. 2014).

Is traditional agricultural knowledge a commons?

While the term "commons" has been alternatively used to refer to resources or goods, to a social process, or even to a worldview (Bollier and Helfrich 2014; Kostakis and Bauwens 2014), one of its most standard uses is to refer to the institutional approach that governs the production, use, management, and/or preservation of shared resources according to which people manage such resources by negotiating their own rules through social or customary traditions, norms, and practices (Ostrom 1990; Frischmann et al. 2014). As resources managed under a public approach (i.e., public goods), resources managed under the commons approach are collectively owned, but – differently than public goods – they are managed by self-organized communities for their collective benefit and not by government institutions (Quilligan 2012). A distinctive aspect of the commons approach is that it emphasizes that social dilemmas, or situations in which there is a conflict between immediate individual self-interest and long-term collective interest, can be solved through resource-use management rules self-defined by the users (Ostrom 1990; MacKinnon 2012; Siefkes 2012). Another distinctive characteristic of the commons approach is that, under this governance system, resources are managed to ensure long-term maintenance of the resource use value, for which it is important that management is oriented to prevent resource degradation (Kostakis and Bauwens 2014).

The term "knowledge commons" refers to the application of the commons approach to governing the production, use, management, and/or preservation of knowledge or information (Frischmann et al. 2014; Hess and Ostrom 2007). Within this context, we aim to explore whether TAK can be considered as a commons, in the sense of being shared and collectively governed by a group who receives non-monetary utility from its existence and reproduction. To do so, we follow Kostakis and Bauwens (2014) who proposed that one needs to examine four interlinked components to understand any commons scheme: 1) the resource; 2) the community who shares it; 3) the rules and property regimes that govern people's access to the resource; and 4) the use value created through the social reproduction or preservation of the resource. In the remainder of this section, we analyse these four components in relation to knowledge systems in general and TAK in particular, before analysing them in more detail in relation to our two case studies.

The resource: The institutional approach of the commons was originally developed in relation to material, degradable resources. This approach uses the rival nature of material goods (i.e., goods could only be possessed or consumed by a single user) to explain the emergence of users' self-defined management rules, in an attempt to avoid resource depletion. Differently to material resources, but similar to other types of socially transmitted knowledge (Boyd, Richerson, and Henrich 2011), TAK is an immaterial non-rival resource developed through social – not individual – processes, i.e., through the cumulative effort of generations of farmers experimenting, improving, and adapting crops and techniques through trial and error and on farm verification. Why should users develop management rules for a non-rival good (i.e., a good that can be used or consumed by one person without reducing the amount left for others)?

A potential answer to this question lies in the inextricable link between TAK and tangible resources (i.e., seeds, agricultural landscapes, tools and practices), that create opportunities for enclosing TAK, and the dramatic process of privatization those tangible resources (primarily seeds) have experienced over the past 100 years (Halewood 2013). Thus, as information stored in a book or a CD, TAK is often embodied in tangible resources such as seeds. Firms can protect the seeds they create either through plant breeder's rights (PBR), covered in the UPOV Conventions, or through patents as stipulated in the Paris Convention of 1883, or – since the signing of the TRIPS Agreement in 1994 – through a combination of both (Shiva 2004; Ghijsen 2009). Furthermore, they can do so without recognizing the farmers' role in the development of the original varieties (Ceccarelli et al. 2009; Kloppenburg 2010; Luby et al. 2015). While initiatives such as the convention on biological diversity (CBD) and the international treaty for plant genetic resources of food and agriculture (IT-PGRFA) aim to subject the access to germplasm and associated knowledge to particular rules for benefit sharing, the terms for access continue to be uncertain (Ghijsen 2009; Frison and Coolsaet, this volume), and often farming communities have no recognized rights on the plants they grow (Brush 2004; Thomas et al. 2011). Given the scarce recognition of TAK under the current legal framework, management rules that protect the free exchange of seeds and associated knowledge seem to be relevant. Indeed, initiatives such as the Open Source Seed Initiative (OSSI) (www.osseeds.org) seek to provide an alternative to pervasive intellectual property rights agreements that restrict freedom to use plant germplasm so to ensure that germplasm, and associated knowledge, can be freely exchanged now and into the future (see Kloppenburg 2010; Luby et al. 2015).

The community: The second component that constitutes knowledge as a commons is the community who shares it. Although, at least in theory, knowledge is a non-rival good which could be shared with and used by other people without reducing the amount left for others, in practice, knowledge is mostly shared within specific communities (see for example Salpeteur et al. 2015). Thus, TAK has typically been mostly shared within a community of users linked by tight social bonds who often share the same geographical environment. In this sense, a growing body of empirical research has been able to monitor the flow of seeds among farmers showing that the exchange of agricultural knowledge and crop propagation materials is interviewed with social aspects such as ethnicity, social differentiation, or gender relations (e.g., Labeyrie et al. 2014; Kawa et al. 2013; Reyes-García et al. 2013). Because TAK is embedded in local cultures, resources, and practices, and because – until recently – it was rarely transmitted in written form but rather stored in social memory and shared through daily practices, its spread outside the close community was rare. Geographical and social barriers (e.g., language, mores, and ethnicity) acted as boundaries delimiting the community sharing TAK, which cannot be understood, performed, or managed without considering its intrinsic social character.

The rules and property regimes: The third component relates to the rules that govern the access and management of the resource production and use within, and potentially beyond, the community. Several studies suggest that local and rural communities often manage their agrobiodiversity collectively, i.e., through collective use and knowledge exchange practices (e.g., Aw-Hassan et al. 2008; Abay et al. 2011; Labeyrie et al. 2016). In that sense, a growing body of research shows that the exchange of seeds is governed by collectively constructed rules regulating access to seed and associated TAK. For example, a recent study on sorghum seed exchange networks in Mount Kenya found a tendency to exchange seeds with members of the same residence and ethnolinguistic group, with measurable effects on crop genetic diversity across communities (Labeyrie et al. 2016). Similarly, Thomas and Caillon (2016) show that the circulation of plant materials and associated knowledge is affected by the social status of farmers, with higher-rank individuals giving more plants and therefore contributing more to the resilience of the community. These studies suggest that, although access to TAK could be theoretically unlimited, indeed social rules define such access and the way in which TAK is disseminated, enjoyed, and reproduced.

It should be noted, however, that while the social characteristics of TAK make it costly for non-community users to access the resource, in the absence of strict mechanisms allowing the community to exclude outsiders from using the resource, accessing it is not difficult. Under the commons framework, access to TAK by external actors, who appropriate, extract exchange value, and potentially exclude community members from use rights, could be considered a form of free-riding, where one reaps benefits from the commons without contributing to their maintenance. In that sense, the movement to enclose and commodify TAK through patent laws, plant variety protection laws, contractual restrictions accompanying seed sales, or bilaterally oriented access and benefit shared laws could be considered a form of free-riding from agricultural companies, as they restrict the use of TAK developed by a community in order to generate private benefits (Halewood 2013). Indeed, this type of appropriation constitutes one of the most important social dilemmas for management of TAK as commons.

The use value: The last component proposed by Kostakis and Bauwens (2014) to understand the commons scheme relates to the use value created through the social reproduction and preservation of the resource, rather than through the exchange value that is normally generated when commodities are created and transferred according to market rules. Kostakis and Bauwens (2014) argue that the governance of resources under the commons framework is not based on the production of exchange value but rather on the production of use value. There is enough evidence to maintain that TAK systems not only have an immediate use value for individual farmers, but that they are also oriented towards the production, hybridization, use, and reuse of knowledge to support environmentally sustainable and culturally adapted food systems in a way that cannot be monetized without subverting the idea of the commons itself (Malezieux 2012; Gilles et al. 2013; Vandermeer and Perfecto 2013).

In a large variety of settings, from places with no access to contemporary agricultural technologies (Brookfield et al. 2003) to rural areas of industrialized countries (Calvet-Mir et al. 2016), farmers rely on the TAK to make decisions regarding subsistence-oriented agricultural production. Maintaining this body of knowledge, however, not only provides private benefits but more generally also generates social and environmental benefits, as the maintenance of the knowledge system strengthens the collective resilience, contributes to the reproduction of the community itself (e.g., reproductive labour), and contributes to the maintenance of local agrobiodiversity (Vogl and Vogl-Lukasser 2003; Reyes-García et al. 2007; Federici 2010). Even if individual farmers can make profit from holding TAK because they can sell their produce to the market, the fact that this knowledge returns always to the community guarantees a collective benefit that exceeds that of the individual members. TAK and its shared nature and practice add the maintenance and the exchange of knowledge to the production of use value, and therefore go beyond the monetary possibilities that are conceived by the notion of market value.

The examination of the four interlinked components proposed by Kostakis and Bauwens (2014) to understand the governance of a resource under the commons framework suggests that TAK (despite being normally governed by small and closed communities) shares similarities with other knowledge commons in the sense that they can be collectively governed by a group that is constructed around and performs certain management rules and that profits from non-monetary value obtained from the existence and reproduction of TAK.

Local governance of TAK as a commons: the seed exchange network in Vall Fosca

In this section, we use information from a previous study (Calvet-Mir et al. 2010; Calvet-Mir et al. 2011; Calvet-Mir et al. 2012a; Calvet-Mir et al. 2012b) to analyse the management of TAK as a commons by a close community with tight social bonds. Specifically, we focus on landrace knowledge (resource) held by gardeners of Vall Fosca, a rural Pyrenean valley of Northeastern Spain (community), and managed through an informal social network of seeds exchange (rules and participatory property regime) promoting cultural identity and social cohesion within the community and likely enhancing agrobiodiversity conservation (use value).

Vall Fosca is a Pyrenean valley, of about 200 km^2 and 1000 inhabitants. It has traditionally been a farming area, where most households practiced stockbreeding and home gardening. Home garden products were mostly grown for household needs and normally not commercialized. As part of household activities, women customarily managed home gardens and were the main seed and TAK holders. According to our informants, before the 1970s, when accessibility to the market town improved, seed self-storage and exchange were the most common ways to procure seeds. Nowadays, as much as 80% of plants in the studied gardens have a commercial origin (Calvet-Mir et al. 2011), but since landraces seeds are not commercialized, they can only be acquired via self-storage or exchange, a practice not regulated under current Spanish legislation.

Traditional landrace knowledge includes information on the appropriated sowing, planting, and harvesting calendar, the type of manure and rotations adequate for each landrace, and instructions for storing or using the plant (Calvet-Mir et al. 2010). Information from a ten-month fieldwork in the area suggests that, typically, landraces knowledge is not transmitted in a vacuum, but in association with the landraces themselves. Moreover, the transmission of landraces and associated knowledge mostly occurs within the inhabitants of the valley who manage a home garden (n=76) and who spontaneously link to one another through the exchange of landraces seeds to form a social network of seeds exchange (*sensu* Borgatti et al. 2009), in which some gardeners living outside the research area also take part (n=35)

(Calvet-Mir et al. 2012b). The close examination of network members suggests that the community within which seeds and TAK exchange happens is restricted to people with close social (i.e., family, friends) and geographical bounds (i.e., neighbours). In other words, the community of Vall Fosca landraces users is based on informal rules of landraces seed exchange shaped by geographical and social proximity.

Regarding the rules that govern landraces use and production, we found that there were no institutionalized decision makers, although the most active gardeners within the network of exchange were also those who cultivated more landraces and had more traditional agricultural landrace knowledge (Calvet-Mir et al. 2012b). These knowledge holders were mainly women and people that had cultivated a home garden for more than 25 years (i.e., experienced gardeners). We also found that the network had a low density of exchanges and was fragmented (implying that not all the people of the network are connected via seed exchange) (Calvet-Mir et al. 2012b). For example, we observed a marked geographical distribution of the network ties, with most network members settled in the valley and clustered around five small networks, three of them corresponding to exchanges among people from the most geographically isolated villages who mainly exchange seeds among themselves. Fragmentation hampers the possibility of an individual to access all the landrace knowledge circulating in the network, while low density implies few interactions between gardeners.

Finally, landraces knowledge might have important social and environmental use values. Our findings suggest that landraces knowledge is associated with the preservation of cultural identity, as gardeners interviewed mentioned the maintenance of local traditions as an important reason to conserve landrace knowledge (Calvet-Mir et al. 2011). Landraces knowledge is also associated with the creation, maintenance, and strengthening of social relations. For example, gardeners in Vall Fosca multiply their seeds not only for their own planting needs, but also so they have enough to offer to friends and relatives. Gifts of local landraces and the sharing of knowledge about them are locally highly appreciated. It is also likely that, as in other areas of the industrial world (Vogl and Vogl-Lukasser 2003), the exchange of landraces in Vall Fosca favours overall agrobiodiversity given that, as landraces are reproduced and circulated outside the market system, they provide an alternative source of agrobiodiversity.

In sum, landraces knowledge in Vall Fosca seems to circulate through a network of users, who ensure the preservation of landraces and associated knowledge (Reyes-García et al. 2013). Such a network is – mostly – locally based and constructed around a social fabric that has created informal governance mechanisms for landraces and associated knowledge management. These management rules promote culturally adapted food systems and respect the right of farmers to produce and exchange their own seeds and knowledge.

P2P governance of digitalized TAK commons: the CONECT-e platform

When referring to traditional knowledge (and TAK specifically), we often think of closely bounded communities living in rural areas, such as gardeners in Vall Fosca. However, the technological revolution opens the possibility of making TAK accessible to all kinds of users around the globe. Indeed, there have already been some attempts to digitalize TAK, mostly with the objectives either to preserve it in databases or to promote its use in alternative food production systems (Lakshmi Poorna et al. 2014; Cox 2015). The digitalization of the knowledge commons has implications regarding its governance (Boyle 2003). In this section, we explore one initiative aiming to digitalize TAK in order to prevent both its erosion and its enclosure. As in the previous section, the resource is landrace knowledge; however, in this case it is shared via a user-based internet platform (the community) and managed through P2P governance mechanisms (rules).

CONECT-e (www.conecte.es) is a Wikipedia-like citizen science initiative in Spain aiming to gather and share traditional ecological knowledge. According to its founders, this project was born out of two coexisting needs: 1) to encourage citizen's contribution to the Spanish Inventory of Traditional Ecological Knowledge (a static bibliographic compilation of ethnobotanical referenced knowledge; Pardo-de-Santayana et al. 2014) and 2) to create a dynamic inventory of traditional ecological knowledge associated with them in order to contest misappropriation/enclosure issues (Calvet-Mir et al. 2018). Thus, the platform gathers not only TAK (knowledge on local landraces' names, uses, and management, among other information) but also traditional knowledge regarding wild plants, animals and ecosystems, and soon on animals and climate change indicators. The structure of the digital platform regarding landraces knowledge has been designed following both scientific and civil society inputs (with various seed exchange network NGOs closely collaborating with the project) (Calvet-Mir et al. 2018).

Despite the fact that the digitalization of knowledge commons loosens the boundaries of excludability and the idea of a community linked by close social ties (such as in Vall Fosca), there is little doubt that there still exists a community of users in CONECT-e. In this case, the community is formed by the registered users that contribute with their TAK to the platform. Although the project is financed by the Spanish government, information displayed is accessible to the whole online community, and thus it does not belong to the state. Moreover, the platform operates outside market rules of use and distribution. However, as in other cases of digital commons, such as Linux or Wikimedia Foundation, there is a core community formed by those who have created the tool and who make decisions on the governance of this digitalized common.

Most of the decisions taken by the platform management group support the implementation of typical P2P governance mechanism: promoting equipotentiality, heterarchy, holoptism, openness, networking, and transparency (Kostakis 2010). In CONECT-e, membership is open and widespread, encouraging the free collaboration of all users (potentially anyone with an internet connection and TAK), thus supporting openness and equipotentiality. Although users are geographically dispersed, some of them are organized in physical networks (e.g., local seed networks, students from agrarian schools). Moreover, features such as "following" or "commenting" promote the creation of a digital social network that can potentially be stronger than the broader community. The platform is also characterized by a high degree of transparency, with users being able to track all editions and comments, as well as to get a horizontal view of all the components of the project (holoptism). Thus, CONECT-e does not function hierarchically but rather allows diverse teams or individuals to participate in the platform at the same time and in different directions (heterarchy).

Thanks to its structure, CONECT-e contributes to the expansion of the benefits and beneficiaries (in terms of use value) of traditional landrace knowledge by harnessing knowledge from its close and territorial boundaries and making it available to a geographically dispersed extended community. It also contributes to preserving TAK as a commons at the same time that it offers the opportunity of cross-pollination of information, trials, and experimental learning (Reyes-García et al. 2018).

As the project was launched at the time of writing (March 2017), it is difficult to foresee which social dilemmas may be associated with its functioning and expansion (but see some first results in Calvet-Mir et al. 2018). However, drawing on previous examples, such as Wikipedia, we can foresee some social dilemmas related to the implementation of the P2P governance in CONECT-e. Firstly, CONECT-e has "contributors" and "editors", who acquire this status through meritocracy (either based on academic or contribution records). This distinction could lead to the emergence of benevolent dictatorships or to conflicts between inclusionists and deletionists (Kostakis 2010). For example, under current rules, editors can decide to delete a specific TAK contribution made by a user, leading to possible biases and bottlenecks regarding the expansion of the platform's content.

Another potential social dilemma associated to CONECT-e relates to the idea that low excludability (and thus potential free-riding) may lead to a lack of incentives to contribute new knowledge (Boyle 2003). In that sense, the lack of economic incentives and the gap between knowledge holders (mainly elders) and the digital environment in which CONECT-e is set might result in a lack of participants sharing knowledge in CONECT-e. Although CONECT-e only gathers previously generated knowledge (and does not necessarily require creative or innovative efforts), one could argue that, as happens with any classic public good that can be copied freely without paying the creator (free-riding), it needs economic incentive mechanisms for sustaining collaboration. It is too early to know to what extent this social dilemma might affect CONECT-e, as it is also true that despite potentially being a threat, examples such as Wikipedia or Encyclopaedia of Life suggest that the project could still work without these incentives.

Discussion

Although operating at different scales, the seed exchange network of Vall Fosca and the CONECT-e platform could be seen as multi-layered or even as a nested series of governance that guarantees overall preservation of TAK. Ostrom (1990) identifies nested enterprises of governance as one of the design principles of robust, long-enduring, common-pool resource institutions. Such resource systems often exist in complex institutional settings in which smaller commons are nested within larger ones (Ostrom 2005). The concept has evolved from the natural resource commons framework to the knowledge commons framework, taking into account the local–regional–global scales, envisioned as different groups functioning at various levels within locally provided but globally shared resources (Hess and Ostrom 2007). This principle of a nested series of governance contributing to the sustainability of the commons resembles the concept of federated networks (Networkcultures 2016) or heterarchical hybrid modes of organization (Kostakis 2010) that are used in the digital commons, where knowledge is circulated via a distributed network (P2P) of one or various centralized networks (entities that aggregate content). The idea is that autonomous sub-entities (such as the network of gardeners in Vall Fosca) can aggregate to form one single larger entity sharing knowledge (i.e., the CONECT-e platform). When applied to the real world, the nested/federated mode of organization could guarantee the exchange of different types of TAK within the multi-layered network, engaging people into a continuous learning process whereby TAK can be updated and adapted from the bottom (Bodin et al. 2006), while assuring coordination to cope with social dilemmas (Prell et al. 2009). Moreover, this type of governance could help avoid the loss of knowledge that is not useful for people (and is not exchanged anymore) within the sub-entity of the P2P network, thanks to the possibility of storing and making it available for others within a superior governance unit (Reyes-García et al. 2016). For example, some of the traditional landrace knowledge in Vall Fosca is being lost due to gardeners' lack of time for making a seed bank: if stored in the CONECT-e platform, this knowledge could be updated, shared, and used by others outside Vall Fosca, maintaining TAK and expanding its boundaries.

However, it is also true that globally sharing locally provided TAK can generate social dilemmas which are not present or not acknowledged in governing TAK with only one level. For example, the tight connections that ensured appropriate management of TAK in Vall Fosca through peer-pressure and monitoring might become looser when the community of people sharing knowledge enlarges to include people without any social ties (as in the case of CONECT-e). The low level of excludability that characterizes digitalized knowledge commons may lead to conflicting issues in terms of protection of local TAK that is gathered and stored in CONECT-e. Although some authors argue that the mere existence of databases could already protect knowledge in case of misappropriation by actors interested in acquiring private intellectual property rights over it (Lakshmi Poorna et al. 2014), others argue that in order to fully protect TAK and its commons nature, a licensing that follows a copy left approach (such as the General Public Licensing, used commonly to secure copy left over open source software) should be implemented (see for example the ideas on Open Variety Rights; Deibel 2013). Following this approach, TAK content in CONECT-e is protected under the Creative Commons Attribution – ShareAlike 4.0 International License,[2] which allows reproduction but only under the same licensing system. In our opinion, this solution may offer a powerful tool to act against knowledge commons enclosure.

Conclusion

We have explored the degree to which traditional agricultural knowledge is constructed and managed under the commons framework, offering two examples where TAK is shared by communities that operate at contrasting scales (local and global). Taking each case study independently, we have examined how various components of TAK (i.e., the community sharing the resource, the rules governing access, and the value created by it) provide arguments to support that 1) TAK has been traditionally managed under a commons framework (in the case of Vall Fosca) and 2) even when TAK shifts from the local to the global digital world (the CONECT-e platform), it can still be managed as a commons. In the last part, we discussed potential links between the two cases and provided insights on how a nested commons governance framework could facilitate the interaction between local and global knowledge commons, thus helping us contextualize commoning as a social process (Kostakis and Bauwens 2014). As such, the approach has the potential to overcome the various intellectual property rights limitations that, in many situations, turned seeds and their associated knowledge into objects of exclusionary property (Dafermos and Vivero Pol 2015).

By uncovering the features that justify considering local and global TAK as a knowledge commons, we have ruled out the idea that TAK is an unmanaged resource, an idea in line with what Hardin (1968) presented to describe the tragedy of the common pool resources and that is typically used as an argument in favour of the commons enclosure (Boyle 2003). In both Vall Fosca and CONECT-e, we found a "regime for ensuring that the artifacts of community based productive efforts remain under the control of that community" (Kostakis and Bauwens 2014). Considering that the commons framework offers an alternative paradigm and creates a space of interaction that is different from that of the market or the state (Bollier and Helfrich 2014; Kostakis and Bauwens 2014), including TAK under the commons framework is more than an intellectual exercise: it is a political stand. If considered, perceived, practiced, and governed as a commons, TAK should not be subjected to the rules of the market nor to state-fostered initiatives, but rather be part of the "commons".

Acknowledgements

Research leading to this chapter has received funding from the Spanish government through a grant of the Economy and Competitiveness Ministry (CSO2014-59704-P) and a PhD studentship to P. Benyei. We thank members of the Laboratori d'Anàlisi de Sistemes Socio-Ecològics en la Globalització (LASEG) for their comments and suggestions. Reyes-García thanks the Dryland Cereals Research Group at ICRISAT-Patancheru for providing office facilities. This work contributes to the "María de Maeztu Unit of Excellence" (MdM-2015-0552).

Notes

1 Despite the controversy around it, we use the term "traditional" (rather than "local" or "customary") when referring to the knowledge systems presented here to emphasize the long-term historical continuity of these bodies of knowledge and the importance of social processes in their transmission and maintenance. The term "traditional" does not imply being archaic or pre-modern, as traditional knowledge systems are highly dynamic and adaptive (Reyes-García et al. 2014).

2 https://creativecommons.org/licenses/by-sa/4.0/

References

Abay, F., W. de Boef, and Å. Bjørnstad. 2011. Network analysis of barley seed flows in Tigray, Ethiopia: Supporting the design of strategies that contribute to on-farm management of plant genetic resources. *Plant Genetic Resources* 9 (4): 495–505.

Aw-Hassan, A., A. Mazid, and H. Salahieh. 2008. The role of informal farmer-to-farmer seed distribution in diffusion of new barley varieties in Syria. *Experimental Agriculture* 44 (3): 413–431.

Berkes, F., J. Colding, and C. Folke. 2000. Rediscovery of traditional ecological knowledge as adaptive management. *Ecological Applications* 10 (5): 1251–1262.

Bodin, O., B. Crona, and H. Ernstson. 2006. Social networks in natural resource management: What is there to learn from a structural perspective? *Ecology and Society* 11 (2).

Bollier, D., and S. Helfrich, eds. 2014. *The wealth of the commons: A world beyond market and state.* London: Levellers Press.

Borgatti, S., A. Mehra, D. J. Brass, and G. Labianca. 2009. Network analysis in the social sciences. *Science* 323: 892–895.

Boyd, R., P. J. Richerson, and J. Henrich. 2011. The cultural niche: Why social learning is essential for human adaptation. *Proceedings of the National Academy of Sciences of the United States of America* 108: 10918–10925.

Boyle, J. 2003. The second enclosure movement and the construction of the public domain. *Law and Contemporary Problems* 66: 33–74.

Brush, S. B. 2004. *Farmers' bounty: Locating crop diversity in the contemporary world.* Yale: Yale University Press.

Calvet-Mir, L., M. Calvet-Mir, and V. Reyes-García. 2010. Traditional ecological knowledge and landraces in situ conservation in high mountain home gardens of Vall Fosca, Catalan Pyrenees, Iberian Peninsula. In *Tradiciones y transformaciones en etnobotánica*, edited by M. L. Pochettino, A. H. Ladio and P. M. Arenas. Buenos Aires, Argentina: CYTED.

Calvet-Mir, L., M. Calvet-Mir, L. Vaqué-Nuñez, and V. Reyes-García. 2011. Landraces in situ conservation: A case study in high-mountain home gardens in Vall Fosca, Catalan Pyrenees, Iberian Peninsula. *Economic Botany* 65 (2): 146–157.

Calvet-Mir, L., M. Calvet-Mir, J.L. Molina, and V. Reyes-García. 2012a. Seeds exchange as an agrobiodiversity conservation mechanism: A case study in Vall Fosca, Catalan Pyrenees, Iberian Peninsula. *Ecology and Society* 17(1): 29.

Calvet-Mir, L., E. Gómez-Bagetthun, and V. Reyes-García. 2012b. Beyond food production: Ecosystem services provided by home gardens. A case study in Vall Fosca, Catalan Pyrenees, northeastern Spain. *Ecological Economics* 74: 153–160.

Calvet-Mir, L., P. Benyei, L. Aceituno-Mata, M. Pardo de Santayana, D. López-García, M. Carrascosa, A. Perdomo-Molina, and V. Reyes-García. 2018. Contributions of traditional agroecological knowledge as a digital commons to agroecological transitions. The case of the CONECT-e platform. *Sustainability* (Special Issue, Agroecology for the Transition towards Social-Ecological Sustainability) 10: 3214. Doi:10.3390/su10093214.

Ceccarelli, S., E.P. Guimarães, and E. Weltzien, eds. 2009. *Plant breeding and farmer participation.* Rome: FAO.

Cox, D. 2015. Farm hack: A commons for agricultural innovation. In *Patterns of commoning*, edited by D. Bollier and S. Helfrich. The Commons Strategy Group and Off the Common Press.

Dafermos, G., and J. L. Vivero Pol. 2015. Sistema agro-alimentario abierto y sostenible para Ecuador. In *Buen Conocer-FLOK Society: Modelos sostenibles y políticas públicas para una economía social del conocimiento común y abierto en Ecuador*, edited by D. Vila-Viñas and X. E. Barandiaran. Quito Instituto de Altos Estudios Nacionales.

Deibel, E. 2013. Open variety rights: Rethinking the commodification of plants. *Journal of Agrarian Change* 13: 282–309.

Engels, J.M., M. Hannes-Dempewolf, and V. Henson-Apollonio. 2011. Ethical considerations in agro-biodiversity research, collecting, and use. *Journal of Agricultural & Environmental Ethics* 24 (2): 107–126.
Federici, S. 2010. Feminism and the politics of the commons in an era of primitive accumulation. In *Uses of a whirlwind: Movement, movements, and contemporary radical currents in the United States.*, edited by T. C. Collective. Oakland, CA: AK Press.
Frischmann, B. M., M. J. Madison, and K. J. Strandburg, eds. 2014. *Governing knowledge commons*. Oxford: Oxford UP.
Ghijsen, H. 2009. Intellectual property rights and access rules for germplasm: Benefit or straitjacket? *Euphytica* 170 (1–2): 229–234.
Gilles, J.L., J. L. Thomas, C. Valdivia, and E. S. Yucra. 2013. Laggards or leaders: Conservers of traditional agricultural knowledge in Bolivia. *Rural Sociology* 78 (1): 51–74.
Halewood, M. 2013. What kind of goods are plant genetic resources for food and agriculture? Towards the identification and development of a new global commons. *International Journal of the Commons* 7.
Hardin, G. 1968. The tragedy of the commons. *Science* 162: 1243–1248.
Heinrich, M. 2015. New medicines based on traditional knowledge: Indigenous and intellectual property rights from an ethnopharmacological perspective. In *Ethnopharmacology*. John Wiley & Sons, Ltd.
Hess, C., and E. Ostrom, eds. 2007. *Understanding knowledge as a commons. From theory to practice*. The MIT Press.
Kawa, N. C., C. McCarty, and C. R. Clement. 2013. Manioc varietal diversity, social networks, and distribution constraints in rural Amazonia. *Current Anthropology* 54 (6): 764–770.
Kloppenburg, J. 2010. Impeding dispossession, enabling repossession: Biological open source and the recovery of seed sovereignty. *Journal of Agrarian Change* 10 (3): 367–388.
Kostakis, V. 2010. Identifying and understanding the problems of Wikipedia's peer governance: The case of inclusionists versus deletionists. *First Monday* 15: 1–11.
Kostakis, V., and M. Bauwens. 2014. *Network society and future scenarios for a collaborative economy*. Palgrave Macmillan.
Labeyrie, V., B. Rono, and C. Leclerc. 2014. How social organization shapes crop diversity: an ecological anthropology approach among Tharaka farmers of Mount Kenya. *Agriculture and Human Values* 31 (1): 97–107.
Labeyrie, V., M. Thomas, Z.K. Muthamia, and C. Leclerc. 2016. Seed exchange networks, ethnicity, and sorghum diversity. *Proceedings of the National Academy of Sciences* 113 (1): 98–103.
Lakshmi-Poorna, R., M. Mymoon, and A. Hariharan. 2014. Preservation and protection of traditional knowledge – diverse documentation initiatives across the globe. *Current Science* 107 (8): 1240–1246.
Luby, C. H., J. Kloppenburg, T. E. Michaels, and I. L. Goldman. 2015. Enhancing freedom to operate for plant breeders and farmers through open source plant breeding. *Crop Science* 55 (6): 2481–2488.
Macilwain, C. 1998. When rhetoric hits reality in debate on bioprospecting. *Nature* 392 (6676): 535–536.
MacKinnon, R. 2012. *Consent of the networked. The worldwide struggle for internet freedom*: Basic Books.
Malezieux, E. 2012. Designing cropping systems from nature. *Agronomy for Sustainable Development* 32 (1): 15–29.
Moran, K., S. R. King, and T. J. Carlson. 2001. Biodiversity prospecting: Lessons and prospects. *Annual Review of Anthropology* 30: 505–526.
Networkcultures. 2016. http://networkcultures.org/unlikeus/resources/articles/what-is-a-federated-network/. Last accessed: 23/12/2016.
Ostrom, E. 1990. *Governing the commons. The evolution of institutions for collective action. 1281*. Cambridge: Cambridge University Press.
Ostrom, E. 2005. *Understanding institutional diversity*. Princeton: Princeton University Press.
Pardo-de-Santayana, M., R. Morales, L. Aceituno-Mata, and M. Molina. 2014. *Inventario Español de los Conocimientos Tradicionales relativos a la Biodiversidad*. Madrid, Spain: MAGRAMA.
Prell, C., K. Hubacek, and M. Reed. 2009. Stakeholder analysis and social network analysis in natural resource management. *Society & Natural Resources* 22 (6): 501–518.
Quilligan, J. B. 2012. Why distinguish common goods from public goods? In *The wealth of the commons: A world beyond market & state*, edited by D. Bollier and S. Helfrich. Amherst: Levellers Press.
Reyes-García, V., V. Vadez, S. Tanner, T. Huanca, W. R. Leonard, and T. McDade. 2007. Ethnobotanical skills and clearance of tropical rain forest for agriculture: A case study in the lowlands of Bolivia. *Ambio* 36 (5): 406–408.
Reyes-García, V., J. L. Molina, L. Calvet-Mir, L. Aceituno-Mata, J. J. Lastra, R. Ontillera, M. Parada, M. Pardo-de-Santayana, M. Rigat, J. Valles, and T. Garnatje. 2013. "Tertius gaudens": germplasm exchange

networks and agroecological knowledge among home gardeners in the Iberian Peninsula. *Journal of Ethnobiology and Ethnomedicine* 9.

Reyes-García, V., L. Aceituno-Mata, L. Calvet-Mir, T. Garnatje, E. Gomez-Baggethun, J. J. Lastra, R. Ontillera, M. Parada, M. Rigat, J. Valles, S. Vila, and M. Pardo-de-Santayana. 2014. Resilience of traditional knowledge systems: The case of agricultural knowledge in home gardens of the Iberian Peninsula. *Global Environmental Change-Human and Policy Dimensions* 24: 223–231.

Reyes-García, V., A. L. Balbo, E. Gómez-Baggethun, M. Gueze, A. Mesoudi, P. J. Richerson, X. Rubio-Campillo, I. Ruiz-Mallén, and S. Shennan. 2016. Multilevel processes and cultural adaptation: examples from past and present small-scale societies. *Ecology and Society* 21 (4).

Reyes-García, V., M. Guèze, I. Díaz-Reviriego, R. Duda, Á. Fernández-Llamazares, S. Gallois, L. Napitupulu, M. Orta-Martínez, and A. Pyhälä. 2016. The adaptive nature of culture: A cross-cultural analysis of the returns of local environmental knowledge in three Indigenous societies. *Current Anthropology* 57 (6): 761–784.

Reyes-García, V., L. Aceituno-Mata, P. Benyei, L. Calvet-Mir, M. Carrascosa, M. Pardo-de-Santayana, J. Tardío for CONECT-e, IECTBA and Red de Semillas. 2018. Governing landraces and associated knowledge as commons. From theory to practice. In F. Girad and C. Frison, *The Commons, Plant Breeding and Agricultural Research. Challenges for Food Security and Agrobiodiversity*. New York: Routledge, Earthscan Food and Agriculture series: 197–209.

Riu-Bosoms, C., L. Calvet-Mir, and V. Reyes-García. 2014. Factors enhancing landrace in situ conservation in home gardens and fields in Vall de Gosol, Catalan Pyrenees, Iberian Peninsula. *Journal of Ethnobiology* 34 (2): 175–194.

Salpeteur, M., H. Patel, A. L. Balbo, X. Rubio-Campillo, M. Madella, P. Ajithprasad, and V. Reyes-García. 2015. When knowledge follows the blood. Kin groups and the distribution of ecological knowledge in a community of semi-nomadic pastoralists Gujarat (India). *Current Anthropology* 56 (3): 471–483.

Shiva, V. 1997. *Biopiracy. The plunder of nature and knowledge*. Boston, MA: South End Press.

Shiva, V. 2004. Trips, human rights and the public domain. *The Journal of World Intellectual Property* 7 (5): 665–673.

Siebenhuner, B., T. Dedeurwaerdere, and E. Brousseau. 2005. Introduction and overview to the special issue on biodiversity conservation, access and benefit-sharing and traditional knowledge. *Ecological Economics* 53 (4): 439–444.

Siefkes, C. 2012. Beyond digital plenty. Building blocks for physical peer production. *Journal of Peer Production*.

Smale, M., M. R. Bellon, D. Jarvis, and B. Sthapit. 2004. Economic concepts for designing policies to conserve crop genetic resources on farms. *Genetic Resources and Crop Evolution* 51 (2): 121–135.

Soejarto, D. D., H. H. S. Fong, G. T. Tan, H. J. Zhang, C. Y. Ma, S. G. Franzblau, C. Gyllenhaal, M. C. Riley, M. R. Kadushin, J. M. Pezzuto, L. T. Xuan, N. T. Hiep, N. V. Hung, B. M. Vu, P. K. Loc, L. X. Dac, L. T. Binh, N. Q. Chien, N. V. Hai, T. Q. Bich, N. M. Cuong, B. Southavong, K. Sydara, S. Bouamanivong, H. M. Ly, T. Van Thuy, W. C. Rose, and G. R. Dietzman. 2005. Ethnobotany/ethnopharmacology and mass bioprospecting: Issues on intellectual property and benefit-sharing. *Journal of Ethnopharmacology* 100 (1–2): 15–22.

Thomas, M., and S. Caillon. 2016. Effects of farmer social status and plant biocultural value on seed circulation networks in Vanuatu. *Ecology and Society* 21 (2).

Thomas, M., J. C. Dawson, I. Goldringer, and C. Bonneuil. 2011. Seed exchanges, a key to analyze crop diversity dynamics in farmer-led on-farm conservation. *Genetic Resources and Crop Evolution* 58 (3): 321–338.

Vandermeer, J., and I. Perfecto. 2013. Complex traditions: Intersecting theoretical frameworks in agroecological research. *Agroecology and Sustainable Food Systems* 37 (1): 76–89.

Vogl, C. R., and B. Vogl-Lukasser. 2003. Tradition, dynamics and Sustainability of plant species composition and management in homegardens on organic and non-organic small scale farms in alpine Eastern Tyrol, Austria. *Biological Agriculture & Horticulture* 21:349–366.

Whitt, L. A. 1998. Biocolonialism and the commodification of knowledge. *Science as culture* 7: 33–67.

12

SCIENTIFIC KNOWLEDGE OF FOOD AND AGRICULTURE IN PUBLIC INSTITUTIONS

Movement from public to private goods

Molly D. Anderson

Introduction

What is scientific knowledge? Who owns it? In particular, who owns scientific knowledge about food and agriculture, which often has contributions from non-scientists? Can the people who need it access it readily? Such knowledge has become increasingly privatized through the growing proportion of research that is privately funded and corporate influence on public institutions that conduct research on food and agriculture. This is part of a larger issue: Western capitalist countries have commodified knowledge in many ways to twist profits from it, just as land, labor, and humans themselves have been commodified (Elliot and Hepting, 2015; Amin and Howell, 2016). The generation and dissemination of agricultural knowledge, products, and services is big business, with vast sums involved in research, patents, technical services, and teaching, and generous profits involved in the transfer of products. So perhaps it is not surprising that efforts to derive financial returns from these activities are expansive.

Yet there is a fundamental contradiction in the privatization of scientific knowledge: science in its "pure" form is a collaborative process in which scientists test and build on previous knowledge. This is not to say that scientists are always open to new ideas: there is an inherent conservatism in scientific communities, with people defending their own ideas sometimes far past the point at which flaws are obvious to others. The ideal of a community of scholars engaging in a selfless search for truth is noble, however, and science at its best can approximate this as long as information is open.

Beyond the generation of knowledge, privatization restricts dissemination and use. Although public funds support state universities, national research centers, and international agricultural institutions such as United Nations agencies and international research centers, the knowledge produced by these institutions does not always automatically enter the public domain, especially in regions that are advocating for greater intellectual property rights. It may be published in journals to which only subscribers or those affiliated with universities and research institutions have access. Or it may be subject to veto power by corporations that have contributed to specific projects, even though public funds created and continue to maintain the facilities and institutions necessary to conduct the research. And even when it gets beyond university or research center walls, the dissemination of scientific knowledge to farmers and other producers has been privatized, so that users in many countries must pay for access, while funding for public agricultural extension has been cut.

However, scientific knowledge about food and agriculture may be created in many places and by different people: it may come from field research by producers or experiments by chefs and other people who prepare food, as well as being created in public institutions. This raises the question of what distinguishes "scientific" from other forms of know-how, or knowledge that is passed from peer to peer or generation to generation. The line is blurry and thus open to exploitation in the search for ways to privatize and profit from knowledge. Vernacular knowledge seldom has clear "owners" or single creators, so theft is relatively easy. A few well-known cases of the attempted appropriation and patenting of traditional knowledge, such as the generations of seed selection and breeding that created basmati rice (Rai, 2001) or knowledge of the multiple uses of neem (BBC News, 2005), have been thwarted successfully. But these success stories are a tiny fraction of all cases of biopiracy.

In this chapter, "scientific" knowledge is considered to be knowledge derived from systematic experimentation through replicable methods and reason; it is an essential part of a liberal democracy because it allows for the falsification of magical thinking, ideology, and deliberate falsehoods (Samir, 2016; Collins and Evans, 2017). The focus here is on knowledge that is generated in public institutions, since arguments against the privatization of this knowledge are particularly compelling.

This chapter argues that privatization hurts the conduct of agricultural and food science because open access is necessary for high-quality scientific work. Furthermore, knowledge generated in public institutions in particular is a global public good and should be considered and managed as a commons: one person "consuming" knowledge doesn't diminish it for others. Moreover, not only have public funds been used to establish and maintain these institutions, but more users increase the likelihood that the knowledge will be tested and refined. As Norbert Weiner expressed the idea, "the value of a piece of scientific work only appears to the full with its further application by many minds and with its free communication to other minds" (Weiner, 1954, p. 153). This is not to say that knowledge generated by the private sector is always rightly private: its quality often would be enhanced by allowing open access, and such knowledge usually builds on and extends knowledge that has been generated in public institutions. The chapter documents the growing privatization of food and agricultural research, and the influence of private money on public agricultural research institutions. It raises the question of how these trends affect and determine the topics that research addresses (or fails to address). Although systematic comparisons across multiple countries and research regimes have not been conducted, case studies under different governments and institutional arrangements show both the consequences of privatization on farmers and other potential users, and the processes by which this has happened.

Just as privatization has taken many pathways, so has the open sharing of scientific knowledge: communities of practice, peer-to-peer sharing, and open access of different kinds. This chapter goes on to articulate why scientific knowledge is properly considered as a global public good. Bringing scientific knowledge into the commons has already started via models in industrialized countries which are lifting restrictions and increasing access; these show that the ongoing pattern toward privatization can be reversed. In addition, some scientific knowledge has never been commodified because it was produced outside capitalist societies. This also may serve as a model for commoning.

Privatization of scientific knowledge: how and how much?

Privatization of scientific agricultural knowledge has occurred through multiple mechanisms. In terms of the generation of knowledge, a United States (US) government study (Fuglie et al., 2011) reported in an analysis of global investment that the global private sector spent $19.7 billion on food and agricultural research in 2007, which made up about half of the total public and private spending on food and agricultural research and development (R&D) in high-income countries (see Figure 12.1 for US trends). According to the same report, for biofuel R&D, the private sector spent $1.47 billion worldwide by 2009.

Discussing the relationship between private and publicly supported agricultural research, the Economic Research Service (2016) of the US Department of Agriculture notes that

> the two sectors have overlapping research interests in areas related to farm production, crop, animal, and farm machinery innovation. However, in areas where reaping benefits from research results is more difficult for private firms, such as in the environmental impacts of agriculture, and human nutrition/food safety, public sector research dominates (Figure 12.2).

When low-income countries are included in the global tally, public research spending increases to 79% of the total of 40.1 billion Purchasing Power Parity (PPP) dollars in 2008. Among developing countries, China, India, and Brazil accounted for one-quarter of the public agricultural R&D. In addition to national agricultural research investments, the CGIAR Consortium increased its R&D spending in developing countries by 41% between 2006 and 2011 (up to a total of $700 million). In contrast, public spending in high-income countries decelerated after the 1980s, with about one-third of the OECD countries spending less on public agricultural R&D in 2008 than in 2000 (Bientema et al., 2012). Global private investment in R&D focusing on agriculture and food processing increased from $12.9 billion in 1994 to $18.2 billion in 2008, with about 45% of that focused on agricultural inputs and the rest on food processing and manufacturing (Bientema et al., 2012). Updated figures indicate that private sector R&D in agriculture alone rose to $15.4 billion by 2014 (Fuglie, 2016). Most private-sector R&D was undertaken by companies in OECD

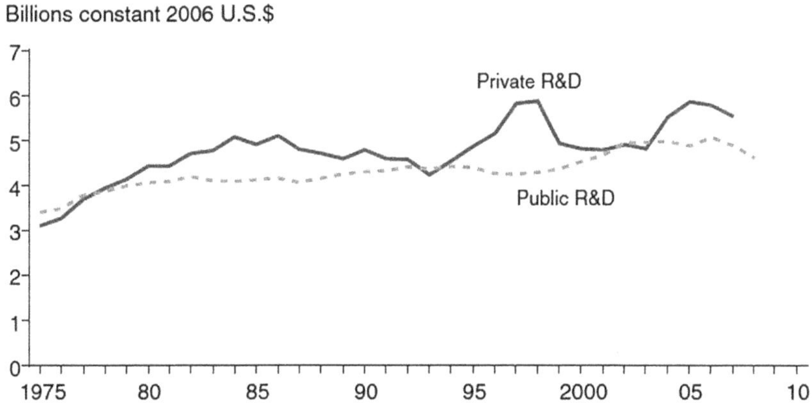

Figure 12.1 US trends in public and private food and agricultural research spending (Fuglie et al., 2011, p. 10).

Source: U.S. public agricultural research and development (R&D) spending is from USDA, Economic Research Service. U.S. private R&D spending is derived from the data in table 1.5, with interpolations for missing data. Nominal research expenditures are adjusted for inflation by the agricultural R&D price index developed by ERS. This price index takes into account changes in the cost of research inputs (scientist salaries, equipment, etc.).

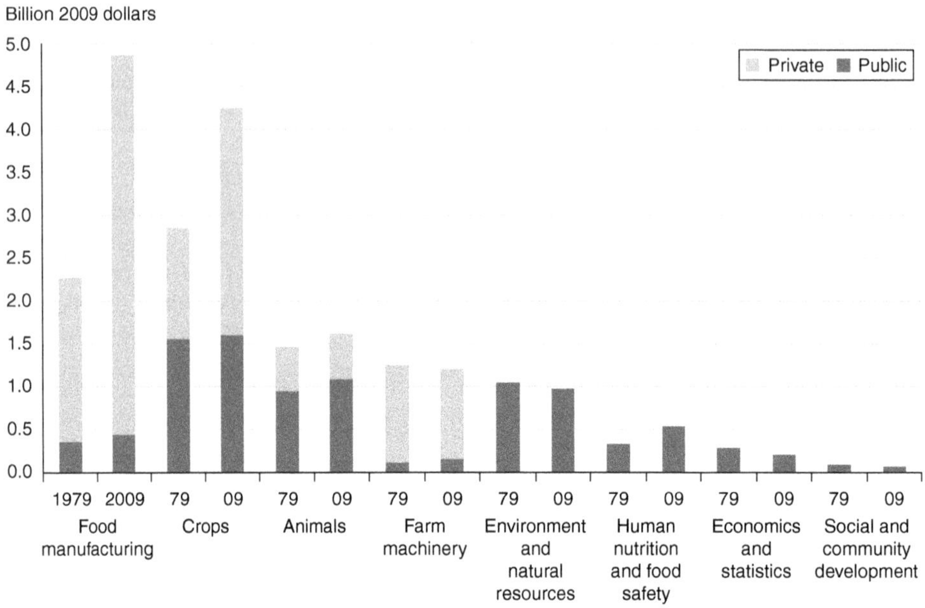

Figure 12.2 Investment in different topic areas by public and private sources (ERS, 2016).

Note: Data are adjusted for inflation using an index for agricultural research spending developed by ERS.
Source: USDA, Economic Research Service based on data from National Science Foundation, USDA's Current Research Information System (CRIS), and private sector data sources.

countries, although many of these countries have experiment stations in developing countries to facilitate the transfer of technologies to these markets (Fuglie et al., 2011).

These trends were confirmed by a more recent analysis of the past 50 years of data (Pardey et al., 2016), which documented that middle-income countries were investing more than high-income countries in 2011 (50% vs. 47% of $38.1 billion spent on global R&D). Furthermore, 52.5% of research in higher-income countries R&D on crop breeding, informatics, fertilizers, pesticides, and food technologies was conducted by the private sector (up from 42% in 1980). Food and beverage research was the focus of 44% of the private investment in R&D in rich countries in 2011. Private firms spent 35.5% of funding in middle-income countries, up from about 16% in 1980. Private R&D growth was especially marked in China. To some extent, these trends are due to firms based in rich countries off-shoring research in middle-income countries (e.g., new R&D facilities in China opened by Nestlé, Syngenta, PepsiCo, and General Mills), but it also reflects new investment by middle-income countries in their own research or, in China's case, privatizing formerly state-managed facilities (Pardey et al., 2016).

The amounts of money invested by the public and private sector tell us relatively little about the terms under which this investment happens, but this is where most concerns arise about privatization of knowledge, especially within public universities. As the proportion of public investment in agricultural research drops, the private sector's power to set the agenda of public institutions grows. Sometimes this happens with full consent of public institutions:

> In recent years, many universities seem to have adopted the attitude that "what is good for General Motors [in this case, consistent with the Bayh–Dole Act] is good for the country." They recognize and exploit the increasing commercial value of the

> intellectual property developed in universities as an important part of their mission. This has infected the research university with the profit objectives of a business, as both institutions and individual faculty members attempt to profit from the commercial value of the products of their research and instructional activities.
>
> *(Duderstadt, 2004, p. 57)*

The Bayh–Dole Act was enacted in December 1980 in the US and allowed small businesses and non-profit organizations, including universities, to retain title to inventions developed through federal research funding. It encouraged universities to collaborate with commercial concerns to promote the utilization of inventions arising from federal funding and to file patents on inventions, thus enabling universities to "take the lead in patenting and licensing groundbreaking discoveries" by retaining patents on inventions and discoveries, even when funded through federal grants. At least 15 other countries now have similar legislation (AUTM, n.d.). The *Economist* (2002, p. 3) hailed the Bayh–Dole Act as "perhaps the most inspired piece of legislation to be enacted in America over the past half-century".

Inspired, perhaps, but by a spirit that is opposed to commoning and open access to knowledge. With shrinking public money available for agricultural research and increased pressure on academicians to secure funding to hold their jobs, gain tenure, and support university overhead costs, more and more academicians turned to private money. For some scientists, this was accompanied by personal anguish about "selling out" and reluctant acquiescence to pressure (Hakim, 2016); but for many other scientists and administrators, private money was simply one more welcome source of revenue. It bought endowed professorships, buildings, conferences, and other items that helped to raise the prestige of their universities. However, private influence also introduced bias into science, with the preponderance of results favoring the product that the industry is selling or undermining questions about its nutritional value (Lesser et al., 2007; Mandrioli et al., 2016; Nestle, 2016).

Other problems are that industry may want to delay or vet results before they are published, attempt to personally discredit researchers who come up with unfavorable findings and want to publicize them, or create front groups and websites posing as trusted information sources to counter reputable research and promulgate results that industry wants the public to hear (Rampton and Stauber, 2001; Bollier, 2003; Washburn, 2005; Mirowski, 2011; Food and Water Watch, 2012; Hamerschlag et al., 2015). At best, such efforts create confusion and engender distrust of science and research establishments in the public. This distrust is well placed at times: recent disclosures by the *New York Times* of academicians who were paid to affix their names to supposedly objective information promoting biotechnology as safe and necessary surely undermined public faith in food and agricultural scientists, which is already at a low point (Lipton, 2015; Funk and Kennedy, 2016). At worst, efforts to bias scientific findings lead to expensive lawsuits to protect professional reputations, poor personal decisions and bad policy based on flawed science, and worsened public health.

The transition towards privatized public laboratories is facilitated by the patent system and the legal construction of the right to patent life-forms, which was granted for the first time in 1980. That year, the US Supreme Court decided by one vote in the *Diamond v. Chakrabarty* case that a strain of bacteria modified by inserting genes could be patented because it was not naturally occurring. The foreign genes gave the bacteria the ability to break down hydrocarbons, and its "inventors" hoped it might be useful for cleaning up oil spills. Since then, patents have been issued on microbes, plants, animals, and their parts through the International Convention for the Protection of Varieties of Plants (UPOV Convention) in 1961. In addition, the Trade-Related Aspects of Intellectual Property Rights (TRIPS) Agreement of 1994 and, more recently, a growing number of free trade agreements promoted by the US, the European Free Trade Association, and the EU have pushed the expansion of intellectual property protection to more biological materials, particularly plants. Since 1995, 40 countries have adhered to the UPOV Convention; before that point, membership had been limited primarily to industrialized countries (Markus and Reichman, 2005). This increasingly harmonized framework of international and domestic law has become an impediment to the free exchange of knowledge, which places poor and developing countries at a disadvantage even when (perhaps especially when) they have abundant biodiversity of potential commercial value.

In terms of public dissemination of knowledge to users, numerous barriers exist between those who generate knowledge and those who need it. Perhaps the best-studied agricultural trend has been the privatization or contracting of extension services in many countries, including Chile, Ecuador, Estonia, Germany, Pakistan, South Africa, Uganda, England, Wales, Honduras, Venezuela, Mali, and Niger (Rivera and Alex, 2004). While Cooperative Extension Services[1] still operate out of public agricultural universities in the US, their funding and staff have been downsized drastically (Wang, 2014). Klerkx and Leeuwis (2008) argue that policy discourse and scientific literature now focus attention on "entrepreneurship development" among farmers, with "entrepreneurship" defined as "the personalized drive and capability to commercialize a product, service, process, or business idea" (Knudson et al., 2004, p. 1333). A concomitant shift has occurred to an "agricultural knowledge market" that has been privatized worldwide to support agricultural innovation, rather than public provision of agricultural extension services, i.e., assistance to farmers and rural citizens in applying food and agricultural knowledge generated in public universities. This shift is consistent with neoliberal attention to private solutions and markets to solve public problems and the withdrawal of government support for public goods, including food.

Of course, some farmers are much less suited to becoming "entrepreneurs" than others—through temperament, ability, or interest—and some countries will benefit more than others from this shift of emphasis. Uddin et al. (2016), while generally supportive of privatization of agricultural extension services in Bangladesh to improve their quality, increase efficiency, and reduce government expenses, caution that several challenges must be overcome for privatization to work well. These include the lack of institutional capacity to coordinate private sector services, the large number of poor and marginalized farmers who cannot afford to pay fees, poor communication facilities in remote areas, corruption and loss of public trust in extension services, and resistance among the current extension service staff and administrators. In addition, private companies are expected to be more interested in selling their own products than in environmental sustainability, and more interested in short-term productivity than long-term issues such as resilience under climate change. As a consequence, it should come as no surprise that poor farmers and women farmers (who tend to have less access to funds to pay for services) have been cut out of privatized extension services in industrialized as well as developing countries (Labarthe and Laurent, 2014; Prager et al., 2016; McIntyre et al., 2009).

Most people, including farmers and other producers, get food and agricultural knowledge from sources other than agricultural extension. Some of these sources have been shut down: Heather Menzies (2015, C6) claims that "the past 10 years has seen the closure of 200 scientific research institutions and at least 16 government libraries, the elimination of some important public-interest government websites and the loss of some vital research by Statistics Canada". She goes on to lament the gagging of government and government-funded scientists, the 21% success rate of Access to Information requests, and "cuts to the National Library and Archives Canada [that] resulted in half the staff that help the public access archived print and other analog material losing their jobs" (ibid.). The United States also saw overt politicization of the dissemination of scientific information by government agencies during the Bush Administration, including government scientists unable to speak to the public without censorship (Shulman, 2006). The newly empowered Trump Administration set off a wave of alarm in late 2016 by asking for names of all government staff who conduct research or attend meetings on climate change (Davidson, 2016). In England, a controversial law was proposed in early 2016 (then retracted before enactment) that would have prohibited government-funded employees from questioning public policy or participating in public debates on issues such as transport, genetic modification, stem-cell research, climate change, and energy (McKie, 2016). In addition to direct contact with government officials and State-supported websites, free access to the Internet (for those who can afford the hardware needed and live in places where cable, satellite, or phone access is possible, that is) has been threatened many times by attempts to force payment by users (Bollier, 2003).

Another significant way in which access to scientific knowledge has been privatized is through the increasing inaccessibility of scientific journals. The costs of subscription services have risen 5–9% per year over the past few years, leaving only scientists and students working at elite colleges and universities with full access (Bosch and Henderson, 2016). While more and more institutions of higher education use e-journal packages to avoid the cost of ever-more-expensive print journals, their cost also has gone up more than inflation per year: in 2016, the average cost of e-journal packages increased 5.8% to 6.3%, down slightly from the 2015 average of 6.6% (ibid.). Subscriptions to scientific journals are especially high: the average price per agricultural journal was US$1,501 in 2014 and US$1,687 in 2016 (increasing 6% per year). The average price per food science journal was US$2,382 in 2014 and US$2,729 in 2016, increasing 7% per year. These prices are modest compared with chemistry and physics journals, however, which were on average US$5,105 and US$4,108 respectively in 2016 (Bosch and Henderson, 2016).

We must add the rising cost of higher education to the ways in which agricultural and food knowledge has been privatized and thus made more exclusive. Colleges and universities not only generate scientific knowledge but are also the primary place for training the next generation of scientists. If people cannot afford to attend, they will lose the opportunity to get jobs that require a degree. In the United States, the average annual out-of-state cost for a bachelor (undergraduate) program in agriculture or related fields is US$34,180 with an estimated average four-year degree total cost of US$136,720 (according to the US Department of Education's 2016–2017 IPEDS survey; CollegeCalc, 2017). Many people who work in agriculture or food studies do not enroll in agricultural degrees, however; they may study ecology, sustainable food systems, international development, or other subjects at public or private colleges. According to the US College Board, the average cost of tuition and fees for the 2016–2017 school year in the US was US$33,480 at private colleges, US$9,650 for state residents at public colleges, and US$24,930 for out-of-state residents attending public universities. The Institute for College Access and Success reports that about 70% of students graduated in 2014 with debt, and their loans averaged about $28,950.

In comparison, costs of higher education in European and middle-income countries are much lower. For example, most of England's universities can't charge more than £9,000, or about US$13,800, in annual tuition until the 2017–2018 school year and many degrees take only three years. The EU average tuition is £9,000, or about US$12,700, for domestic (EU) students for an undergraduate degree (Collier, 2016). Germany announced in 2014 that it would eliminate tuition at its universities, but its cost of living is relatively high compared with other European countries. Brazil's universities don't typically charge tuition, but they do require registration fees. Chinese students' annual tuition is usually between US$3,300 and US$9,900, and in India's Institute of Technology system, tuition averages around US$1,400 a year. Denmark, Finland, and Iceland offer free higher education to local residents, who get a stipend while in school (Glum, 2015). Overall, countries vary tremendously in the amount of funding they make available to students matriculated in institutions of higher education; the debt load for students in the US has gone up sharply over the last two decades as the costs of education have increased, average family incomes have remained flat, and government sources of financial aid have not kept paces with the costs of education (Kantrowitz, 2016).

Another interesting aspect of the privatization of food and agricultural education is visible in Japan and is described by Aya Kimura (2010). She analyzed "food education experts", mainly women, who had obtained their certification from private enterprise institutions offering certificates in food education. She writes that Japan has emphasized food education in its recent policy; but the kind of knowledge that is imparted in these certificate programs is superficial mastery of facts and skills needed for "right eating", rather than the ability to critique the agri-food system. This resonates with the long-standing conservatism of the Academy of Nutrition and Dietetics (AND, formerly the American Dietetic Association), the largest organization of nutrition professionals in the US, which finally stopped accepting sponsorship from Coca-Cola in 2015 (but on Coca-Cola's initiative, not because AND was uncomfortable with the close relationship). Between 2001 and 2011, the number of industry sponsors of AND rose from 10 to 38, about 23% of speakers at the 2011 annual conference had industry ties, and corporate contributions were the single largest source of revenue in 2011: $1.3 million out of a total of $3.4 million (38%). Corporations, including Coca-Cola, provided continuing education courses for AND members that included messages that sugar was not harmful to children (Simon, 2013). A more public-spirited contingent of AND, the Health and Environmental Nutrition Dietetic Practice Group, came together within the last decade to "empower members to be leaders in sustainable and accessible food and water systems", according to its website. This is a welcome development, but reforming the ways that food professionals are taught in universities and ensuring that they have strong critical thinking skills will take a long time.

Impacts of the privatization of agricultural knowledge

As the cases discussed above demonstrate, the impacts of these trends of increasing privatization fall in at least five areas: what is researched, how it is disseminated, who is able to use the knowledge generated, how much they pay, and the quality of that knowledge. There is a clear shift in the focus of research funded by the private sector into topics that will bring profits back to the company and its investors, away from basic research and projects that result in knowledge other than products or services that can be sold. Furthermore, researchers may shift their focus to avoid intellectual property lawsuits and other actions. A study in 1998 found that 25% of 100 research facilities surveyed affirmed that they had shifted their focus for these reasons (Bollier, 2003).

Two well-known examples of shifting emphasis to profitable ventures rather than ones needed for a sustainable future are the neglect of research on so-called orphan crops—such as tef, finger millet, yam, roots, and tubers that are regionally important but not traded around the world—and the almost-exclusive use of biotechnology in agriculture to develop crops planted in industrialized and middle-income countries that express pesticide-resistant traits rather than pest and disease resistance, drought tolerance, or flood tolerance. Aspects of food and agriculture related to public health, environmental quality, human rights, and a host of other issues get short shrift in privately funded research and are chronically underfunded because of the budget constraints of public research. Yet these short-changed issues are precisely the topics that most need research to combat global challenges such as creating and encouraging sustainable diets, increasing access to affordable seeds, protecting and enhancing biodiversity, growing sufficient food without fossil fuels or back-breaking labor, or restoring degraded soil and water so that food can be produced with them again.

Because of its intrinsic orientation towards profit, the private sector will invest in these areas only if there is an angle for commercialization, not because they are intrinsically important and especially critical to poor and small-scale farmers in poor countries. King et al. (2012) emphasize the "complementary" nature of public and private agricultural research, but the relationship might as easily be termed parasitical as symbiotic. Private firms reap the gains by selling products and services that were developed only after public investment in "pre-technology" and basic knowledge. While the Bayh–Dole Act and similar legislation in other countries was intended to allow more of the profits to be retained by scientists and universities, the public is paying twice: first to support public universities, and second to buy the products and services.

Second, commodification of knowledge means that people who already have wealth and power are more easily able to access agricultural knowledge that is useful to them than poor people. Given that agriculture is proportionally more significant to livelihoods and national revenues in developing countries compared with industrialized ones, it follows that people in poor countries are least able to access new agricultural knowledge relevant to new market conditions or new environmental problems. For example, growers in poor countries who are selling to wealthy countries must meet the quality standards set by industry buyers, or certification standards that are often set by boards either dominated or strongly influenced by industry. This happens even with organic and fair-trade products, which are especially enticing for corporate involvement because of their rapid growth as food market segments (Jaffee and Howard, 2010). Furthermore, growers and fishers in poor countries are increasingly affected by global environmental change generated by the Global North. Thus, they are hurt twice over: first by rapid changes in the production environment and second by lack of access to knowledge and technology necessary for adaptation.

Next, privatization of knowledge leads to higher costs to access that knowledge: when research is taken from the commons realm and subjected to market conditions, scarcity becomes a way to increase price. Indeed, this is the main reason to exclude some users. Reichman and Samuelson (1997) describe the privatization of Landsat satellite images, which resulted in prices jumping from US$400 to US$4,400 per image. When researchers complained, Congress restored access at close to the previous price to government researchers, but non-government researchers still had to pay the higher price, which was infeasible for most research budgets. Similarly, privatization may have similar impacts on the rapid increase in cost for scientific journals: Van Noorden (2013) reports profit margins of 20–30% for the publishing industry with an average cost of producing an article around US$3,500–4,000. The Cambridge Economic Policy Associates in London estimated that society publishers have profit margins of 20% and university publishers 25%; Elsevier's profit margins are estimated by financial analysts to be 40–50% for science, technology, and mathematics journals (ibid.).

Finally, the fundamental task of science and scientists to challenge received wisdom is made immeasurably more difficult when information is privatized, kept secret, or distributed on the basis of its exchange value. This means that challenges to privately owned knowledge are not informed by full understanding of how results have been derived and thus cannot be replicated or falsified. Furthermore, information that is released without full details of how it was generated is inherently untrustworthy. Moreover, it generates a vicious circle by which only certain laboratories and well-funded research centers can benefit from what has already been discovered and added to the scientific realm. These quality issues increase and are compounded by the private-sector influence on universities that teach about food and agriculture, on professional societies of their graduates (as described briefly below for the Academy of Nutrition and Dietetics), and on the media. In combination, these impacts create overwhelming reinforcement of an industry world-view that squeezes out independent science.

The trends and consequences described above are consistent with neoliberal trends in government and economics across the globe, which have encouraged a shift from public welfare based on the notion of collectivity and use value to the creation and support of market mechanisms constructed around demand, offer, and exchange value. Lave et al. (2010) published a list of recurrent themes of neoliberal science management, including the rollback of public funding for universities, "ghost-writing" that appends the reputation of scientists to industry-slanted text, the narrowing of research agendas to focus on the needs of commercial actors, and the fortification of intellectual property in order to commercialize knowledge. This decimation of the full range of research that the public needs to understand and work productively with food and agriculture is ominous, but there are alternatives.

Why scientific knowledge belongs in the commons

Neoliberalism looks to market solutions for all of society's problems, and indeed is blind to mechanisms for meeting needs and solving problems that exist outside markets. But markets do not deal well with goods that are non-rivalrous and non-excludable, i.e., those that belong to "global public goods". Of course, by setting up ways to enclose public goods and thus limit their access and use, markets can appear to benefit some people, at least for a while. I argue here that scientific knowledge is a global public good, in that sciences require the free exchange of information to flourish.

As demonstrated by the meticulous work of Elinor Ostrom, managed commons have been effective ways for sharing global public goods for as long as humans have lived in societies. Ostrom and Charlotte Hess hosted a meeting on scholarly communication as a commons in 2004, coming out of growing interest in transferring ideas from the management of natural resources commons to knowledge and digital media (Hess and Ostrom, 2009). This workshop and the volume produced from it introduced several of the ways that knowledge commons were being protected from enclosure and created. Since that time, commons literature has blossomed; examples from digital media, natural resources, and collective living or providing for human needs have proliferated more than cases of science commons. Yet the arguments for commoning used elsewhere (e.g., Bollier and Helfrich, 2015) clearly apply to scientific knowledge as well.

Despite the current trends, there are multiple reasons why scientific knowledge must be considered a commons and why our aim should be to re-common what has been privatized already. First, scientific knowledge cannot be privatized or owned because it builds on other knowledge that is already in the commons; it always builds on previous work and discoveries and is cumulative. Even inventions result from multiple sources coming together in a new way. Lewis Hyde (2010, p. 112) characterizes Benjamin Franklin, lauded as the "quintessential self-made man" and contributor of foundational ideas to electrical theory, as more of a pirate. He constantly collaborated with a network of others at home and abroad in his scientific inventions and saw no reason to patent a wood-stove that he designed and which took his name. He argued:

> [t]hat as we enjoy great Advantages from the Inventions of Others, we should be glad of an Opportunity to serve others by any Invention of ours, and this we should do freely and generously.
>
> *(Franklin, Autobiography p. 98, as quoted in Hyde 2010, p. 120)*

He shared the view that his work belonged to the public with other scientists such as Jonas Salk, Albert Sabin and John Enders (who did polio research in the 1940s and 1950s, leading to a vaccine), and Cesar Milstein, Nobel Prize winner in 1975 (Bollier, 2003). According to Bollier (2003, p. 137):

> The scientists of previous generations who refused to patent their breakthrough discoveries were neither naïve nor saintly. They were members of a flourishing gift culture, the academy, which presumed (and still presumes, for the most part) that data, research tools, and other scholarly resources should be widely shared and openly scrutinized.

He then goes on to quote law scholar Robert P. Merges (1996):

> Those who produce [scientific knowledge] understand that the community always has extensive claims on it, because without shared knowledge, research techniques and even biological materials, there would often be no results, no progress, and hence nothing to argue about.
>
> *(as quoted in Bollier, 2003, p. 137)*

That is, privatization destroys a rich academic gift culture which depends on give and take to continue. The academic gift culture is the root of public respect for science and institutions of higher education; scientists are not just working for money but in service to the public, even if the direct applications of basic science are not always apparent. This is the justification for the non-profit status of public and most private colleges and universities, which excuses them from paying property tax and sales tax.

In addition to the scruples of individual scientists about the disposition of their work and the kind of culture within which they want to work, another argument for scientific knowledge as a commons is that this is necessary to avoid the wrong kinds of incentives. Hyde (2010) argues that self-interest must be channeled to serve the public good, but near-term rewards of monopoly control over one's creations are hardly necessary. The original argument for copyright was that the inventor should have limited control over his or her invention, and then it would pass into the public domain after a certain number of years. This argument has been distorted by numerous extensions of copyright and patent law, based on the premise that the inventor or creator should have unlimited rights to his or her products, even extending to rights of the estate after the creator's death. But, Hyde (2010, pp. 159–160) points out,

> [s]ometimes we work merely for gain, to be sure, but we also work out of simple curiosity … or for the fun of it, or for the aesthetic pleasure of finding an elegant solution, or to while away the idle hours, or out of gratitude, spite, or patriotic zeal. In reputation systems, we may work for any of these and for status as well, or for the delayed gains that status can bring. … [A]ny incentive system that emphasizes only present financial reward may have the side effect of crowding out these others.

A third argument for keeping food and agricultural knowledge in the commons is that such knowledge is necessary to fulfill human rights, such as the rights to food and decent compensation for freely chosen work. That is, people need this information to be healthy, perhaps by producing the food they will eat or by choosing which foods to buy. People who earn a living in food production need this knowledge to work productively. Article 27 of the Universal Declaration of Human Rights states that, "Everyone has the right freely to participate in the cultural life of the community, to enjoy the arts and to share in scientific advancement and its benefits". Echoing arguments for copyright, the Article continues: "Everyone has the right to the protection of the moral and material interests resulting from any scientific, literary or artistic production of which he [*sic*] is the author", but this has a very different flavor when juxtaposed with the previous clause. In a similar way to Hyde's arguments for how reputation systems can repay contributions to shared scientific knowledge in ways beyond financial gain, even when it is commoned, the human rights argument comes from a different domain of reference that is oppositional to market-based solutions to food and agricultural challenges (Anderson, 2008). The Declaration on the Rights of Peasants and Other People Working in Rural Areas, which has been under negotiation in the Human Rights Council since 2008, uses human rights arguments to shore up many of its knowledge-related demands, such as peasants' rights to seeds and the protection of traditional knowledge.

Bringing agricultural science back to the commons: what is possible and necessary

Once we start to look for places where people have never allowed the privatization of scientific knowledge on food and agriculture, where movements and individuals are resisting its commodification, or where farmers are experimenting with forms of more open access, examples are abundant. All over the world, scholars and scientists continue to collaborate with communities of practice or involve themselves in informal sharing arrangements operating outside the institutionalized framework of laws and contracts. They put their work in open-source repositories that are maintained by increasing numbers of colleges and universities. In addition, solidarity networks support scientists who are being persecuted by corporations for publicizing results that throw doubts on products of the corporation. For example, the reaction was prompt after Monsanto mounted a campaign of defamation against Professor Gilles-Eric Séralini and his research team in 2012. Séralini had discovered signs of toxicity in raw data of a rat-feeding study of genetically engineered (Roundup Ready) corn done by Monsanto, which had not been reported by the corporation (in fact, Monsanto continues to insist that Roundup Ready corn is perfectly safe). Rats fed Roundup Ready corn or drinking water with levels of Roundup permitted in drinking water in the US developed cancers faster and died earlier than rats fed on a standard diet. They also suffered breast cancer and severe liver and kidney damage. The journal *Food and Chemical Toxicology* retracted Séralini's article under pressure from lobbyists. Defamatory remarks were first publicized in *Forbes Magazine* by Henry Miller and Bruce Chassy, without any kind of conflict-of-interest statement (Miller and Chassy, 2012) despite the fact that Miller was a former lobbyist for the tobacco industry who tried to discredit the association between tobacco, cancer, and heart disease. The defamation was picked up by *Marianne Magazine* in France, which wrote about "scientific fraud in which the methodology served to reinforce pre-determined results" (Séralini, 2015). In response, scientists from around the world rallied in support of Séralini: a letter signed by 140 French scientists was published in *Le Monde*, while scientists from 33 countries signed an open letter of protest, and Séralini et al.'s research was supported in individual letters sent by 160 scientists (Sustainable Pulse, 2012). Although this level of support from other scientists is encouraging, it took three years to bring *Marianne Magazine* to task for libel. It was only in late 2015, in fact, that the 17th Criminal Chamber of the High Court of Paris fined *Marianne Magazine* and its journalist for public defamation of the researchers and Séralini's research institute (ibid.). And the chilling effect on other scientists, especially junior scientists, is stark.

Another way in which scientific knowledge is "opening" is through multi-organizational teams that work together on research projects. This goes beyond the interdisciplinary or single-discipline teams that are more and more common in research, especially on global challenges, because it consists of creating spaces where non-governmental organization, producers, government agencies, academicians, and others can collaborate with formal agreements that protect the space of each partner. Examples include potato research in several countries leading to new management systems, including the famous Potato Park in Peru which establishes a protected agroecology region (Ortiz et al., 2013; Bollier, 2015); participatory action research that engages academicians and producers (e.g., Mendez et al., 2015); and research on the management of natural resources (e.g., Brewer, 2014; Walker and Salt, 2012). Transdisciplinary research teams that bring together people within academic institutions or government and people from social movements and NGOs are another example, when results are open access. Opening up the formulation of research questions, design, protocols, and dissemination to different people and ways of knowing not only makes the results much richer but also ensures that the research will serve people who are the intended beneficiaries.

Similarly, the open-access movement in publishing is growing, although it has not found a perfect solution to expensive publishing costs. At present, there are several examples where the costs of publishing have just shifted from the subscriber to the person submitting a manuscript for publication. The cost of submitting to open-access journals ranged from US$8 to US$3,900 in 2010, although most open-access journals have waivers for people in developing countries or small institutions. At the least, the availability of open access journals is making researchers question the added value they get from academic publishers and whether it is commensurate with the costs (Van Noorden, 2013).

Open-access publishing is a part of a broader access to knowledge movement. Consumers International sponsors the Access to Knowledge Network (A2Knetwork.org), which defines the movement's aims:

> to create more equitable public access to the products of human culture and learning. The ultimate objective of the movement is to create a world in which educational and cultural works are accessible to all, and in which consumers and creators alike participate in a vibrant ecosystem of innovation and creativity. These goals are of interest to a broad coalition of consumer groups, NGOs, activists, Internet users and others.

Consumers International sees intellectual property rights as a consumer issue because it can inhibit the sharing and development of culture by denying consumers the freedom to use goods as they would reasonably expect, or make learning materials unaffordable or unavailable to consumers in developing countries. Massive on-line open-access courses (MOOCs) offered by a few universities for free are a great example of how learning materials might be shared freely (Kop and Fournier, 2015).

The global status of agricultural knowledge in the commons is also characterized by multiple intriguing examples of communities that have resisted commodification of knowledge because this is intrinsically opposed to cultural or political beliefs. However, little has been published about them. For example, biotechnology research in Cuba has been supported by the government and serves very different purposes than in the US: it is focused on creating products that farmers need to resist pests and diseases, as well as on affordable medicine and vaccines (Randal, 2000; Starr, 2004). Open biotechnology is in its beginning stages and might be the best hope of overcoming staunch public resistance to genetic engineering of food (Lemmens, 2014).

The best example of resistance to the commodification and commercialization of knowledge in industrialized countries is perhaps Indigenous research, rooted in Indigenous worldviews that humans exist in relationship, between themselves and with the members of the natural world. This creates a very different view of accountability and responsibility than a non-Indigenous person, perhaps accountable to a government agency for reporting on a research grant, would have (Wilson, 2008; Kimmerer, 2013).

So, what is needed to bring scientific knowledge (back) into the commons? We can look at some of the examples of where commoning is already happening, of course, and seek to expand these. But some solid groundwork would be useful. First, David Bollier (2003) argues for a better articulation of the public interest in scientific knowledge. Second, separating the concept of the commons from the concept of "community" would be useful. Amin and Howell (2016) are concerned that language about the commons is sometimes heard as a call for homogenizing differences, in a nostalgia for community as an antidote to neoliberal emphasis on the individual. They quote Jeremy Gilbert (2014, p. 165):

> What is particularly useful about the idea of the commons as distinct from the idea of community is that it does not depend upon any presumption that the participants in a commons will be bound together by a shared identity or a homogeneous culture. Rather, they will be related primarily by their shared interest in defending or producing a set of common resources, and this shared interest is likely to be the basis for an egalitarian and potentially democratic set of social relationships.

This respectful difference (agreeing to disagree) seems especially vital to science, with its constant play of setting up arguments only to have others shoot holes in them. Without differences among the collaborators, there would be no advance in thinking.

Another clear need is to publicize and disseminate the idea of scientific commons. One of the worst aspects of the commodification of almost everything we do and use has been the obscuring of alternatives, so that it is sometimes difficult to even know that it is possible to live and work in a different way. Once the idea of the scientific commons has been firmly embedded in how society views science itself, next steps will involve advancing it through accessible and high-quality higher education that encourages science and critical thinking, better funding for research on public goods (and transparent funding processes), and a firm prohibition on the use of public money for research that aims to create commodified knowledge and that will benefit the private sector first. This might include truly free and mandatory open-access publication of any research that is supported by public money.

Beyond these steps to (re)establish a scientific commons, there are certain habits of mind and responsibility that are necessary to maintain it and keep it alive. Among these are the constant defense of the commons, similar in concept to "beating the bounds", or the annual patrols around the border of English commons in which commoners destroyed fences, ditches, gates, or buildings that had been erected without permission (Hyde, 2010). Solidarity networks to vociferously protest privatization of scientific knowledge would serve this purpose.

A second element of changing mental habits and attitudes about responsibility would be the recognition of vulnerability of science to enclosure, and the identification of the responsibility to protect it that goes along with this. Amin and Howell (2016, p. 1) see the commons as a "process, a contest of force, a reconstitution, a site of convening practices"; others argue for thinking of "common" as a verb or a set of patterns, to emphasize that it must be continually recreated. Commoning would include not only maintaining open boundaries of knowledge but also inhabiting the commons space by publishing through open access or "copylefting" one's work rather than copyrighting it. The concept of copylefting (GNU Project, 2017) comes from the software domain, in which many software developers have learned that making their products freely accessible, with the stipulation that the same rights must be preserved for any derivative products, both increases the number of people using the software and its quality.

It may be that opening to knowledge as relationship, following in the footsteps of Indigenous researchers, is also necessary to create a commons of scientific knowledge. This recognizes a different type of accountability, what Gibson-Graham et al. (2016, p. 195) define as "establishing rules and protocols for access and use, taking care of and accepting responsibility for a resource, and distributing the benefits in ways that take into account the well-being of others". These authors suggest that the politics of commoning and the knowledge needed by it could lead to a

> different mode of humanity [in which] humans might take their place as only one in a multi-species community of life on this planet, abandoning illusions of mastery to become 'team players' with non-human earth others.
>
> *(Gibson-Graham et al., 2016, p. 207)*

Commodification and privatization require a disarticulation of knowledge from its context and history, whether this is, for example, the history of how seeds were domesticated over millennia, the current use of the plethora of varieties of seeds by farmers all over the world, or why keeping multiple varieties in circulation is desirable. Only when a tiny fragment of knowledge has been parsed out (the genetic code of the manipulated seed, a tiny library in itself) can it be patented; but in parsing it from its context, it is deprived of the multiplicity of values and the cues that show how that knowledge does something more than generating profit and why it should be used responsibly, which is one of the differences between knowledge and wisdom. Brewster Kneen (2015, p. xiv) claims that "Western reductionist science is all about control and management. What it seeks to know about the organism is how to make it do what the 'owner' (or manager) wants". This perspective on science is consistent with neoliberal aims of narrowing and restricting the focus. A viewpoint that sees the organism first in relation to everything around it and considers that the observer/scientist is obligated to respect and protect those relationships would illuminate the world in which humans live with bright threads of mutual obligation. If food and agricultural knowledge were generated and shared through a managed commons, perhaps we could learn together how to hold other life and the planet itself in trust for future generations, rather than destroying it through attempts to control and own it.

Note

1 The Cooperative Extension Service was established in the United States through the Smith Lever Act of 1914, which created a partnership for research and education between the Department of Agriculture and Land-grant Colleges, which had been established in each state through the 1860 Morrill Act. The Extension system addressed rural agricultural issues by helping farmers and rural citizens to apply research results.

References

Amin, A. and Howell, P. (2016) 'Thinking the commons', in A. Amin and P. Howell (eds.) *Releasing the Commons: Rethinking the Futures of the Commons*, Routledge, London, Oxfordshire, UK and New York

Anderson, M.D. (2008) 'Rights-based food systems and the goals of food system reform', *Agriculture & Human Values*, vol. 25, no. 4, pp.593–608

Association of University Technology Managers (n.d.) 'Bayh–Dole Act', https://autm.net/about-tech-transfer/advocacy/legislation/bayh-dole-act, accessed 17 October 2018

BBC News (2005) 'India wins landmark patent battle', 9 March, http://news.bbc.co.uk/2/hi/science/nature/4333627.stm, accessed 17 October 2018

Beintema, N., Stads, G.-J., Fuglie, K., and Heisey, P (2012) 'ASTI global assessment of agricultural R&D spending', International Food Policy Research Institute, Washington, DC, and Global Forum on Agricultural Research, Rome
Bollier, D. (2003) *Silent Theft: The Private Plunder of Our Common Wealth*, Routledge, New York and London
Bollier, D. (2015) 'The potato park of Peru', in D. Bollier and S. Helfrich (eds.) *Patterns of Commoning*, Levellers Press, Amherst and Florence, MA
Bollier, D. and Helfrich, S. (2015) *Patterns of Commoning*, Levellers Press, Amherst and Florence, MA
Bosch, S. and Henderson, K. (2016) 'Fracking the ecosystem – Periodicals price survey 2016', *Library Journal*, http://lj.libraryjournal.com/2016/04/publishing/fracking-the-ecosystem-periodicals-price-survey-2016/, accessed 17 October 2018
Brewer, J.F. (2014) 'Harvesting a knowledge commons: Collective action, transparency, and innovation at the Portland Fish Exchange', *International Journal of the Commons*, vol. 8, no. 1, pp.155–178
CollegeCalc (2017) 'Cheapest colleges for agriculture, agriculture operations, and related sciences, other', http://www.collegecalc.org/majors/agriculture-agriculture-operations-and-related-sciences-other/, accessed 12 December 2017
Collier, S. (2016) 'How much does it cost to study in Europe?' https://www.topuniversities.com/student-info/student-finance/how-much-does-it-cost-study-europe, accessed 12 December 2017
Collins, H. and Evans, R. (2017) *Why Democracies Need Science*, Polity Press, Cambridge, UK and Malden, MA
Davidson, J. (2016) 'Energy Dept. rejects Trump's request to name climate-change workers, who remain worried', *The Washington Post*, 13 December
Duderstadt, J. J. (2004) 'Delicate balance: Market forces versus the public interest', in D.G. Stein (ed.). *Buying In or Selling Out?: The Commercialization of the American Research University*, Rutgers University Press, New Brunswick, NJ
Economic Research Service, USDA (2016) 'Incentives drive public vs. private agricultural research and development expenditure mix', https://www.ers.usda.gov/data-products/chart-gallery/gallery/chart-detail/?chartId=78592, accessed 14 December 2017
The Economist (2002) 'Innovation's golden goose', *The Economist*, vol. 8303, p.3
Elliott, P.W. and D.H. Hepting (eds.) (2015) *Free Knowledge: Confronting the Commodification of Human Discovery*, University of Regina Press, Regina, Saskatchewan
Food and Water Watch (2012) 'Public interest, private gain: Corporate influence over university agricultural research', http://www.foodandwaterwatch.org/insight/public-research-private-gain, accessed 17 October 2018
Fuglie, K. (2016) 'The growing role of the private sector in agricultural research and development world-wide', *Global Food Security*, vol. 10, pp.29–38
Fuglie, K, Heisey, P., King, J., Pray, C.E., Day-Rubenstein, K., Schimmelpfennig, D., Wang, S.L., and Karmarkar-Deshmukh, R. (2011) 'Research investments and market structure in the food processing, agricultural input, and biofuel industries worldwide', Economic Research Report No. (ERR-130), US Department of Agriculture, Washington, DC
Funk, C. and Kennedy, B. (2016) 'The new food fights: U.S. public divides over food science', Pew Research Center, http://assets.pewresearch.org/wp-content/uploads/sites/14/2016/12/19170147/PS_2016.12.01_Food-Science_FINAL.pdf, accessed 14 December 2017
Gilbert, J. (2014) *Common Ground: Democracy and Collectivity in an Age of Individualism*, Pluto, London
Glum, J. (2015) 'Amid free education protests, college costs around the world reveal countries' differences', *International Business Times*, 4 November
Gibson-Graham, J.K., Cameron, J. and Healy, S. (2016) 'Commoning as postcapitalist politics', in A. Amin and P. Howell (eds.) *Releasing the Commons: Rethinking the Futures of the Commons*. Routledge, Abington and Oxfordshire, UK, and New York, pp.192–212
GNU Project (2017) 'What is copyleft?', https://www.gnu.org/copyleft/, accessed 13 December 2017
Hakim, D. (2016) 'Scientists loved and loathed by an agrochemical giant', *New York Times*, 31 December
Hamerschlag, K., Lappé, A. and Malkam, S. (2015) *Spinning Food: How Food Industry Front Groups and Covert Communications Are Shaping the Story of Food*, Friends of the Earth, Washington, DC
Hess, C. and Ostrom, E. (2007) *Understanding Knowledge as a Commons: From Theory to Practice*, MIT Press, Cambridge, MA
Hyde, L. (2010) *Common as Air: Revolution, Art and Ownership*, Farrar, Straus and Giroux
Jaffee, D. and Howard, P. (2010) 'Corporate cooptation of organic and fair trade standards', *Agriculture & Human Values*, vol. 27, pp.387–399
Kantrowitz, M. (2016) 'Why the student loan crisis is even worse than people think', *Time Magazine*, 11 January

Kimura, A.H. (2011) 'Food education as food literacy: privatized and gendered food knowledge in contemporary Japan', *Agriculture and Human Values*, vol. 28, pp.465–482
Kimmerer, R.W. (2013) *Braiding Sweetgrass: Indigenous Wisdom, Scientific Knowledge, and the Teachings of Plants*, Milkweed Editions, Canada
King, J., Toole, A. and Fuglie, K. (2012) 'The complementary roles of the public and private sectors in U.S. agricultural research and development', Economic Brief 19, Economic Research Service, USDA
Klerkx, L. and Leeuwis, C. (2008) 'Matching demand and supply in the agricultural knowledge infrastructure: Experiences with innovation intermediaries', *Food Policy*, vol. 33, pp.260–276
Kneen, B. (2015) 'Prologue: Free knowledge, seeds, and other beings', in P.W. Elliott and D.H. Hepting (eds.). *Free Knowledge: Confronting the Commodification of Human Discovery*, University of Regina Press, Saskatchewan, pp. ix–xviii
Knudson, W., Wysocki, A., Champagne, J. and Peterson, H.C. (2004) 'Entrepreneurship and innovation in the agri-food system', *American Journal of Agricultural Economics*, vol. 86, no. 5, pp.1330–1336
Kop, R. and Fournier, H. (2015) 'Peer2peer and open pedagogy of MOOCS to support the knowledge commons', in C.J. Bonk, M.M. Lee, T.C. Reeves and T.H. Reynolds (eds.) *MOOCs and Open Education Around the World*, Routledge, New York, pp. 303–331
Labarthe, P. and Laurent, C. (2013) 'Privatization of agricultural extension services in the EU: Towards a lack of adequate knowledge for small-scale farms?', *Food Policy*, vol. 38, pp.240–252
Lave, R., Mirowski, P. and Randalls, S. (2010) 'Introduction: STS and neoliberal science', *Social Studies of Science*, vol. 40, no. 5, pp.659–675
Lemmens, P. (2014) 'Re-taking care: Open source biotech in Light of the need to deproletarianize agricultural innovation', *Journal of Agricultural and Environmental Ethics*, vol. 27, no. 1, pp.127–152
Lesser, L.I., Ebbeling, C.B., Goozner, M., Wypij, D., and Ludwig, D.S. (2007) 'Relationship between funding source and conclusion among nutrition-related scientific articles', *PLoS Med*, vol. 4, no. 1, pp.e5
Lipton, E. (2015) 'Food industry enlisted academics in G.M.O. lobbying war, emails show', *New York Times*, 5 September, https://www.nytimes.com/2015/09/06/us/food-industry-enlisted-academics-in-gmo-lobbying-war-emails-show.html?_r=0, accessed 17 October 2018
Mandrioli, D., Kearns, C.E. and Bero, L.A. (2016) 'Relationship between research outcomes and risk of bias, study sponsorship, and author financial conflicts of interest in reviews of the effects of artificially sweetened beverages on weight outcomes: A systematic review of reviews', *PLoS ONE*, vol. 11, no. 9, pp.e0162198
Marcus, K.E. and Reichman, J.H. (2005) 'The globalization of private knowledge goods and the privatization of global public goods', *Journal of International Economic Law*, vol. 7, no. 2, pp.279–320
McIntyre, B.D., Herren, H., Wakhungu, J., and Watson, R.T. (2009) *Global Report. Agriculture at a Crossroads: International Assessment of Agricultural Knowledge, Science & Technology for Development*, Island Press, Washington, DC
McKie, R. (2016) Scientists attack their 'muzzling' by government, *The Guardian*, 20 February
Mendez, E., Bacon, C., Cohen, R., and Gliessman, S.R. (2015) *Agroecology: A Transdisciplinary, Participatory and Action-oriented Approach*, CRC Press
Menzies, H. (2015) 'Culture is essential to democracy', *Ottawa Citizen*, 25 September, p.C6
Merges, R.P. (1996) 'Property rights theory and the commons: The case of scientific research', *Social Philosophy and Policy*, vol. 13, no. 2, pp.145–167
Miller, H.I. and Chassy, B. (2012) 'Scientists smell a rat in fraudulent genetic engineering study', *Forbes Magazine*, 25 September, http://academicsreview.org/2012/09/scientists-smell-a-rat-in-fraudulent-study/, accessed 14 December 2017
Mirowski, P. (2011) *Science-Mart. Privatizing American Science*, Harvard University Press, Cambridge, MA
Nestle, M. (2016) 'Food industry funding of nutrition research: The relevance of history for current debates', *JAMA Intern Med.*, vol. 176, no. 11, pp.1685–1686
Ortiz, O., Orrego, R., Pradel, W., Gildemacher, P., Castillo, R., Otiniano, R., Gabriel, J., Vallejo, J., Torres, O., Woldegiorgis, G., Damene, B., Kakuhenzire, R., Kasahija, I. and Kahiu, I. (2013) 'Insights into potato innovation systems in Bolivia, Ethiopia, Peru and Uganda', *Agricultural Systems*, vol. 114, pp.73–83
Pardey, P.G., Chan-Kang, C., Dehmer, S.P. and Beddow, J.M. (2016) 'Agricultural R&D is on the move', *Nature*, vol. 537, pp.301–303
Prager, K., Labarthe, P., Caggiano, M. and Lorenzo-Arribas, A. (2016) 'How does commercialisation impact on the provision of farm advisory services? Evidence from Belgium, Italy, Ireland and the UK', *Land Use Policy*, vol. 52, pp.329–344
Rai, S. (2001) 'India–U.S. fight on basmati rice is mostly settled', *New York Times*, 25 August

Rampton, S. and Stauber, J.C. (2001) *Trust Us, We're Experts: How Industry Manipulates Science and Gambles with Your Future*, Putnam, New York
Reichman, J.H. and Samuelson, P. (1997) 'Intellectual Property Rights in data?', *Vanderbilt Law Review*, vol. 50, p.121
Randal, J. (2000) 'Despite embargo, biotechnology in Cuba thrives', *Journal of the National Cancer Institute*, vol. 92, no. 13, pp.1034–1037
Rivera, W. and Alex, G. (eds.) (2004) *Privatization of Extension Systems Case Studies of International Initiatives. Extension Reform for Rural Development*, Volume 2, Agriculture and Rural Development Discussion Paper 9, The World Bank, Washington, DC
Samir, O. (2016) *Philosophy of Science: A Very Short Introduction*, University of Oxford Press, Oxford
Séralini, G.E. (2015) 'Toxicity of GMOs and pesticides: Séralini's team wins defamation and forgery court cases', http://www.globalresearch.ca/toxicity-of-gmos-and-pesticides-seralinis-team-wins-defamation-and-forgery-court-cases/5492480, accessed 13 December 2017
Shulman, S. (2006) *Undermining Science: Suppression and Distortion in the Bush Administration*, University of California Press, Berkeley and Los Angeles, CA
Simon, M. (2013) *And Now a Word from Our Sponsors: Are America's Nutrition Professionals in the Pocket of Big Food?*, http://www.eatdrinkpolitics.com/wp-content/uploads/AND_Corporate_Sponsorship_Report.pdf, accessed 13 December 2017
Starr, D. (2004) 'The Cuban biotech revolution', *Wired Magazine*, 1 December, https://www.wired.com/2004/12/cuba/
Sustainable Pulse (2012) 'French scientists slam orchestrated attacks on Seralini GMO study', *Sustainable Pulse*, 18 November, http://sustainablepulse.com/2012/11/18/french-scientists-slam-orchestrated-attacks-seralini-gmo-study/#.WHvjXpKkfoU, accessed 13 December 2017
Uddin, E., Gao, Q. and Rashid, M.U. (2016) Crop Farmers' Willingness to Pay for Agricultural Extension Services in Bangladesh: Cases of Selected Villages in Two Important Agro-ecological Zones', *Journal of Agricultural Education and Extension*, vol. 22, pp.43–60
Van Noorden, R. (2013) 'Open access: The true cost of science publishing', *Nature*, vol. 495, pp.426–429
Walker, B. and Salt, D. (2012) *Resilience Practice: Building Capacity to Absorb Disturbance and Maintain Function*, Island Press, Washington, DC
Wang, S.L. (2014) 'Cooperative extension system: Trends and economic impacts on U.S. agriculture', *Choices*, vol. 29, no. 1, pp.1–8
Washburn, J. (2005) *University, Inc. The Corporate Corruption of American Higher Education*, Basic Books, New York
Weiner, N. 1954 (1993). *Invention: The Care and Feeding of Ideas*, MIT Press, Cambridge, MA
Wilson, S. (2008) *Research Is Ceremony. Indigenous Research Methods*, Fernwood Publishing, Halifax and Winnipeg

13

WESTERN GASTRONOMY, INHERITED COMMONS AND MARKET LOGIC

Cooking up a crisis

Christian Barrère

Introduction

Cuisine is a cultural process transforming natural resources into foodstuffs and, according to technical recipes, into more or less elaborated dishes. Its main goal is feeding people. Nevertheless, under specific circumstances, the *feeding function* of cuisine diminishes for the benefit of its *pleasure function*; food is prepared not mainly to feed people but to delight them. Gastronomy appears thus as a segment of cuisine. In these circumstances, the preparation and consumption of food acquires autonomy from the nutritional content of the raw ingredient, with the search of pleasure prevailing over any other component of what is eaten. For some people (Gods or chiefs, for instance) and/or under specific circumstances (feasts, ceremonies, etc.) the act of eating is not only connected with necessity but with pleasure and honour. Along those lines, in Western societies, when we think of gastronomy we think of meals that are consumed for the sake of pleasure rather for utilitarian purposes; they become 'hedonic goods' (Hirschman and Holbrook, 1982; Holbrook and Hirschman, 1982; Addis and Holbrook, 2001). Tastes and preferences then move beyond strict rationality since they result from emotional behaviour, which needs to consider rationality and emotions together, unlike the mainstream hypothesis of economic analysis.

Gastronomy, as a very old invention, takes different forms according to the culture and the social structure of societies. Generally, we can distinguish popular gastronomies, dedicated to the 'lower' classes, and aristocratic or elitist gastronomies reserved to 'upper' ones. Popular gastronomies are often defined by domestic principles and by domestic dynamics. However, in contemporary Western societies, market relations shape an important part of the gastronomic supply. Similarly, market forces represent the dominant driver of the transformation of gastronomy and the main source of gastronomic creativity. This chapter focuses on the market sector and, mainly, on the relation between its intrinsic logic and the use of commons. Contrary to appearances, this sector does not only use market resources; it would not work without a systematic use of commons!

Market gastronomy works by using creativity and inherited knowledge commons. On the basis of old dishes and recipes, and by reinterpreting them, chefs create new ones. The development of gastronomy is a ceaseless movement of combination of creativity and heritage. Thus, the history of Western gastronomy can be analysed through the lens of the commons and in relation to a specific development path. Gastronomy, as a set of open knowledge and cultural norms, is clearly a commons.

However, if we travel back in time to the European Middle Ages, we could witness how the strict segmentation between a dominant group, the aristocracy, and a dominated one, the common people, produced the splitting up of the culinary and gastronomic commons into two different categories: *aristocratic* and *popular*. Modern gastronomy, developed for the social elite and driven by market forces, re-articulated this separation by building a hierarchical model of gastronomy from the top (the elitist niche) to the bottom (the traditional bourgeoise cuisine). As a consequence, traditional popular gastronomy was undervalued and marginalized. It was the post-war period that led to the over-valuation of the elitist gastronomy and to the development of its sophistication and costly character. Today, however, this model encounters growing difficulties. The extension and intensification of competition, the costly investments that are needed and the growing financial returns lead to the hyper-sophistication of cooking and its over-commodification, undermining the economic and social sustainability of the Western market model.

Within this framework, the second section of this chapter is dedicated to the definition of 'inherited commons' as the analytical tool deployed to understand the transformation of gastronomy. In the third section, the chapter focuses on the study of what is called gastronomic commons. The fourth section engages with the French gastronomic model as the basis of the dominant model of market gastronomy all over the Western world. In the fifth section, the effects of market-based relationships on the organisation and evolution of this model are considered, along with the way in which they led up to its present difficulties. This raises the question of a new gastronomic model which is anchored in a gastronomic pluralism that drinks from all the inherited commons of aristocratic and popular heritage.

An economic and social analysis resorting to the notion of inherited commons

Cuisine and gastronomy consume scarce resources whose production and transport comes at a price. Inputs (raw ingredients, energy, gender disparity, etc.) are transformed through productive processes that combine labour and equipment. Their economic dimension cannot be overlooked. Some gastronomic establishments make astonishing revenues; famous chefs export their restaurants all over the world (Joël Robuchon owned thirteen restaurants with a total of twenty-five Michelin stars; Paul Bocuse, a three-star chef since 1965, had seventeen; Gordon Ramsay has eleven) and others build profitable groups (under the Ducasse group name, Alain Ducasse has thirty restaurants throughout the world, and a total of eighteen Michelin Red Guide stars). Furthermore, specialists in tourism are shedding light on the growing role of gastronomy in the touristic appeal of certain localities and specialists in gastronomy note the shifting of restaurants towards touristic areas.

Despite the continuous attempts to transform gastronomy into an economic and business subject, there are at least three reasons why we cannot analyse it according to the standard apparatus of microeconomics or industrial organisation:

- Cuisine and gastronomy connect the individual with the collective dimensions. As such, the study of cuisine and gastronomy cannot be limited to market equilibrium between individual demand and supply; social context must also be taken into account. Cuisine has long since been organised mainly by collective and non-profit institutions as family and community and often still is. Even though individual talents, behaviours and strategies play a crucial role, a focus on non-individual factors remains absolutely necessary in order to grasp the relation between the collective and individual dimensions of cuisine and gastronomy and to examine how market and non-market processes cohabit in hybrid modes of organisation. Besides markets and private properties, the functioning of the gastronomic field is populated by commons, heritages and hybrid forms of property and use that cannot be trivialized (i.e. who owns the cooking recipes?).

- Cuisine and gastronomy depend on the cultural and semiotic dimension of food: if fifty years ago eating horse meat in Western Europe was normal (and often recommended for medical reasons), today most people consider this meat inedible because horses are too close to human beings to be transformed into food. In the same way, the distinction between luxury and ordinary food and between standard and gastronomic cuisine, depends on food goods' semiotic value. Major changes are discernible over many years. In the 19th century, the captains of fishing boats on the Gironde and Adour rivers in south-west France used to give their employees the catch of wild sturgeon, as non-noble, un-saleable fish; employees would eat the fish, but only after feeding the eggs to their pigs. Today these eggs constitute the caviar of Aquitaine – which is as expensive as the Iranian product. Taste and food attitudes result in, and are produced by, individuals' integration into a situated community, place and time.
- Finally, cuisine and gastronomy connect past (mainly the culinary legacy) and present (mainly via innovation and creativity). Because social norms and routines are embedded within a historical and geographical context (Italian gastronomy being fundamentally different from American gastronomy), the historical and territorial points of view are necessary to fully understand the current problems and transformations of gastronomy. By including the past in our analysis, we step right away from the formalist approach of mainstream economics. Gastronomy should thus be studied from the perspective of commons and heritages analysis because this new framework allows consideration of social, collective, diverse and past dimensions.

Therefore, this chapter proposes a heterodox economic and social analysis based on the notion of commons and, more precisely, on the notion of inherited commons (see Section 3 below), that catches the collective, cultural and historical dimensions of gastronomy and contrasts it with contemporary practices of appropriation and commodification.

The foundations of gastronomy: from food and culinary commons to gastronomic commons

Whether we are aware or not, eating is a social and cultural activity. It supposes production and selection of foodstuffs and their combination with a cultural substratum so as to build cuisine as a cultural and social practice. Nutritional values could refer only to the individual, but the same could not be said for food – upon which all of this confers a semiotic dimension that is fundamentally dependent on social values and norms. People eat differently in every society, thus developing particular cultural uses of any potential foodstuff; however, not every produce is good to eat, even if it has relevant nutritional value. According to Lévi-Strauss, "natural species are chosen not because they are 'good to eat' but because they are 'good to think'" (Lévi-Strauss, 1962, 1964: 89). In addition, cultural norms such as the distinction between *le cru* (the raw), *le cuit* (the cooked) and *le pourri* (the putrefied) serve to classify foodstuffs and the different ways they are used (Lévi-Strauss, 1964).

Thus, transforming natural resources into foodstuffs according to social norms and technical recipes is a cultural and collective process that is produced by all, benefits all and is not enclosed by absolute proprietary regimes. Therefore, it is a cultural commons in all human societies. Similarly, gastronomic commons as the top segment of culinary commons must be considered as cultural commons, i.e. "a cultural resource shared by a group where the resource is vulnerable to enclosure, overuse and social dilemmas" (Hess, 2012: 25). Such a cultural commons has at least four components.

First of all, it concerns the selection of natural resources used in the food process. For instance, Fischler (1993) argues that the present distinction between eatable and uneatable varies from country to country and from one culinary culture to another: in forty-two cultures dog meat is commonly eaten; in many countries rat meat is much appreciated; ants are cooked in Colombia, as are bees, wasps and cockroaches in China. That means that the cultural commons shared by the people in a specific context defines the framework within which tastes and preferences can be displayed, but the cultural commons also contributes to build an identity. At the same time, it includes the ways of transforming resources by developing new resources on the basis of nature and culture: farming, selection of plants, breeding and so on. Thirdly, it concerns the use of resources: ways of preparing food, recipes, suggestions of foodstuffs or meals, preserving, and cooking food, but also the manner of eating and drinking. Finally, it includes the rules, constraints and taboos that organise food activity.

As a community changes through time, it inherits and reproduces cultural rules around the organisation of eating. While it keeps some, it may abandon or modify others. Culinary commons become *inherited commons,* i.e. culinary heritages (Di Giovine and Brulotte, 2014; Barrère et al., 2012). They are social constructs passing through time. They frame the present context and their influence is particularly strong in the field of cuisine because they define foodstuffs, preparations, uses, presentations and dining manners. They are not homogeneous and they are not stable; on the contrary, they include more or less coherent sets of principles, attitudes and rules and are in continuous transformation. Since they are embedded within a given culture, they express the identity of a community (making it specific and different from any other). Inherited commons can be said to be the reason behind these specificities. For instance, Poulain (2002) analyses the specificity of the French food model and its three structured meals. Similarly, Fischler and Masson (2008) and Rozin et al. (2011) study fundamental differences in attitudes towards food held in different countries – mainly between French and American adults. They connect these to cultural trends: "compared to the French" (Rozin et al., 2011: 1). They conclude by stating that "Americans emphasise quantity rather than quality in making choices, Americans have a higher preference for variety, and Americans usually prefer comfort things (that make life easier) over joys (unique things that make life interesting)" (Rozin et al., 2011: 1).

The transformation and changes in gastronomy produce new commons within the culinary commons. Rules, norms, recipes and eating manners dedicated to gastronomy increase their specificity from ordinary cuisine. According to the observation of Ostrom (1990), "the devil is in the details"; the specificity of culinary and gastronomic commons has to be understood and respected. Culinary and, mainly, gastronomic commons are specific commons that operate far from the economic image of commons as a set of gathered and undifferentiated resources (Barrère et al., 2012). This happens for a series of reasons, starting from the fact that cuisine and gastronomy are collective processes that are socially organised (i.e. organised by a community). A given cuisine arises out of a social and cultural building of local communities and societies according to the semiotic value of food. It connects the commons to the group's identity in relation to other groups and expresses its specificity through collective idiosyncrasy (Di Giovine and Brulotte, 2014). Secondly, cuisine and gastronomy are socially structured and differentiated because they define the uses of food (according to social conditions, status, age and gender) that draw together different situations and different social status: food intended for the Gods cannot be the same as food for the people; food for dominant groups is not the same as food for ordinary people; food for feasts and religious ceremonies is not the same as everyday food. As a consequence, food, culinary and gastronomic systems are not inherently equal, just and democratic. On the contrary, they constitute sets that serve to organise and consolidate differences; they are structured sets that reproduce hierarchy and fragmentation. Then, culinary commons appear as a large set composed of different and generally intertwined subsets: commons coming from ordinary cuisine, commons coming from popular gastronomies and commons coming from aristocratic or elitist gastronomies.

Finally, culinary and gastronomic commons pass through time by means of cultural transmission. Since they include rules of composition (structures) and are connected to places, they constitute *heritages* (for a development see Barrère, 2016, 2018). These sets can be local, regional or national heritages and can be more or less rigidly structured. The spatial dimension of heritage derives from the connection between heritage and communities. However, a world set of shared gastronomic resources does exist and everybody can use these resources; for instance, in any place in the world, one can concoct a bouillabaisse using the recipe that originated in Marseilles or sell a pizza, even far from Naples. Nevertheless, a large part of these cultural practices is strongly linked to local cultures and communities and belong to specific heritages, to regional or local ways of cooking, largely dictating how dishes are made, how flavours are blended, how textures are combined and so on. The world set of gastronomic commons collects resources worldwide: recipes, knowledge, know-how and the organisation of meals, manners and so on. This large commons includes different subsets that are more or less specific - and thus, more or less compatible. For instance, for a long time, Western gastronomy utilised spices and other condiments (though not indiscriminately) obtained from Asian gastronomies and geographies. Moreover, as the use of culinary and gastronomic cultural resources is generally non-rival, a great social dilemma is not to avoid waste and overuse – that generally affects natural commons - but to produce and develop them (Madison et al., 2010). In our case, that means developing gastronomic creativity and adding it to increase the inherited gastronomic commons.

The Western gastronomic model

The organisation of Western gastronomy is rooted in the historical division between popular and aristocratic gastronomy. However, it represents an original combination of their inherited food and culinary commons, over-valuing the aristocratic heritage and under-valuing the popular one. In my reading, Western gastronomy was formalised by the French model of gastronomy as a top-down process, directed by the elitist tastes and desires that were produced within and around the French Royal Court. Because of historical contingencies and long-term pan-European processes, this model was then exported through the whole continent, and, later, to the whole Western world. For this reason, talking about French gastronomy is, although indirectly, a way to talk about Western gastronomy.

1) Popular and aristocratic gastronomies

Historians have shown that the upper classes of ancient societies – and sometimes the lower classes too – were familiar with gastronomy. In the Christian Europe of the Middle Ages, the development of gastronomy was limited by a religious ideology that broke with the cult of pleasure that existed in Greek and Roman civilisations; people were on the Earth awaiting the Judgement Day rather than for enjoyment; the Devil, omnipresent, lurked behind every pleasure – as can be seen in the religious frescos that painted 'gluttony' as a deadly sin. Yet the continuous repetition of this propaganda shown that the desire for culinary pleasure was ever-present and that popular gastronomies persisted.

Using regional resources (cereals, game, truffles, fish or mushrooms, for instance), locally selected and prepared, these popular gastronomies defined regional recipes, mainly for special occasions (Easter, Christmas or the end of harvest - see the peasant feasts painted by Brueghel). Recipes and local norms circulated within the community, they were commons in the sense of a cultural resource shared by a group. The public component of gastronomic production was dominant; no individual contribution could be credited (who invented *bouillabaisse, Christmas turkey* or *truffle risotto*?). Indeed, early recipe books were collections of traditional recipes (generally no specific author of recipes could be identified and such books were often published anonymously) and made no claim to innovation. These commons reproduced and changed through time by a process of cultural transmission, most of the time through the gendered mother–daughter relationship within families. As such, cooking was mainly a conservative act of repetition of the known; innovation and experimentation were very low. Yet recipes were able to change, particularly when cooks in the West encountered new knowledge, new recipes and new resources. For example, the emblematic dish of French south-western cuisine – *cassoulet*, which is based on haricot beans, used to be made using broad beans and chickpeas until the 17th century, when these were supplanted by the haricot bean, following its arrival in Europe from Mexico.

Since the European Middle-Ages, aristocracy wanted to be distinguished from the people. In terms of narrative, they imposed a conception of society as a set of two closed subsets radically distinct – the first one having blue blood while the second had only red one. One way aristocrats had to separate themselves was to develop a specific gastronomy that was impossible to achieve outside of the wealth of palace. Good taste was considered to be specific to aristocracy – and thus expensive luxury goods should be reserved for it. The vulgar people were considered to be incapable of appreciating these foods for which a taste had to be acquired: "it would be as to throw pearls before swine", it used to be said.

In the field of nutrition, the hierarchical construction of food and society led to rank food according to status and classes. The best variant of this conception was the paradigm of the Great Chain of Being (Lovejoy, 1936) according to which the universe was considered on the basis of a hierarchical principle from Earth to sky, from material to spiritual, from sin to perfection, from the Devil to God. The same principle governed the hierarchy of food: at the bottom, bulbs (garlic or onion) and roots (carrots, beets, or turnips), which were under the ground; then vegetables whose leaves started from the root (salads, spinach); followed by vegetables whose leaves started from the stem (peas, cabbage); then foodstuff located in the air (cereal); then fruit coming from trees (pears had a strong value but strawberries, which were on the ground, were looked down on); the birds, higher still, and so on. This 'natural' order implied parallels between foods and eaters. Upper beings had to eat upper food while lower beings ate lower food; for instance, peasants ate leek and onion but never peacock or swan – while the reverse was true for aristocrats.

Therefore, the aristocratic product was portrayed as exceptional in comparison to the everyday product; it was (it had to be) a 'higher' good marking the absolute superiority of aristocracy; in the case of food: fine wines, 'noble' game – wild boar or stag but not rabbit. Even on extraordinary occasions, popular consumption was different from aristocratic consumption: aristocratic celebrations were not the same as village ones and balls and dances were different too – the popular banquet painted by Breughel was very different from the paintings of court banquets.

In the Middle-Ages, the aristocratic banquet was characterised by surfeit – an abundance of dishes, scarce and expansive foodstuffs, waste – and aristocratic cuisine was quite conservative.

Yet French Court society dramatically changed things. Rare and costly foodstuffs, coupled with perfect execution, were not enough. The rivalry between courts to seem most magnificent led to continuous outbidding in terms of originality, novelty and creativity. By making cuisine a social issue that was as crucial to the court as fashion (Beaugé, 2010: 5), Louis XIV reinforced this movement. To dazzle the *Grands*, the *Grand Cuisinier* resorted to creativity: new recipes, new presentations and new sauces. No longer an artisan, he (because women were not permitted to be *Grandes Cuisinières*) became an artist, so that cooking was no longer a material activity but rather a formalised, aestheticized and thus intellectualised pursuit. The chef became the core of aristocratic gastronomy and an object of envy among monarchs and nobles, who competed among themselves to attract the best.

Throughout centuries, the aristocratic gastronomy model imposed a costly and complex cuisine with a very long supply chain aimed at satisfying high tables with scarce, high-quality and 'noble' foodstuffs. For example, lobster, *foie gras*, sweetbreads, caviar, oysters, asparagus, turbot, etc. were sources throughout France and, later, all over the world. The preparation of dishes and the transformation of raw materials required (and requires) highly qualified labour in the context of lengthy preparations. As such, aristocratic cuisine was clearly very different from popular cuisine, including in its gender dynamics. The *cuisinier* (chef) was defined in contrast to the *cuisinière* (housewife) in exactly the same way as the *grand couturier* was defined in opposition to the *couturière* (seamstress). This aristocratic gastronomic social model made room for individuals (mainly chefs) yet at the same time defined both their function and the pattern of their work.

Aristocratic gastronomy introduced precise sets of principles, rules and institutions; for instance, instead of stressing the product, aristocratic cuisine was aestheticized, prioritising the presentation of the dish. They included recipes, but mainly manners of eating and drinking, the cult of novelty and creativity and a normalised process of producing gastronomy. Similarly to popular gastronomy, aristocratic gastronomy also reproduced along time as a heritage.

2) The development of market relations and evolution towards an elitist gastronomy

In the post-French revolution Western world, the development of market relations in the gastronomic field came about with the emergence of the bourgeoisie as the dominant class and capitalism as the way of framing socio-economic relationships. Aristocratic cuisine was replaced by the elitist version that was served up at the great restaurants.

Similar to the aristocracy, the upper bourgeois class had money and could buy goods – even expensive ones – and had access to luxury goods. They replaced aristocratic splendour with luxury – and in so doing encouraged the expansion of the luxury sector in general and gastronomy in particular. They created strong ties with the old aristocracy and contributed to the establishment of a social elite separated from the 'vulgar' people by means of wealth, power and education. Old aristocracy and the high bourgeoisie flocked to new and ostentatious leisure venues (Veblen, 1899): seaside holidays, horse races, concerts, art galleries, dancing – and gastronomic restaurants.

Competition and the search for profitability led to both the rationalisation and standardisation of the old aristocratic gastronomy. Whereas aristocratic behaviour was characterised by a distance from economic calculus and economic rationality, the development of market logic implied an adaptation of aristocratic heritage in order to take costs and benefits into consideration and build the cuisine's rationality. That led to a new model, which was supposed to be rational and coherent. It was the end of the extravagant sophistication of aristocratic cuisine.

Creativity and novelty remained dominant – but sophistication diminished. According to Parkhurst-Ferguson (2004:10), Carême (1783–1833) achieved "the reconfiguration of the aristocratic cuisine of the Ancien Régime into the elite and assertively national cuisine of the nineteenth century". Carême broke with the extravagance of court cuisine (henceforth too costly) yet continued to draw inspiration from the old heritage of the aristocratic cuisine. In addition, he developed the intellectualisation and scientification of cuisine. Inspired by enlightenment and the idea that everything could be controlled, cooks stopped using traditional recipes, preparation methods and routines and were interested in understanding the 'laws' of cuisine, the processes of transformation of inputs and the mechanisms of taste.

At the time of the industrial revolution, technological innovation in the kitchen was encouraged; cooking was increasingly defined as a professional activity demanding specific and lengthy training, applying professional norms and routines within the context of a rigid organisation (the *brigade* with its divisions, the *sauciers*, the *légumiers*, the bakers and so on). Plus, the market-based organisation adopted by the elitist gastronomy and professionally executed gradually took it further from popular gastronomy, which was mainly organised within the domestic arena.

A further aspect of standardisation was the process of selecting ingredients and recipes that belonged to other commons and in particular, to popular and regional cuisines. In France – unlike other countries as, mainly, Italy – the powerful process of building a national identity and developing central power against local autonomies was accompanied by the definition of a national cuisine – including some 'reconfigured' regional dishes and recipes, though others were excluded as vulgar. Bourgeois cuisine was thus a rationalised and euphemized form of aristocratic cuisine, but at the same time, and more interestingly, a wilful attempt to appropriate and utilise elements of local and 'lower' gastronomy. In doing so, *chefs* created a continuum from elitist gastronomy via bourgeois gastronomy to popular gastronomy which contrasted with the old separation between aristocracy and common people; however, it was a top-down process that extracted from below in order to please the palate of those above.

This hierarchical model over-evaluated elitist cuisine coming from the aristocratic commons, even if they were used in a rationalised way. The *Grand Restaurant*, as the epitome of elitist gastronomy, at the top of the pyramid of restaurants, ranked according to their evaluations by experts, guides and customers, constituted the ideal type. In its menus and space, it concentrated the creativity of the gastronomic field and the most talented chefs. Its recipes showed off sophistication and the time that had been lavished on the food preparation. Every element was luxurious, including the crockery, glassware and tablecloths; the staff was large and competent. The restaurant presented a profusion of dishes (*mignardises, amuse-bouches, trous normands*, etc.), offering a wide variety of choices as well as many complementary extras (famous wines, cigars and alcohols) – and doing so in a sumptuous setting. Costs and prices were, needless to say, a representation of the status associated with the restaurant and with the idea that taste and gastronomy were a matter of class.

3) Exporting the elitist model and the diversity of Western gastronomy

The French elitist model of gastronomy became a European term of reference. This happened because of several historical moments: the link between aristocracy and the French Court culture; the disregard for the gastronomy of English and Spanish puritan societies; the influence exerted internationally by the French luxury firms; and the reputation of the French *bon gout*, which was reproduced in literary accounts and vulgar knowledge. Furthermore, other factors favoured the export of this model and drove the organisation of Western gastronomy along its lines. First of all, because the model was defined as rational and scientific, it could be disconnected with the specific national context and associated with status and practices rather than geographies and culture. What was good to eat for French elite would also be good for other countries' elites. French 'culinary imperialism' held, then, that the French knew what 'good cuisine' was and thus could and should pass it on to the rest of the world.

Secondly, since Carême's cuisine no longer relied on empirical knowledge based on repetition and errors, it was possible to develop a coherent system of principles and applicative techniques. With the technical systemisation of cooking, it became easier to reproduce recipes in other geographies. In the words of Parkhurst-Ferguson: "Carême's French cuisine is not tied to or rooted in a particular place" (2004: 71).

Thirdly, the popularity of French gastronomy across the world elite grew to the point that foreigners visited France to eat in the most famous restaurants. These businesses benefited from a superstar effect, but at the same time they operated as attractions for many who loved and admired the French way of life. Finally, experts and guides in the most important cities of the Western world mimicked the Michelin Guide (the Red Guide) and favoured the consolidation of sophisticated gastronomy. Chefs, for their part, mimicked French chefs and competed for Michelin stars (experts think a new star increases the sales revenues by anywhere from 20 to 30% in France). The growing role of capital investment in the gastronomic field sped up the process.

Thus, Western gastronomy, more or less quickly, adopted a model based on the over-valuation of elitist principles and on the disregard for popular and local gastronomies, which tended to be considered an impoverished version of the elitist gastronomy; that does not mean popular or local gastronomies disappear, but that they are under-valuated. In some countries, as Italy or England, the old aristocratic families continued to honour the old aristocratic principles of gastronomy, but it was happening more within the domestic area than in the market of elitist restaurants. One of the consequences was the reduction of the speed at which elitist restaurants diffused in these countries, at least until recently. On the other hand, in some regions local cuisines continued to play a strong role, mainly when they express regional identities. This was even more the case in countries where different regions were competing, as in Italy (there remained a Sicilian cuisine, a Tuscan and Piedmontese ones) and in Spain (the Catalan and the Basque, for instance). These cuisines remained strong when they were founded on local produce and local institutions, constituting local districts and local commons, based on local agrifood systems and benefiting from the advantages of the economy of proximity (Torre and Rallet, 2005). Instead of developing a 'cuisine of techniques', they favoured a 'cuisine of produce'.

The present limits of the domination of elitist gastronomy

During the 20th century, the transformation of Western gastronomy has been driven by market competition and by the search for profitability. This has led to the establishment of elitist gastronomy as a narrow, transnational niche. Compared to the past, Western gastronomy has become increasingly autonomous and different from popular ways of eating and cooking. Today, it might seem that the triumph of the elitist gastronomy is accomplished. Chefs are stars who populate TV channels and newspapers; nevertheless, a closer look shows that the extension and intensification of competition stretch to the limit this elitist type of organisation and finds itself questioning its own economic and social sustainability.

1) The extension and intensification of competition

One of the main drivers and implications of the contemporary expansion of the profit logic in the Western gastronomy is that eating and enjoying food has increasingly become commoditised.

The new term of reference is the Michelin Red Guide "starred" restaurant, which reproduces the remaining features of the old aristocratic model: a famous chef, extraordinary products (e.g. truffles, caviar, lobster, etc.), sophisticated dishes, sumptuous setting, service and expansive cooking. The competition between restaurants implies a race for stars and profit; when Alain Ducasse was hired as chef at Le Louis XV in Monte Carlo, his contract stipulated that he had precisely one year to obtain the coveted third Michelin star; had he failed to do so, he would have been dismissed! This anecdote shows the strong connection between gastronomy and capital that develops when elitist gastronomy becomes a product for high-end consumption; producing that commodity is costly, in particular in terms of products and setting and thus needs strong investments, implying moving closer to financial and luxury groups.

Experts and guides justify this model, defining membership of each segment within the pyramid of legitimated gastronomy, from the not-starred-but-selected restaurants to the narrow niche of the three-starred ones. They also contribute to the development of this model and the corresponding segmentation in up-and-coming 'culinary' countries and cities (e.g. Spain, Italy, Japan, London, New York, etc.), where gastronomic consumption is on the rise.

One of the most significant transformations of the contemporary gastronomic framework is that globalisation extends elitism of production and consumption worldwide. On one hand, French elitist restaurants are opening up throughout the world (Bocuse, Ducasse and Robuchon hold transnational franchises, to name a few). On the other hand, the rush towards fame and money is such that several local chefs enter into the competition to create new elitist cuisines, strengthen their reputation, open new restaurants throughout the world and participate in the global market for elite consumers.

Moreover, as in the leader-follower models, followers have to over-invest in innovation in order to challenge those who are ahead. They in turn tend to develop an increasingly creative and sophisticated cuisine, but also to strengthen the vicious circle of investment, financialization and commodification. An iconic example is the new Spanish creative cuisine, representing a strong challenge to French cuisine. Ferran Adrià (followed by numerous Spanish chefs) embarked upon a creative bidding race, applying new technologies (based on liquid nitrogen and centrifugation), inventing new textures and developing molecular cooking as a new semiotic cooking style featuring dishes presented as 'paintings'. Adrià's claim to be an artist was eventually legitimated in 2007, when he was invited to the documenta exhibit in Kassel.

A new challenge, for Western elitist gastronomy, derives from the industrialisation of cuisine. This process gathered pace during the second half of the 20th century, when culture and entertainment fields had to give way to market and capital. The Frankfurt School (Benjamin, 1936) interprets the fundamental changes brought about in the cultural field (and that were later to affect the fields of cuisine and gastronomy) as the birth of a mass culture adapted to suit a mass society. The process had two sides: substituting capitalist relations for domestic and market ones; industrialising the production process of goods and services.

On the one hand, capitalism invaded areas yet managed by non-capitalist relations, mainly in the field of household consumption. That modified the food systems in relation to the development of women's paid work and the increase of urbanisation. But demands inspired by desires or pleasures, mainly in the middle and upper classes, but also among industrial workers, became increasingly significant in family spending, either superseding or adding to those strictly related to basic and material needs and opening new opportunities for capitalist investment.

On the other hand, technological developments allowed the mass production, mass distribution and mass consumption of cultural, artistic and craft goods in societies in which the middle and upper classes were experiencing rising living standards. In terms of food, it led to strong changes in the way food was produced, selected, sold, packaged, preserved, prepared and eaten.

Within the gastronomic market, a pseudo-gastronomy (low quality, based on industrial cuisine) emerged. With the development of new technologies allowing the standardisation and industrialisation of cooking (vacuum cooking, deep-freezing, and other mass-production technologies), industrial cuisine was able to supply gastronomic or pseudo-gastronomic services to large numbers of consumers and throughout the planet. Households represented the initial targets for these products but, increasingly, many restaurants started serving both home-made dishes and industrial dishes purchased from new food companies. Industry became able to provide food businesses with standardised quality for sophisticated dishes. The great chefs' recipes were copied and dishes that looked very much like the gastronomic ones became available; though not comparable with those of the famous restaurants, their prices were far lower.

Nevertheless, the top gastronomy had to justify the difference in prices between low and industrial gastronomy and the old elitist one, a difference which was strongly growing. Then, top elitist gastronomy overreacted to this competition by stepping up its innovation, creativity and (above all) sophistication. They looked for something that industrial cuisine was unable to supply.

2) A market crisis: is elitist gastronomy financially sustainable?

A point that is worth particular attention is the fact that the financialization of elite gastronomy and the growing global competition are connected with steeply-rising financial costs. Baumol and Bowen have introduced the idea of what they call the Baumol effect (Baumol and Bowen, 1966); production costs are ever higher and productivity gains (for cuisine and restaurant staff, for instance) often impossible to achieve. Moreover, investing in setting increases expense, weighing heavily on profitability. From the fifties to today, prices increase and those that rise most during the period belong to the most gastronomic services: 'à la carte' rather than 'prix fixe' menus and high-end prices rather than low-end.

From the year 2000, this movement accelerated (cf. Barrère et al., 2014 for Parisian gastronomy) and led to an increasing disruption of elitist gastronomy. On the one hand, as already noted, some very expensive establishments that serve highly sophisticated cuisine make huge amounts of money and export their restaurants worldwide. Benefiting from a superstar effect, their reputation is such that they are able to offer sophisticated cuisine at very high (and continuously rising) prices. For example, in Paris, dinner for two at a two- or three-starred restaurant costs in excess of 1,000€ (more than $1,100). These establishments concentrate on places in which they can reach a very wealthy clientele – and rich tourists in particular. A growing number of these are connected to luxury hospitality financial groups and are used to attract the most affluent customers so that their low profitability (sometimes non-profitability) is compensated for by the earnings of luxury hospitality. On the contrary, other restaurants that aim to reach the top of their league are unable to reach this clientele (often due to location) and begin experiencing serious financial difficulties. As a matter of fact, gastronomic restaurants, then, can only survive when their customers are numerous and rich. In France, Pierre Gagnaire, who had a three-starred restaurant in a mid-range city (Saint-Etienne), was not able to earn enough to cover his costs, so he went bankrupt, closed the restaurant and moved to Paris.

The necessity of achieving superstar status in order to survive exacerbates the difficulty of assuring the succession of great chefs. In the elitist regime, grand chefs' main assets are their reputation and social capital, which are inextricably connected to their talent, their personal investment in the activity, their person and thus, their name. Once they leave the job, what can possibly guarantee that the quality and creativity of the cooking will be maintained? Their succession is very uncertain. When a son or a daughter steps into their shoes, consumers may suppose some level of continuity. An alternative is, as in *haute couture*, the transformation of the *griffe* (the creator's own name) into the trademark (as a guarantee of quality); everyone knows, of course, that Alain Ducasse is not able to be simultaneously present at all eighteen of his starred restaurants (indeed, he now states that he no longer cooks) and the same is true of Gordon Ramsay, with his New York and Los Angeles establishments – but it is accepted that they leave their restaurants in the capable hands of top-notch chefs (just as consumers know that Christian Dior, who died in 1957, no longer creates dresses). Yet in gastronomy, as elsewhere, establishing a trademark implies heavy investment, limiting this possibility to the lucky few who have generated enough financial capital to guarantee the continuity of the business. As a consequence, chefs spend more and more time in public relations; this leads to a purely commodified and strange system that over-values the name and talent of chefs who do not cook anymore!

The increase in prices and the consolidation of a global, gastronomic elite contrasts (or is the reflection) of the growing inequalities of the present world. Since the 1980s, the wealth of rich and super-rich people has been growing and from 2010, this movement is accelerating; in 2017, the number of High Net Worth Individuals (HNWI) who have investable assets of US $1 million or more is at 16.5 million, more than a 50% increase compared to the 10.9 million in 2010, with wealth of US $63.5 trillion (42.7 in 2010; Capgemini, 2018). When it comes to food, poverty and hunger are growing and the middle class cannot afford market gastronomy and high-end restaurants any longer. In Paris, for example, the average prices of the best class of Michelin-selected restaurants, according to the setting, grew from 50€ in 1960 to 160€ – constant euros – in 2012. Similarly, the average prices of the top restaurants in the Michelin guide grew from 50€ in 1960 to 250€ in 2012 (Barrère et al., 2014). At a time of austerity and social inequality, the social issue adds to the economic issue.

3) A social crisis: is elitist gastronomy socially sustainable?

These days, cultural economics is no longer restricted to the traditional efficiency criterion (defined as efficiency of allocation of scarce resources). On the contrary, it aims to consider the social consequences of economic workings and quality of life in particular (Santagata, 2010). People benefit from this social and cultural heritage and, in return, they participate in its preservation, evolution and enlargement. The formal and informal components of this social heritage form the basis of the individual's social life while simultaneously building social cohesion. Moreover, current patrimonial inflation reveals that both individuals and communities are increasingly seeking out their roots and links to the past. Culinary and gastronomic heritages are part and parcel of social heritages and also contribute to the quality of social life. Individual access to these heritages is therefore a crucial issue in improving the welfare of all. The problem is twofold and involves both the access to gastronomy and the access to the cultural knowledge included in the gastronomic commons.

As previously seen, a growing section of the population, even within the bourgeoisie, is now beyond market elitist gastronomy and condemned to the pseudo-gastronomy supplied by industrial cuisine groups. The issue of accessing gastronomic commons is a difficult one and must be taken seriously. On one hand, gastronomy is a social and collective creation, but, on the second one, some individuals strongly contribute to the enlargement of the inherited commons and the contribution of famous chefs cannot be denied; nevertheless, they create new dishes, new recipes and new types of presentation on a double basis. First of all, they inherit commons by revisiting old recipes and benefiting from experimental rules and techniques that have been passed down from generation to generation of chefs and non-professional cooks. Moreover, they benefit from the present cultural atmosphere: if in the past the presentation of plates was inspired by the impressionist painting, today it is by the performance and exhibitions of modern art. Consequently, despite the attempts of some chefs to transform the commons into private property by copyrighting recipes, it has been impossible to implement a definition and a protection of individual recipe contributions based on the idea of appropriation and exclusion. As a matter of fact, the market logic is unable to build an individualistic and exclusionary system of property rights in a sector where creativity and heritages inextricably mix (Barrère and Chossat, 2004).

However, the interchange between elite and people is not reciprocal. If haute-cuisine can draw from the inherited and popular gastronomic commons, individuals and communities cannot access commodified gastronomy that is the privilege of the few. Surely the desire for gastronomic experiences expands beyond the niche thanks to market mechanisms. Famous chefs make magazine headlines; TV programmes show luxury restaurants, and provide chefs' recipes; channels are sometimes exclusively devoted to gastronomy. Millions of potential consumers are exposed to these images and develop an appetite for these experiences, which are linked with their culinary tradition, but portrayed as superior. However, the distance between top professional and amateur cuisines is growing and creates financial and social obstacles in what is a continuum of gastronomies. Elitist gastronomy extracts from, but does not nourish, the commons. As a result, more and more people turn away from the present elitist gastronomy.

Conclusion: For gastronomy, times are changing...

Our analysis shows that the financial success of a few superstar chefs hides the deep problems of commodified gastronomy. The historical path of elitist gastronomy depends less and less on incentives coming from intrinsic reasons (i.e. the development of creativity) and more and more on extrinsic reason such as the need to maintain or increase the profitability of invested capital. This leads to a headlong rush towards super-sophistication, which increases prices and, consequently, contracts the size of the market; so, top gastronomy becomes less and less economically sustainable and encounters growing difficulties to be the leader of the market gastronomic system. As it has never been socially sustainable, its crisis-in-the-making calls for strong changes.

Modern Western societies present themselves as democratic and, along those lines, pretend to export their model of gastronomy worldwide, even in countries that have mainly been characterised by popular gastronomies; however, the analysis of commodified gastronomy presented here denies this democratic pretention. Market gastronomy is, as of today, anything but democratic. It continues to be founded on an aristocratic framework under-evaluating popular gastronomies and honouring certain specific values emanating from the old aristocratic *Société de Cour*: sophistication of recipes, scarcity and high value of foodstuffs, richness of setting, etc. As a result, *haute* gastronomy is implemented and separated from the lifestyle of 99% of the world's population.

It is time to imagine a new pathway for gastronomies which breaks with joint market-elitist gastronomy and recognizes the shared bases of gastronomy and cuisine. Many people are in search for natural products and non-sophisticated culinary practices which are accessible and sustainable, but also creative and original. New ways of producing and consuming gastronomy can be opened. Gastronomy may become less comfortable or hedonistic, but could be experienced in a relaxed and friendly atmosphere. Some gastronomic actors are adapting their offer to this shift in values, trying to engage with these needs and benefit from the opportunities.

The elitist restaurant is no longer the only dominant model; a sort of gastronomic pluralism is in the making. New categories of restaurants emerge; extraordinary ways of cooking are no longer the sole symbol of luxury and taste. Creativity mixed with ordinary foodstuffs and setting attracts a wider, less wealthy audience in search of new codes and settings. This "low cost gastronomy" (in the words of famous French chef, Alain Senderens) is actively involved in the democratisation and mass consumption of gastronomy. The Michelin Red Guide is also evolving and creates new awards: Bib for the French Red Guide and "Small Plates" in New York, selecting establishments chosen for the originality of their menus, mood and service. The emergence of this category reflects the growing popularity of restaurants providing gastronomic quality at reasonable prices. The same is true of Tokyo; since 2011, the Tokyo Michelin Guide has a new pictogram indicating a starred restaurant offering a menu priced fewer than 5,000 yen ($50) for lunch and/or dinner. According to the guide: "Value for money is one of five criteria [used] to select star restaurants, and the new pictogram serves readers [by helping them] to find local eateries at affordable prices" (2017: XI).

Gastronomy could even go down the streets, according to the model of Asian street food and the recent developments of high-quality food trucks. Ecological values stand in opposition to extraction, appropriation and excess and are increasingly influencing food experiences – as shown by the Slow Food, Km Zero and locavore movements. The Old French model of gastronomy, based on the leading role of the *Grand Restaurant* can no longer claim to be the pillar of European gastronomy. The gastronomy of the future has to rediscover its roots and redefine the relationship between elitist and popular gastronomy, opening it up to other knowledge commons.

The globalisation process mixes cultures and heritages. Culinary heritages can be enjoyed beyond their places of origin. Multiculturalism is growing and world fusion cuisine is spreading, drawing from culinary commons and heritages. In the new globalised world, different gastronomies coexist and interact daily. New models of gastronomy have been emerging and have been strengthening the idea of gastronomic pluralism that cuts across geographies, classes and the past and present of inherited culinary commons. Today, things are changing. It's time to invent a new model of gastronomy: democratic, ecological and pluralist!

References

Addis, M. and M.B. Holbrook (2001) "On the Conceptual Link between Mass Customisation and Experiential Consumption: An Explosion of Subjectivity", *Journal of Consumer Behaviour*, 1, 1, 50–66.

Barrère, C. and V. Chossat (2004) "Intellectual Property Rights and Cultural Heritage; The Case of Non-Cumulative and Non-Degenerative Creation", *Review of Economic Research on Copyright Issues*, 1, 2, 97–117.

Barrère, C. (2016) Cultural Heritages: From Official to Informal, *City, Culture and Society*, 7, 2, 87–94.

Barrère, C. (2018) "Cultural Heritage: Capital, Commons, and Heritages", in Labrador, A.M. and N.A. Silberman (eds.), *The Oxford Handbook of Public Heritage Theory and Practice*, Online Publication Date: March 2018, DOI:10.1093/oxfordhb/9780190676315.013.3.

Barrère, C., Bonnard, Q. and V. Chossat (2012) "Food, Gastronomy and Cultural Commons", in Bertacchini, E., Bravo, G., Marrelli, M. and W. Santagata (eds.), *Cultural Commons, A New Perspective on the Production and Evolution of Cultures*, Cheltenham: Edward Elgar, p. 129–150.

Barrère, C., Bonnard, Q. and V. Chossat (2014), "Are We at a Turning Point in the Evolution of Gastronomy? Paris: An Exemplary Case", *Applied Economics*, 46, 12, 1409–1419. DOI:10.1080/00036846.2013.875110.

Baumol, W.J. and W.G. Bowen (1966) *Performing Arts: The Economic Dilemma*, New York: The Twentieth Century Fund.

Benjamin, W. (1936) *The Work of Art in the Age of Mechanical Reproduction*, Translated by H. Zohn, Berlin: Schocken; New York: Random House. http://www.marxists.org/reference/subject/philosophy/works/ge/benjamin.htm.

Beaugé, B. (2010) "Sur l'idée de nouveauté en cuisine", *Mode de recherche*, Institut Français de la Mode, Paris, 13, 3–11.

Capgemini (2018) *World Wealth Report 2017*, www.worldwealthreport.com.

Di Giovine, M.A. and R.L. Brulotte (eds.) (2014) *Edible Identities: Food as Cultural Heritage*, Burlington: Ashgate.

Elster, J. (1996) Rationality and the Emotions, *The Economic Journal*, 106, 438, 1386–1397.

Ferguson, P. (2004) *Accounting for Taste: The Triumph of French Cuisine*, Chicago: University of Chicago Press.

Fischler, C. (1993) *L'homnivore*, Paris: Odile Jacob.

Fischler C. and E. Masson(2008) *Manger: Français, Européens et Américains face à l'alimentation*, Paris: Odile Jacob.

Hess, C. (2012) "Constructing a New Research Agenda for Cultural Commons", in Bertacchini, E., Bravo, G., Marrelli, M. and W. Santagata (eds.), *Cultural Commons, A New Perspective on the Production and Evolution of Cultures*, Cheltenham: Edward Elgar, p. 19–35.

Hess, C. and E. Ostrom (eds.) (2007) *Understanding Knowledge as a Commons : from Theory to Practice*, Cambridge: MIT Press.

Hirschmann, E.C. and M.B. Holbrook (1982) "Hedonic Consumption: Emerging Concepts, Methods and Propositions", *Journal of Marketing*, 46, 3, 92–101.

Holbrook, M.B. and E.C. Hirschman (1982) "The Experiential Aspects of Consumption: Consumer Fantasies, Feelings, and Fun", *Journal of Consumer Research*, 9, 2, 132–140.

Lévi-Strauss, C. (1962, 1964) *Totémisme*, Paris: Plon. English Translation (1964); *Totemism*, Translated by R. Needham, London: Merlin Press.

Lévi- Strauss, C. (1962, 1966) *La pensée sauvage*, Paris: Plon. English translation (1966); *The Savage Mind*, London: Weidenfeld and Nicholson.

Lévi- Strauss, C. (1964, 1966) *Le Cru et le Cuit*, Paris: Plon. English translation: *The Raw and the Cooked: Introduction to a Science of Mythology*, Translated by D. Weightman, New York: Penguin.

Lovejoy, A.O. (1936) *The Great Chain of Being: A Study of the History of an Idea*, Cambridge: Harvard University Press.

Madison, M.J., Frischmann, B.M. and K.J. Strandburg (2010) "Constructing Cultural Commons in the Cultural Environment", *Cornell Law Review*, 95, 657–710.

Michelin (2017) *The Michelin Guide: Tokyo*, Paris: Michelin Guide.

Ostrom, E. (1990) *Governing the Commons: The Evolution of Institutions for Collective Action*, New York: Cambridge University Press.

Poulain, J.-P. (2002) *Sociologies de l'alimentation, les mangeurs et l'espace social alimentaire*, Paris: PUF.

Rozin, P., Remick, A.K. and C. Fischler (2011) "Broad themes of difference between French and Americans in attitudes to food and other life domains: personal versus communal values, quantity versus quality, and comforts versus joys", *Frontiers in Psychology*, 2, http://dx.doi.org/10.3389/fpsyg.2011.00177

Santagata, W. (2010) *The Culture Factory: Creativity and the Production of Culture*, New York: Springer.

Torre, A. and A. Rallet, (2005) Proximity and Localization, *Regional Studies*, 2005, 39, 1, 47–59.

Veblen, T. (1899) *The Theory of the Leisure Class*. New York: MacMillan. Republished in 1970, London: Unwin Books.

14

GENETIC RESOURCES FOR FOOD AND AGRICULTURE AS COMMONS[1]

Christine Frison and Brendan Coolsaet

Introduction

Access to the diversity of genetic resources for food and agriculture (GRFA) is vital for human survival (Altieri, 1999; Frison et al., 2011; FAO, 2015), notably given the current context of climate change (IPES-Food, 2016; Beddington et al., 2012). However, access to GRFA is under threat for at least two main reasons: (1) biodiversity erosion, caused by a multitude of drivers, including climate change, land use changes and the introduction of invasive species (FAO, 1998, 2010; Esquinas-Alcazar, 2005; Butchart et al., 2010), as well as micro-level farming decisions and macro-level policies (Pascual and Perrings, 2007; 2012), including agricultural intensification, shifting cultivation and indiscriminate crossbreeding; and (2) hyperownership over genetic resources (Safrin, 2004; Chiarolla, 2006; Aoki, 2008 and 2010), understood as "exclusive ownership and restrictions on the sharing of genetic material" (Safrin, 2004: 641). Both elements limit the use and sharing of GRFA among food and agriculture stakeholders and constrain international GRFA management within an appropriation paradigm, which stands at odds with the relative free flow of germplasm inherent to agricultural practices since the first forms of plant and animal domestication. This appropriation paradigm considers resources as commercial objects to be appropriated and sold on the global market as a commodity, reducing resources to merely an economic value and erasing all the other (cultural, social, religious, etc.) values of the resources (Bavikatte, 2014: 232; Posey, 1999). This commodification process eludes any other contributions and values related to resources and confines GRFA to a vicious circle of appropriation and exclusion which contradicts the intrinsic interdependence of such resources and their exchange systems.

These ways of governing GRFA have made access more difficult. Indeed, market mechanisms, which are regulated by a regime complex, control the access to genetic resources (Raustiala and Viktor, 2004; Gerstetter et al., 2007; Andersen 2008) and have gradually reinforced market control over homogenised GRFA. This has increasingly impeded the use and exchange of GRFA diversity (Fowler and Hodgkin, 2004; Fowler and Mooney, 1990; Kloppenburg, 2004; Pautasso et al., 2013). Acknowledging that feeding the world's population (2015 Sustainable Development Goal n°2 "Zero Hunger"; see UN, 2015) requires changing our agricultural practices (IPES-Food, 2016), there is an urgent need to unlock access to these essential resources (Pistor and De Schutter, 2015) by stepping out of the dominant appropriation paradigm. This means facilitating the use, conservation and exchange of the diversity of local and traditional GRFA and protecting it from appropriation and homogenisation. Indeed, one may question the usefulness of biotechnologically improved seeds as a solution to hunger (Saab, 2015). Moreover, protecting the diversity of local and traditional GRFA goes hand in hand with the protection of traditional and innovative farming systems, particularly well adapted to local needs, culture and social behaviours. Engaging in a virtuous circle – whereby GRFA conservation and use in a holistic system where social, cultural and spiritual needs are taken into account besides the food, nutrition and economic aspects – promises to enable local populations to produce sufficient and nutritious food for all in a resilient manner (Altieri and Merrick, 1987; IPES-Food, 2016).

Recent literature has explored the idea that managing seeds as a commons would unlock access to seed diversity and promote their efficient conservation and sustainable use (Helfer, 2005; Halewood and Nnadozie, 2008; Dedeurwaerdere, 2012; Frison, 2016). Extending this hypothesis to all GRFA, we argue that GRFA should be governed as a commons in order to ensure both food security and sovereignty, as well as to support the transition towards more sustainable agriculture. The theory of the commons sheds light on the efficiency of local collective management systems over natural resources, where private (market) or public (State) control failed to sustainably manage specific resources (Ostrom, 1990 and 2010b). However, Ostrom's work remains within the boundaries of the above-mentioned appropriation paradigm by considering resources as economic objects to be governed for the sustainability of the community's living (Schlager and Ostrom, 1992). It fails to explore the governance of resources in a more holistic manner (Capra and Mattei, 2015), i.e. by focusing on the relationship between culture and nature rather than on the institutional arrangements designed mainly to manage an object considered exclusively as a "resource". As a commons, GRFA exchange schemes, we argue, are considered networked knowledge-goods with non-exclusive access and use conditions, which are governed, produced and consumed by communities.

To illustrate this argument, two case studies, at two different policy levels and dealing with two different types of GRFA (plant and animal), are examined in this chapter: (a) the Global Seed Commons established under the International Treaty on Plant Genetic Resources for Food and Agriculture; and (b) the reintroduction and "commonification" of a traditional pig breed by a local community enterprise in Schwäbisch Hall, Germany. While ticking off many of the institutional boxes identified in the commons literature, these cases call for a "re-politicisation" of the conservation of GRFA. By drawing on the underlying social, political and economic values at the heart of commons-based collective governance, they highlight the need to bridge the nature-culture divide in order to achieve a genuine transition towards sustainable agriculture.

We conclude that innovative legal frameworks and governance arrangements inspired from the philosophy of the commons may help us to go beyond Ostrom's views and facilitate access to and sharing of GRFA. Doing so would not only ensure food security and sovereignty but also contribute to the transition towards sustainable agriculture by ecologically, socially and culturally sustaining agri-food systems.

The International Treaty on Plant Genetic Resources for Food and Agriculture: re-commonising seeds

The International Treaty on Plant Genetic Resources for Food and Agriculture (hereafter the Treaty) entered into force in 2004 (October 2016: 142 Member States). Its objectives are the conservation, sustainable use and fair and equitable access to agricultural seeds through a

Multilateral System (MLS) where seeds are exchanged through a virtual common pool and where benefit-sharing obligations rest on its users. Access to a list of 64 crops and forages (constituting 80% of the world's staple food) is facilitated for research, breeding and training purposes to all Treaty members. When accessing seeds through the MLS using a Standard Material Transfer Agreement (SMTA: i.e. a standard contract), recipients agree that they will freely share any new developments with others for further research through the MLS. If recipients protect the new material using intellectual property rights (IPRs) and sell it on the market (i.e. exclude MLS members from accessing it for free), they agree to pay a percentage of commercial benefits derived from it into the Benefit-sharing Fund (BSF). The BSF (established in 2008) is a common fund aimed at supporting conservation and the further development of agriculture in developing countries. It was created to increase resources to be invested in crop diversity with the overall objective of increasing world food security.

Ostrom's work is crucial to understand the Treaty as a collective management system for seeds (Halewood et al., 2012; Halewood, 2013). However, recent analyses show that merely applying Ostrom's theory as such to the Treaty is insufficient to unlock access to seeds and fulfil food security and sustainable agriculture goals (Frison, 2016). Indeed, the way the Treaty is designed and implemented remains within and reinforces the current appropriation paradigm. Intellectual property and access to seeds are bargaining stakes directing the globalised food and agricultural market and creating major conflicts of interests between the seed industry and farmers (Kloppenburg, 2004). Seeds are regulated by a regime complex (Raustiala and Viktor, 2004; Andersen, 2008) made of the Treaty's obligations but also by IPRs regulations reinforcing market control over seeds (Gerstetter et al., 2007): the International Union for the Protection of New Varieties of Plant (UPOV) revised in 1991 – creating plant variety protection – and the Trade-Related Aspects of Intellectual Property Rights (TRIPS Agreement) in 1994 – ruling on the patenting of "plant genes" (Cullet, 1999; Helfer, 2004; Chiarolla, 2011).

Hence, the Treaty is not functioning well (i.e. facilitated access is limited to specific recipients, lack of funding impedes benefit-sharing and increasing distrust between stakeholders renders implementation difficult) and is currently undergoing a review process to enhance the functioning of its MLS and to identify more secure funds for its BSF. For now, facilitated access to seeds functions mainly for gene bank curators, researchers and breeders, but does not formally allow direct access to seeds for farmers, who are the first "users" of seeds. In addition, the BSF has received very limited funding and only supports a small portion of all the potential conservation and sustainable use projects in developing countries. Moreover, the Global Information System (GLIS), which aims at facilitating the exchange of information, targets information relevant only for specific types of users (i.e. breeders and researchers), but not for the majority of smallholder farmers in the world. Finally, the general governance mechanism (i.e. the Treaty's Governing Body) lacks participation from the biggest "seed user group", i.e. farmers, on which obligations are imposed without taking into account their voices, needs and interests.

These flaws constitute a quick summary of the deeper and more systemic issues that the Treaty has been raising in terms of obligations, tools, instruments and the implementation process from 2004 to 2016 (Frison, 2016). Overall, a thorough analysis demonstrates that the Treaty expresses an imbalance of rights penalising farmers, notwithstanding the fact that they are the ones feeding the world's population. This imbalance of rights is twofold. First, Farmers' Rights (FRs) are not recognised at the same level as IPRs. Indeed, building the funding strategy and its BSF around IPRs de facto imposes the IPR system as being the "default" system, which in turn fragilises and marginalises further farmers' (informal) system of exchanging seeds and in which recognition and support are left to national recognition and enforcement. Second, small-scale farmers are excluded from the governance of the MLS and their ancestral role of innovators is not recognised, excluding them from research and breeding processes, at the exception of very few participatory research projects under the BSF.

Undeniably, the Treaty has created innovative tools and instruments under international law, which emphasise the collective interest in the international seed management. For this reason, analysing the Treaty using the lens of the commons is more than relevant. However, according to us, this has to be done in a more "holistic manner" (this book; Capra and Mattei, 2015; Frison, 2018; Girard and Frison, 2018; Girard, 2018), i.e. focusing on the complex, interdependent and multi-layered relationship between human and nature rather than on the institutional arrangements designed to manage an object considered exclusively as a (potentially commercial) "resource". Therefore, expanding to more recent developments of the commons theory (this book), the emphasis is set on the underlying values supporting collective governance (social, cultural, ecological values etc.) and on the political philosophy of the commons (Dardot and Laval, 2014) as potential way forward to mitigate the limits of the Treaty.

Going beyond Ostrom's work with her eight design principles, we can identify six invariable principles for a Global Seed Commons (GSC principles) to effectively function beyond this appropriation paradigm. These are: (a) sustainability; (b) interdependence; (c) the anti-commons dilemma (i.e. the underuse of seeds as a main factor in erosion); (d) physical and informational components inextricably bound to the use of seeds; (e) the global seed community; (f) diversity, heterogeneity and complexity. Realising these GSC principles has the potential to transform the Treaty – currently functioning as a management system focused on the collective use of seeds by specific users – into a sustainable global multi-stakeholder management regime focused on the collective governance of seeds by and for the collective benefit of all (Frison, 2016).

(a) Sustainability

Sustainability is seen as a "dynamically maintained system condition rather than a static equilibrium" (Agrawal, 2002:59), i.e. users of the community manage a resource with the perspective of duration and renewal in an adaptive relationship with each other and with the resources. While the conservation and sustainable use objectives of the Treaty undeniably fit with this invariable principle, the Treaty's implementation tools and instruments do not sufficiently take this objective into account. Long-term objectives for the benefit of the community should be translated into concrete rules within the Global Seed Commons, which integrate a "more interactive and participatory process between scientists, policy makers and stakeholders" (Dedeurwaerdere, 2014:24). To be sustainable, commoning should embed the act, mechanism and movement of creating, preserving and reproducing the commons.

(b) Interdependence

Interdependence goes hand in hand with sustainability. In the food and agriculture field, interdependence is the result of long-run human cooperation and collaboration in the exchange of food and feed plants across the world. Farmers and breeders have selected, exchanged and bred seeds to develop certain characteristics over millennia that respond to specific needs and adaptation (FAO, 1998; 2010; Khoury et al., 2016). There is therefore an ongoing need to exchange plant genetic resources from countries all over the world. This calls for the sustainable management of the resource in the collective interest. Interdependence thus becomes a philosophical and political goal to be attained by all countries to reach global food security. In the same line, Capra and Mattei state that "recogniz[ing] the interconnectedness of our global problems [would] enable us to find appropriate, mutually supportive solutions that [...] would mirror the interdependence of the problems they address" (Capra and Mattei, 2015: 159; see also Capra 1996; Moore 2015).

(c) The anticommons dilemma

Heller defines the anticommons dilemma as occurring when "there are too many owners holding rights of exclusion, [then] the resource is prone to underuse" (Heller, 1998:624). Indeed, the dilemma is not that overconsumption leads to the depletion of seed diversity[2] but rather that under-use leads to erosion (i.e. the drastic loss – approx. 75% – of the diversity of seed varieties). This is why over the last decades, with access to seeds that has become more and more restricted due to the appropriation paradigm, erosion of seed diversity has never been so wide. Halewood emphasises that this "aspect of [seeds] informs the need for collective action institutions that are necessary to support their continual creation/evolution as well as ensuring that they are conserved and available for use" (2013: 291). Therefore, the only sustainable way of managing seeds and avoiding the anticommons dilemma that erodes seeds diversity is to facilitate access to them and use by all users, not only breeders and researchers as provided for by the Treaty, but at the global level for every farmer feeding the world.

(d) Physical and informational components

Hess and Ostrom contend that advances in law and technology have enabled profit-oriented firms to extract value from resources previously held in common and to establish property rights over them (Hess and Ostrom, 2003). Ostrom reminds that

> [f]or most of human history, the [global commons] remained unclaimed due to a lack of technology for extracting their value and for establishing and sustaining property rights. To our peril, the technology to extracting value from [the global commons] has developed more rapidly than the appropriate legal mechanism for establishing an effective property regime. The treasured resources for all mankind are threatened by the very technological abilities that we have mastered during recent eras.
>
> *(Ostrom's foreword in Buck, 1998: xiii)*

Hence, IPR and new technologies rather go hand in hand in enclosing information, technologies and access to knowledge and material, which were traditionally available.

As regards seeds, the physical and informational components are inextricably bound to their use. Dedeurwaerdere clearly makes this point when he states that seeds are somewhere in between the exclusive "natural resource commons" and the exclusive "knowledge commons", containing both a physical component and an informational component (Dedeurwaerdere, 2012). This dual component as a physical and informational asset should be taken into account when considering the institutionalisation of a Global Seed Commons. It would require specific governing rules, which change and adapt with the evolution of the intellectual property (IP) protection scheme. Initiatives to facilitate access to breeding information, such as the DivSeek initiative and the Global Open Genome Sequence Data Framework, constitute false sharing and "reopening" initiatives; they will only serve specific categories of seed users, de facto excluding users who do not have the adequate training and technology to benefit from them.

(e) Global Seed Community

Bollier reminds us that "community" is one of the three constitutive elements of a commons, along with a "resource" and a set of "social protocols" (2014:15). A commons is not only about the resource but also about surrounding practices and behaviours, i.e. the establishment of a fair, equitable and sustainable management that guarantees access to and use of the resource in the collective interest (Bollier and Helfrich, 2015:1–12). Baland and Platteau define a community as an arena where "all members of a social group have an access to the local resources" (Baland and Platteau, 1998:644). This definition implies belonging to an identified social group and a notion of scale. In the Treaty arena, the scale and diversity of the community are crucial issues. Defining who is part of the community is crucial as it sets the legitimacy for rights in managing the resources, i.e. only those members that are part of the community may collectively manage (and benefit from) the resource. In the Treaty, farmers are clearly identified as a (passive) target group for benefit-sharing (the first benefit of which being facilitated access to seeds). Breeders, researchers and trainees for food and agriculture are those identified as part of the MLS community (Article 12.3a of the Treaty). To be congruent with the objectives of the Treaty, it would be essential to recognise the fundamental role of farmers in the sustainable use and conservation of seeds, in innovating new seed varieties and in producing food and thus to automatically integrate them in the "management team" of the Treaty, i.e. the Governing Body and its implementation tools and instruments, to constitute the Treaty's global community.

(f) Diversity, heterogeneity and complexity

Ostrom's design principle on "nested enterprises" (Ostrom, 2009) premised that larger commons are more complex to govern than smaller ones. Further studies showed that heterogeneity, diversity and complexity (Cox et al., 2010; Hughes, 1997; De Burca, 2012) were important aspects to consider in collective governance institutions. In studying the character of an adaptive system to a changing context, Dedeurwaerdere pointed to the importance of the modular character of organisational architecture in institutional management systems (Dedeurwaerdere, 2012). As Ostrom and Basurto put it,

> we do not seek to be complex for the sake of being complex, but we have to overcome our obsessive tendency to simplification [...] taking into account the complex and nested character of the systems of the biophysical world, one needs to develop a social science of complexity and nested systems.
>
> *(2013:16; authors' translation)*

Unfortunately, globalisation and the homogenisation of biodiversity governing regimes hinder the emergence of institutional diversity, regime heterogeneity and systems complexity. It is important not to forget, as Roa-Rodríguez and Van Dooren remind us, that

> [t]he dynamics unleashed by IP and sovereign regimes are transforming the varied common spaces, with their multiple modalities of access, use and alienation of resources, into a de facto homogeneous common space where the negative and exclusive characteristics are predominant. This is a highly undesirable outcome if our true goal is the conservation and sustainable use of [seeds] for the well-being of society at large.
>
> *(2008:193–194)*

Although the Treaty and its instruments (MLS, BSF, GLIS, etc.) are innovative from an international law perspective, an in-depth study of their implementation reveals major dysfunctions and incongruences with a political and progressive understanding of the commons and with an ecologically oriented vision of sustainable food systems (Frison, 2016). One cross-cutting aspect is the lack of recognition of the role and rights of smallholder farmers. A *UN Declaration on the rights of peasants and other people working in rural areas* is currently being negotiated to address this gap. Recognition of Farmers' Rights at the international level is promoted as a compulsory step in order to overcome the imbalance of rights pertaining to seeds and to reach food security and sovereignty, as well as the sustainable agriculture overall goals of the Treaty. The theory of the commons in its philosophical- and political-wide dimension is identified as a useful theoretical framework to address these constraints. Transforming the current intergovernmental multilateral legal instrument into an effective and collectively governed political Global Seed Commons would thus propose an alternative path to the current seed regulatory setting entangled in an out-of-date public/private good dichotomy appropriation scheme. It would allow stepping out from the dominant appropriation paradigm and solving major conflicts of interests and power that would inevitably arise between stakeholders, as well as moving from stakeholders managing resources to members of a global community commoning seeds and their ecosystems.

Commons-based animal genetic resources governance: the case of the Swabian-Hall swine

Commons-based strategies and ideas for food and agriculture at the international level need to be implemented through and complemented by both formal (top-down, State-led) and/or informal (bottom-up) collective action (Ostrom, 2004). This section offers an opposite perspective to the previous one and provides an illustration of how grassroots collective action for food and agriculture can provide for public goods through bottom-up innovation.

The Swabian-Hall swine (*Schwäbisch-Hällische Landschwein*) is a local pig breed from Schwäbisch Hall, a small town and capital of the eponymous district, in the state of Baden-Württemberg in southwestern Germany. The swine is the result of a crossbreed between the Chinese Meishan pig, imported by King William I of Württemberg in 1821, and a German landrace. The locally adapted landrace gained enormous popularity in the 19th and first half of the 20th century, with a market-share of over 90% by 1959. Despite its popularity, the swine almost disappeared 25 years later, with the introduction of fast-growing Dutch "high-performance" breeds, suitable for mass production and with low fat content, but ill-adapted to their environment. Livestock of the local landrace declined sharply, and by 1984 the Swabian-Hall swine was considered to be extinct.

The critical condition of the local landrace, owing to the extreme commodification of pig breeding in the region, led a small group of farmers–breeders to launch a conservation campaign to save the Swabian-Hall swine. In the 1980s, they created the Schwäbisch Hall Producers' Community (*Bäuerlichen Erzeugergemeinschaft Schwäbisch Hall*, or BESH), defending a "holistic approach to rural development". The initiative turned into a genuine success story (Coolsaet, 2015): the local landrace was rebuilt and its genetic potential sustained, the breeding footprint was improved and local farmers–breeders were made the main stakeholders of farming and breeding activity. Although the swine is still considered to be at risk of extinction, the community now counts over 1400 farmers breeding the Swabian-Hall swine.

The governance arrangements set out by the BESH tick off many of the institutional boxes identified in the literature as leading to success: well-defined boundaries, predictability of the resource flow, group homogeneity, monitoring system, accountability, articulation with markets, legal recognition, etc. (Agrawal, 2001). However, understanding the success of the BESH requires going beyond a technicist, depoliticised and institutional approach and looking at the governance arrangements through the underlying values supporting collective governance in general and this initiative in particular. Three broad overlapping dimensions underlie the initiative: (a) economic distribution (strengthening farmers economically); (b) cultural recognition (valuing traditional knowledge, cultural traditions and local identities); and (c) political representation (improving farmers' autonomy and self-determination) (Coolsaet, 2015).

(a) Economic distribution

First of all, the BESH established a community-based pricing system. Both meat prices and production amounts are fixed communally and in advance and the association guarantees buying of these amounts. To account for higher production costs of the Swabian-Hall swine (approximately 12% higher than for "high performance" breeds (Leipprand et al., 2006)), part of the network's profits is redistributed as financial support for adaptation. More specifically, BESH breeders get a 0.33 euro "adaptation premium" per kg of carcass of pork on top of the purchase price to allow for the adaptation of their breeding practices to the community's standards. While this allows for controlling the predictability of the market and of the resource flow, hence providing stability to the production process, it also allows for avoiding over-production and internal competition between farmers.

With over 1400 pig farmers–breeders, the BESH is by no means a small group. But the group's size gradually increased over time and does not constrain its activities. On the contrary, one could argue that given the peculiar nature of the resource (i.e. genetic resources), a rather large group is necessary to avoid genetic degradation. The BESH employs permanent staff who support the farmers in tasks as diverse as commercialisation, product marketing, logistics support, internal communication, research subsidies and recreation.

These different elements function as a (re)distribution mechanism that guarantees a stable income and a fair share of the profits for farmers. This financial redistribution offers both autonomy from an increasingly centralised semen industry and greater independence from the retail sector, with farmers having more security with regards to the volatility of the market and global competitors.

(b) Cultural recognition

Another characteristic of the group is its homogeneity regarding both production and retail. This homogeneity is a result of the group's ambition to account for both cultural and natural issues through the breeding process. Joining the association is only possible for breeders located in the traditional breeding area of the Swabian-Hall swine. This ensures an overlap between the farmers–breeders' residential and cultural location and the resource location. As such, the landrace is not only adapted to its natural environment, hence requiring less external farming inputs, but also to local culture and traditional knowledge.

But this reliance on situational knowledge does not amount to the glorification of traditions or to some form of extreme localism. Although the project initially faced criticism and defiance from the scientific community, the BESH has been increasingly collaborating with scientists through the establishment of shared learning spaces. For example, it has recently teamed up with 24 European partners to launch a joint project under the EU Horizon 2020 research program, studying connections between traditional feed (e.g. grass) and improved meat quality.[3] Moreover, through its internal training centre, the BESH facilitates the exchange of knowledge between farmers. These instruments allow for the farmers and their knowledge to be recognised. In so doing, they promote a form of status equality between those who practise farming (i.e. the farmers) and those recognised as holding authoritative representational and interpretative knowledge of farming (i.e. the scientists).

Finally, the BESH also organises cultural events aimed at strengthening farmers' identity and promoting sustainable agriculture. The regional "Rock for Nature" festival combines rock music and debates on the future of agriculture (in 2008 it hosted both Joe Cocker and Vandana Shiva). The motto of the festival is "*Gen Tec – Nein Danke!*" (Gen[etic] [bio]tec[hnology] – no thanks!).

Strengthening farmers' identity and recognising the validity of their traditional knowledge and local breeding techniques were necessary to depart from high-input breeding. In the contemporary context of the EU's Common Agricultural Policy, the dominance of industrial farming systems and the hierarchisation of knowledge systems leave little space for the emergence of alternative ways of farming (Coolsaet, 2016). Central to the activities of the BESH was thus the idea that "the genetic code is not fixed in time, it evolves according to the environment and the farmer" (participating farmer; authors' interview, 2013). This evolution, however, is only possible when genetic resources are free from access and use arrangements which freeze the genetic code according to characteristics defined by external, non-producing actors. As a participating farmer puts it: "rare breeds are not for the museum" (authors' interview, 2013).

(c) Political representation

Governance arrangements aiming to economically and culturally strengthen the farmers have led the BESH to rethink representational spaces and conditions. Community-based pricing systems, for instance, only work if the whole community is represented and the whole point of collaborative learning spaces is to enlarge the evidence base through broad(er) participation.

But the issue of representation was taken a step further by specifically thinking about how to reconnect individual farmers, consumers and the broader community with a larger collective rural movement, empowering participants, both producers and consumers, as agents of change and encouraging self-determination. To allow for a strong articulation with consumers and to guarantee a stable return to all the parties involved in the chain of production, the BESH created its own market. When a local slaughterhouse threatened to close, the BESH teamed up with surrounding communities and local authorities to collectively buy the slaughterhouse. It then gradually established a network of community-owned (butcher)shops and established a series of partnerships with regional hotels and restaurants, which exclusively supply BESH's meat. This allowed the broader community to gain control over the whole value-chain, redirecting added value to the farm instead of the industry. But it also allowed them to break loose from the conditions imposed by the retail sector by redefining the conditions of market access.

While working within limited geographic and cultural boundaries, the BESH also gradually opened its production process to external actors to help improve farming practices, ensure monitoring and accountability and create a more direct relationship between producers and consumers. While farmers keep their autonomy and define their own farming practices, an independent environmental NGO ensures the definition and improvement of the community's breeding guidelines and production standards, as well as their monitoring and compliance throughout the production chain. Scientific partners and agricultural engineers help develop technical breeding criteria for meat quality, vitality, stress resistance, fertility, etc. And the inclusion of end-user and transformers has allowed to directly promote and encourage the use and consumption of the local landrace.

Beyond strengthening the conditions for farmers–breeders, these three dimensions have also allowed for reducing the environmental impact of the breeding activity. Not only was the locally adapted landrace saved from extinction, hence reintroducing more genetic diversity and associated local knowledge in the industry, the breeding standards gradually incorporated strict environmental objectives. A stronger focus on animal welfare led to doubling the mandatory space per pig, developing better-adapted barns and banning the use of antibiotics and tranquilisers. Moreover, if not produced personally, BESH farmers must buy pig feed from genetically modified (GM)-free suppliers, which are monitored and pre-selected by the BESH. While the whole enterprise obviously still produces greenhouse gases, the breeding standards also include emissions reduction objectives. As such, at least 80% of the feed must be bought from regional suppliers (max. 500 km from the farm) and the whole chain of production is shortened, leading to less transport. To limit emissions and minimise nutrient runoff, the BESH imposes sufficient fallow land in the members' farms to allow for surface spreading of their pigs' manure. Finally, while official organic labelling is not imposed on the farmers, the BESH has created specific support programs to help the farmers adapt to environmental standards and provides nature conservation training to its members.

The BESH's holistic approach hence combines a series of institutional arrangements to successfully improve access to, conservation of and use of the genetic diversity of the local landrace. But these arrangements offer more than just a standardised governance system. By encompassing strong socio-cultural, economic and political objectives, the BESH replaced and complemented the tradeable dimension of food and farming with a much larger understanding based on agency and empowerment of its members, while at the same time reducing the environmental impact of pig farming.

Conclusions

Access to GRFA for farmers is an essential component of more sustainable farming. It helps transform local agri-food systems by enabling environmentally appropriate and culturally adapted food production. However, this access has become increasingly difficult due to the way in which GRFA are being governed. Current approaches to their management tend to overemphasise their economic dimension, reducing them to tradable commodities managed through market approaches, reinforcing their appropriation by certain actors and fostering hyperownership.

We have argued in this chapter that governing GRFA as commons can improve access and mitigate failing management. Doing this, however, requires departing from a narrow institutional understanding of commons-based governance. The conservation and use of GRFA must be re-politicised by drawing on some of the underlying social, political and economic values at the heart of commons-based collective governance.

Transforming the current intergovernmental multilateral system of the Treaty into an effective and collectively governed political Global Seed Commons constitutes an example of a possible global "re-commonisation" of seeds. The review process of the MLS could allow for stepping out from the dominant appropriation paradigm and solving major conflicts of interests arising between seed stakeholders. Implementing the identified six invariable principles has the potential to transform the Treaty into a sustainable global multi-stakeholder management regime focused on the collective governance of seeds in the interest of all, including smallholder farmers. In the same vein, the Schwäbisch Hall case provides an illustration of how GRFA can shift from being a commodity to being a commons. The successful conservation and breeding of the local swine was both the driving force and the consequence of the strong sociocultural, political and environmental principles underlying the initiative.

By governing GRFA as politically constructed commons, both cases offer effective and sustainable solutions that differ from the usual market- or state-based approaches which led to the current appropriation paradigm. A political rediscovery of the commons can thus help in constructing innovative legal frameworks and governance arrangements that facilitate preservation, access to and sharing of GRFA and the transition towards more sustainable agriculture.

Notes

1 Most of this chapter is a summary of Christine Frison's doctoral thesis, referred to in the bibliography.
2 Although it is obvious that when someone eats a potato, the potato is no longer available for growing or for somebody else's consumption.
3 See 'TREASURE – Diversity of local pig breeds and production systems for high quality traditional products and sustainable pork chains' – https://treasure.kis.si/

References

Agrawal, A. 2001 "Common Property Institutions and Sustainable Governance of Resources" *World Development*, 29(10), 1649–1672.

Altieri, M. A., and Merrick, L., 1987 "In Situ Conservation of Crop Genetic Resources through Maintenance of Traditional Farming Systems" *Economic Botany*, 41(1), 86–96.

Altieri, M., 1999 "The Ecological Role of Biodiversity in Agroecosystems" *Agriculture, Ecosystems and Environment*, 74(1), 19–31.

Andersen, R., 2008 *Governing Agrobiodiversity: Plant Genetics and Developing Countries, Global Environmental Governance*, Farnham: Ashgate.

Aoki, K., 2010 "Seeds of Dispute: IPRs and Agricultural Biodiversity" *Golden Gate University Environmental Law Journal*, 3(1), 79–160.

Baland, J.-M. and Platteau, J.-P., 1998 "Division of the Commons: A Partial Assessment of the New Institutional Economics of Land Rights" *American Journal of Agricultural Economics*, 80(3), 644–650.

Batur, F., 2014 *Agrobiodiversity Conservation and Plant Improvement: Adjustments In intellectual Property Rights Reclaiming the Public Domain towards Sustainability and Equity* (PhD thesis), Université Catholique de Louvain, Belgium.

Bavikatte, K. S., 2014 *Stewarding the Earth: Rethinking Property and the Emergence of Biocultural Rights*, New Delhi: Oxford University Press.

Beddington, J., et al., 2012 "Achieving Food Security in the Face of Climate Change: Final Report from the Commission on Sustainable Agriculture and Climate Change" Copenhagen: CGIAR Research Program on Climate Change, Agriculture and Food Security (CCAFS), available at https://cgspace.cgiar.org/handle/10568/35589.

Blomquist, W. and Ostrom, E., 1985 "Institutional Capacity and the Resolution of a Commons Dilemma" *Review of Policy Research*, 5(2), 383–394.

Capra, F., 1996 *The Web of Life: A New Synthesis of Mind and Matter.* New York: HarperCollins.

Capra, F. and Mattei, U., 2015 *The Ecology of Law – Toward a Legal System in Tune with Nature and Community*, Oakland: Berrett-Koehler.

Chiarolla, C., 2011 *Intellectual Property, Agriculture and Global Food Security: The Privatization of Crop Diversity*, Cheltenham: Edward Elgar.

Coolsaet B., 2015 "Transformative Participation in Agrobiodiversity Governance: Making the Case for an Environmental Justice Approach" *Journal of Agricultural and Environmental Ethics*, 28(6), 1089–1104.

Coolsaet, B., 2016 "Towards an Agroecology of Knowledges: Recognition, Cognitive Justice and Farmers' Autonomy in France" *Journal of Rural Studies*, 47, Part A (October 2016), 165–171.

Cox, M., Gwen, A., and Sergio Villamayor, T., 2010 "A Review of Design Principles for Community-Based Natural Resource Management" *Ecology and Society* 15(4), 38.

Cullet, P., 1999 "Revision of TRIPS Agreement Concerning the Protection of Plant Varieties" *Journal of World Intellectual Property*, 2(4), 617-656.

Dardot, P. and Laval, C. 2014 *Commun: Essai Sur La Révolution Au XXIe Siècle*, Paris: La Découverte.

Dedeurwaerdere, T., 2012 "Design Principles of Successful Genetic-Resource Commons for Food and Agriculture" *International Journal of Ecological Economics and Statistics*, 26(3), 31–46.
De Schutter, O., 2011 "The Right of Everyone to Enjoy the Benefits of Scientific Progress and the Right to Food: From Conflict to Complementarity" *Human Rights Quarterly*, 33(2), 304-350.
Frison, C., 2016 *Towards Redesigning the Plant Commons* (PhD thesis) Université Catholique de Louvain and Katholieke Universiteit Leuven, Belgium.
Frison, C., 2018 "Planting the Commons: Towards Redesigning an Equitable Global Seed Exchange" in Girard and Frison (eds) *The Commons, Plant Breeding and Agricultural Research. Challenges for Food Security and Agrobiodiversity*, London: Routledge.
Frison, C., et al. (eds) 2011 *Plant Genetic Resources and Food Security – Stakeholder Perspectives on the International Treaty on Plant Genetic Resources for Food and Agriculture*, London: Earthscan/Routledge.
FAO, 2010 "The Second Report on the State of the World's Plant Genetic Resources for Food and Agriculture: Synthetic Account" Rome: FAO, available at http://www.fao.org/docrep/013/i1500e/i1500e00.htm.
FAO, 2012 "Towards the Future We Want – End Hunger and Make the Transition to Sustainable Agricultural and Food Systems" Rome: FAO, available at http://www.fao.org/docrep/015/an894e/an894e00.pdf.
FAO, IFA and WFP 2014 "The State of Food Insecurity in the World 2014" Rome: FAO, available at http://www.fao.org/3/a-i4030e.pdf.
FAO, 2015 "The Second Report on the State of the World's Animal Genetic Resources for Food and Agriculture" (eds) B.D. Scherf and D. Pilling. Rome: FAO, available at http://www.fao.org/3/a-i4787e/index.html.
Gerstetter, C., Görlach, B., Neumann, K., and Schaffrin, D., 2007 "The International Treaty on Plant Genetic Resources for Food and Agriculture within the Current Legal Regime Complex on Plant Genetic Resources" *Journal of World Intellectual Property*, 10(3/4), 259–283.
Girard, F., 2018 "Composing the Common World of the Local Bio-Commons in the Age of the Anthropocene" in Girard and Frison (eds) *The Commons, Plant Breeding and Agricultural Research. Challenges for Food Security and Agrobiodiversity*, London: Routledge.
Girard, F., and Frison, C., (eds) 2018, *The Commons, Plant Breeding and Agricultural Research. Challenges for Food Security and Agrobiodiversity*, London: Routledge.
Halewood, M., 2013 "What Kind of Goods Are Plant Genetic Resources for Food and Agriculture? Towards the Identification and Development of a New Global Commons" *International Journal of the Commons*, 7(2), 278–312.
Halewood, M. and Nnadozie, K., 2008 "Giving Priority to the Commons: The International Treaty on Plant Genetic Resources for Food and Agriculture" in Tansey and Rajotte (eds) *The Future Control of Food*, London: Earthscan.
Halewood, M., Noriega López, I. and Louafi, S., 2012 *Crop Genetic Resources as Global Commons: Challenges in International Law and Governance*, Abingdon: Routledge.
Helfer, L., 2005 "Using Intellectual Property Rights to Preserve the Global Genetic Commons: The ITPGRFA" in Maskus and Reichman (eds) *International Public Goods and Transfer of Technology under a Globalized Intellectual Property Regime*, Cambridge: Cambridge University Press.
Hughes, B., 1997,"Local Commons and Global Interdependence: Heterogeneity and Cooperation in Two Domains – Keohane,Ro, Ostrom,E", *American Political Science Review*, 91(1), 236.
iPES Food, 2016 "From Uniformity to Diversity: A Paradigm Shift from Industrial Agriculture to Diversified Agroecological Systems", iPES Food, available at http://www.ipes-food.org/images/Reports/UniformityToDiversity_FullReport.pdf.
Jungcurt, S., 2007 "Institutional Interplay in International Environmental Governance: Policy Interdependence and Strategic Interaction in the Regime Complex on Plant Genetic Resources for Food and Agriculture" in Beckmann and Hagedorn (eds), *Institutional Change in Agriculture and Natural Resources*, Aachen: Shaker Verlag.
Kloppenburg, J., 2004 "First the Seed. The Political Economy of Plant Biotechnology, 1492–2000" Madison: University of Wisconsin Press.
Köhler-Rollefson U. I., Mathias, E., Singh, H., Vivekanandan, P. and Wanyam, J., 2010 "Livestock Keepers' Rights: The State of Discussion" *Animal Genetic Resources*, 47(1–5).
Moore, J. W., 2015 *Capitalism in the Web of Life: Ecology and the Accumulation of Capital*. New York: Verso Books.
Ostrom, E., 1990 *Governing the Commons: The Evolution of Institutions for Collective Action*, Cambridge: Cambridge University Press.

Ostrom, E., 2004 "Understanding collective action" in Meinzen-Dick R, Gregorio M Di (eds) *Collective action and property rights for sustainable development*, Washington, DC: IFPRI International Food Policy Research Institute.
Ostrom, E., 2009,"Design Principles of Robust Property-Rights Institutions: What Have We Learned?" in Ingram and Hong (eds) *Property rights and land policies*, Cambridge: Lincoln Institute of Land Policy.
Ostrom, E., 2010a "The Institutional Analysis and Development Framework and the Commons", *Cornell Law Review*, 95, 807–816.
Ostrom, E., 2010b "Beyond Markets and States: Polycentric Governance of Complex Economic Systems" *Transnational Corporations Review*, 2(2),1–12.
Ostrom, E., and Basurto, X., 2013,"Façonner Des Outils D'analyse Pour Étudier Le Changement Institutionnel", *Revue de la régulation. Capitalisme, institutions, pouvoirs*, 14, 16.
Ostrom, E., Burger, J., Field, C. B., Norgaard, R. B. and Policansky, D., 1999 "Revisiting the Commons: Local Lessons, Global Challenges" *Science*, 284(5412), 278–282.
Pascual, U. and Perrings., C., 2007 "Developing Incentives and Economic Mechanisms for In Situ Biodiversity Conservation in Agricultural Landscapes" *Agriculture, Ecosystems and Environment*, 121(3), 256–268.
Pascual, U. and Perrings, C., 2012 "Developing Mechanisms for In Situ Biodiversity Conservation in Agricultural Landscapes" in Ninan (ed) *Conserving and Valuing Ecosystem Services and Biodiversity: Economic, Institutional and Social Challenges*. London: Earthscan,151–174.
Pautasso, M. et al., 2013 "Seed Exchange Networks for Agrobiodiversity Conservation. A Review" *Agronomy for Sustainable Development*, 33(1), 151–175.
Pistor, K. and De Schutter, O., 2015 "Governing Access to Essential Resources" New York: Columbia University Press.
Posey, D.A. (ed), 1999 *Cultural and Spiritual Values of Biodiversity*, Nairobi: London: UNEP and IT.
Raustiala, K. and Victor, D. G., 2004 "The Regime Complex for Plant Genetic Resources" *International Organization*, 58(2), 277–309.
Roa-Rodríguez, C and Van Dooren, T., 2008,"Shifting Common Spaces of Plant Genetic Resources in the International Regulation of Property" *The Journal of World Intellectual Property*, 11(3), 193–194.
Saab, A., 2015 "International Law and the Pyramid of Assumptions: Do We Need More Food to Tackle Hunger in the Face of Climate Change?" *Sustainable Consumption*, 8, 50.
Safrin, S., 2004 "Hyperownership in a Time of Biotechnological Promise: The International Conflict to Control the Building Blocks of Life" *American Journal of International Law*, 98(4), 641–685.
Schlager, E. and Ostrom, E., 1992 "Property-Rights Regimes and Natural Resources: A Conceptual Analysis" *Land economics*, 67(3), 249–262.
Six, B., Van Zimmeren, E., Popa, F., Frison, C., 2015 "Trust and Social Capital in the Design and Evolution of Institutions for Collective Action" *International Journal of the Commons*, 9(1), 151–176.
United Nations 2015 "Millennium Development Goals Report 2015" New York: UN, available at http://www.un.org/millenniumgoals/2015_MDG_Report/pdf/MDG%202015%20rev%20(July%201).pdf.

15

WATER, FOOD AND CLIMATE COMMONING IN SOUTH AFRICAN CITIES

Contradictions and prospects

Patrick Bond and Mary Galvin

Introduction

Social change has occurred unevenly in South Africa, with adverse implications for the strategy of 'commoning.'[1] The framing of a commons is not as popular in this extremely unequal society as are various versions of 'Right to the City' narratives, or simply the informal and mainly illegal appropriation of state-supplied services, especially water and electricity, sometimes in the wake of the thousands of 'service delivery protests' that occur each year. The narrow, constitutionalist framings of rights are most often articulated by lawyers supporting low-income people in these struggles, while other organizers (e.g. Ngwane, 2009) have taken up a more expansive argument consistent with arguments made by Henri Lefebvre (1996) or David Harvey (2012). The direction the latter may go, if the 'popcorn protests' can be linked up more effectively, could be towards a new version of mutual-aid philosophy often considered within the 'eco-socialist,' feminist and decolonizing traditions of radical South African politics. To understand the concrete form these are taking, it is useful first to frame these as contestations of the commons.

Progressive movements have regularly expressed a desire to expand various kinds of commons, especially those that are connected with nature (water, air, land, sub-soil resources), ideas (humanity's intellectual and cultural traditions), society (the mixing of peoples through regional migrations) and state services (water/sanitation, electricity, social services, healthcare, education, etc.). The most crucial South African example is represented by the successful commoning of intellectual property over Anti-Retroviral Medicines (ARVs), which led to free provision of AIDS medicines through the public service since 2004. Earlier, ARVs were too costly for anyone aside from a few thousand individual healthcare customers (nearly all white) in the private sector. By 2018, with four million receiving the ARVs for free, life expectancy soared from 52 (in 2004) to 64 (Bond, 2014).

There have also been illustrative commons struggles for water decommodification, free tertiary education financing, access to land and nature and resistance against society's xenophobic tendencies. The neoliberal era's enclosing movements often were accompanied by counter-movements (as predicted by Polanyi, 1957). In South Africa, they could claim partially successful efforts to decommodify, defend or expand state- or mutually-owned or managed goods and services, including free basic water and electricity.

Two standard economistic ideas that are typically applied to commons processes, namely 'merit goods' and 'public goods' (e.g. Ostrom, 1990), are not only dangerously localistic and small-scale in orientation (Harvey, 2012). They are also weak versions of the concept, for they hardly capture the radical political essence of South Africa's struggle over the commons. Over millennia in this region, commons have been constructed through mutual aid termed 'Ubuntu' – meaning 'we are who we are through each other' – and then deconstructed through slavery, colonialism, apartheid, patriarchy, neoliberal commodification, social atomization and the broader process of uneven development (Smith, 1984; Bond and Ruiters, 2017).

This chapter focuses on urban commoning in South African cities as both a survival strategy and potential eco-socialist project. Through this example it draws out aspects of a potential counter-movement against neoliberalism that will evolve through small-scale experimentation, social-democratic public policy and – in some cases already – a broader framing of radical politics. We read this as an expression of political commoning: building commons spaces as a transformative politics (Ferrando and Vivero-Pol, 2017). However, the cases considered below can also be considered as a contradictory mode of social commoning, which Vivero-Pol (2017) describes as following a pathway of either political disaffection in which people are not yet engaged in broader political struggles, or of 'political activism and self-awareness of working at the community level but with a greater (counter-hegemonic) global objective.'

In this chapter, we examine the background to the commoning of water, food and climate adaptation. When combined – as Naomi Klein (2014) recommends in *This Changes Everything* – as an eco-socialist strategy, there are inklings of a new social movement ideology in the otherwise fragmented South African urban context. At the same time, however, we also explore contradictions of this approach within the present socio-political context: fragmentation of struggles, geographical and scale tensions and differential radical and liberal approaches especially applied to juridical, rights-based strategies. The idea of the commons could transcend these contradictions, as shown in the struggles for water, food and air as manifestations of prefigurative politics.

The chapter is based upon both authors' experiences as scholar-activists over the last thirty years in Johannesburg and Durban, but mainly relies upon the organic expressions of township residents who exhibit both socialist and capitalist behaviour. Together, these undermine the state's strategies for what are essentially neoliberal modes of water delivery, a non-existent food policy and minimalist climate adaptation. The lessons from these commons strategies point us to both the pathway and the potholes en route to building eco-socialism, especially in the world's most unequal cities.

History, struggles and contemporary conditions in urban South Africa

Since its origin in the 1920s, the anti-apartheid movement's strengths have always been linked with explicitly urban (and urbanizing) social and labour collectivities. Anti-apartheid resistance was unlike most other anti-colonial movements in Africa, which had a stronger rural than urban base. Although the 1955 Freedom Charter led by the African National Congress (ANC) called for a modernizing social democracy, there was also a strong communist presence within the ANC, along with a powerful labour contingent. After the state's 1960s imprisonment of liberation movement leaders and repression of civil society, urban trade unions began re-organizing during the early 1970s (Baskin, 1991).

By the mid-1980s, most cities also witnessed the rise of powerful 'civic associations' based in the black townships. In 1992, they forged a network called the South African National Civic Organization (SANCO), joining the broader front known as the Mass Democratic Movement. Most community leaders were also labour leaders who, due to apartheid segregation, lived side-by-side in townships with poor people and an aspiring middle class. In that context, SANCO's resistance principles, analyses, strategies, tactics and allies reflected a collectivist, proto-socialist identity, often pitted directly against the neoliberal capitalist approach of the state and its corporate allies (Bond, 2000, 2014; Mayekiso, 1996). SANCO's loose federal form experienced perpetual problems of movement coordination, yet nearly all the urban civics pursued an agenda that conjoined democratization, deracialization and developmental demands (Mayekiso, 1996).

These demands were in part reflected in the 1994 *Reconstruction and Development Programme* adopted by the African National Congress (ANC) and its Alliance partners. The programme included detailed visions of access to decommodified housing and associated services. The various promises relating to urban restructuring were progressive and ambitious (Bond, 2000) – but were nearly universally broken once the ANC took power (Bond and Khosa, 1999). Within a few years, SANCO degenerated into a junior partner of the ANC, with little impact on the wave of urban revolts against neoliberal public policy that began in the late 1990s (Heller and Ntlokonkulu, 2001; Mayekiso, 1996; Seekings, 1997; Zuern, 2004).

Those policies have generated dire socio-economic conditions. According to Josh Budlender, Ingrid Woolard and Murray Leibbrandt (2015), the poverty rate is 63 percent, far higher than the 45 percent level of 1994. As for inequality, over the first two decades of freedom, the market-income Gini coefficient level was 0.77, the world's highest (World Bank, 2014). Unemployment soared from 16 to 25 percent from 1994 to 2014 and, adding those who gave up looking for jobs, brought the rate to 35 percent (Bond, 2014). At the same time, extremely high increases in fees for consuming basic state services (especially electricity and water) began to kick in, creating the conditions for intense urban unrest. Johannesburg is typically considered (Razvadauskas, 2017) the world's 'most unequal major city,' with Durban and Cape Town not far behind. In this context, the water, food and climate inequalities necessarily called for movements for social justice.

Divergent double-movement responses in South African cities

The neoliberal era is characterized by the 'movement' of capital into every form of life. In reaction, a 'double-movement' – as Karl Polanyi (1957) termed such resistance – can be identified in several sectors that were especially important in South Africa's cities. Although food commoning has been limited, several struggles for decommodified water and clean air suggest the enormous potential of constructing resistance around the notion of the commons. But there are also quite profound contradictions to confront, including ideological contradictions. Urban social movements did not respond to the post-apartheid neoliberal policy terrain with a consistent 'Polanyian' double-movement, mainly because of confusing political subjectivities (Bond, 2014; Duncan, 2016; Ngwane, 2017). In addition, if the Polanyian schema is applied to water, the complex South African situation in which it is partially commodified and partially decommodified allows for diverse types of struggle to qualify as double-movement activism (Galvin, 2016).

As SANCO's mid-1990s demobilization progressed, disruptive and often violent 'service delivery protests' emerged in the vacuum. These became ubiquitous, starting in Johannesburg and quickly moving as far afield as several small Eastern Cape towns during the late 1990s then rising into thousands of demonstrations measured by police and researchers annually (Duncan, 2016; Ngwane, 2017; Runciman et al., 2016). Starting in the late 1990s, 'new social movements' rose in the main cities. There was a general expectation that they would muster the strength to network nationally with increasingly cross-sectoral connections (e.g. combining the drive to decommodify AIDS medicines, water and electricity in a unifying way).

The most impressive social movement was the Treatment Action Campaign (TAC) which, from 1999, rose against the state's refusal to provide AIDS medicines to six million HIV+ citizens. At the time anti-retrovirals cost $10,000 per person annually. After an intense struggle that entailed battles against the World Trade Organization, Big Pharma and the US and South African governments, AIDS treatment was, from 2004, provided for free in state clinics, using generic medicines (Mbali, 2013). This can be considered South Africa's main example of social commoning, for it entailed the decommodification of medicines, the deglobalization of their production (in India's and then Africa's generic drugs factories) and the globalization of people involved in solidarity with TAC, including the US group ACTUP!, African activists, Mèdicins sans Frontières and Oxfam. While there were new opportunities herein for South Africa's main generics company, Aspen, to accumulate capital, making its chief executive a billionaire, the activists' ability to shift the terms of debate on intellectual property, so as to common life-saving ideas and technologies, was nothing short of miraculous, as reflected in the high life-expectancy increase in South Africa and all other countries with high HIV+ incidence (Mbali, 2013).

Other social movements arose and took stands that were highly visible but did not obtain notable successes. Fighting rural inequality, the Landless People's Movement emerged in 2001 to demand land redistribution in the wake of Zimbabwe's 'jambanja' occupations of most white farms (Ntsebeza, 2011; Shonhe, 2018). In a similar anti-colonial and anti-land concentration vein, the critique of apartheid-era debt repayment and profit repatriation was put at the center of the demands advanced by the Jubilee 2000 movement, Khulumani Support Group and their allies. Their ultimately unsuccessful claim for reparations from multinational corporations went as far as the US Supreme Court (Bond, 2003). Moreover, police repression (Bond, 2014) had a lot to do with the difficulties faced by these movements.

In other cases, proponents of environmental justice embarked upon a series of discrete campaigns that were occasionally networked, increasingly against mining industry exploitation and urban pollution (Bond, 2002; Cock, 2011; Womin, 2018). These could be interpreted as attempts to achieve ecological local paradigms through the commoning of nature and society. However, many social movement activists and supporters who expected mass support for social movements to grow and to translate into a socialist political project found their expectations dashed (Bond et al., 2013). In this environment of disappointment, debates emerged about the role of romantic academic activists in these movements, some of whom were prone to ventriloquism, substitutionism and careerism (Mdlalose, 2014; Bond, 2015).

Still, substantial achievements were recorded by the two most fiercely anti-neoliberal urban movements: the Anti-Privatization Forum (APF), focused largely on Gauteng and comprised of 19 affiliate members as of 2004, and the Abahlali baseMjondolo (AbM) shackdwellers movement, which began in 2005 during anti-eviction battles in the Durban shack settlement Kennedy Road and continued experiencing life-and-death conflicts with municipal neoliberals and security forces into 2018. In Cape Town and other major cities, the resistance movements were more fragmented, and this is one reason it has been difficult for visionary organizers to link climate politics, even in sites (like Cape Town) suffering extreme drought and water shortages.

The movements prided themselves on having a stronger political orientation than typical service delivery protesters. In Johannesburg, the APF affiliates initially defined themselves as ideologically heterogeneous and dismissed predetermined political or ideological programmes. Their objective was:

> to bring together the collective struggles of poor/working class communities against the devastating effects of capitalist neoliberalism in South Africa … (so as) to effect

> fundamental shifts in the basic service/needs policies of the state so that the majority of South Africans can enjoy the full realization of their basic human needs and rights.
> *(cited in McKinley, 2016)*

Later, APF adopted 'socialism' as its constitutional objective, which was an unusual ideological commitment. As Ashwin Desai (2006) explains, South Africa's 'movements of the poor must be celebrated for being what they are: relatively small groupings of awakening antagonism in a sea of political apathy, nationalist ignorance and informal repression.' The APF was pragmatic and – like the TAC – raised its members' awareness about the national Constitution's socio-economic rights clauses. They fought for the recognition of rights as part of their overall strategy, including recourse to the courts for injunctions and even major history-making claims upon the state (Runciman, 2012). In the case of the decommodified AIDS medicines won by TAC, this proto-socialist success was consistent with a saying attributed to the assassinated South African Communist Party (SACP) leader Chris Hani: 'Socialism is not about big concepts and heavy theory. Socialism is about decent shelter for those who are homeless. It is about water for those who have no safe drinking water. It is about healthcare' (*Sunday Independent*, 2016).

However, crucial to commons strategies is that these basic services are delivered in a decommodified, destratified manner by either the state or by mutual-aid systems (such as cooperatives, ideally worker-controlled) that get sufficiently generous state subsidies. Without these, there is the danger that housing markets dependent upon bank financing, water privatizers, for-profit health care, pre-paid (commercially-tariffed) electricity meters, private schools and similar intrusions of capitalism will continue weakening – not strengthening – the commoning project. Hence some activists contend that these battles to common – by decommodification and destratification – the basic requirements of daily reproduction are part of a slow-but-sure movement toward socialism; the SACP's slogan is, 'The future is socialism, build it today.'

While it is often prophesied that anti-neoliberal mass-democratic urban movements could not be sustained (Bond et al., 2013), the APF deserves closer attention in part because of the path-breaking work its affiliates – especially the Soweto Electricity Crisis Committee – did in order to decommodify water, which is central to the commons project and also closely related to food commoning (McKinley, 2016; Miraftab and Wills, 2015; Runciman, 2012). After a decade of struggle, which is in itself a success for any social movement, the APF achieved levels of political activism and conscientization which stand out in post-1994 South Africa. Moreover, both organizers and members had a critical aim: not simply to support localized struggles but to develop the linkages between these struggles and between activists. They also built a cadre of activists with a vision of what local struggles mean at a national and global level in an increasingly commodified world and reached out regularly to international allies.

Crises emerged, however, which led to the APF's demise not long after Jacob Zuma became president in 2009, according to its former treasurer Dale McKinley (2012): 'Zuma's politics created both short-term confusion and a variegated "turn" away from independent movement-community politics and struggle towards institutionalized party politics and a creeping (Zuma-inspired) social conservatism, individualism.' As discussed below in a Durban case study (Mzinyathi), individualism is the dominant strategy for survival, even where people continue to protest collectively. In short, the private politics of survival crowded out the potential for commoning in even the strongest urban social movement.

In Durban, Joel Kovel (2007: 251) initially identified an AbM strategy 'to recreate commons' as 'a modern simulacrum of the Paris Commune,' both in the shackdwellers' demands for housing and for access to water (Galvin, 2016). However, the vocabulary of the commons was often hidden or translated into the paradigm of human rights, which itself experienced limitations when water was tested in the Constitutional Court in 2009. The differences between a narrow, constitutionalist 'Right to the City' and the more expansive possibilities of eco-socialist commoning were explored in Johannesburg, especially, in a Constitutional Court case that ended in failure.

Water rights litigating versus activist commoning

Whether in shacks, townships, inner city slums or South Africa's other stressed urban and peri-urban settings, in the early 2000s a turn to constitutional strategies become attractive to donors, NGOs, social movements and their legal advisors. Given the long history of civil disobedience and intense confrontations with the apartheid state, the use of rights discourse became one of the main narratives dissidents used to challenge government policy and practice after 1994. However, Erik Swyngedouw (2014) has correctly pointed out that the uncritical adoption of a litigative approach to rights enforcement readily fits into a 'neoliberal tyranny of participatory governance,' even in a state with a liberal democratic constitution like South Africa. If the liberals' rights strategies were effective, that would make it more difficult for more radical commons approaches.

The limits of merely constitutionalist framings of rights – instead of practical commoning – became evident in the case *Mazibuko v Johannesburg Water*. Supported by the APF and its allies in the broader Coalition Against Water Privatization (2009), the legal battle is illustrative of the limitations of liberalism. The lower courts declared unconstitutional Johannesburg's commodified water strategy – originally established by the French firm Suez during its 2000–06 management contract – on the grounds that pre-payment meters and the meagre 25 litres per person per day allotment of Free Basic Water were insufficient for a dignified life.[2]

In 2008, the High Court agreed with APF that 50 litres were required and the judge also banned pre-payment meters, a decision confirmed in the Supreme Court (though with 42, not 50, litres specified). In the Constitutional Court (2009), however, Judge Kate O'Regan confirmed the conservative character of juridical sensibilities in refusing to make detailed policy recommendations, and in finding that Johannesburg Water had exercised its duties of expanding rights on an incremental, affordable basis (Roithmayr, 2011). It was not only this outcome, but also the process of litigating rights that angered APF activists like Trevor Ngwane (2003). What happened to the struggle for water in Johannesburg can be seen as a case of the 'domestication' of the politics of need, which according to Tshepo Madlingozi (2007) entails taking militants off the street and putting them into courts. There, activist arguments had to be panel-beat, removing any progressive and quasi-socialist intent. Even attempts to draw connections to ecological factors such as the overconsumption of water by wealthy residents would not have served the purposes of narrow legal argumentation, and so they were omitted.

Another critical legal scholar, Marius Pieterse (2007), complained that 'the transformative potential of rights is significantly thwarted by the fact that they are typically formulated, interpreted, and enforced by institutions that are embedded in the political, social, and economic *status quo*.' Daniel Brand (2005) added, 'The law, including adjudication, works in a variety of ways to destroy the societal structures necessary for politics, to close down space for political contestation.' Brand specifically accuses courts of depoliticizing poverty by casting cases 'as private or familial issues rather than public or political.' In sum, following the Critical Legal Scholarship tradition, rights talk is only conjuncturally and contingently useful (Roithmayr, 2011).

The case of *Mazibuko v Johannesburg Water* raises the question of whether to dismiss rights as a narrative and courts as a terrain and to instead pursue commoning alone (e.g. illegal reconnection of water supplies) or whether to use juridical strategies when the appropriate conjuncture arises, while steering clear of rights as a foundational argument. The APF and its Johannesburg allies ultimately wasted time, energy and vast resources in pursuing juridical justice under the impression that they had genuine water rights (Bond, 2013). In contrast, movements, such as the TAC and AbM, developed diverse juridical strategies that used socio-economic rights in both an offensive way – when TAC won nevirapine access (to prevent HIV transmission between a pregnant woman and new-born child) in a 2002 Constitutional Court decision – and for defensive purposes – when AbM had the provincial Slums Act declared unconstitutional in 2009 because it had no provisions for rehousing displaced communities. Similarly, rights-based injunctions were sought by some local groups such as the Westcliff Flat Residents Association in a way to combine litigation with protest. In a seminal 2001 case, their attempt to prevent water disconnections was initially supported by a lower court but then overturned on appeal by the state. Nevertheless, the Association gained from the experience and retained confidence to regularly challenge the municipality for further socio-economic concessions (Galvin, 2016). What these cases suggest is that strategies for each conjunctural circumstance must be considered on their own merits.

Scaling-up the commons: the localist dangers

Another challenge to constructing a genuine 'Right to the City' (Harvey, 2012; Lefebvre, 1996) is the persistently localistic focus of most urban activists. Their failure to develop wider linkages with other likeminded groups in South Africa was, for obvious reasons, related to the way apartheid had used zoning boundaries as race-based barriers to mobility. So, their own township or the shack settlement remain the immediate terrain of struggle for most community activists, even though there is little doubt that the stingy Treasury central budget and national-level neoliberal policies can be blamed for so many of the country's urban problems.

The question of how, whether and to what extent local protests are linked has become a topic for theoretical engagement and debate. Whether as 'popcorn protests' which spontaneously erupt or as more consciously organized revolts sometimes termed a 'rebellion of the poor' (Alexander et al., 2013), service delivery protests take place more or less a thousand times each year, with varying reports as to whether they are increasing in number and intensity. According to Carin Runciman et al. (2016: 44), community protests comprised about 22 percent of all police-recorded protests in South Africa from 1997 to 2013. From 2009 to 2013 there were typically 2–3 community protests per day and a peak of 5 per day in 2012 (Runciman et al., 2016: 48–49, Appendix 4). Protesters typically call on a municipal councilor or official to provide better water, sanitation, electricity, roads, stormwater drainage, clinics and other municipal services and often demand jobs in the process. However, the main problem remains the 'scale jumping' to national government, because it is in the Treasury that the central-to-local subsidies are determined and is the parastatal Eskom that sets electricity tariffs. It is in national public policy that service standards and subsidization strategies are set, which municipalities have to pursue. The failure of protesters to think nationally and locally, and to also act nationally while continuing to mobilize locally, characterizes these protests and helps explain their failures to generate widespread commons politics.

The central government typically ignores the protests, although in extreme cases, a national politician will visit the scene and typically provide platitudes, without proving a real commitment to changing the underlying conditions. Such a commitment would interfere with national fiscal constraints and neoliberal policies. This is especially true when it comes to water and electricity access. When the opposite takes place, protesters are surprised. This is, for example, the case of late 2017 when the students demanding free tertiary education pleasantly received the information that then President Zuma had agreed that 90 percent of students (those from the middle class and below) would not have to pay fees in future. This was the second major commoning victory in public policy – the decommodification of education – following TAC's medicines decommodification a dozen years earlier.

What is also missing in South Africa's urban civil society sphere is a coherent ideology: specifically an anti-racist, feminist, eco-socialism that can transcend intra- and inter-urban competitive tendencies and generate the kinds of social movements and political parties that, for example, Southern European city protests seem to have embraced since 2011, especially in Athens, Barcelona and Madrid. In the years ahead, there is nevertheless a distinct possibility for utilizing struggles for the commoning of water and electricity – especially in Soweto, the huge township on the edge of Johannesburg – for the purpose of developing models for broader political strategy. In the Gramscian sense, commoning could become an implicit ideology to transcend the prevailing tendencies to commodification, atomization and individualism fostered by neoliberal hegemony.

Air, water and food commoning as prefigurative politics

Even in the absence of an explicit ideology, there are important local signs of commoning around environmental justice and urban farming. In the name of environmental justice, activist groups are taking action locally to protect water and the environment as a commons and to challenge local air, water and soil pollution by industry and mines and government inaction. Increasingly, these draw on 'citizen science' (Patel, 2009), in which residents collect air or water samples and have them tested. In one of the most polluted areas of South Africa, one hour's drive south of Johannesburg, the Vaal Environmental Justice Alliance works with Save the Vaal to monitor and report river pollution. This treatment of the commons exists across class and racial divides, a notable achievement in the highly fractured South African context. This type of citizen science then places local activists' organizations in the position of informal regulator, able to demand that the state take action against polluting corporations or parastatal agencies, by providing proof of pollution and abuse of natural resources (Steward et al., 2007).

In terms of rural, community-based climate change adaptation, small-farming activists are developing surface water alternatives, monitoring their own weather data and changing crop selection to cope with climactic changes. In the Western Cape village of Goetverwacht, a local network of the community, university and local government has emerged to promote progressive land tenure agenda and make other far-reaching climate adaptations. Yet their access to water is still limited by the land being owned by the church. In this rural, isolated location, activists are compelled to work independently, providing hope that bureaucratic networks that would co-opt these groups into the state do not exercise the power to do so (Galvin et al., 2015; Galvin, unpublished).

In terms of water, a higher scale form of commoning has already been constructed by the national authority through the broader institutional frameworks established by the government, including Catchment Management Forums (CMFs), which agree on the allocation of water resources to farmers, industry, mining and other large users. Legislation specifies that 10 percent of a water course must be allocated to the 'ecological reserve,' to maintain the commons and ensure the integrity of the water supply in eco-social regards. However, in spite of their merits in theory, in practice CMFs are not formulated in a way to properly involve small users and local groups. Those that have tried to participate in CMFs report being treated in a condescending manner and sidelined (Munnik, 2015). In a more optimistic reading, van Koppen (2014) argues that Multiple Use Water Services can be scaled up. Likewise, Woodhouse et al. (2016) show that local farmers are using irrigation systems that they designed autonomously.

In peri-urban areas such as Mzinyathi in Durban, women's garden groups support one another in a classic case of mutual aid serving as a means of coping. Their gardens offer a means of supplementing purchased foods and, if possible, selling the surplus locally. These groups have not made broader linkages and are not well profiled in the community, arguably due to their gender and class composition (Galvin, 2010). In contrast, in Soweto, urban farming is the focus of the Izindaba Zokudla network, which enjoys support from University of Johannesburg food activists. In the easily accessible markets within this dense, income-diverse township, Izindaba Zokudla has had success in developing gardens, linking to markets and engaging with the local state. Their aim is to change agricultural and food systems, but they operate with an entrepreneurial spirit and lack a concrete ideological agenda. However, the expectation in this instance is that once local farmers have a reliable platform, such as regular Soweto buyers, new opportunities can arise to challenge South Africa's notoriously concentrated retail structures and for farmers themselves to become recognized stakeholders engaging with government structures (Malan, 2015; Gwamba, 2018).

In Seawinds, in the Cape Flats outside Cape Town, outside activists with funding from conservation groups (the Table Mountain Fund and the Rufford Foundation) have built bridges with local activists. Together, they have appropriated unused land along the roadside to farm plants to be used in traditional medicine. While their work appears to challenge the state and the traditional notion of property, it has carved out space that is non-contentious and they are able to work closely in developing a win-win strategy with the state (Sustainable Livelihoods Foundation, undated). This evokes the 'guerrilla urbanism' that food commons analyst Robert Biel (2016: 113) endorses and that has the potential to '[emphasize] that the city is a human system and its emergent properties develop from its people: we cannot simply address self-organization at a technical level without also embracing struggles for emancipation and environmental justice.'

Biel (2016: 109) argues that such food struggles 'encompass both the issues of immediate material livelihood, which all revolutions must address, and the big strategic issues going beyond immediate survival: dis-alienation, human rights and real democracy; all of which tend to converge in today's land and food struggles.' Given the history of land and food in South Africa, and the dominance of privately own lands and commercial farming, land remains a struggle around which people make demands for reparations and even, in the early 2000s, built the Landless People's Movement around. But food farming itself rarely moves past material concerns into ideological ones.

From community organizing to climate adaptation

From water, food and livelihood, the notion of commoning and the paradigm of the commons could also be utilized to reinforce and define the ongoing struggles for climate change adaptation, i.e. the fights to protect low-income residents from events such as the extreme flash flooding and droughts that affected South Africa in 2016–17. To date, this has been treated as risk and disaster management, but, as a result, de-politicization of the issue accompanies state failure in even the most rudimentary support for victims of extreme weather events. Although cities have formulated adaptation plans at a city-wide level, they rarely address directly what climate change means for the poorest areas. In response, local activists, such as community organizations that are part of the GroundWork network, are challenging the strategy of transition that is being imposed from above. Instead, they are pursuing a strategy often termed 'Just Transition,' which challenges underlying power relations and makes demands upon the state. These are still being formulated, but two examples are illustrative.

First, in the Eastern Cape's 'Wild Coast,' largely inhabited by the low-income, rural Pondo ethnic group who possess very strong traditions of fighting apartheid, a campaign has been waged since 2008 by the Amadiba Crisis Committee (ACC) and allies in the green network 'Sustaining the Wild Coast' (Bennie, 2017). They are opposed to the coastal extraction of titanium at the world's tenth largest deposit, in the community of Xolobeni. Instead, the ACC promotes Wild Coast eco-tourism and traditional small-scale farming. Members of the network ask simple questions that are based on the close connection between land, water, food and ecological balance: 'How can we be poor when we have land? We grow maize, sweet potatoes, taro, potatoes, onions, spinach, carrots, lemons and guavas, and we sell some of it to the market. We eat fish, eggs and chicken. This agriculture is what should be developed here' (Washinyira, 2016).

Andrew Bennie (2017) shows how villagers in search of food sovereignty have deployed agriculture as a tactic of resistance, giving literal expression to the notion that 'resistance is fertile.' Tactics like these represent a particular form of stern defiance against the plans of a fledgling business class and state elite that, to them, has long severed itself from their humble desires: to keep their land, to decide what to do with it, to welcome others to appreciate it with them and to be assisted in a forward-looking approach that seeks to build on it rather than destroy it. The ACC and its lawyers have brought the case to court, arguing that titanium mining would destroy 'the biome and ethnobotanical elements of the area. This includes reliance on the ocean, and the socio-cultural and economic value derived from the land and ocean' (ACC, 2016). A titanium smelter would also draw upon coal-fired electricity or a proposed new nuclear generator. The ACC (2016) also expressed concern that mining would disrupt plants used in traditional medicine, destroy water sources and grazing land, disturb scores of burial sites – hence 'breaking of links with ancestors' – and delimit 'self-sufficient development… Financial compensation and the provision of alternative housing for those who are physically displaced cannot adequately compensate for the destruction of a community, its culture and traditions and its members' way of life.' This narrative and the intense local struggles in which more than a dozen ACC activists have died under mysterious conditions over the past decade, testify to the fusion of desperation and power in many such settings (Bennie, 2017).

Another example is represented by the South Durban Community Environmental Alliance (SDCEA), which has made various post-carbon development demands for the South Durban Basin, mainly in opposition to the area's massive port-petrochemical expansion (SDCEA, 2008; SDCEA, 2011; Bond, 2016a). These include defending the 'airport farmers,' a group of more than 100 land managers and workers, against displacement as the old airport becomes a potential site of redevelopment. In its campaign, SDCEA challenges the liberalized zoning that currently allows freight transport to creep into historic Clairwood, displacing thousands of black households. The reversal of the zoning would also protect and regenerate green spaces in the already toxic-saturated industrial and petrochemical areas of the city, revealing the close interconnection between the protection of livelihoods and the preservation of the environment. SDCEA's (2008) 30-page 'Spatial and Development Vision' includes demands such as 'a halt to the privatization of ocean, Bay and shore resources that belong to all the people of this country.' A lengthy follow-up statement just prior to SDCEA's (2011) co-hosting of the counter-summit to the UN climate summit in 2011 – 'Feeling the heat in Durban' – included this language: [productive-sector economic] 'localization is essential to any serious programme of mitigation and requires that national resources should be focused on supporting people's capacities to direct local development…We call for people's energy sovereignty founded on democratic and local control.'

Contradictions in commoning

In every case of community-based adaptation to climate change, it is critical to interrogate the way in which the framing and the strategies are presented and articulated. It is rare for civil society groups to make connections between the less tangible impacts of climate change and systemic aspects of a capitalist system that thrives on the extraction, production and emission of hydrocarbons. Because of the different perspectives, vocabularies, interests and confrontations, opportunities to 'connect the dots' between struggles are often not taken up. In other cases, connections are made, albeit with a strong ideological orientation. This is the case, for example, of the rural Goedverwacht community in the Western Cape (Rodina et al., 2017). The challenge is for urban groups to assert their agency through climate commoning, while beginning to draw out larger lessons. The so-far disconnected struggles over water, food and climate are illustrative of the potentials, but also the pitfalls, of bottom-up mobilization.

Two critical contradictions immediately emerge. First, the concept of water as a commons as with both a political and managerial role has been successfully applied in contexts (such as Soweto) that are not particularly water-stressed. Residents are not competing for scarce water resources supplied by inadequate bulk systems; they do not suffer limits on intermediate infrastructure such as pipe width and pumping station size.

Second, the commoning of water or electricity in the form of 'self-reticulation' or 'illegal connections' has certain negative impacts on people's quality of life. There is no universally agreed redefinition of access, so idiosyncratic action can have a negative impact on the water or electricity supply system as a whole. For example, debilitating water leaks due to faulty informal plumbing connections and electricity outages due to inadequate circuit breakers are unintended consequences of activists' reconnection strategy. Mitigating against this danger requires an understanding within and between local areas of the wider political strategy of commoning. The Soweto Electricity Crisis Committee is one of the institutions that regularly addresses these problems, as well as the continual danger of electrocution when the power grid is informally accessed – as applies to 80 percent of Soweto – with the occasional tragic result of deaths, especially of children who step onto loose, live wiring.

Taking water as the topic of concern, the critical question is whether the level of social organization and trust within the broader society allows the commoning and reconnection approach to go beyond the individual household scale and be adopted into the broader hydrological system. There is a need to 'jump scale' for the sake of expanding genuine commoning of water, from the atomistic household to a catchment-level residential population. This partly depends on whether social organization and trust within a specific and well-organized community can extend to catchment-wide water planning.

To take one example, in some parts of North America, citizens have banded together as 'Riverkeepers' (in spirit of Robert Putnam's 'social capital') and protect the water commons and the health of river systems by extending their local actions to the level of the whole river. Without the pressure of scale constraints (thanks to strong communications systems) and without the concern with basic survival that makes so many South Africans think locally and act locally (only), activists in the Global North have the luxury of organizing together horizontally and vertically across space and scale, so as to protect their water resources at the source and consumption site together. In addition, within what might be termed the 'Global South within the North' (e.g. Flint, Michigan), water activists did make this jump along the socio-hydrological system, but too late – only after a river source was switched to save money and hence poisoned the mainly low-income users. In the Global North, activists can normally use trust in broader-based geographically expansive alliances and in very few cases there is no immediate threat to water coming out of their tap. Activists have thus the luxury to work together using a citizen-science approach to monitor water quality, drawing immediate attention to pollution and challenging polluters whether upstream or downstream.

Some sites in South Africa have this capacity, such as South Durban, where rapid-alert systems have been established to protect the air from excessive pollution, or in the Vaal, where environmentalists have begun to monitor water quality. But in most parts of the country, activists are far from being able to achieve a commons-like approach based on bottom-up coordination. Societal trust has been eroded into collective individualism. This is evident in peri-urban areas like Mzinyathi, a black residential area within Durban city limits, but located outside the 'Urban Development Line' (UDL). That location demarcates areas with formal sewer lines available to those with closer-in housing from those further away who reside in what was a former 'KwaZulu' homeland area. Since these lines were drawn, Mzinyathi has witnessed extremely rapid urbanization, with middle class households moving into what were formally rural areas. There they suffer parallel 'traditional governance' (i.e., ethnic-patriarchal) power structures but, at the same time, they receive free water. The water utility indicates that it plans to provide metered water infrastructure with greater flow capacity, so as to charge for the increased supply. However, at present, all water in this area is free, since administratively the city has been unable to charge individual households due to the collective land tenure, in which ethnic overlords manage residential allocations.

With the influx of middle-class (black) households who have a higher consumption level, the Mzinyathi water system is unable to cope with much larger flows of water demanded by a rising population. Household taps provide high-pressure water, rather than the 300 liters per household per day provided in containers (at low pressure) that poor households receive. But as a result of the increased consumption, areas further north of Mzinyathi on the same water pipeline have suffered for lengthy periods with no water. These lower-income citizens are thus forced to rely on water tankers from the city due to the easy (and free) access that is enjoyed by citizens who are economically and socially better off. Everyone in the area understands that it is this 'overuse' of free, unlimited water by new households that prevents other areas further down the pipeline from having any water. Under ideal circumstances, in which commoning was widely accepted, a community meeting and an agreement would be all that is needed to share the available water fairly and to halt the overconsumption by some that prevents other people's access. Yet without that level of simple social organization and ethos of sharing equally in Mzinyathi, it is 'every household for itself.' The ethnic leaders – Zulu chiefs – have little to contribute to a solution.

As a result, consumer expropriation using illegal connections to access unlimited free water is a short-term tactic that may help achieve positive outcomes for households, whether in Detroit or Soweto, but with potential, unintended, adverse consequences. Activists see this as decommodifying the water commons and winning their 'Right to the City.' But in the case of Mzinyathi and countless others, this approach often leads to long disruptions of the water system due to leaks that are the consequence of people's self-connection and excessive use upstream that leaves very little left over downstream along the pipe (often with high levels of wastage). The same is true of illegal electricity connections, in which the individual household's access may put at risk the entire neighborhood's supply, given the frequency of brown-outs that correlate to the amount of stress on the power system when the circuit breakers trip. In the context of systemic deprivation, it requires enormous skill and organization to generate a horizontal and diverse movement that both fights against excessive payment by poor people and reinforces respect for natural resources and social connections to the Earth.

In parts of the world where there is a high level of trust and social capital, communities may manage and protect their own resources, even though they struggle for water. This typically applies to indigenous communities, whether First Nations in Canada or the collective water systems of Latin America, that are known for having a deeper connection to the land and water and a history of managing these resources. There are untapped potentials in South Africa, for example, in water court cases lodged by Lawyers for Human Rights and the Legal Resources Centre in the city of Carolina, in the province of Mpumalanga, in 2012 and by AfriForum in the Vhembe District Municipality, Limpopo. Both were successful in forcing municipalities to supply water to residents. Nevertheless, they failed to use their successes to link with other groups and escalate their grievances past the local level. Often a state agency is required as an intervening force and, in a case like South Africa where both neoliberal policy and corruption are barriers, additional community organizing is needed to gain sufficient power.

The 'Right to the City' *and* to water commons in South Africa

Resistance strategies and tactics develop over time. Throughout the townships of South Africa, activists interested in guaranteeing access to water and electricity have been attempting to evolve what was already a popular survival tactic at the time of apartheid: reconnecting water and electricity illegally, once it was disconnected by state officials due to nonpayment. In 2001, 13 percent of Gauteng's water connections were deemed to be illegal (Bond, 2002) and by 2016, Eskom announced that in Soweto the share of illegal electricity connections had risen to 80 percent (Le Cordeur, 2016). By 2018, the cumulative cost had risen to $1 billion, which was as much as the rest of the country owed Eskom (Mahlangu, 2018). As discussed above, the most serious problem with what could be called 'informal commoning' is that once the water infrastructure is tapped by township plumbers (many working for a small fee), leaks are exacerbated and water quality is sometimes compromised. Likewise, as electricity lines are commoned using illegal connections, the capacity of the entire township system is stressed and the power supply regularly trips. Accidental electrocutions become more frequent as live wires criss-cross pedestrian pathways.

The typical short-term response from a Soweto Electricity Crisis Committee activist is to embark upon technical training, stressing insulation and effective pipe repair. Autonomist and localist activists are generally comfortable with the insurgent spirit represented in such strategies, in contrast to socialists who typically argue for a future metropolitan-scale planning and redistribution so as to avoid system degradation due to illegal connections. According to the latter, bottom-up commoning should be replaced with top-down tariff reforms providing a minimal decent supply of water and electricity to all residents but penalizing high-volume (hedonistic) users so as to provide surpluses for cross-subsidization. In this conception, the class struggle occurs within the public authority over the shape and slope of the tariff curve. Here, the eco-socialist committed to both social justice and conservation (i.e. reducing wasteful demand for water and electricity) is fully aware that the state must be controlled by socialists in order for such tariffs and other reforms to be implemented.

However, hacking the water network is not the only space of intervention. Together with these survival tactics, community-based leaders such as Ngwane (2009) adopt a strategic approach that links the locality of community with the global character of ecological processes, mainly through questioning the bulk supply of water to Johannesburg from Lesotho mega-dams and of electricity from coal-fired power stations, because 'The climate crisis can only be solved if the profit motive is severely restricted or eliminated altogether. Capitalism is incapable of solving the ecological crisis because it is the main culprit.'

Our argument about commoning is premised on the reality that there are limits to the use of constitutional rights. If the objective of those promoting the 'Right to the City' includes making water primarily an eco-social and decommodified good, these constitutional limits will have to be transcended. Yet there are differences of opinion about the next logical step on how to move beyond commoning to eco-socialism. Some argue that the scope for change is limited to engaging with the present state. The strengthening of free basic water provision is one strategy derived from such an approach. But doing so requires raising the scale of politics from the local to the national level (Galvin, 2016). It is at that level that commoning can be achieved, both horizontally across the populace and vertically from the raindrop above or borehole below, all the way to the sewage outfall and the sea. But to get to the next mode of financing, extraction, production, distribution and disposal of water requires a formidable social force to take us through and beyond rights, to the commons.

Conclusion

We have observed in this South African exploration of commoning's potentials and contradictions that the crucial missing link in many sites of informal water, food and climate politics is an eco-socialist (and also hopefully feminist and anti-racist) ideology. The ideological deficit is not a problem unique to South Africa. As Frantz Fanon (1969: 186) once remarked, 'For my part the deeper I enter into the cultures and the political circles, the surer I am that the great danger that threatens Africa is the absence of ideology.' After all, neoliberals are extremely forceful, moving their arguments deep into the terrain of commoning. One multilateral agency profoundly committed to urban private property rights, Habitat (2014: 7) argues that 'The "Commons" reinforce the social function of property and that of the city as a whole, while recognizing the dynamism of private assets.'

In contrast, the challenge for South Africans committed to a different society, economy and city is to humbly combine the limited gains that social movements have won so far (in many cases matched by regular defeats on economic terrain) with the soaring ambitions that are needed to match the scale of the systemic crisis and the current extent of social protest. The irony is that the upsurge of recent protest of a 'popcorn' character – i.e. rising quickly in all directions but then immediately subsiding – screams out for the kind of organization that once worked so well in parts of Johannesburg, Durban and Cape Town. Moreover, there are ideological, strategic and material problems that South Africa's independent left has failed to overcome, including the division between autonomist and socialist currents and the lack of mutual respect for various left traditions, including traditional communism, Trotskyism, anarchism, syndicalism, Black Consciousness and feminism. A synthetic approach from the top down still appears impossible.

Dignity is ultimately the outcome of a struggle in which not simply individual demands for rights, but collective solutions – e.g. the full-fledged bulk infrastructure required for sanitation – are achieved, once political parties (perhaps the Economic Freedom Fighters, the leftist movement that began in 2013) gain sufficient state power to answer the demands of social movements (Bond, 2016b). The next generation of South African urban activists will have learned the prior movements' lessons and will have less and less satisfaction with constitutionalism, as it becomes clearer how courts protect property in times of stress and as the police react to protest with more violence. *En route*, society is girding for degeneration into far worse conditions than now prevail: a post-apartheid South Africa that is more economically unequal, more environmentally unsustainable and more justified in fostering anger-ridden grassroots expectations than during apartheid itself. One of the central questions, once dust settles following battle after battle and activists compare notes, is whether the cadres find they can use commoning as a strategy towards an ideology and practice more robustly eco-socialist in character.

Notes

1 The literature about the South African transition and its limitations to progress in various sectors includes works by Alexander (2002); Ballard et al. (2006); Bell and Ntsebeza (2003); Desai (2002); Cock (2011); Hart (2002); Hassim (2006); Marais (2001); Mhone and Edigheji (2003); Naidoo (2007); Ntsebeza (2005); Padayachee and Habib (2000); Shivambu (2015); Terreblanche (2002); van Driel (2003); and Webster and Adler (1995).

2 Traditionally white areas had credit meters, revealing the municipality's higher levels of trust in the higher-income residents.

References

Alexander, A. (2002) *An Ordinary Country*, University of KwaZulu-Natal Press, Pietermaritzburg.

Alexander, P., Runciman, C. and Ngwane, T. (2013) 'South Africa's rebellion of the poor,' University of Johannesburg Centre for Social Change, Johannesburg, https://ddd.uab.cat/pub/caplli/2016/157741/I9UAlexander_Peter_et_al_OK_.pdf.

Amadiba Crisis Committee (2016) 'About the Amadiba Crisis Community and the region,' Unpublished report, Xolobeni, South Africa, http://www.bench-marks.org.za/press/amendments_accepted_in_annexure_a.doc.

Ballard, R., Habib, A. and Valodia, I. (2006) *Voices of Protest*, University of KwaZulu-Natal Press, Pietermaritzburg.

Baskin, J. (1991) *Striking Back*, Raven Press, Johannesburg.

Bell, T. and Ntsebeza, D. (2003) *Unfinished Business*, Verso, London.

Bennie, A. (2017) 'Resistance is fertile,' *Daily Maverick*, 20 July 2017, https://www.dailymaverick.co.za/article/2017-07-20-op-ed-resistance-is-0fertile-amadiba-agriculture-challenges-elite-mining-agenda/#.Wl7usHlx3Z4.

Biel, R. (2016) *Sustainable Food Systems*, UCL Press, London.

Bond, P. (2000) *Cities of Gold, Townships of Coal*, Africa World Press, Trenton.

Bond, P. (2002) *Unsustainable South Africa*, Merlin Press, London.

Bond, P. (2003) *Against Global Apartheid*, Zed Books, London.

Bond, P. (2014) *Elite Transition*, Pluto Press, London.

Bond, P. (2015) 'The intellectual meets the South African social movement subject,' *Politikon*, 42, 1, pp.117–122.

Bond, P. (2016a) 'Red-green alliance-building against Durban's port-petrochemical complex expansion,' in L. Horowitz and M. Watts (eds) *Grassroots Environmental Governance*, Routledge, London.

Bond, P. (2016b) 'South Africa's next revolt,' in L. Panitch and G. Albo (eds) *Socialist Register 2017*, Merlin Press, London.

Bond, P. and Khosa, M. (1999) *A Reconstruction and Development Programme Policy Audit*, HSRC Press, Pretoria.

Bond, P. and Ruiters, G. (2017) 'Uneven development and scale politics in Southern Africa,' *Antipode*, 49, S1, pp.171–189.

Bond, P., Desai, A. and Ngwane, T. (2013) 'Uneven and combined Marxism within South Africa's urban social movements,' in C. Barker, L. Cox, J. Krinsky and A. Nilsen (eds) *Marxism and Social Movements*, Routledge, London.

Brand, D. (2005) 'The politics of need interpretation and the adjudication of socio-economic rights claims in South Africa,' in A.J. van der Walt (ed) *Theories of Social and Economic Justice*, Stellenbosch University Press, Stellenbosch.

Budlender, J., Woolard, I. and Leibbrandt, M. (2015) 'How current measures underestimate the level of poverty in South Africa,' *The Conversation*, 3 September 2015, https://theconversation.com/how-current-measures-underestimate-the-level-of-poverty-in-south-africa-46704.

Coalition Against Water Privatization (2009) 'Phiri Water case: Constitutional Court fails the poor and the Constitution,' Johannesburg, 2 October.

Cock, J. (2011) *The War Against Ourselves*, Wits University Press, Johannesburg.

Constitutional Court of South Africa (2009) *Mazibuko & Others v the City of Johannesburg & Others*, Johannesburg.

Desai, A. (2002) *We are the Poors*, Monthly Review Press, New York.

Desai, A. (2006) 'Vans, autos, kombis and the drivers of social movements,' Harold Wolpe Memorial Lecture Series, Centre for Civil Society, Durban, 28 July, http://wolpetrust.org.za/dialogue2006/DN072006desai_paper.pdf.
Duncan, J. (2016) *Protest Nation*, University of KwaZulu-Natal Press, Pietermaritzburg.
Fanon, F. (1969) *Toward the African Revolution*, Grove Press, New York.
Ferrando, T. and Vivero-Pol, J.L. (2017) 'Commons and 'commoning': A 'new' old narrative to enrich the food sovereignty and right to food claims,' P2P Foundation, Amsterdam, https://blog.p2pfoundation.net/commons-and-commoning-a-new-old-narrative-to-enrich-the-food-sovereignty-and-right-to-food-claims/2017/10/05.
Galvin, M. (2010) 'Rural survival, development or advocacy?' in B. Maharaj, A. Desai and P. Bond (eds) *Zuma's Own Goal*, Africa World Press, Trenton.
Galvin, M. (2016) 'Leaving boxes behind,' *Transformation* 92, 1, pp.112–134, https://ujcontent.uj.ac.za/vital/access/services/Download/uj:22606/SOURCE1.
Galvin, M. (forthcoming, 2019) 'Making community based adaptation a reality: Different concepts, different politics,' *Human Geography*.
Galvin, M., Wilson, J., Sabine Stuart-Hill, S., Pereira, T., Warburton, M., Khumalo, D., Robinson, G., Chibvongodze, D. and Lewis, M. (2015) 'Planning for adaptation: applying scientific climate change projections to local social realities,' Water Research Commission, Pretoria, http://www.wrc.org.za/Knowledge%20Hub%20Documents/Research%20Reports/2152-1-15.pdf.
Gwamba, K. (2018) 'Growing participation in urban agriculture: Technology development in the Izindaba Zokudla Project in Soweto, South Africa,' University of Johannesburg, MA thesis submitted for examination.
Hart, G. (2002) *Disabling Globalization*, University of KwaZulu-Natal Press, Pietermaritzburg.
Harvey, D. (2003) *The New Imperialism*, Oxford University Press, New York.
Harvey, D. (2012) *Rebel Cities*, Verso Press, London.
Hassim, S. (2006) *Women's Organizations and Democracy in South Africa*, University of Wisconsin Press, Madison.
Heller, P and Ntlokonkulu, L. (2001) 'A civic movement, or a movement of civics?,' Johannesburg, Centre for Policy Studies.
Klein, N. (2014) *This Changes Everything*, Simon & Schuster, New York.
Kovel, J. (2007) *The Enemy of Nature*, Zed Books, London.
le Cordeur, M. (2016) 'Eskom to waive Soweto users' debt – on condition,' *Fin24*, 25 May, https://www.fin24.com/Debt/News/eskom-to-waive-soweto-users-debt-on-condition-20160525.
Madlingozi, T. (2013) 'Post-Apartheid social movements and legal mobilization', in M. Langford, B. Cousins, J. Dugard, and T. Madlingozi (eds) *Socio-economic Rights in South Africa*, Cambridge University Press, Cambridge.
Mahlangu, I. (2018) 'Soweto residents owe Eskom R15bn in unpaid bills.' *Sowetan*, 18 September, https://www.sowetanlive.co.za/news/south-africa/2018-09-18-soweto-residents-owe-eskom-r15bn-in-unpaid-bills/
Malan, N. (2015) 'Design and social innovation for systemic change,' in L. Collina, L. Galluzzo, and A. Meroni (eds) *The Virtuous Circle*, McGraw-Hill Education, Milano.
Marais, H. (2001) *South Africa Limits to Change*, Zed Books, London.
Mayekiso, M. (1996) *Township Politics*, Monthly Review Press, New York.
Mbali, M. (2013) *South African Aids Activism and Global Health Politics*, Palgrave Macmillan, London.
McKinley, D. (2012) 'Lessons of struggle,' *The South African Civil Society Information Service*, 2 February 2012, http://sacsis.org.za/site/article/1197.
McKinley, D. (2016) 'Lessons in community-based resistance?' *Journal of Contemporary African Studies*, 34, 14, pp.268–281.
Mdlalose, B. (2014) 'The rise and fall of Abahlali baseMjondolo, a South African social movement,' *Politikon*, 41, 3, pp.345–353.
Mhone, G. and Edigheji, O. (2003) *Governance in the New South Africa*, Juta Press, Cape Town.
Miraftab, F. and Wills, S. (2005) 'Insurgency and spaces of active citizenship,' *Journal of Planning Education and Research*, 25, 2, pp.200–217.
Munnik, V. (2015) 'Making space for cognitive justice,' Unpublished paper, University of the Witwatersrand, Department of Sociology, Johannesburg.
Naidoo, P. (2007) 'Struggles around the commodification of daily life in South Africa,' *Review of African Political Economy*, 34, 111, pp.57–66.

Ngwane, T. (2009) 'Ideology and agency in protest politics,' University of KwaZulu-Natal School of Development Studies, Master's research thesis proposal.
Ngwane, T. (2017) '"Amakomiti" as "democracy on the margins,"' University of Johannesburg, Department of Sociology, D.Litt. et Phil. thesis, https://ujcontent.uj.ac.za/vital/access/manager/Repository/uj:22963.
Ntsebeza, L. (2005) *Democracy Compromised*, HSRC Press, Pretoria.
Ostrom, E. (1990) *Governing the Commons*, Cambridge University Press, Cambridge.
Padayachee, V. and Habib, A. (2000) 'Economic policy and power relations in South Africa's transition to democracy,' *World Development*, 28, 2, pp.245–263.
Pieterse, M. (2007) 'Eating socio-economic rights,' *Human Rights Quarterly, 29*, pp.796–822.
Polanyi, K. (1957) *The Great Transformation*, Beacon Press, Boston.
Rodina, L., Baker, L., Galvin, M., Golden, J., Harris, L., Manungfala, T., Musemwa, M., Sutherland, C., and Ziervogel, G. (2017) 'Water, equity and resilience in Southern Africa,' *Current Opinion in Environmental Sustainability*, 26-27, pp.143–151.
Roithmayr, D. (2011) 'Lessons from *Mazibuko*,' *Constitutional Court Review*, 3, pp.317–346.
Runciman, C. (2012) 'Mobilization and insurgent citizenship of the Anti-Privatization Forum, South Africa,' University of Glasgow, PhD thesis, http://theses.gla.ac.uk/3706/.
Runciman, C., Alexander, P., Rampedi, M., Moloto, B., Maruping, B., Kumalo, E. and Sibanda, S. (2016) 'Counting police-recorded protests,' Social Change Research Unit, University of Johannesburg, https://www.researchgate.net/profile/Peter_Alexander9/publication/304076282_Counting_Police-Recorded_Protests_Based_on_South_African_Police_Service_Data/links/5765768708aedbc345f380d5/Counting-Police-Recorded-Protests-Based-on-South-African-Police-Service-Data.pdf
Seekings, J. (1997) 'SANCO: Strategic dilemmas in a democratic South Africa,' *Transformation* 34, pp.1–29.
Shivambu, F. (2015) 'South Africa's negotiated transition from apartheid to an inclusive political system,' University of the Witwatersrand, Master's thesis, http://wiredspace.wits.ac.za/handle/10539/17051.
Shonhe, T. (2018) *Reconfigured Agrarian Relations in Zimbabwe*, Africa Books Collective, London.
Smith, N. (1984) *Uneven Development*, Basil Blackwell, Oxford
South Durban Community Environmental Alliance (2008) *Spatial and Development Vision for the People of South Durban*, South Durban Environmental Alliance, Durban.
South Durban Community Environmental Alliance (2011) 'Feeling the heat in Durban,' *Capitalism Nature Socialism*, 22, 4, pp.50–73.
Steward, C., Aguiar, M., Aziz, N., Leaning, J., and Moss, D. (2007) 'Towards a Green Food System,' Grassroots International, Boston, Food and Water Watch, Washington, http://safsc.org.za/wp-content/uploads/2015/09/Towards-a-green-food-system.pdf.
The Sunday Independent (2016) 'Honour Chris Hani,' *The Sunday Independent,* 10 April, https://www.pressreader.com/south-africa/the-sunday-independent/20160410/281844347791978.
Swyngedouw, E. (2014) 'Where is the political?,' *Space and Polity* 18, 2, pp.122–136.
Terreblanche, S. (2002) *A History of Inequality in South Africa, 1652-2002*, University of KwaZulu-Natal Press, Pietermaritzburg.
United Nations Habitat (2015) *Global Public Space Toolkit*, Nairobi, UN-HABITAT, https://unhabitat.org/wp-content/uploads/2015/10/Global%20Public%20Space%20Toolkit.pdf.
van Driel, M. (2003) 'Unions and privatization in South Africa, 1990-2001,' in T. Bramble and F. Barchiesi (eds) *Rethinking the Labour Movement in the 'New' South Africa*, Ashgate, Aldershot.
van Koppen, B., Smits, S., Rumbaitis del Rio, C. and Thomas, J.B. (2014) *Scaling up Multiple Use Water Services*, Practical Action Publishing, Rugby.
Vivero-Pol, J.L. (2017) 'Epistemic regards on food as a commons: plurality of schools, genealogy of meanings, confusing vocabularies,' Unpublished paper, Catholic University of Louvain, 5 April, https://papers.ssrn.com/sol3/papers.cfm?abstract_id=2947219.
Washinyira, T. (2016) 'We will die for our land, say angry Xolobeni villagers as dune mining looms,' *Amabhungane*, Johannesburg, 12 February 2016, http://amabhungane.co.za/article/2016-02-12-we-will-die-for-our-land-say-angry-xolobeni-villagers-as-dune-mining-looms.
Webster, E. and Adler, G. (1995) 'Challenging transition theory,' *Politics and Society* 23, 1, pp.75–106.
Womin (2018) 'African women unite against destructive resource extraction' website, http://womin.org.za/
Woodhouse, P., Veldwisch, G.J., Venot, J.P., Brockington, D., Komakech, H. and Manjichi, A. (2016) 'African farmer-led irrigation development,' *The Journal of Peasant Studies*, 44,1, pp.213–233, http://dx.doi.org/10.1080/03066150.2016.1219719.

World Bank (2014) 'South Africa economic update; fiscal policy and redistribution in an unequal society,' World Bank Group, Washington, DC,, http://documents.worldbank.org/curated/en/2014/11/20339043/south-africa-economic-update-fiscal-policy-redistribution-unequal-society.
Zuern, E. (2004) 'Continuity in contradiction? The prospects for a national civic movement in a democratic state: SANCO and the ANC in post-apartheid South Africa,' University of KwaZulu-Natal, Centre for Civil Society, Durban.

PART IV

Commoning from below: Current examples of commons-based food systems

16

THE 'CAMPESINO A CAMPESINO' AGROECOLOGY MOVEMENT IN CUBA

Food sovereignty and food as a commons

Peter M. Rosset and Valentín Val

Introduction

Agroecology has played a key role in helping Cuba survive the acute crisis caused by the collapse of the socialist bloc in Europe in 1989-90 and the subsequent chronic crisis due the U.S. trade embargo, which have prevented economic normalization on the island ever since socialist trade relations were lost. Cuban peasants have been able to boost food production without scarce and expensive imported agricultural chemicals by first substituting more ecological inputs for the no longer available imports and then by making a transition to more agroecologically integrated and diverse farming systems.

This has been a process of collective transformation, based on the high level of organization of the Cuban peasantry through their national organization, the National Association of Small Farmers (ANAP), a member of the transnational peasant movement, La Via Campesina (LVC). This process has been stimulated through a horizontal, peasant-to-peasant learning and sharing methodology called 'Campesino a Campesino' (CAC), which helped create a national grassroots agroecology movement among peasants. This has been assisted by the perspective in Cuba in general and in ANAP and LVC, in particular, of food production as a social, public good and, in the sense of this book, of food as a commons or *bien común* produced through collective social process, rather than as a commodity (Rosset 2006a). In this sense, Cuba, ANAP and LVC are an example of what Vivero-Pol (2017 and this volume) calls the epistemological school of thought that sees the commons – food in this case – as a social construct that is defined by a collective practice by specific communities – peasants in this case. As we discuss below, this counter-hegemonic vision of food as a commons and not as a commodity (Vivero-Pol 2017) has permeated Cuba since the 1959 Revolution (Funes *et al.* 2002; Benjamin *et al.* 1984; Enríquez 1994).

An historical perspective on Cuban agriculture

Before the 1959 Cuban Revolution, the island was characterized by a typical *latifundio-minifundio* system of land distribution and tenure, with a strong presence of US capital, the production of sugar for export and a marginalized peasantry (Nova 2002). In the early years of the Revolution, the government invested heavily in improving conditions in the countryside and carried out an extensive agrarian reform over several progressive phases. While initial policy was directed at diversifying away from sugar and export dependency, extreme hostility by the US and the opportunity to join the international socialist division of labor (Comecon) on favorable terms of trade ended up strengthening the export monocrop emphasis as well as dependency on imported food, agricultural inputs and implements (Nova 2002; Machín Sosa *et al.* 2010). By 1989, 30% of agricultural land was devoted to a single export crop, sugarcane, which generated 75% of export revenues, while 57% of all food was imported (Rosset and Benjamin 1994).

In the export-led wave, Cuban agriculture was a world-class case of modernization and of the Green Revolution (Machín Sosa *et al.* 2010), with the most tractors per person and per unit of area and the second highest average grain yields of Latin America (Rosset and Benjamin 1994). Agriculture made heavy use of chemical inputs such as fertilizer, 48% of which was imported (with a 94% import coefficient for the fertilizer that was manufactured domestically), and pesticides, 82% of which were imported (Rosset and Benjamin 1994). While this model was able to guarantee a relatively high level of food security and standard of living to the Cuban population thanks to the continuation of the favorable terms of trade with the socialist bloc, in the long run it turned out to generate a dangerous dependency on foreign trade, providing temporary food *security* but not long-term food *sovereignty*. It also proved not to be very sustainable from an ecological and productive viewpoint, as the chemical-intensive industrial monocultures experienced ever increasing pest problems; after decades of increases, yields of some key crops, like rice, began to decline in the 1980s due to soil degradation and pests (Machín Sosa *et al.* 2010). This pattern of long-term yield leveling and/or decline is found in Green Revolution-lead areas around the world (Pingali *et al.* 1997; Radford *et al.* 2001; Kundu *et al.* 2007; Mulvaney *et al.* 2009) and Cuba was no exception (Rosset *et al.* 2011).

As a consequence of this conventional model, when the collapse of the socialist bloc in Europe came in 1989 and the United States tightened the trade embargo (called 'the blockade' by Cubans), Cuba lost 85% of its trade relations and was no longer able to import sufficient food, or the machinery, inputs and petroleum to grow it, under the capital-intensive production model (Rosset and Benjamin 1994; Funes-Monzote 2008; Wright 2008). The 1990s saw the Cuban population face an economic and food crisis while attempts were made to recover and boost national food production. In 1990, the Cuban government declared the 'Special Period in Peacetime', a war-style economic policy based on austerity measures, to survive the crisis. Part of that involved the breaking up of large state farms into Basic Units of Cooperative Production (UBPCs), basically cooperatives made up of former employees with usufruct privileges on the former state enterprise land.[1] One of the motivations was that, while peasant cooperatives were quick to adopt new low external input practices, the state farms seemed incapable of such rapid change (Rosset 1997).

But perhaps the most important changes occurred in the peasant sector itself. Virtually all peasants in Cuba belong to ANAP, and almost all of them belong to one of two types of cooperatives. Credit and Service Cooperatives (CCSs) are made up of peasant families who own their own farms and work them individually, but group together in the CCS to achieve economies of scale in marketing harvests, obtaining credit, sharing farm machinery, etc. Agriculture Production Cooperatives (CPAs) are collective farms in which the land and all productive assets, like machinery, warehouses, etc., are owned collectively. In 1989, on the eve of the Special Period, 78% of arable land was in the hands of the state, 10% belonged to CPAs, and 12% to CCSs (Machín Sosa *et al.* 2010, 24).

Under the imperative to boost the production of food as a public good in the early part of the Special Period, Cuba obtained some successes with alternative farming technologies such that, by the end of the decade, the acute food crisis was in the past and food was being produced with a fraction of the inputs and equipment previously imported (Rosset and Benjamin 1994; Funes *et al.* 2002; Wright 2008; Funes-Monzote 2008, 2010). While we agree that the Cuban experience in the 1990s with alternative agriculture was remarkable compared to other countries around the world, our vantage point in 2017 gives us a more nuanced perspective.

First, when Cuba faced the shock of lost trade relations in the early 1990s, food production initially collapsed due to the loss of imported fertilizer, pesticides, tractors, parts, petroleum, etc. The situation was so bad that Cuba posted the worst mark in all of Latin America and the Caribbean in terms of the annual per capita rate of growth of food production (−5.1% for the period from 1986 through 1995, against a regional average of −0.2%). But as the country re-oriented its agriculture to depend less on imported chemical inputs, Cuba rebounded to show the best performance in all of Latin America and the Caribbean over the following time period, a remarkable rate of 4.2% annual growth in per capita food production from 1996 through 2005 (the most recent year for which statistics are available). Interestingly, in the same period, the regional average was 0.0% (FAO 2006). However, this transition was not enough to transform Cuba from a net food importer into a net exporter, as the gap was too large to overcome.

Second, the better performance in the late 1990s was largely based on input substitution practices, like biopesticides, biofertilizers and animal traction, rather than on advanced agroecological integration. Although the initial adoption by Cuban farmers of these and other alternatives was fairly rapid, by the end of the decade it was clear to the leadership of ANAP that things were stagnating. Further breakthroughs were urgently needed, both technological and methodological, in order to speed the transition. While hindsight now shows us that the technological breakthrough that was needed was greater agroecological integration, it was the adoption by Cuba of a methodological innovation that in our view has proved key to its contemporary success. We believe that, in the typical case, in most countries most of the time, there are abundant and productive ecological farming practices 'on offer', but low adoption of them is the norm, because what is lacking is a methodology to create a social dynamic of widespread adoption.

Horizontal communication vs. conventional extension

There is an extensive debate concerning the effectiveness and appropriateness of conventional agricultural research and extension systems for reaching peasant families in general (Freire 1973) and, more specifically, for promoting agroecology rather than the Green Revolution (see, for example, Chambers 1990, 1993; Holt-Giménez 2006). The fact that agroecology is based on applying principles in ways that depend on local realities means that the local knowledge and ingenuity of farmers must necessarily take a front seat, as farmers are not blindly following pesticide and fertilizer recommendations prescribed on a recipe basis by extension agents or salesmen. Methods in which the extensionist or agronomist is the key actor and farmers are passive are, in the best of cases, limited to the number of peasant families that can be effectively attended to by each technician. In those cases, there is little or no self-catalyzed dynamic among farmers themselves to carry innovations well beyond the last technician. Thus, these cases are finally limited by the budget, that is, by how many technicians can be hired. Many project-based, rural development NGOs face a similar problem. When the project funding cycle comes to an end, virtually everything reverts to the pre-project state, with little lasting effect.

The history of agroecological transition in the Latin American region suggests that the most successful methodology for promoting farmer innovation and horizontal sharing and learning is the 'Campesino a Campesino' (farmer-to-farmer or peasant-to-peasant) methodology (CAC). While farmers innovating and sharing goes back to time immemorial, the more contemporary and more formalized version was developed locally in Guatemala and spread through Mesoamerica beginning in the 1970s (Holt-Giménez 2006). CAC is a *Freirian* horizontal communication methodology (*sensu* Freire 1970). It, in fact, is a counter-hegemonic social process methodology that is based on farmer-promoters who have innovated new solutions to problems that are common among many farmers or have recovered/rediscovered older, traditional solutions, and who use popular education methodology to share them with their peers. A fundamental tenet of CAC is that farmers are more likely to believe and emulate a fellow farmer who is successfully using a given alternative on their own farm than they are to take the word of an agronomist of possibly urban extraction. This is even more the case when they can visit the farm of their peer and see the alternative functioning with their own eyes. In Cuba, farmers say, '*cuando el campesino ve, hace fe*',[2] which translates roughly to 'seeing is believing'.

Whereas conventional extension can be demobilizing for farmers, CAC is mobilizing, as they become the protagonists in the process of generating and sharing technologies. In comparing CAC with conventional extension, the key question to ask is, who is the passive actor and who is active? Note that there is still a role for technical staff in CAC, but it is a *different* role. Rather than bringing knowledge to the (presumably) ignorant, the extensionist now concentrates on facilitating and supporting a process of farmer exchanges. Furthermore, CAC is a communal process that makes possible the collective transformation of reality, while the relationship in conventional extension is typically a more individual relation between technician and peasant family.

It is worth noting that Cuban peasants, like peasants everywhere, have always employed some traditional agroecological practices that are commonly held by the collectivity. They were preserved even during the heyday of the Green Revolution and made it through informality and connections to today's Cuba (Rosset *et al.* 2011). This pool of traditional knowledge proved to be a key resource for CAC and the associated "CAC Agroecology Movement" (MACAC) in Cuba.

'Campesino a Campesino' in Cuba

Through a series of somewhat fortuitous events, ANAP in Cuba learned of, and learned from, the experience with CAC in Nicaragua during the mid-1990s, just about the same time as it became clear that the spread of alternative practices to produce food during the Special Period needed a boost.

After hosting a meeting of CAC delegates from Mexico and Central America in 1996, ANAP decided to try the methodology on a 'trial' project basis with external donor funding in the province of Villa Clara (Machín Sosa *et al.* 2010). In November of 1997, the first workshop was held in Villa Clara to train local members of the organization in the CAC methodology. The early methodology and structure were the same as in Mesoamerica. The keys actors in this phase thus consisted of promoters, facilitators and peasant families who belonged to the ANAP. Success was fast and by 1999 CAC had spread to the nearby provinces of Cienfuegos and Sancti Spíritus.

Promoters are recruited from farmers who are recognized by their peers for the successful innovations and agroecological practices employed on their own farms and their desire and ability to teach others. Successful means that these innovations and practices work well in terms of productivity and in terms of the economy of the peasant family, without using toxic chemicals or other off-farm, purchased inputs. Their farms are their classrooms and other farmers visit them to learn. A principle of CAC in Cuba is that they receive no compensation other than the satisfaction of helping others and the status of being considered a good role model. If they were to be paid, people say, then other farmers would not believe in their technologies, finding it easier to think they just use them so they will get a salary. Facilitators are in charge of the logistics of matching and arranging visits for farmers in need of solutions to promoters who have them, organizing workshops and generally keeping things running. Some of them are trained agronomists or technicians, while some are peasants and co-op members, though they share a commitment to the ecological transformation of farming, that is, they are *activists*. They are hired and paid by each cooperative that chooses to have one or more facilitator. Emphasis is placed on this latter point; if cooperative members do not feel they gain anything worthwhile from having a particular facilitator, or any facilitator at all, then they will fire them. This, people say, ensures that they do a good job for their farmer-employers.

By 2000, CAC was clearly successful at accelerating the transition to productive agroecological farming much faster than conventional extension had been able to. It was now firmly established in Villa Clara, Cienfuegos and Sancti Spíritus and had begun in the provinces of Holguín, Ciego de Ávila, Matanzas and La Habana. But the ANAP leadership was frustrated at the time it took to get CAC established in each new province, especially as the implementation was still dependent on external funding from donor agencies, which made the grant cycle the key limiting factor (CAC was being run as a *project* or *program* inside of ANAP). Although the food crisis had, by this time, eased quite a bit, there was still a strongly felt need to boost national food production more rapidly and imported inputs were still not abundantly available. In February of 2001, the First National Encounter of the Campesino-to-Campesino Program of ANAP was held. At this meeting, Orlando Lugo Fonte, the president of ANAP, put forth the radical idea that CAC should become a *movement* and stop being a *project* or *program*. This meant it could no longer depend on external financing (though such would always be welcome), but rather must cut the reins that were holding it back and unleash campesino energy and creativity to rush forward at its own pace. He said,

> The vanguard movement of our organization has to be the movement of Campesino promoters. We want a thousand promoters, but beyond this first thousand, we want a thousand more aspiring to become promoters, and so on, with new *compañeros* joining the movement all the time. And speaking thusly, of a movement, in a short period of time we should see thousands of men and women working for this noble idea [agroecology].
>
> (*Machín Sosa et al. 2010, 41*)

Reflecting on what happened at that time, Lugo Fonte later said, 'If we couldn't find external financing, the Cuban agroecological movement was going to have to emerge with our own resources, even though we had very little' (Machín Sosa *et al.* 2010, 41). While promoters were not going to be paid at all and facilitators were going to be paid by the cooperatives themselves, significant resources were still needed, as the basis of CAC rests on exchange visits, meaning expenses such as transport, fuel, food, lodging, etc. However, the ANAP was determined to cover that mainly from their own resources, plus whatever they could obtain from government agencies.[3]

From this point on, ANAP assumed the promotion of the henceforth-named 'Campesino a Campesino Agroecology Movement' (MACAC) as an 'organic task' at each level and in every structure of the national organization. Every cadre and every militant of the organization was to take on responsibility for facilitating and promoting the movement within their area of authority or work. As a revolutionary mass organization, ANAP had inherent strengths in movement building. It had a political organizing methodology for 'mass mobilization', a methodology which had been used successfully in earlier times to promote other internal mobilizations.

During our field work across the Cuban countryside, we could feel the high level of political consciousness of the ANAP grassroots membership, a testimony of the on-going 'ideological work' carried out inside of ANAP. An example of this is the general belief among members of the organization that the 'historic mission' of the peasant sector is to feed the Cuban people. In other words, food is a commons, a public good for all that should be provided by Cuban peasants for the Cuban people.

ANAP exhibits an unusual degree of *organicity*.[4] Virtually all peasants are members of cooperatives, which are the basic units of ANAP membership. Each cooperative has a general assembly and officers, and ANAP has a leadership structure in every municipality of the country, as well as at the provincial and national level. This essentially means that the organization can call on cadres with leadership experience in literally every corner of rural Cuba.

There were thus powerful forces ready to be turned to the task of promoting agroecology. In this environment MACAC rapidly took on a 'mass character', in which agroecology was blended with socialist, communitarian and environmental values. In the anonymous written words of a participant in a workshop that we held in Granma Province,

> To *massify* is to move all the methods and forms possible to promote and multiply any task. Taking the practices of peasants and promoters and spreading them in training workshops, seminars, and conversations on the farm. Learn the practices by doing them. Do them in schools, with the children, in the barrio, with the community, so that all these people carry the word from mouth to mouth, to the men or women they are closest to… The need to build a great movement at the district, municipal, and national level. To consolidate the practices in an organized fashion; demonstrate that something good is happening, is being experimented with, on the farm. That nothing shall be left which hasn't been taught to others; that all of us can learn and can also teach, each according to our role.

From 2000 to 2003, MACAC spread to *all* Cuban provinces, taking the form of a movement, and ANAP began to tinker with the methodology inherited from Mesoamerica. As the farmer exchanges began taking place between provinces and over longer distances, the organizational complexity grew. It was difficult for a facilitator in a cooperative in Pinar del Río province, for example, with members who needed to solve a particular weeds problem, to know that a promoter in Cienfuegos had a good solution and then organize an exchange visit. ANAP thus created a new role in the organization and started training new actors, the coordinators. These are typically professionals, sometimes from agricultural sciences, but also include professionals in everything from public relations to administration, who, like the facilitators, are first and foremost activists. They identify and coordinate exchanges and trainings at higher levels or on broader scales. Gradually coordinators have been hired at the municipal and provincial levels, and a national coordinator was added as well. ANAP pays their salaries.

Impacts and achievements

By 2008–2009, 12 years after CAC came to Cuba, the results were quite impressive in terms of the membership growth of MACAC, the productivity of agroecological farms and of the peasant sector in general, and other variables (Rosset *et al.* 2011). In little more than a decade, MACAC had grown dramatically in terms of numbers of families who have formally joined the movement and numbers of promoters, facilitators and coordinators. From just over 200 families in 1999, the movement had grown to 110,000 families ten years later. By way of comparison, in 2009 there were less than 350,000 families in the peasant sector (CCSs and CPAs) of Cuba, so this number represents about one third of families joining in a relatively short period of time, giving CAC a much faster growth rate than anywhere in Mesoamerica, both in relative and in absolute terms.[5] There were some 12,000 farmer-promoters, 3,000 facilitators and 170 coordinators.

It is important to note that not just the families who have joined the movement are influenced by it. In fact, a lot of neighbors-emulating-neighbors takes place in rural areas and within cooperatives and non-MACAC members also adopt some of the practices that they see their more agroecological neighbors using successfully. Although it cannot totally be attributed to MACAC, this partial 'spill-over' effect nevertheless can be seen in Cuba, where typical practices promoted inside the movement are now used by more than the one-third of all peasant families who are movement members. These include the use of organic soil amendments on 64% of all peasant farms and ecological pest management methods on 82% (Machín Sosa *et al.* 2010, 51).

Productivity of agroecological farms and of the peasant sector

In data assembled by Rosset *et al.* (2011) we can see that the greater the level of agroecological integration, the greater the total value of production, measured in Cuban non-convertible pesos per year, both *per worker* and *per hectare*. This would seem to suggest that, at least in Cuba, agroecology is an effective way to intensify production and generate income for farmers. Contrary to popular belief or myth, the model does not suffer from low labor productivity. These findings are in broad agreement with those of Badgley *et al.* (2007) on a global scale and Martínez-Torres (2006) in her study of organic and conventional coffee in Mexico.

Another way to tease out the relationship between peasants, food production and agroecology is to look at production data and use of agrochemicals (Rosset *et al.* 2011). For example, the production of vegetables, which are typical peasant crops, fell by 65% from 1988 to 1994, but by 2007 had rebounded to 145% over 1988 levels. This increase came despite using 72% fewer agricultural chemicals in 2007 than in 1988. Similar patterns can be seen for other peasant crops like beans (down 77% in 1994, but at 351% over 1988 levels by 2007, with 55% less use of agrochemicals) and roots and tubers (down 42% in 1994, at 145% of 1988 levels by 2007, with 85% fewer agrochemicals). This contrasts dramatically with sugarcane, *not* a peasant crop, which saw yields fall in 1994 to 25% below 1988 levels and fall another 3% by 2007, precisely the same time period during which production of peasant crops leaped, and this even though the reduction in agrochemical use in sugar (down just 5% by 2007) was insignificant (Machín Sosa *et al.* 2010, 52).

In summary then, the data shows that more agroecological farms produce more than less agroecological farms and that the peasant sector as a whole has made dramatic strides in food production both in absolute terms and relative to other sectors over the same time period, while consuming much less agrochemicals. In more general terms, we can argue that the more collective process of agroecology based on CAC, which takes "food as a commons" as a given, is actually more effective at feeding people than the more conventional, individualistic and capitalist alternatives.

An illustrative case: CAC in the Escambray Mountains

We can illustrate some arguments with a specific example, through the voice of a key actor in the local sphere. The Escambray mountains in Villa Clara were a pioneer area in the implementation of agroecology in Cuba and an emblematic example of the close interaction between individuals, collectivity, food and ecology. This was one of the earliest places where the CAC methodology was put into practice. The "Ignacio Pérez Ríos" Credit and Services Cooperative is a representative case of success. The members farm high-quality organic peasant coffee with the complementary production of root and tuber crops, fruit trees and farm animals, the combination of which provides food for the peasant families and products for local and even regional markets (Val 2017).

The spirit of the members of the cooperative is manifest in the words of Genaro Rafael González Baltrón, "El Cojo", the main agroecology promoter in the area:

> Here where we are, in these hills, here, the only thing I would have to purchase from outside would be a little rice, that doesn't grow that well here, and a little salt, in terms of food. We are creating sustainable farms to preserve the biodiversity of plants and animals, to give something to us and to not harm others, not a neighbor, not a plant, nor a little animal. We seek the coexistence of all and for all. That is my idea of what agroecology is.

Food is part of a constant exchange in family and community networks of solidarity and reciprocity. Via MACAC, knowledge, practices, seeds and food circulate. Each activity, whether workshops, meetings or events, provides a special communal space to share food as a form of collective integration, a commons, that enhances the exchange and consolidates the relationships of solidarity and reciprocity (Val 2012). When they get together, the members of the community are living and experiencing the commons (i.e. *commoning*), but at the same time producing and protecting the commons as tangible and intangible goods. Framed in this process, the cooperative has become a provincial, national and international reference in the promotion of ecological agriculture that is locally adapted and environmentally sustainable.

It was precisely the strength of this type of solidarity structure that tempered the impact of the food crisis of the Special Period in the Cuban countryside. Esther, a peasant woman associated with the cooperative, reflected on that moment:

> It left us a very great lesson ... There may be a very special period, there may be a very big blockade, but if a person has ideas and initiatives, the very commitment to move forward, then no blockade or special period can prevent it ... Because if you are facing a crisis, like the whole world is now in crisis, then you can grow food in every corner, in the tiniest space, food that is for your own benefit, for your family, and for your community. With every crop I sow, I am providing food for my community.

In this context, the concept of food sovereignty is manifested in the articulation of these exchange networks at the local (and sometimes regional) level of food in the present, through direct exchanges of complementary products, as well as in the future, through exchanges of knowledge, seeds and seedlings. This is not only the right to self-sufficiency, autonomy of production and availability of food for the population, but also involves the social awareness of producers and consumers for the production, demand and consumption of healthy, nutritious and accessible food, as a common good.

Likewise, food sovereignty is understood as a process of liberation and autonomy, as El Cojo tells us:

> If we had thought before about how to do things in the way that we are working today, we would have much more food, much more. Beyond that, we produce healthier food

> and we are more independent. We already feel freer because we have it right there, we do not have to look for it ... One has already become freer, more a master of one's own life! In Cuba almost everyone realizes that you need to grow food in every little space, be it a yard, a plot or a farm, everyone on their little bit of land, no matter how tiny. And you have to know how to care for, and how to produce. You have to try to have food to be independent. "Food sovereignty" is not that you have a pigsty or a dairy, which is what gives the most... no, it is not that. It is about how you handle everything, as you take care with everything, how you teach what you know to others. Because after you have an equilibrium established, you already have sovereignty. And that is what we have to achieve here in Cuba and everywhere.

A very important dimension of Cuban food production is that of products intended for social purposes (schools, hospitals, community centers, etc.). This work is generally taken on as a commitment of peasants to society, generating a great satisfaction to producers in terms of the destination of their crops and making them, in turn, recipients of prestige and social recognition within their communities. In Cuba, this a key way in which food transcends the limits of the individual and the local collective and becomes a commons that is produced by the many and is accessed by the many.

Once again in the words of El Cojo: "There is always food, for me and for the family. I also give to the hospital, to the day care center and I sell to the government food program. I take part to the market on the plaza. We almost always sell through the cooperative. This gives us enough to maintain the family and to deliver food to the people; healthy food, and food that is from here, without having to bring it from somewhere else." In addition to the family (in a broad sense), the village (generally referring to the non-food producing part of the population) and their animals, many Cuban peasants argue that 'the land also has to eat', thus anchoring many of the agroecological soil management practices, like the use of organic and green manure, cover crops, etc. (Val 2012).

Thus, in the hills of the Escambray, food sovereignty, agroecology and sustainability are amalgamated in a way that produces and reproduces life. Manuel "Manolo" González, president of the cooperative, clearly points out the logic that agroecological production has for these peasants: "I produce agroecologically not because I am rich or fill my pockets from agroecological production, but rather for the benefit of my family, of all the people that I sell to, that I give to as a gift, or that comes looking food. I do not want to get rich with agroecological food, on the contrary I sell it even cheaper. It's because you eat something healthy. That will not harm any human being. That is why I work for the overall welfare of the population, and not for economic well-being."

We find that the agroecological peasants of the Escambray offer a much broader, more complex and holistic conception of food than simply a production for self-sufficiency and the market that may be associated to the notion of food sovereignty. This perspective of food as a commons certainly questions the hegemonic model of agribusiness, but also proposes an interesting alternative for the production of healthy and accessible food, as well as a valuable reflection and praxis around the relationship between humans and nature (Val 2012).

This alternative order disputes meanings and discourses with the developmentalism of the Green Revolution and the postulate of food as a commodity, entails elements that represent a radical critique that transcends the immediate dimension of productive alternatives and questions the hegemonic capitalist modernity that underlies it. These elements of radical ontological, epistemic and political criticism nourish Latin American peasant struggles and form a horizon of alternative possibilities to that posed by the imperial hegemony of agribusiness, the paradigm of development and green capitalism (Val 2012; Martínez-Torres and Rosset 2014).

Lessons and challenges

We can distill some tentative lessons by comparing the more and less successful local cases that unfolded in Cuba, especially with regard to the more challenging aspects of applying CAC on a large scale and to the possibility of enhancing a food system inspired by the principles and visions of the commons. A principal challenge has been to achieve gender equality in the movement. While Machín Sosa *et al.* (2010) found that agroecology may dilute patriarchy within the family, that is not the same as gender balance in the movement itself. Although women made up 40% of coordinators in 2009, only 12% of facilitators and 8% of promoters were women (Machín Sosa *et al.* 2010, 70). In light of the implications and of the importance that equality (of gender and other forms), plays in the construction of the commons, it is clear that the movement needs to make a more concerted effort to recruit and train women activists, especially as many members of MACAC lauded the skills that women bring to promotion and facilitation.

We observed that the CAC process develops best when special attention is devoted to training and privileging the centrality of peasants (rather than technicians, political leaders, etc.) in all aspects of the process. This means that a careful balance has to be achieved between the vertical and horizontal elements of the structure of the movement. Where peasant protagonism is overly diluted by other actors, the process slows to a crawl. There have also been some cases where peasant promoters developed 'know-it-all' superior attitudes reminiscent of technicians and extension agents, with similar effects in reducing the dynamism of the overall process.

The implementation of CAC in a cooperative or municipality should be based as much as possible on resources that are already available locally. That means both human and material resources. Minimizing external dependency is the best way to build sustainable processes; where the local process has been overly dependent on the outside it has typically failed to develop. However, this does not mean that the organization (i.e. ANAP) does not need to play a large role in planning and in obtaining needed resources. When peasant promoters have been overly saddled with bureaucracy, like paperwork for reporting, the process has typically ground to a halt. Nevertheless, it is crucial that promoters, facilitators and coordinators work together closely in planning, monitoring and evaluation.

Another key to success is the achievement of absolute respect for local culture and customs in each locality as integral parts of each experience and reality. This process should better emphasize the recovering, valuing, recognizing and promoting of local knowledge, complementing it with external elements, but not overwhelming it. It is critical to avoid imbalances between technological aspects, which have a rapid dynamic, and the social methodology process, which takes time to develop. The most successful cases involved and built on the skills of and respect for local leaders, took advantage of local structures, like the cooperative assembly, and involved potential local allies, ranging from school teachers and physicians to local officials.

Another success is that of the consolidation of the political nature of agroecology and the commons. MACAC has proven to be a 'hot house', to use a phrase that was often repeated, for identifying and developing new grassroots leadership within ANAP. Peasants who become promoters receive training in popular education methodology and experience success in helping other farmers transform their production systems, gain self-confidence and gain respect from their peers. Many are soon elected to leadership positions in their cooperatives and some rise further to municipal, provincial or national leadership positions in ANAP. We could literally feel the bottom-up rise of a whole new generation of peasant leaders as a result of MACAC, some of whom eventually leave ANAP and come to occupy political offices, start to work for government agencies, etc. ANAP and MACAC activists see this as both a plus and a minus. A plus because this is providing ANAP with a dynamic new cohort of leadership, but a minus because it means constantly identifying and training new promoters to replace those who are 'lost upward and outward'. But even this loss carries within it a plus, as the former MACAC promoters in leadership positions inside ANAP reinforce the importance given to the movement by the organization itself, and those who now occupy leadership positions outside of ANAP have proven to be key institutional allies for MACAC and for agroecology in general, (re)shaping government policies to support MACAC and agroecology.

Reflections on Campesino a Campesino, agroecology and food as a commons

The story of MACAC in Cuba provides a lot of material for reflection on a variety of issues, from a variety of perspectives. From a natural science perspective, it speaks to the productivity of more complex and more integrated agroecosystems. In this case, there is a correlation between the transition from conventional farming to simple input substitution to agroecological integration and an increase in total productivity both of land and of labor. But it is also a warning to natural scientists, technicians and extensionists; more and better technology will not alone lead to widespread ecological farming. Typically, many agroecological practices are available but not widely adopted because of the lack of a social process that encourages and drives their adoption. Thus, the limiting factor is most often not technical but social and methodological and the latter are most often under-addressed. Furthermore, even a good social process may not be successful unless structural barriers to agroecology and food sovereignty can be at least partially overcome.

From a policy perspective, it speaks to questions of achieving national food sovereignty in the face of the global economic, climate and food crises. The Cuban experience would tend to support the arguments of La Via Campesina (2010, see also Rosset 2006b) that building food sovereignty requires putting land in the hands of peasants, through genuine agrarian reform, fair prices through protection from dumping of cheap food from abroad and a transition to agroecological farming. And it confirms the arguments advanced in this volume that a true food transition requires the adoption of the paradigm of the commons. Agroecological farming breaks dependence on imported inputs in times of economic crisis (in Cuba it helped boost national food production just when the global food crisis had driven the foreign exchange cost of imported food to unacceptable levels) and increases the resiliency of the economy to ever more frequent climate shocks.

Some observers raise the issue of 'Cuban exceptionalism',[6] arguing that experiences from the island do not apply to other countries that had not had social revolutions or had not faced food crises as severe as that faced by Cuba during the Special Period. It is, of course, important to be cautious about universalizing and generalizing particular experiences. But we would first observe that the growth of MACAC in Cuba occurred after the most difficult moments of the Special Period had passed, when the economy was experiencing some level of recovery. But, of course, there is no denying that MACAC in Cuba, and Cuban peasants in general, have greatly benefited from a facilitating and counter-hegemonic rather than a hostile state. As a matter of fact, it can be argued that the Cuban state adopts a policy that treats food as a commons, with relatively high food prices that translate into fair crop prices, land already in the hands of peasants (organized peasants) and a high 'scarcity cost' for imported farm inputs. Rather than accepting such conditions as impossible outside of Cuba, LVC and many other social movements actively struggle around the world for genuine agrarian reform, banning dangerous pesticides, protection of the national economy from dumping and speculation by transnational corporations, and other food sovereignty policies (Rosset 2006b; Martínez-Torres and Rosset 2010; Borras and Franco 2010). When we can demonstrate that certain policies function in Cuba, this is not to stress its uniqueness but to offer a powerful argument for use in other countries.

A key lesson to draw from our study is that to scale up agroecology requires a peasant organization and a socially dynamic methodology like CAC, as has been argued by La Via Campesina (2010). Peasant self-organization must be supported and encouraged, and conventional agricultural extension from the state, NGOs or the private sector is no substitute. The question of how to scale up agroecology is under debate in the literature (von der Weid 2000; Altieri and Nicholls 2008) and our results fall squarely in support of the position of Holt-Giménez (2001, 2006), that the CAC methodology is the most effective way found to date, and of Altieri (2009), that rural social movements hold the key.

From the perspective of a peasant organization searching for a way to support its member families in a transition from conventional to ecological farming, the experience of ANAP presented here is unequivocal. When conventional extension was used, the results were slow and haphazard. But a dramatic speed-up occurred through the adoption of the socially dynamic CAC methodology, with another important leap taking place when this was combined with a grassroots social movement-building methodology.[7] From the perspective of LVC, a key lesson is that the campesino a campesino methodology can and should be applied at the international level. This would be a 'campesino organization'-to-'campesino organization' method based on exchange visits and this is something we are already beginning to carry out.

Of course, the fact that a national peasant organization using the CAC methodology under favorable structural conditions was able to achieve so much success does not guarantee that an international peasant movement will be able to use the same methodology to advance agroecology worldwide under decidedly less favorable structural conditions. That the CAC methodology is now in the hands of an international peasant federation with increasing 'organicity' would seem to be a necessary, but not sufficient, condition. First, not many organizations inside or outside of LVC boast the degree of organicity that ANAP has. Second, while in some countries conditions are becoming more supportive, those countries may still lack such a well-organized peasant organization and/or the supportive conditions may still be partially lacking. Moreover, their governments most decidedly are neither counter-hegemonic nor do they treat food as a commons.

It is clear to LVC that the internal work of strengthening member organizations is a critical priority (Martínez-Torres and Rosset 2010) and, in fact, is probably a precondition for achieving further structural and policy changes, as well as for developing CAC and agroecology on a broad scale in other countries. However, the tasks of internal strengthening and the promotion of CAC can be mutually supportive in terms of developing a grassroots leadership cadre and credibility inside organizations, as the example of ANAP has shown us. In many countries, organizations find that food, agroecology and food sovereignty are much more disputed terrains than they are in Cuba. Typically, the countryside is awash with NGOs, reformist and reactionary farmers organizations, foreign foundations and government and inter-governmental programs, which all treat food as a commodity (Vivero-Pol 2017) and tout a sometimes intentionally confusing mixture of a re-packaged Green Revolution, sustainable agriculture, organic farming, etc.[8] Can the CAC methodology be a tool to help LVC member organizations navigate this complicated landscape and build internal strength? LVC is betting on this, carrying a wide array of exchange visits to ANAP and Cuba.

Recuperating the reflections, theorizations and practices of the Cuban agroecological peasantry leaves us with valuable lessons to advance towards a more inclusive food production that addresses the needs of all the inhabitants of this planet and of those who will come in the future: a food production that responds to people's basic needs, cultural practices and the ecological boundaries of the planet, for the participated construction of long-term economic, political, social, cultural and environmental sustainability; a proposal that not only feeds the world, but feeds it better and in a way that is compatible with the survival of human and non-human beings; an alternative model that proposes a collective exit from the civilizational crisis (as the summation of all crises) and introduces a paradigm shift in the forms of production and reproduction of life in the logic of the food system as a common good of humanity (Houtart 2011; Val 2012; Vivero-Pol 2017; Chapter 3, this volume).

Notes

1 Laura Enríquez (2003) has called this repeasantization, though as stated below in the text, the transition to becoming peasants has been uneven.
2 In fact, this saying is the subtitle of the book by Machín Sosa *et al* (2010).
3 Contrary to common belief, ANAP is not funded by the Cuban government, but rather by a voluntary self-tax on farm sales by member cooperatives. While the Cuban state has historically provided a much greater degree of support (credit, marketing, crop insurance, extension, etc.) to the peasant sector than other Latin American governments, it is also true that long-term and larger investments were more directed to the state farm sector than to the peasant sector.
4 Among Latin American social movements, 'organicity', or organicidad in Spanish, refers to the degree of internal organization that a movement or an organization has.
5 In 2017, we now estimate that fully half of the Cuban peasantry participates in MACAC.
6 See Hoffmann and Whitehead (2006) for a discussion of Cuban exceptionalism.
7 Like many other farmer organizations, ANAP has a national farmer training school. A key lesson of the ANAP experience is that the school can play an integral role in supporting MACAC. Promoters, facilitators, and coordinators all take short courses at the school to learn methods (i.e. pedagogical and organizing methods) specifically tailored to their roles. Cooperative presidents and other ANAP cadre and leaders from all levels receive courses to sensitize them to agroecology and to the CAC methodology (Machín Sosa *et al.* 2010, Val 2012).
8 See Borras (2010) for an example of how 'cluttered' the landscape can be.

References

Altieri, M.A. 2009. Agroecology, small farms, and food sovereignty. *Monthly Review*, 61(3), 102–13.

Altieri, M.A. and C. Nicholls. 2008. Scaling up agroecological approaches for food sovereignty in Latin America. *Development*, 51(4), 472–80.

Badgley, C. et al. 2007. Organic agriculture and the global food supply. *Renewable Agriculture and Food Systems*, 22(2), 86–108.

Benjamin, M., Collins, J. and M. Scott. 1984. *No Free Lunch: Food & Revolution in Cuba Today*. New York: Grove Press.

Borras, S.M. 2010. The politics of transnational agrarian movements. *Development and Change*, 41(5), 771–803.

Borras, S.M. and J.C. Franco. 2010. Contemporary discourses and contestations around pro-poor land policies and land governance. *Journal of Agrarian Change*, 10(1), 1–32.

Chambers, R. 1990. Farmer-first: a practical paradigm for the third agriculture. In: M.A. Altieri and S.B. Hecht, eds. *Agroecology and Small Farm Development*. Ann Arbor, MI: CRC Press, pp. 237–44.

Chambers, R. 1993. *Challenging the Professions: Frontiers for Rural Development*. London: Intermediate Technology Publications.

Enríquez, L. 1994. *The Question of Food Security in Cuban Socialism*. Berkeley, CA: Institute of International and Area Studies, University of California.

Enríquez, L. 2003. Economic reform and repeasantization in post-1990 Cuba. *Latin American Research Review*, 38(1), 202–18.

Food and Agriculture Organization of the United Nations (FAO). 2006. The state of food and agriculture 2006. Rome: FAO.

Freire, P. 1970. *Pedagogy of the oppressed*. New York: Seabury Press.

Freire, P. 1973. *Extension or Communication?* New York: McGraw.

Funes, F. *et al*, eds. 2002. *Sustainable Agriculture and Resistance: Transforming Food Production in Cuba*. Oakland, CA: Food First Books.
Funes-Monzote, F.R. 2008. Farming like we're here to stay: the mixed farming alternative for Cuba. Thesis (PhD). Wageningen University. Available from: http://edepot.wur.nl/122038 [Accessed October 2010].
Funes-Monzote, F.R. 2010. Cuba: a national-level experiment in conversion. In: S.R. Gliessman and M. Rosenmeyer, eds. *The Conversion to Sustainable Agriculture: Principles, Processes and Practices*. Boca Raton, FL: CRC Press, pp. 205–27.
Hoffmann, B. and L. Whitehead. 2006. Cuban exceptionalism revisited. *Giga Working Papers* 28, 1–24.
Holt-Giménez, E. 2001. Scaling-up sustainable agriculture. *LEISA*, 3(3), 27–9.
Holt-Giménez, E. 2006. *Campesino a Campesino: Voices from Latin America's Farmer to Farmer Movement for Sustainable Agriculture*. Oakland, CA: Food First Books.
Houtart, F. 2011. De Los Bienes Comunes al "Bien Común de la Humanidad"., Brussels: Fundación Rosa Luxemburgo.
Kundu, S. et al. 2007. Long-term yield trend and sustainability of rainfed soybean–wheat system through farmyard manure application in a sandy loam soil of the Indian Himalayas. *Biology & Fertility of Soils*, 43(3), 271–80.
La Via Campesina. 2010. Sustainable peasant and family farm agriculture can feed the world. Jakarta: La Via Campesina. Available from: http://viacampesina.org/en/index.php?option=com_content&view=section&layout=blog&id=8&Itemid=30 [Accessed October 2010].
Machín Sosa, B. *et al*. 2010. Revolución agroecológica: el movimiento de campesino a campesino de la ANAP en Cuba. Cuando el campesino ve, hace fe. Havana: ANAP and La Vía Campesina. Available from: http://www.viacampesina.org/downloads/pdf/sp/2010-04-14-rev-agro.pdf [Accessed September 2010].
Martínez-Torres, M.E. 2006. Organic coffee: sustainable development by Mayan farmers. Athens, OH: Ohio University Press.
Martínez-Torres, M.E. and P.M. Rosset. 2010. La Vía Campesina: the birth and evolution of a transnational social movement. *Journal of Peasant Studies*, 37(1), 149–75.
Martínez-Torres, M.E. and P.M. Rosset. 2014. Diálogo de saberes in La Vía Campesina: food sovereignty and agroecology. *Journal of Peasant Studies*, 41(6), 979–97.
Mulvaney, R.L., Khan, S.A and T.R. Ellsworth. 2009. Synthetic nitrogen fertilizers deplete soil nitrogen: a global dilemma for sustainable cereal production. *Journal of Environmental Quality*, 38, 2295–314.
Nova, A. 2002. Cuban agriculture before 1990. In: F. Funes, *et al.*, eds. *Sustainable Agriculture and Resistance: Transforming Food Production in Cuba*. Oakland, CA: Food First Books, pp. 27–39.
Pingali P.L., Hossain, M. and R.V. Gerpacio. 1997. *Asian Rice Bowls: The Returning Crisis*. Wallingford: CAB International.
Radford, B.J., *et al.* 2001. Crop responses to applied soil compaction and to compaction repair treatments. *Soil and Tillage Research*, 61(3–4), 157–66.
Rosset, P.M. 1997. Alternative agriculture and crisis in Cuba. *Technology and Society*, 16(2), 19–25.
Rosset, P. 2006a. *Food is Different: Why the WTO Should Get Out of Agriculture*. London: Zed Books.
Rosset, P. 2006b. Moving forward: agrarian reform as part of food sovereignty. In: P. Rosset, R. Patel and M. Courville, eds. *Promised Land: Competing Visions of Agrarian Reform*. Oakland: Food First Books, pp. 301–21.
Rosset, P. and M. Benjamin. 1994. *The Greening of the Revolution: Cuba's Experiment with Organic Agriculture*. Melbourne: Ocean Press.
Rosset, P.M., Machín Sosa, B., Roque Jaime, A.M. and D.R. Ávila Lozano. 2011. The Campesino-to-Campesino agroecology movement of ANAP in Cuba: social process methodology in the construction of sustainable peasant agriculture and food sovereignty. *Journal of Peasant Studies*, 38(1), 161–191.
Val, V. 2012. Sembrando alternativas; cosechando esperanzas. *(Re)campesinización agroecológica* en las lomas del Escambray, Provincia de Villa Clara, Cuba. Thesis (Master's). Centro de Investigaciones y Estudios Superiores en Antropología Social. Available from: http://repositorio.ciesas.edu.mx/handle/123456789/176.
Val, V. 2017. Hacer, conocer y *ser*. Agroecología en primera persona [Web Log Post]. Retrieved from https://sites.google.com/site/agroecologiadesdesur/autores/el-cojo.

von der Weid, J.M. 2000. Scaling up, and scaling further up: an ongoing experience of participatory development in Brazil. Saõ Paulo: AS-PTA. Available from: http://www.fao.org/docs/eims/upload/215152/AS-PTA.pdf [Accessed October 2010].

Vivero-Pol, J.L. 2017. Food as commons or commodity? Exploring the links between normative valuations and agency in food transition. *Sustainability*, 9, 442; doi:10.3390/su9030442.

Wright, J. 2008. *Sustainable Agriculture and Food Security in an Era of Oil Scarcity: Lessons from Cuba*. London: Earthscan.

17

THE COMMONING OF FOOD GOVERNANCE IN CANADA

Pathways towards a national food policy?

Hugo Martorell and Peter Andrée

Introduction

A 'food movement' effectively working to further the values of food as a commons and as public good and explicitly seeking to strengthen the role of civil society organizations (CSOs) in food policy-making and governance is gaining ground in Canada. This social movement assembles a disparate array of CSOs, Indigenous food sovereignty advocates, fair-trade businesses, and sustainable and just food system advocates (including academics) from each of Canada's thirteen provinces and territories (Wittman et al., 2011; Levkoe, 2015). Koç et al. (2008) define the CSOs involved as primarily "community-based not-for-profit organizations working for the public interest independently of governments and the market place" (p. 125), though many of these CSOs are actually funded by various levels of government and often seek to influence government policies. As of 2005, Canada's food movement connects nationally under the auspices of a "pan-Canadian alliance" called Food Secure Canada (FSC).[1] From 2007 to 2011, this movement organized under the banner of food sovereignty to develop a "People's Food Policy".

The food movement's most recent victory was the 2015 commitment by Canada's newly elected federal government to deliver a healthy-eating strategy, reduce poverty, and invest in food access in Northern and remote areas of the country. The new Prime Minister also mandated the development of a National Food Policy in a process that convenes ten federal departments and is intended to gather widespread citizen input (Meredith, 2016). FSC and its movement partners advocated for these commitments during the 2015 election campaign, and now see them as an opportunity to connect the issues of hunger, health, and sustainability in a way that strengthens the role of citizens in the bottom-up governance of their own food systems.

The intent of this chapter[2] is to recognize, celebrate, and strengthen the 'commoning' of food in Canada. Following Linebaugh (2010), we understand commoning to be the activity that brings together people's labour to own and govern a good or resource collectively. This collective ownership and governance can take various forms. The idea of governing food as a 'commons' is a direct challenge to its increasingly commodified form. Valuing food only as a commodity, whereby money is the regulating objective of thought and action, erodes relationships between people and ecology. In contrast, valuing food as a commons prioritizes life as the source of value (McMurtry, 2002). It involves recognizing food as a multidimensional good that has benefits for the common good (Vivero-Pol, 2017a) and thereby affirms the efforts of people, and the organizations created by them, to build systems of food governance from the bottom-up to meet collective needs (Renting et al., 2012). We see the idea of food as a 'common-pool resource' (Ostrom, 2010) – discussed below in the context of the management of wildlife by Canada's Indigenous and Inuit peoples – as a way of thinking about the governance of food resources with specific management attributes to ensure their sustainable use and the equitable sharing of their benefits. Common-pool resources are thus a subset of the broader category of the extant food 'commons' which also includes recipes (Walljasper, 2011), agricultural knowledge (Chapter 12), and unpatented genetic resources (Chapter 14; Vivero-Pol, 2017a).

Seeing something as a 'public good' is another form of collective ownership and governance. It depends on the vehicle of the state to institutionalize collective ownership and control. The products of such institutionalization (e.g. public libraries, bridges and environmental legislation) are together sometimes referred to as the 'civil commons' (Sumner, 2011). In this chapter, the 'commoning of governance' is understood to involve the myriad ways in which citizens, organized through CSOs, are (re)claiming food provisioning systems as their own, including the public laws and policies that regulate them, in efforts to ensure these systems benefit the common good above all else. Such efforts involve both de-commodification, or what Ferrando and Vivero Pol (2017, p.50) refer to as the "contemporary food commons," and ongoing resistance to commodification through the land-based, Indigenous-led "customary food commons."

How could the values of food as commons *and* as public good translate into national food policy? This chapter is not explicitly focused on the institutionalization of principles like the 'right to food' or 'food sovereignty' in a constitution (as undertaken by countries like Bangladesh and Bolivia respectively; Knuth and Vidar, 2011). Instead, it focuses on institutional arrangements that improve universal access to healthy foods. We seek to identify novel approaches already existing or in emergence at sub-national levels in Canada upon which a national food policy could be built on. We are especially interested in approaches that involved collaboration across government silos and levels of government (federal, provincial, municipal, and Aboriginal) and in co-governance between civil society, governments, and the business sector.

This chapter is framed in terms of the commoning of food governance in Canada. We see the values of access to healthy food as a common-pool resource and as public good, as well as the commoning of food governance processes, to be exemplified in the cases presented. It is important to recognize, however, that as of yet few food movement actors (including FSC) explicitly frame their work in terms of furthering "food as commons," nor are Canada's public policies yet articulated in these terms.

Three case studies are presented in this chapter, each of which represent stories of success at the sub-national level in Canada that could be scaled to the national level. The first, focused on Aboriginal self-determination in Northern and remote areas (IV.1), is about institutionalizing *access to healthy traditional food as common pool resource.* The second is about the *commoning of health promotion policy-making* through stakeholder engagement processes led by CSOs (IV.2). The third is about improving *access to healthy food as a public good* through joint programming between civil society, businesses and governments (IV.3). In our analysis, we discuss the possibilities for the commoning of food governance represented by these cases – such as stakeholder engagement, co-governance, and joint programming – as well as some of the tensions and constraints of these approaches.

This chapter is based on a research collaboration with Food Secure Canada. "Mapping the food policy landscape in Canada" included a review of national, territorial, and provincial policies. Twenty-seven provincial and national CSOs and members of the research community were also interviewed in a semi-open questionnaire format between May and September 2016.

All of these interviews inform this paper and seven are quoted directly. Research results, compiled in reports, underwent a peer review process that included both food movement practitioners and academics before release.[3]

The next section (II) sets the stage with a review of theories of co-governance, polycentrism, and regulatory pluralism. Section III presents a brief historical overview of the food policy context and the rise of the food movement in Canada. Section IV presents our case studies from three different policy arenas, while Section V offers our conclusions.

Theorizing the commoning of food governance

Sumner (2011) contextualizes food as public good in Canada by drawing a parallel with another universal public service: "Like the Canadian health-care system, a made-in-Canada sustainable food system would be supported by four pillars: universality, accessibility, portability and public administration. Based on a single payer, we would all contribute so everyone could eat" (p. 69). This would represent a significant shift of power away from the market and the state back to the civic food networks though "new configurations of agri-food governance mechanisms" (Renting et al., 2012). These include recognizing food and culture as human rights (De Schutter, 2012), delivering a universal school food program (Oliveira and Frazáo, 2009), guaranteeing "a minimum of food for all" (Vivero-Pol, 2017b), and sensibly taxing a private sector to return benefits to the citizens whose resources they ultimately extract surplus from (ibid.). These ideas align with literature on bottom-up and collaborative approaches to sustainable development (e.g. Voss et al., 2006) and the governance of natural resources (Nagendra and Ostrom, 2012). This literature highlights the concepts of co-governance and polycentrism.

Co-governance (aka 'collaborative governance') can be defined as multiple actors working together to meet shared governance goals (Kooiman, 2003). Co-governance is about more than simple co-ordination. It refers to a middle ground between the hierarchical nature of most governmental processes and the self-organization of social movement or market processes (Somerville and Haines, 2008). Participants are co-producers of governance outcomes and feel equal stewardship of the process (Paquet and Wilson, 2011). In practice, co-governance schemes are typically designed to be flexible, reflexive, and adaptive so that social learning can take place (Voss et al., 2006).

Polycentrism is a related concept. First defined by Elinor Ostrom in the early 1970s, polycentric governance refers to organizational structures in which a number of independent actors coordinate their relationships with one another under an agreed upon set of rules (Araral and Hartley, 2013). When brought to the realm of public policy, polycentrism does not simply imply de-centralization; rather, it implies first the recognition of existing self-governing processes (Nagendra and Ostrom, 2012). Ostrom's life work has involved specifying the conditions under which polycentric systems emerge, so that public policy could "develop institutions that bring out the best in humans" (Ostrom, 2010, p. 25). While never suggesting polycentrism as a panacea for addressing all challenges, her research emphasizes the value of exploring diverse polycentric institutions because of their potential to help "innovativeness, learning, adapting, trustworthiness, levels of cooperation of participants," and the ability to achieve "more effective, equitable and sustainable outcomes at multiple scales" (Ostrom, 2010, p. 25).

Co-governance and polycentric governance are the theoretical anchors for the exploration that follows. The concept of polycentrism articulates the potential benefits of building governance processes from the bottom- up (i.e. the 'commoning' of food policy-making and governance), while co-governance theory articulates the potential value of collaborative processes that involve state, civil society, and private sector actors.

In the context of efforts to influence state policies, it is important to also introduce the concept of regulatory pluralism. Regulatory pluralism is part of how co-governance might be defined by (and justified to) a hierarchical state. It refers to "a governance regime that embraces a wide range of coordinated and integrated instruments (including some traditional command and control regulations), well matched to the desired effect and implemented by an equally wide range of state and non-state actors" (MacRae and Winfield, 2016, p. 21). Gunningham and Sinclair note that regulatory pluralism occurs when "governments harness the capacities of markets, civil society, and other institutions to accomplish its policy goals more effectively, with greater social acceptance, and at less cost to the state" (quoted in Seed et al., 2012, p. 465). Regulatory pluralism is not the same as co-governance or polycentrism, but we believe collaboration with self-governing systems may be helpful for regulatory pluralism to function.

The potential for new approaches to food policy in Canada

Canada currently has no national food policy to improve the health and sustainability of the food system. The 2015 promise from Canada's federal government is thus an opportunity to 'co-govern' a national food policy that builds on experiences at the sub-national level. It is also an enormous challenge, because the federal government's history of policy interventions related to food appears to represent an obstacle to innovation. Consider Canada's productivist orientation to agriculture from the 1950s to the 1970s, which translated into narrow policy goals with a dominant focus on cheap prices and exports and limited departmental mandates (e.g. Canada has no 'department of food'; MacRae, 2011). Then, in the neoliberal era (post 1980s) of multilateral trade agreements, further restrictions came into place. As a champion of global free trade, Canada aligned many of its policies (e.g. food safety) with key trading partners like the United States and gave up some of its capacity to determine national priorities (ibid., 2011, p. 428).

Notwithstanding these challenges, there are also enablers of change. First, Canada's highly decentralized system of governance may offer a starting point for polycentrism. The ten provinces and three northern territories, plus Aboriginal governments, are key jurisdictional arenas in the policy-making process (Skogstad, 1987). They have responsibility over several policy domains related to food (social policy, health care, education, resource management, etc.) and are empowered to negotiate and deliver a range of federal initiatives. These governments are critical players in food policy, each involving a set of institutions shaped according to their respective regulatory history. Policy experimentation has been taking place at the provincial and territorial (PT) levels (as well as the municipal level; MacRae and Donahue, 2013), which could be scaled up to a national food policy.

Second, Canada has a growing 'food movement.' Today's movement is the latest wave in a history of community organizing and a variety of forms of resistance to the dominant food system. These earlier forms include five hundred years of Aboriginal resistance to the changes brought to Indigenous food systems through reliance on colonial systems (Morrison in Wittman et al., 2011).

Earlier forms of resistance and organizing also include the agrarian and cooperative movements of the late 19th–early 20th century documented by Lévesque in Quebec (2007) and Winson (1994) in Ontario and Saskatchewan. These movements led to policy innovations like Canada's model of supply management in the dairy and poultry sectors, which pays producers based on a formula related to their costs of production (Winson, 1994). The latest wave of social movement activities include a range of diverse, place-based actions that are seeking to reduce the negative externalities of the food system and foster positive social, environmental, and economic benefits (Wittman et al., 2011).

To articulate the emergence of the national food movement into the policy sphere, we refer to three successive moments in the past two decades: the steps that led to the emergence of FSC from 1996 to 2001, the People's Food Policy project (PFPP) of 2007–2011, and the Eat Think Vote campaign during the 2015 federal elections.

In 1996, the World Food Summit organized by the Food and Agriculture Organization, provided a short-lived impetus for the Department of Agriculture and Agri-Food Canada (AAFC) to activate an interdepartmental committee in which representatives of various government branches would periodically meet with invited civil society representatives. Koç et al. (2008, p. 131) argue that "the interdepartmental committee's only achievement was convincing the federal government to fund a food security conference to listen to civil society organization voices." That conference was held in 2001 and sowed the seeds for CSOs and others to come together and form, in 2005, the membership-based, non-profit organization known as Food Secure Canada (FSC).

Later, in 2007, representatives of several Canadian CSOs (including the founding director of FSC, Cathleen Kneen) participated in the Food Sovereignty Forum in Nyeleni, Mali[4]. This forum was organized by La Via Campesina, along with a wide range of social movement partners from around the world, to broaden the food sovereignty conversation beyond only agrarian concerns (Kneen in Wittman et al., 2011). Once back in Canada, the leadership of FSC began working alongside progressive farm organizations, anti-biotechnology activists, international solidarity organizations, and others to re-define their policy positions through the lens of food sovereignty (ibid.). They instigated a three-year "kitchen table talk" consultation, involving over 3,000 citizens across the country in the process and ultimately leading to the positions encapsulated in 'Resetting the Table' (PFPP, 2011). The process was sufficiently credible among food advocates to bring the voices of the food movement to the national scene and Food Secure Canada adopted this document as its policy platform.

Central to the PFPP was a process of deep consultation. La Via Campesina speaks of the "right of peoples to define their agricultural and food policy" (Desmarais, 2007, p. 45). Similarly, the PFPP was focused on democratic engagement. At its core, the PFPP (2011) recognized the way that many Canadians, through community organizing, were already building and governing food systems to meet their own needs. It also called on the federal government to offer more opportunities for people to control their food systems, taking that power back from corporations and their lobbyists. It is in these ways that the PFPP and the work of FSC align so well with the notion of the 'commoning of food governance' around which this chapter is framed. In the end, the PFPP was about furthering decision-making structures rooted in the principles of deliberative democracy and in particular the inclusion of the perspectives of those most marginalized by dominant structures and systems.

The 2015 Canadian election was the most recent moment for food movement actors to bring their collective voices into the public debate. FSC and its partners converged on four 'common ground' policy demands[5]: (1) to assess the feasibility of a basic guaranteed income for addressing food insecurity in Canada, (2) to overhaul the Nutrition North program (which subsidizes retailers to bring food into difficult to reach parts of Canada's north), (3) to increase support for the next generation of farmers, and (4) to invest in a national healthy school food program. Food movement actors argued that these inter-related issues should be addressed through a comprehensive national food policy developed through participatory processes and designed to support existing grassroots and CSO initiatives. FSC's Eat Think Vote campaign during the 2015 election campaign was a call to action to organize community-based events with local political candidates and communicate a common message. The campaign was a success, and Canada's new Liberal government committed to a national food policy in late 2015.

With this opportunity in mind, this chapter presents three case studies in the area of access to healthy food that we argue could be scaled up into a polycentric national food policy.

Access to healthy food in Canada's food policy landscape

In Canada, the strong tension between the consideration of access to healthy food as a common (or public) good and food as a commodity is one of the major reasons behind the food movement's entry into the political arena. The government of Canada says that it seeks to "improve access to nutritious food" (Public Health Agency, 2016) and values the making of "healthy food choices" by Canadians (Health Canada, 2007). However, indicators of both food insecurity and obesity – the two faces of malnutrition – are getting worse (Tarasuk et al., 2014; Twells et al., 2015).

The efforts of the government, CSOs, and – to a lesser degree – the private sector[6] to address malnutrition are leading to institutional arrangements worth analyzing. The case studies presented here focus on sub-national levels of government (Aboriginal and PT levels) to articulate possibilities that a national food policy can learn from and build upon. These are: (1) hunter support programs in Northern and remote regions (reflecting food as common pool resource); (2) multi-stakeholder engagement for health promotion (exemplifying the commoning of food policy-making); and (3) joint food security programming (furthering food as public good through co-governance mechanisms). A view of food as a cultural determinant and a vital need for health, as opposed to simply food as a commodity, is a recurring element of these otherwise different case studies.

Indigenous self-determination through hunter support programs

A major nutrition transition is well underway in Canada's northern regions and Aboriginal communities.[7] Colonialism and the reserve system of land and population management, among other factors, produced a shift in diets from nutrient-dense, 'traditional,' or 'country' foods provided by nature to nutrient-poor, store-bought food delivered by the market (CCA 2014, p. 116). In this context, the de-commodification of diets depends on continued (or renewed) access to traditional foods – foods that have historically been governed as common-pool resources by communities under customary rules and tradition[8] (Berkes et al., 1991). Access to these healthy foods as commons relies on traditional indigenous knowledge of how to harvest and prepare traditional foods and are a cornerstone to Aboriginal people's goals around cultural self-determination (Morrison, 2011). This case study demonstrates how access to healthy, wild food as a common-pool resource has been institutionalized through hunter support programs.

Aboriginal rights to hunt, fish, harvest, and trap are legally protected after amendments were made to the Section 35 of the Canadian Constitution (1982)[9]. Specifically, it recognizes Comprehensive Land Claim Agreements (CLCAs), which have been negotiated between the government of Canada, PT governments, and the Inuit or specific Indigenous nations since the 1970s (INAC 2015). CLCAs are 'modern treaties' intended to delineate ownership rights and the use and management of land and natural resources. Twenty-six have now been signed, but hundreds of negotiations are still taking place as "the governance of vast areas of territory [...] remains unsettled and contested" (CCA, 2014, p. 151).

In the words of one elder, "food security has always been a priority in the settlement of land claims, and included species conservation [provisions] and [ongoing] access to our homeland" (N. Kassi, interview, June 22, 2016). In reference to the James Bay and Northern Québec Agreement, Natcher et al. (2015, p. 2) write, "although the agreement required the Cree and Inuit to cede certain territorial rights they gained new forms of recognition and protection for wildlife harvesting activities, most notably through various Hunter Support Programs (HSPs)". Funded through treaty commitments, HSPs involve subsidizing the purchase of equipment, gas, labour, and other expenses for going onto the land. They vary in how they are designed and implemented. In general, however, resources are transferred to Indigenous governments or PT departments, and then allocated through individual applications or to hunters' and trappers' organizations.

HSPs are not always tied to CLCAs, but are part of the socio-economic and cultural fabric of the North, and are situated within a wider set of institutional arrangements. As a polycentric form of natural resource management, HSPs involve multiple levels of coordination. Hunter-trapper organizations, fisheries councils, and wildlife committees manage animal populations and set harvesting quotas, drawing on direct wildlife observation, traditional ecological knowledge and harvest studies (Berkes et al., 1991; CCA, 2014). Funding for HSPs and the inclusion of traditional knowledge into planning strategies ensure that these traditional food resources are being maintained as common-pool resources.

Further, harvested food is normally widely shared within these isolated communities in order to ensure that the elderly, children, and others who cannot themselves go on the land continue to have access to healthy, nutrient-dense foods (CCA, 2014). One of our interviewees explained the value of HSP in this way:

> In a small community of Ontario of 70 band members with no infrastructure, the band council bought communal equipment (quads, boats, guns). Everyone is employed through the band council, which encourage people to go on the land half of the week. Half of the harvest goes to the community freezers. It's been so successful that the band council send food to neighbouring communities. It's an example of a contemporary model of an operational tribal economic system.
>
> *(J. Leblanc, interview, July 14, 2016)*

Echoing Ostrom's point that polycentrism is no panacea, it is notable that HSPs may not be equally appropriate for all communities. Prioritized by one interviewee was community size and ready access to wild game: "The success of the programs requires free mobility on the land and suitable community size. I've seen smaller communities thriving, whereas larger ones [do not as the] result of [a] forced sedentary lifestyle" (J. Leblanc, interview, July 14, 2016).

Across Canada's provinces, territories and First Nations' governments, varying levels of institutional capacity and legislative provisions represent a challenge to HSPs as a mechanism for access to healthy traditional foods. For example, in the Yukon territories, the Umbrella Final Agreement (a CLCA) "included a provision for a study to examine the possible implementation of an HSP. However, no program has yet been implemented and there is no indication that either the territorial or federal governments are inclined to support a program in the future" (Natcher et al., 2015, p. 5), albeit some of the local governments have initiated their own. Interviewees raised a great deal of concerns about the legal entitlements of Aboriginal peoples to access land, including the weak monitoring mechanisms of wildlife (J. Leblanc, interview, July 14, 2016), as well as the burdensome regulatory hurdles (N. Kassi, interview, June 22, 2016). Finally, the impacts of climate change, which are felt across the North, pose serious challenges to the sustainability of hunting, harvesting, and fishing practices.

Hunter support programs represent an opportunity to improve food security and increase cultural self-determination. A national food policy could support HSPs by ensuring Canada-wide recognition of these programs and stable nation-wide funding for them, whether through the CLCAs or other channels (N. Kassi, interview, June 22, 2016; J. Leblanc, interview, July 14, 2016).

However, any national program must recognize and build on the self-governance mechanisms already on the ground across Canada, thereby ensuring management of specific resources remains locally controlled, while adapting to the rapidly changing conditions caused by climate change.

CSO-led stakeholder engagement for health promotion policy

One of the responses to the problems of both food insecurity and obesity in Canada by Canada's federal, PT ministers of health/healthy living and sport, physical activity, and recreation was to reintroduce their commitment to health promotion through a joint Declaration on Prevention and Promotion in 2010. The declaration was announced in the spirit of the Ottawa Charter for Health Promotion (1986), a landmark global agreement of the World Health Organization (WHO, 2017). This case study shows how the declaration was advanced in Ontario and the province of Newfoundland and Labrador through stakeholder engagement processes led by CSOs.

The case of Ontario – Canada's most populous province – provides an important example of the commoning of food policy-making through the leadership of CSOs. The seeds of the Ontario Food and Nutrition Strategy (OFNS) were sown in a 2009 meeting hosted by a provincial collaboration of not-for-profit, public health, food, and academic organizations. Over a five-year period, the OFNS process was developed at arm's length from government, though inclusive of relevant governmental representatives. It built momentum across a credible and diverse group of organizations mobilized to create a "cross-government, multi-stakeholder coordinated approach to food policy development and a plan for healthy food and food systems in Ontario" (OFNS, 2014, p. 3).

Through the contribution of the provincial networking CSO affiliated with FSC, Sustain Ontario, the OFNS gained additional members from the agricultural sector as well as government representatives (OFNS, 2014, p. 2). In the spirit of advocating for a comprehensive and integrated governmental response to a range of food issues, the group also issued recommendations to a number of relevant government consultations[10]. The OFNS is developing system-level metrics for food access and food literacy, suggesting a possible alignment with the provincial government's Local Food Strategy (2014, p. 54). More significantly for us is the breadth and depth of the cooperative relationships that CSOs built to encourage the provincial government (2013–2018) to improve access to healthy foods.

The experience of Newfoundland and Labrador provides another example of the commoning of policy-making processes. In 2015, Food First Newfoundland (another membership-based provincial networking CSO affiliated with FSC) launched, with the NL Public Health Association (an independent public health advocacy group), the *Everybody Eats* initiative to "raise public awareness and facilitate dialogue on food issues to enhance the level of public discourse" (FSC, 2016). This process started with a discussion paper and an invitation to all citizens to attend forums, host kitchen table talks, and submit input. What is most notable about *Everybody Eats* is the early alignment with key arms of the civil service of Newfoundland. Food First and the Public Health Agency (PHA) invited the co-chairs of the provincial government's interdepartmental food security working group onto the *Everybody Eats* advisory committee. This was deemed to be a major success by CSO organizers who hope this representation translates into provincial policy uptake of *Everybody Eats* recommendations (K. Jameson, interview, May 26, 2016). Will the process influence the government's upcoming Food Security and Agriculture Growth Strategy? It is it too early to say, but CSO representatives expect to align with the government's priority over increasing investment in domestic agriculture production and reducing dependency on imports.

These two examples of stakeholder engagement processes led by CSOs are in contrast to traditional 'consultation' processes led by governments. MacRae and Abergel (2012, p. 3) point out that "the effectiveness of the [government-driven] public consultation process is uncertain, given a lack of resources and skills within the civil service and, depending on the government of the day, the degree of political commitment to it. Inherent in the structure of government-driven consultative processes is their legitimacy in light of ideological goals and international commitments." In contrast, CSO-led processes appear to involve greater opportunity to build a shared vision among the stakeholders involved, including relevant government actors.

The developments in Ontario and Newfoundland are recent and their implications are not fully materialized. Moreover, it is likely that future provincial policy will conceive of food only as a basic need or a socio-cultural determinant, instead of a human right everyone is entitled to and will therefore be partial and imperfect. They do provide, however, examples that we believe the creation of a national food policy can build upon. Drawing on the work of McNicoll (2013), both FSC and industry argue for the creation of a national food policy council as part of a national food policy. Building on similar models at the municipal level, such a council would bring together key stakeholders in an inclusive way to help design responses to the food system challenges that Canada faces.

Joint food security programming

Canada is considered a wealthy country, but hunger persists with social assistance rates remaining insufficient (De Schutter, 2012, p. 11). In the absence of a coordinated federal poverty reduction initiative, only four provinces and territories have implemented their own strategies (PHAC, 2016). We can again look to the PT to identify models that can be scaled-up. In this case study, details about recent efforts to tackle food insecurity in New Brunswick are provided. It then raises similarities between the experiences of Manitoba and British Columbia. These examples demonstrate ways of institutionalizing access to healthy food as a public good through joint programing between the state, private sector, and CSOs.

In New Brunswick, the Economic and Social Inclusion Corporation (ESIC)[11] is empowered by legislation to establish a five-year social inclusion action plan (2014–2019). As part of this action plan, a partnership led by the Department of Social Development with the Food Banks Association and the Food Security Action Network (both networks of CSOs) was created to "promote [the] transition of food banks to community-based food centers" (ESIC, 2014, p. 29). Community food centres differ from food banks because of how they integrate healthy food access, education, and community engagement.[12] "Like libraries, but for food literacy… where people exchange recipes, seeds, stories and support" (Palassio, 2016), community food centres are being established across Canada through a combination of government and private sector funding, with the explicit goals of realizing the ideal of "food as a public good" and "putting the commons back at the centre of communities… [so that] those communities are taking their food decisions into their own hands" (ibid.).

As a first step, the department worked with the staff of the provincial food security network (an affiliate of FSC) to conduct a consultation with food banks (R. Hutchins, August 9, 2016). The consultation resulted in the development of a strategic framework to change long-established provincial mechanisms for funding allocation and food distribution. After visiting fifty-three out of the fifty-nine food pantries in the province, the consultation garnered sufficient input, as well as legitimacy, in the eyes of the key stakeholders to be affected to recommend a shift from a core-funding model to a needs-based framework that assesses local community capacity for joint programming and food provisioning along the lines of the community food centre model (R. Hutchins, August 9, 2016). While the implementation of new funding and food distribution arrangements would remain far from a full-fledged realization of food as a human right in New Brunswick, the example illustrates the value of a collaborative approach to policy assessment with CSOs. It is ultimately the food-insecure who will benefit the most from the shift, where feasible, to a community food centre model.

The provinces of Manitoba and British Columbia have more established joint programming experiences between CSOs and government (Thompson et al., 2011; Seed et al., 2012). Although these emerged in notably different contexts, two similarities are worth mentioning. First, the Northern Healthy Food Initiative (MB) and the Core Food Security Program (BC) are both inter-departmental in scope, with one department empowered by government to provide leadership. The case of BC can illustrate inter-departmental reach, whereby food security as a public health issue was integrated into the Ministry of Employment and Social Assistance and the Ministry of Agriculture, which designed programs to make locally grown fresh foods more universally available (Seed et al., 2012).

Second, these programs would be difficult to realize without active food movement involvement. In Manitoba, three northern regional CSOs and a province-wide food security organization pooled their respective resources to build greater capacity for self-provisioning in remote areas of the province and reduce food insecurity (Thompson et al., 2011). In British Columbia, "both civil society food security networks and health-focused [CSOs] played key roles in laying the foundation for, and support the integration into Public Health" (Seed et al., 2012, 462). These programs have also been a springboard to increase CSO capacity. The establishment of a public-private collective funding partnership, the Northern Manitoba Food, Culture and Community Fund, is one such example (Thompson et al., 2011).

The development of co-governance arrangements that involve and empower CSOs are circulating new considerations of food as a public good into the policy-making spheres and (re-)introducing ways for furthering this goal (e.g. self-provisioning, farmers markets, and schools). What we have termed here the 'commoning of food governance' is starting to make a difference in specific provinces, and now awaits federal involvement and funding to achieve a greater impact across the country. As one interviewee noted, "While there are a lot of good things happening, we noticed school food meals are managed in a ad hoc and voluntary way [and] the issue of funding was recurrent" (K. Atkey, interview June 21, 2016). However, there are also tensions associated with having CSOs in the frontline programming role for addressing food insecurity.

In their work on CSOs in Canada's food movement, Koç et al. (2008) note that they serve a variety of key roles. CSOs are "vital drivers of change and the democratization process; contribute to transparency and accountability of policy-making; introduce new information, experiences, and perspectives; and contribute to the practical implementation of various initiatives, including filling the gaps in service delivery unfulfilled by public programs" (p. 125). These roles ensure governance processes become nimbler and more flexible, but they also raise constraints.

The first tension is that the experiments are financially under-resourced. CSOs are also weary of relying too much on the government. As one of our interviewees pointed out, "on the flip side, I've talked to community providers, they don't want full support from government because it could be jeopardized if there is a change in [the governing] party" (K. Atkey, interview June 21, 2016). In making the case for a universal school meal program, advocates have proposed a cost-shared model between multiple levels of governments, households, and the private sector.[13]

Second, there are typically conflicting mandates or agendas within the government. In particular, there is a split between the government's public health mandate and its food safety and trade priorities (MacRae and Winfield, 2016; Seed et al., 2012). This dilemma reinforces the need for developing strategies across government departments in a more integrated way: "It's not about programming, but about developing a comprehensive strategy" (S. Roberts, May 24, 2016).

In thinking of scaling up these efforts to further food as a public good, what are some of the immediate implications for the federal government? One idea would be to recognize food security as a full-fledged pillar of the upcoming federal Poverty Reduction Strategy. This would lead to greater likelihood that all PT government prioritize food security in their respective plans. This raises a third tension and recurring debate among advocates in Canada, which is whether policy can accommodate both community-based responses and income-based measures (e.g. guaranteed income) (Tarusak et al., 2014).

Notwithstanding these proposals, it is worth noting that some of our interviewees feel that the food movement should not overly rely on the government. According to one informant, "the solutions aren't in programs and policies... Community initiatives need to become financially viable: they are often initiated by grassroots groups with the help of public funding. They are not entrepreneurs, and when funding ends, we see (...) these initiatives seriously compromised" (Pedersen, June 24, 2016).

Conclusions

Diverse models of food policy-making and governance, reflecting the values of food as commons and as a public good, are fostering cooperation and trust to change the way healthy food is promoted, distributed, and accessed in Canada. The material changes remain small and are situated at the edges of food systems otherwise defined primarily by global markets, but they are not insignificant. This chapter drew from a sample of institutional arrangement to explore the 'commoning of food governance', including the hunter support programs in Canada's North, CSO-led health promotion engagement strategies in Ontario and Newfoundland and Labrador, as well as examples of joint, community, food security programming in New Brunswick, Manitoba and British Columbia. There are many other examples of the growing role of CSOs at sub-national levels of governments in food governance in Canada.

These examples are part of a larger narrative. They represent the food movement's project to increase civic participation in the governance of the food system. In Canada, the food movement is gaining institutional recognition and lessons from those efforts should inform the design of the national food policy process, as well as the content of the policy itself. These examples show that provincial departments in Canada, such as public health institutions and social development departments, acknowledge that food is vital for human health and a socio-cultural determinant and that participatory approaches to food governance will help them achieve broader aims associated with public health and social equity. This is a starting point for increased engagement by CSOs in food governance.

The examples provided show how provincial and territorial networking organizations and associations are experimenting with polycentrism and co-governance, affirming Levkoe's (2015) point that they are "uniquely positioned [...] to foster and sustain collaboration" (p. 177). These membership-based, non-profit organizations, working with Food Secure Canada at sub-national levels, provide important exemplars within a national food policy landscape that otherwise has limited feedback mechanisms to facilitate innovation (MacRae, 2011). Provincial networking CSOs and their partners are building bridges upstream between health and agriculture, instilling norms of trust and cooperation and encouraging other food system actors to value food differently.

Food is a 'contested space' (Lang, 2005) and the experiences outlined in this chapter raise structural tensions. There are many challenges ahead, not least the historic role the Canadian federal government has played as a champion of agri-food exports and its limited engagement with food policy on other fronts (e.g. nutrition). Lessons suggest that there are conditions which will need to be met to continue translating food as a public good into policy: the implementation of accountability mechanisms (e.g. goals and metrics) and sustained funding to experiment with policy and consolidate social innovations. Finally, governance models to advise and/or coordinate emerging food policies (such as a National Food Policy Council) still need to be built.

The commoning of food governance in Canada clearly opens possibilities for regulatory pluralism. Whether or not these arrangements lead to true polycentrism, co-governance, and the whole-of-government approach to food policy is unlikely given the tensions this research reveals. Still, experience at sub-national levels of inclusion of plural knowledge systems, deep public engagement, and joint programming suggest that a future national food policy may yet become more than a high-level, aspirational policy framework.

Interviews quoted in chapter

Kayla Atkey. Policy Analyst at the Alberta Policy Coalition for Chronic Disease Prevention. June 21, 2016.

Susan Roberts. Representative of Alberta Food Matters. May 24, 2016.

Stephen Pedersen. Consultant for the Alberta First Nations Food Security Strategy. June 24, 2016.

Rick Hutchins. Executive Director of the New Brunswick Food Security Action Network. August 9, 2016.

Kristie Jameson. Executive Director of Food First Newfoundland. May 26, 2016.

Joseph Leblanc. Executive Director of the Sudburry Social Planning Council. July 14, 2016.

Norma Kassi, Katelyn Frienship, and Jodi Butler Walker. Researchers at the Arctic Institute of Community-Based Research. June 22, 2016.

Notes

1 The goals of FSC are to achieve zero hunger, healthy and safe food, and sustainable food systems.

2 The cases presented came to the fore through a partnership between FSC and *Food: Locally Embedded, Globally Engaged* (FLEdGE), a pan-Canadian research initiative led out of Wilfrid Laurier University's Centre for Sustainable Food Systems. A desk analysis, complemented with interviews, was carried out in 2016.

3 To view the discussion papers, maps, and summary tables, see fledgeresearch.ca/2017/08/11/mapping-the-food-policy-landscape-in-canada/. Details on the community peer review process are described in Levkoe, Schembri and Wilson (forthcoming).

4 See www.nyleni.org.

5 See https://foodsecurecanada.org/policy-advocacy/eat-think-vote.

6 For a private-sector-led initiative, see the Maple Leaf Centre for Food Security launched in late 2016.

7 The term 'Aboriginal people' refers to the descendants of the original inhabitants of North America. In Canada, Aboriginal peoples comprise of First Nations peoples, Métis and Inuit.

8 The governance systems of First Nations, Métis and Inuit, are plural and complex, spanning 98 linguistic communities and ranging from single Band Councils and Tribal Councils representing several First Nations to province-wide and national organizations (Kaufman and Roberge, 2015).

9 See Section 35 of the Canadian Constitution (1982): http://laws-lois.justice.gc.ca/eng/Const/page-16.html#h-52.

10 For instance: Ontario's Local Food Act, the Healthy Kids Panel, and the Poverty Reduction Strategy.

11 The ESIC is funded primarily by the provincial government and governed by a board "representing government, business, non-profit organizations and persons having experienced poverty," and is mandated to lead initiatives "to reduce poverty and assist thousands of New Brunswickers to become more self-sufficient." See http://www2.gnb.ca/content/gnb/en/departments/esic.html.
12 Community Food Centres Canada is a national non-profit that partners with local groups and organizations to provide integrated food security services. See http://cfccanada.ca/.
13 This proposal stems from the Coalition for Health School Food, a coalition of CSOs coordinated by Food Secure Canada: https://foodsecurecanada.org/coalitionforhealthyschoolfood.

Bibliography

Araral, E. and Hartley, K. (2013). Polycentric governance for a new environmental regime: theoretical frontiers in policy reform and public administration. First International Conference on Public Policy. Grenoble, France.

Berkes, F., George, P., and Preston R. (1991). Co-Management: The Evolution of the Theory and Practice of Joint Administration of Living Resources. TASO Research Report. 1st Second Annual Meeting of International Association for the Study of Common-Property.

The Expert Panel on the State of Knowledge of Food Security in Northern Canada (2014). Aboriginal food security in Northern Canada: an assessment of the state of knowledge. Council of Canadian Academies.

De Schutter, O. (2012). Report of the Special Rapporteur on the right to food, Mission to Canada. United Nations General Assembly. Human Rights Council, Twenty-second session.

Desmarais, A.A. (2007). *La Via Campesina: Globalization and the Power of Peasants*. Halifax and London: Fernwoord and Pluto Press.

Dryzek, J. (2010). *Foundations and Frontiers of Deliberative Governance.* London: Oxford University Press.

Economic and Social Inclusion Corporation. (2014). Overcoming poverty together: the New Brunswick economic and social inclusion plan 2014–2019. Government of New Brunswick.

Ferrando, T. and J.L. Vivero-Pol (2017). Commons and 'commoning': A 'new' old narrative to enrich the food sovereignty and right to food claims. *Right to Food and Nutrition Watch* 10: 50–56.

Food Secure Canada (2016). Great Food Organizations We Love to Support: Food First NL. Retrieved July 3, 2016, from http://foodsecurecanada.org/resources-news/news-media/great-food-organizations-we-love-support-food-first-nl.

Health Canada (2007). Canada's Food Guides from 1942 to 1992. Government of Canada. Retrieved July 1, 2016, from http://www.hc-sc.gc.ca/fn-an/food-guide-aliment/context/fg_history-histoire_ga-eng.php.

Indigenous and Northern Affairs Canada. (2015). Comprehensive Claims. Government of Canada. Retrieved July 18, 2016, from https://www.aadnc-aandc.gc.ca/eng/1100100030577/1100100030578.

Public Health Agency of Canada (2016). Food Security. Government of Canada. Retrieved July 18, 2016, from http://cbpp-pcpe.phac-aspc.gc.ca/public-health-topics/food-security/.

Kaufman, K., and F. Roberge (2015). Aboriginal Governance in the Canadian Federal State 2015, Working Paper 2003 (3).

Koç, M., MacRae, M., Desjardins, E., and W. Robert (2008) Getting Civil About Food: The Interactions Between Civil Society and the State to Advance Sustainable Food Systems in Canada. *Journal of Hunger & Environmental Nutrition* 3(2–3), 122–144.

Kooiman, J. (2003). *Governing as governance.* London: Sage.

Knuth, L. and M. Vidar (2011). *Constitutional and Legal Protection of the Right to Food around the World.* Rome: Food and Agriculture Organization of the United Nations.

Lang, T. (2005). Food control or food democracy? Re-engaging nutrition with society and the environment. *Public Health Nutrition* 8(6A), 730–737.

Lévesque, B. (2007). Un siècle et demi d'économie sociale au Québec : plusieurs configurations en présence (1850-2007). *Cahiers du Centre de recherche sur les innovations sociales (CRISES) Collection Études théoriques* ET0703: 1–79.

Levkoe, C, Schembri, V., and A. Wilson (under review). Community Peer Review: Strengthening Scholarship and Practice. *Engaged Scholar Journal.*

Levkoe, C. (2015). Strategies for forging and sustaining social movement networks: A case study of provincial food networking organizations in Canada. *Geoforum 58*, 174–183.

Linebaugh, P. (2010). Some Principles of the Commons. Retrieved January 1, 2017, from http://www.counterpunch.org/2010/01/08/some-principles-of-the-commons/.
MacRae, R. (2011). A Joined-Up Food Policy for Canada, *Journal of Hunger & Environmental Nutrition*, 6(4), 424–457.
MacRae, R. and E. Abergel (2012). *Health and Sustainability in the Canadian Food System*. Vancouver: UBC Press.
MacRae, R. and K. Donahue (2013). Municipal food policy entrepreneurs: A preliminary analysis of how Canadian cities and regional districts are involved in food system change. The Canadian Agri-Food Policy Institute, 1–34.
MacRae, R. and M. Winfield (2016). A little regulatory pluralism with your counter-hegemonic advocacy? Blending analytical frames to construct joined-up food policy in Canada. *Canadians Food Studies*, 3(1), 140–194.
McMurtry, J. (2002). *Value wars: the global market versus the life economy*. London: Pluto Press.
Meredith, G. (2016). Canada's New Food Policy: Opportunities, Challenges and Dilemmas. Presentation in plenary session. Food Secure Canada 9th Assembly. Toronto, Canada.
McNicoll, A. (2013). More Voices at the Table: The Feasibility of a National Food Policy Council for Canada (Unpublished master's thesis). University of York, Toronto, Canada.
Nagendra, H. and E. Ostrom (2012). Polycentric governance of multifunctional forested landscapes. *International Journal of the Commons*, 6(2), 104–133.
Natcher, D., Castro, D., and F. Lawrence (2015). Hunter support programs and the Northern social economy. In Southcott, C. (Ed.), *Northern Communities Working Together* (pp. 183–197). Toronto: University of Toronto Press.
Oliveira, V. and E. Frazão (2009). The WIC Program: Background, trends, and economic issues, 2009 Edition. USDA Economic Research Report Number 73. Washington: USDA.
Ontario Food and Nutrition Strategy Design Team (OFNS). (2014). Ontario Food and Nutrition Strategy: A Plan for Healthy Food and Food Systems. Sustain Ontario. Retrieved from https://sustainontario.com/initiatives/ontario-food-and-nutrition-strategy
Ostrom, E. (2010). Beyond Markets and States: Polycentric Governance of Complex Economic Systems. *American Economic Review*, 100(3), 641–72.
Palassio, C. (2016) Community food centres must become more commonplace across Canada. *This Magazine*. Retrieved November 1, 2016, from https://this.org/2016/11/03/community-food-centres-must-become-more-commonplace-across-canada/.
Paquet, G. and C. Wilson (2011). Collaborative co-governance as inquiring systems. *Optimum Online*, 41(2), Retrieved October 16, 2016 from http://www.christopherwilson.ca/papers/Collaborative_Co-Governance_May16_2011.pdf.
PPFP People's Food Policy Project (2011). Resetting the table: a people's food policy for Canada. Retrieved May 2, 2016, from https://foodsecurecanada.org/people-food-policy.
Prime Minister of Canada (2015). Minister of Agriculture and Agri-Food Mandate Letter. Retrieved May 20, 2016, from pm.gc.ca/eng/minister-agriculture-and-agri-food-mandate-letter.
Renting, H., Schermer, M. and A. Rossi (2012). Building Food Democracy: Exploring Civic Food Networks and Newly Emerging Forms of Food Citizenship. *International Journal of Sociology of Agriculture & Food*, 19(3), 289–307.
Seed, B., Lang, T., Caraher, M., and A. Ostry (2012). Integrating food security into public health and provincial government departments in British Columbia, Canada. *Journal of Agriculture and Human Values*, 30, 457–470.
Skogstad, G. (1987). *The Politics of Agricultural Policy-Making in Canada*. Toronto: University of Toronto Press.
Somerville, P. and N. Haines. (2008). Prospects for local co-governance. *Local Government Studies*, 34(1), 61–79.
Sumner, J. (2011). Serving Social Justice: The Role of the Commons in Sustainable Food Systems. *Studies in Social Justice*, 5(1), 63–75.
Tarasuk, T., Mitchell, A. and N. Dachner (2014). Household Food Insecurity in Canada. Retrieved June 2, 2016, from proof.utoronto.ca.
Thompson, S., Gulrukh, A., Ballard, A., Beardy, B., Islam, D., Lozeznik, V., and K. Wong (2011). Is Community Economic Development Putting Healthy Food on the Table? Food Sovereignty in Northern Manitoba's Aboriginal Communities. *Journal of Aboriginal Economic Development*, 7(2), 14–39.

Twells, L.K., Gregory, D.M. Reddigan, J., and W.K. Midodzi (2014). Current and predicted prevalence of obesity in Canada: a trend analysis. *CMAJ Open*, 2(1), E18–E26.
Vivero-Pol, J.L. (2017a). Food as Commons or Commodity? Exploring the links between normative valuations and agency in food transition. *Sustainability*, 9(3), 442.
Vivero-Pol, J.L. (2017b). The idea of food as commons or commodity in academia. A systematic review of English scholarly texts. *Journal of Rural Studies*, 53, 182–201.
Voss, J.P., Bauknecht, D., and R. Kemp (Eds.) (2006). *Reflexive governance for sustainable development.* London: Edward Elgar Publishing.
Walljasper, J. (2011). Celebrating all we share three times a day. *On the Commons Magazine*, Retrieved January 15, 2017, from http://www.onthecommons.org/magazine/celebrating-all-we-share-three-times-day.
Wittman, H., Desmarais, A.A., and N. Wiebe (2011). *Food Sovereignty in Canada: Creating Just and Sustainable Food Systems* (pp. 97–111). Halifax: Fernwood Publishing.
Winson, A. (1994). *The Intimate Commodity: Food and the Development of the Agro-Industrial Complex in Canada.* Toronto: University of Toronto Press.
World Health Organization (2017). The Ottawa Charter for Health Promotion. Retrieved November 3, 2016, from http://www.who.int/healthpromotion/conferences/previous/ottawa/en/.

18

FOOD SURPLUS AS CHARITABLE PROVISION

Obstacles to re-introducing food as a commons

Tara Kenny and Colin Sage

Introduction

One of the consequences of the 2008-09 economic crisis has been a growing casualization across the labour market through short-time, zero-hour contracts and other flexible working practices. This has resulted in increased pressure on household budgets, such that opportunities to 'save' on food spending – particularly at 'no-frills' discount supermarkets, through 'own-brand' labels and by eating outside the home (with a steep rise in 'takeaway' fast-food outlets) – have been welcomed. It is no coincidence that this period has witnessed, particularly in the UK, an explosion in the number of food banks and other charitable food-provisioning arrangements. Yet alongside changes in the labour market and rising levels of poverty, we also witness an enthusiastic celebration in certain quarters for the emergence of a new 'sharing economy', with social entrepreneurs spearheading logistical solutions for a more sustainable future. The prevailing narratives are that of celebration and transformation: fixing an inefficient system that wastes one third of all food produced into a 'win-win' solution where food surpluses are channelled to the less fortunate and thereby solving the twin burden of food poverty and food waste in the process. Of course, the fundamentally structural unsustainability of the food system remains unchallenged.

In this chapter we attempt to stitch together disparate developments which we believe have implications for understanding the challenges confronting the 'food as a commons' agenda. In particular, we suggest that the widespread acceptance of surplus food redistribution as a solution to either food waste or food poverty represents an obstacle to working towards the implementation of the right to good food – the cornerstone of building a commons-based food system. We argue that this new era of food poverty, characterised as much by *malconsumption* as by under-nutrition, where the abundance of cheap, highly processed food rather than a lack of food prevails, presents significant challenges to those working for a healthy, sustainable and inclusive food system. This requires that we begin by recognising the industrial *massification* of food, the scale whereby ever greater volumes of highly-processed products are driven through a linear supply chain destined for supermarket shelves, accompanied by strong media promotion such that customers are encouraged to buy more than they require. In the make-believe world of consumer sovereignty and free choice, food is always and everywhere a commodity purchased by 'rational economic actors'. This narrative serves, however, to reinforce the individual focus that bedevils food, social and public health policy, crowding out the notion that food is also a public good. If food *is* just about consumer choice then the responsibility for eating healthily falls upon individuals, not the state nor those who supply the food. In the context of charitable food assistance programmes, however, consumer choice does not apply for here there is no consumer: only a 'needy' beneficiary relying upon the food industry, social entrepreneurs and an army of volunteers.

Consequently, the chapter proceeds as follows. First, we note how the contemporary food system, marked by corporate concentration, the excessive supply of energy-dense packaged products and resulting in rising volumes of waste, requires, in the 'age of sustainability', new means of disposal. This is key to comprehending the rise of charitable food redistribution, discussed in section II, for which the new 'sharing economy' provides a logistical platform. Thus we briefly explore the rise of 'smart app tech entrepreneurs' that serve as intermediaries between suppliers of 'surplus' food and charitable partners willing to distribute this amongst their client base, the 'deserving poor'. We then drill into the Irish case, first providing some institutional context, then through analysis of primary data drawn from extensive fieldwork in Cork, offering detailed insights into the implications of charitable food distribution. Finally, we close with a discussion on how these insights will help in developing a more durable case for transitioning towards a commons-based food system.

Big food > cheap food > food waste

The global food industry is estimated to be worth US$8 trillion per year with the packaged food market worth US$2.4 trillion in 2014 and where the top 100 food and beverage firms account for 77% of all packaged foods sold globally (Clapp and Scrinis, 2016). Companies have grown market share around the world not only by taking advantage of liberalised trade agreements, but by encouraging governments to relax regulations while intervening to re-shape consumer trends, particularly in middle-income countries. Besides growing corporate power and increasing consolidation and concentration (IPES, 2017), this food system is also characterised by ever increasing availability of edible oils, sweeteners and meat (Basu, 2015: 248), an excessive supply of nutrient deficient, energy dense "pseudo foods" (Winson, 2004) and a significant increase in calorie supply into the human diet (Carolan, 2011).

However, analysis of the changes in the global food supply suggest more complex processes at play than simply an outcome of increased volumes, rising incomes or attribution to 'economic development'. Basu (2015) highlights the role of a handful of companies that are responsible for the increased sales of sugar-sweetened beverages and packaged foods distributed through domestic systems (often by franchising local production). With healthier food options "inherently less profitable" (Stuckler and Nestle, 2012: 2), companies have strong economic incentives to promote highly-refined products. Ultra-processed foods now dominate food systems (Baker and Friel, 2016; Monteiro et al., 2013) and overwhelming evidence demonstrates that this process of dietary change is compounding ill-health with growing levels of obesity and non-communicable diseases, such as Type II diabetes, cardio-vascular disease, hypertension and various cancers across the globe (Mann et al, 2015; Roberto et al., 2015; Pearson-Stuttard et al., 2017).

Recent years have seen rising political recognition and a more popular consciousness around the issue of food waste (Campbell et al., 2017). Initially addressed within an environmental management framework, food waste prevention was seen in terms of resource efficiency and the need to reduce landfill costs. The publication and screening of a number of exposés, together with campaigns and gleaning activities accompanied by redistribution, began to raise the moral argument about the scale of food waste. This coincided with the period of austerity that followed the financial crisis of 2008 and accompanied by food price volatility. Inevitably, then, the conditions were created not only for more frugal domestic budgeting, in which wasting food was to be avoided, but as a solution to food insecurity and poverty. As Campbell et al. (2017) observe, food waste becomes a compelling new arena of political action, in which it becomes morally positioned as 'bad'. Given this normative shift, those actors, such as the major retailers widely regarded as a major part of the problem, then moved swiftly to position themselves as the solution, leading them to partner with food banks and intermediaries that possessed the capability of disposing of food surpluses.

The rise of food banks as key social actors has been noted for both high and middle-income economies (Riches and Silvasti, 2014; Caraher and Coveney, 2016). Indeed, it would appear that food banks will continue to grow in number and importance across Europe, taking into account the recent introduction of sanctions for failing to redistribute food in France and voluntary agreements in the UK (Mansuy and Ferrando, 2017). Donating surplus food to charities is an attractive option for food retailers as it is cheaper to dispose of in this way than to pay for collection and landfill taxes and charges. Besides the 'halo effect' that such philanthropic gestures allow, and which add to the triple bottom line under corporate social responsibility, such donations also squash critical questions about the failings of a food system that drives such levels of structural oversupply with all its attendant resource costs. The key to ease of operation, however, is the emergence of intermediaries willing to perform the work of matching surplus donations with the capacity of charitable bodies to absorb them. Ironically, these developments are embedded within the rhetoric of sharing, caring, supporting and community – vocabulary traditionally associated with the language of the commons in what could arguably be described as an appropriation of solidarity. Some accounts go as far as to describe these activities as part of a 'food revolution' (McGrane, 2015).

The sharing economy and charitable food

The rise of technological-based business models, typified by Uber, Airbnb and TaskRabbit, have proliferated in recent years. This 'sharing economy' has been defined as the "peer to peer based activity of obtaining, giving, or sharing access to goods and services" (Hamari et al., 2015: 2047). Alternative names include platform economy, access economy, collaborative economy and crowd-based capitalism (Yaraghi and Ravi, 2017). Frenken and Schor (2017: 4-5) define the sharing economy as "consumers granting each other temporary access to under-utilised physical assets" which they refer to as 'idle capacity'. This is exemplified as 'spare rooms or beds' in the case of Airbnb and 'spare cars' for Uber. However, as Frenken and Schor argue, this process is also surrounded by a discourse of progress, technological sophistication and innovation that seeks to differentiate itself from the 'corporate-centred model' (Sundararjan, 2017) and as a potential platform to elevate sustainable consumption practices (Heinrichs, 2013; Botsman and Rodgers, 2011). Others describe this new economy as 'neoliberalism on steroids', commercialising aspects of life previously outside the reach of the market (Morozov, 2013). The role of new information technology applications, involving sensors and other devices that can be utilised remotely using smart phones and tablets, has been critical to this process.

The sharing economy has also found its way into parts of the food system offering 'community'-based innovation in dealing with food waste. Perhaps one of the most successful ventures in this regard has been that of *FoodCloud*, an Irish-based social enterprise which has been accumulating numerous accolades. Founded in 2013 by two female entrepreneurs with the aim of matching surplus food with charities, it utilises a dedicated smart phone app that, in February 2017, was voted one of the top apps (No. 3) by *The Guardian* for tackling food waste globally (Wong, 2017).

While a lot of the work is done though an online platform, it is important to note that the model depends upon over 200 volunteers who perform 'food rescues' (Foodcloud, 2016). While it has been argued that 'perma-temps'[1] (Hill, 2015) are critical to the sharing economy, in the case of food redistribution, 'perma-temps' become volunteers. Despite the many noteworthy benefits of this well intentioned and highly successful social enterprise in highlighting the issue of food waste, using resources more efficiently does not, by itself, equate to ecological or 'sustainable' practices (Martin, 2016). This is especially pertinent when the defining feature is a product of a structurally unsustainable food system driven to over-supply and to the generation of surplus. Indeed, one concern is for the potential rebound effect whereby the key protagonists have no incentive to reduce volumes of food waste and where charitable partners enrol new clients to consume this surplus and, in so doing, conceal the extent of poverty, social exclusion and the lack of a right to food security with dignity.

Food banks serve as a collection and sorting point for food surplus donated by supermarkets and food manufacturers and have been described as a channel between welfare services and the food industry (Butcher et al., 2014) by capturing surplus food that would otherwise be discarded or diverted to landfill. Their precise function may vary between distributing food directly to the end user, distributing food to intermediary charities and organisations who then distribute to the end users (Ronson and Caraher, 2016) or a mixture of both. Across Europe, the amount of food channelled through these secondary food systems has increased annually. In 2014, the European Food Banks Federation (FEBA) redistributed an estimated 441,000 tonnes of food, a figure that increased to 531,537 tonnes in 2015 (FEBA, 2017). Not all food banks are a member of FEBA and not all redistribution is done via food banks (Fusion, 2016). Consequently, these figures do not reflect the actual scale of charitable food provisioning, an issue highlighted in an ongoing food aid mapping project led by the Independent Food Aid Network (IFAN, 2017).

Despite the noteworthy benefits of re-distributing edible food, food banks occupy a highly contested space and have long been a symbol of government failure: "The reliance on food banks is symptomatic of a broken social protection system and the failure of the state to meet its obligations to its people" (De Schutter, 2012: 5). Over recent decades, concerns have been raised regarding the appropriateness of such 'corporatised' and 'institutionalised' responses (Lambie-Mumford and Dowler, 2014; Booth and Whelan, 2014) as well as health, welfare, human rights and social justice dimensions (Riches, 2011;Riches and Silvasti, 2014).

The *corporatisation* of food banks – and arguably of poverty itself – has been explored in Canada (Riches and Tarasuk, 2014) and the USA (Fisher, 2017). Indeed, Fisher (2017) argues that anti-hunger advocates have now become part of the 'hunger industrial complex', also attracting corporations from outside the food industry. For example, in 2017, Enterprise Rent-A-Car Foundation announced its 'Fill Your Tank' initiative, designed to celebrate the company's 60th global anniversary, that will provide US$60 million over the next six years 'to address food insecurity' in countries where it has wholly-owned business operations in Europe and North America. In Europe, this initiative is being undertaken in collaboration with The Global FoodBanking Network (GFN) "that accelerates the development and growth of food banks" (Crosscare, 2017). GFN founders include a selection of the largest food and beverage companies in the world – Cargill, General Mills Foundation, Kellogg's and DLA Piper, a global law firm (GFN, 2017).

Scholars have long argued that charitable food distribution models are part of the problem as opposed to part of the solution (Berry, 1984; Poppendieck, 1999; Riches, 1997, 2011; Riches and Tarasuk, 2014). Such models have increasingly served to de-politicize hunger and solidify the perception of food waste and food poverty as a single issue with a 'win-win' solution (Caraher and Furey, 2017). This narrative has worked effectively to divert attention from the causes of hunger (Caraher and Furey, 2017) and food charity recipients' lack of rights in an ad hoc secondary food system (Tarasuk and Eakin, 2003), while maintaining public support for measures to address food waste (Mansuy and Ferrando, 2017).

Food banks are limited in their ability to provide a healthy diet (Poppendieck, 2014) and do not address the social, cultural and political aspects of food (Caraher and Dowler, 2014). Moreover, the consequences of a minimal and insecure diet in terms of wellbeing and mental and nutritional health are 'potentially severe' (Gairthwaite et al., 2015) and often hidden (Dowler and Lambie Mumford, 2015). Receiving food from charitable organisations is not a dignified solution to food insecurity. Much research (van der Horst et al., 2014; Purdam et al., 2016; Fisher, 2017) has highlighted the shame, degradation, humiliation and embarrassment associated with having to resort to charity to meet the most basic of all needs. Moreover, the much celebrated environmental benefits of redistributing food surplus ignores the numerous externalised costs elsewhere in the agri-food chain where resources (land, water, energy) have been invested and waste streams (greenhouse gases, contaminated water) have resulted. This arrangement does little to support the local economy and does not encourage systematic change. On the contrary, redistribution merely props up an inadequate food system that is incompatible with the concept of the commons.

Food poverty in Ireland

Ireland has a long history of food insecurity which extends beyond the experience of the Great Famine of 1845-52 and reflects a history of colonial control over agricultural land use. For centuries food security at times of want relied upon charitable interventions, a state of affairs that has persisted into the contemporary era. Moreover, Ireland is no stranger to 'silo' solutions to food security – it strongly defends its own ruminant-based agri-food strategy as a way 'to feed the world' (see Kenny et al., 2018) and extends this approach to matters of health, the environment and poverty. Here, the third sector is playing a key role in the large-scale redistribution of 'surplus food' from corporate retailers to those in need, thereby seemingly solving the twin problems of food poverty and food waste. This narrative has gained considerable traction recently across a wide spectrum of society and policy circles such that it is permeating – and obstructing – efforts to develop a wider civic conversation around the meaning of rights to food and how we should be working toward a healthier and more sustainable food system for all.

Poor diet is a significant risk factor for Ireland's total burden of disease. Currently, Ireland has one of the highest obesity rates in Europe and almost one-third of children in Ireland are overweight (IRSPEN, 2017a). An estimated €70 million is spent on *treating* diabetes annually (Shannon, 2017) and the Irish Society for Clinical Nutrition and Metabolism (IRSPEN), the Royal College of Physicians in Ireland (RCPI) and the European Association for the Study of Obesity (EASO) are now calling for a national obesity *treatment* programme (IRSPEN 2017b). It seems that pulling drowning people out of the river is still preferred to asking why they are falling in upstream.

In a country of less than five million people with reputedly the fastest growing economy in Europe, Ireland manifests striking social inequality.[2] Despite recovery from the economic crisis of 2008-12, the under-25 cohort is disproportionately affected by both un- and under-employment, while the incidence of precarious work "has skyrocketed" (Nugent, 2017: 28). Out of the total working population, 18.8% are below the poverty line (SJI, 2016), 30% earn below the Eurostat low-pay threshold (Collins, 2016) and 10% earn less than, or equal to, the minimum wage, 58% of whom are women (Brennan, 2017). Other social justice challenges, such as housing, are equally acute and marred with similar band-aid responses[3] and short-term thinking.

The latest figures indicate that 1.2 million people living in Ireland are experiencing deprivation, 789,855 are living in poverty, with 58.3% of these in consistent poverty (SJI, 2017).

Taken together, these factors are contributing to the growth and entrenchment of charitable food provisioning given inadequate social safety nets and an expanding pool of potential recipients. Undoubtedly, this prevents progress in acknowledging food as a right, which we consider to be the cornerstone of re-introducing the notion of food as a commons. Yet in Ireland's constitution, enacted on the 29th of December 1937, the right to food, encompassing the three key elements of respecting, protecting and fulfilling, is implied in Art. 45.2 and 45.3 (Box 18.1).

Box 18.1 The Irish Constitution (1937)

Art. 45.2 'The State shall, in particular, direct its policy towards securing:

i that citizens (all of whom, men and women equally, have the right to an adequate means of livelihood) may through their occupations find the means of making reasonable provision for their domestic needs.

ii that the ownership and control of the material resources of the community may be so distributed amongst private individuals and the various classes as best to subserve the common good'.

iii that, especially, the operation of free competition shall not be allowed so to develop as to result in the concentration of the ownership or control of essential commodities in a few individuals to the common detriment'.

Art. 45.3 2° 'the state shall endeavour to secure that private enterprise shall be so conducted as to ensure reasonable efficiency in the production and distribution of goods and as to protect the public against unjust exploitation'.

Art. 45.3 2° 'the state pledges itself to safeguard with especial care the economic interests of the weaker sections of the community, and, where necessary, to contribute to the support of the infirm, the widow, the orphan, and the aged'.

Food poverty manifests as poor health outcomes and widening health inequalities, which are critical justifications for tackling the problem. Tarasuk and Davies (1996: 73) note that "the way a problem gets defined or typified shapes responses to it", but this is not the case in Ireland. Here, the Department of Social Protection (DSP) defines food poverty as "the inability to have an adequate or *nutritious* diet due to issues of *affordability* or *accessibility*" (DSP, 2017: 51). Yet, despite the reference to 'healthy' and 'nutritionally adequate' diets in the definition of food poverty (Friel and Conlon, 2004), practical responses involving the redistribution of food pay little attention to its nutritional value; what people eat is not considered.

Distributing surplus food to the 'poor' is not a new concept in Ireland (as elsewhere) and has been done at a national level since 1987 through the 'food aid for deprived persons', a programme that relied upon intervention stocks, such as cheese and butter (Reilly, 2010). The successor programme, the Fund for European Aid to the most Deprived (FEAD), which is designed to "help people take their first steps out of poverty and social exclusion" (Welfare.ie, 2017) now canvasses charitable partners for a list of food items that it then purchases from contracted suppliers. This programme is now administered by FoodCloud Hubs[4] – FoodCloud's partner which, during the pilot phase of 2016, distributed over 162 tonnes of food to almost 55,000 people (personal communication).

FoodCloud – and their partner social enterprise FoodCloud Hubs – has emerged as the single most important charitable entity responsible for food redistribution in Ireland and describe themselves as having a "transformative impact on addressing the problems of food waste and food poverty in Ireland and internationally" (Irish Tech News, 2016). According to the FoodCloud website, in 2015, around 567 tonnes of food were donated to 325 charities across Ireland, equating to over one million meals (a measurement gauged by weight whereby *any* 0.5kg of food is considered one meal) (Wood, 2016). By October 2016, more than 3,320 tonnes of food had been distributed in Ireland and the UK (Murphy, 2016) and, by February 2017, FoodCloud had distributed "8.5 million meals" (Reeve, 2017: 57). As of March 2017, FoodCloud was working with 300 Aldi and Tesco stores in Ireland, 1200 Tesco stores in the UK and had a pilot scheme in operation with Waitrose (Ryan, 2017). However, FoodCloud is not just a matter of redistributing rising volumes of surplus food:

> Their innovative and technology-led approach to surplus food redistribution contributes to a different future, where food waste prevention is recognised as an opportunity that can save resources, create jobs, alleviate hunger, conserve water, and reduce greenhouse gas emissions.
>
> *(Reeve, 2017: 57)*

In May 2017, Tesco UK released their annual food waste figures, showing a 150% increase in charitable redistribution (Fareshare, 2017). In Ireland, the CEO of Tesco Ireland has said:

> One of my proudest acts as CEO has been to tackle food waste by working in partnership with FoodCloud [...] My One Big Idea would be that the Government should focus the Budget to support resources that will create the infrastructure to facilitate good food management. Some charities would greatly benefit from having the ability to both store and freeze the food we have pledged to provide.[5]
>
> *(Sept 26, 2016, independent.ie)*

Tesco Ireland is a founding member of FoodCloud (Tesco Ireland, 2017) and has enabled the social enterprise to "grow and grow" (Murphy, 2017). Yet, in the early part of 2017, workers in 23 Tesco stores in Ireland were on strike, protesting against revised contract terms, proposed wage reductions and "increased flexibility" in their working conditions (O'Halloran, 2017). Further, in March 2017, Tesco were heavily criticised for selling low-cost lamb by farmers' unions (Halpin, 2017), lamb which ends up as food surplus as we will discuss later. It is strikingly clear that the many injustices visible throughout the food system are not bound by geography, but have the potential to become less visible when the problem becomes part of the solution. This is explored in the next section.

The implications of charitable food redistribution in Cork

This part of the chapter rests upon research conducted within the Cork region, comprising of surveys and semi-structured, in-depth interviews with 11 organisations involved in charitable and subsidized food assistance. The decision to only interview charities and not their clients (the ultimate 'beneficiaries' of redistributed food) was based on the fact that charities decide what food they accept – albeit restricted, based on availability. All organisations interviewed are recipients of surplus food from a variety of sources. Based on this research we draw a number of observations regarding the often-overlooked implications of charitable food distribution for efforts to achieve a healthy, equitable and sustainable food system for all.

Consistent supplies

Prior to the advent of FoodCloud and FoodCloud Hubs, charitable food provisioning was indeed a rather ad hoc system. This is no longer the case. Marked by streamlined logistics characteristic of the corporate food system, charitable partners are embedded in a supply chain such that none of the surveyed entities reported ever running out of food or not being able to meet demand, at least in terms of food quantity. On the contrary, one issue that has emerged is that some charities struggle to dispose of the food which they have received. In some cases, food that is channelled through these charities acts more as a supplementary feeding programme insofar that products are handed out irrespective of need because, as with the general population, nobody likes to throw out food. Moreover, charities are also able to distribute seasonally appropriate foodstuffs, assured of a consistent supply of Brussel sprouts and even smoked salmon at Christmas, lamb at Easter and processed BBQ meats during the summer months, all of which are surplus due to retail demand. Given the nature of reliance on surplus food, charities are also enrolled to utilise new products that fail to perform as well in the market as might have been hoped. An already saturated market expected to absorb 2,000 new food products each year (Milone, 2009) may explain the crates of gluten-free cakes and other fashionable food items that arrive from time to time. In the first quarter of 2017, "64 of the top 100 donated items by Tesco were Bakery comprising 59,003 individual bakery items" (Ward, 2017). Box 18.2 provides an example of the top ten bakery items being channelled through FoodCloud to charitable partners from Tesco donations. While 'the consumers" desires for 'treats' is driving bakery innovation' (Bord Bia, 2014), these 'treats' are not always purchased and consequently end up on the tables of those without a choice in the matter.

Box 18.2 Top ten bakery items donated through FoodCloud by Tesco for the first quarter of 2017.

1. Petit Pain Small
2. White Roll
3. Jam Doughnuts 5 pack
4. Custard Doughnuts 5 pack
5. Large Ring Doughnuts 5 pack
6. Tesco Butter Croissant
7. Tesco Black Olive Roll
8. Petit Pain 4 pack
9. Hi Fibre Petit Pain
10. Cinnamon Roll

Source: Ward, 2017

Long-term use

In Ireland, food provisioning through charities is far from short-term emergency support. In the majority of cases, products supplied by charities reflect their (restricted) choices about what individuals and families will eat for weeks, months, years and, arguably in some instances, generations to come. Recipient individuals and families includes those in low-paid and precarious full-time employment, struggling families, people with mental health problems, people and families in emergency accommodation, elderly people and people with serious health issues, such as cancer, who are made aware of existing charitable food services by doctors and other front-line services. Unlike referral systems elsewhere, such as the UK, in Ireland, the process is informal and in most cases, clients are self-referred. While there is acknowledgement that some of the food being provided may not be health-promoting, the prevailing perception is that it is better than nothing.

In other cases, recipient charities are what many of the end users call 'home', a situation that can mean a lifetime diet based on surplus food. In the context of a healthy, sustainable and nutritious system, the appetite for highly processed food should be a cause for concern. However, the desire to feed the poor/elderly/disadvantaged/ill on restricted or, in some cases, no budget specific to food seems to have created an incognisant situation, one incompatible with the concept of food as a commons. In relation to the Fund for European Aid to the most Deprived (FEAD) any registered not-for-profit, community- or voluntary-based organisation is eligible to apply for assistance and their number grew to 120 in the first quarter of 2017 (Personal communication, 23 May 2017). However, in contrast to the products currently distributed through FEAD – comprising of canned, packaged and instant food – a consequence of lowest-price tendering, imagine how FEAD could work by sourcing local, seasonal and healthier foods. FEAD recipient organisations operate in areas such as family resource centres, addiction support programmes, community creches, women's refuges, food banks, homeless support services, youth programmes and senior citizens' associations. Whether or not related to the mass availability of free food, the numbers of organisations in receipt of FEAD food items has increased from 94 organisations in 2016 to 120 in the first quarter of 2017 (Personal communication, 23 May 2017).

Pressure to take what is being offered

The acquisition of foods outside that of the large food retailer surplus redistribution efforts is also prevalent in the Irish context. Established arrangements now exist with food service companies such as Kentucky Fried Chicken (KFC), as well as with independent stores, including those from the latest food craze sector in Cork, doughnut shops. For the most part, charities receiving such donations are content with these arrangements given that recipients – and donors – are contributing to avoiding waste and saving money. Yet, considering Ireland's immense health burden, one cannot help reflecting on the role of charities, state agencies and indeed CSR agendas in facilitating this burden.

There is a complex triangular relationship with acute disparities of power in play between food donors, their charitable partners and recipient households and individuals. First, charities appear to be under pressure to take whatever products that are being offered for fear of negative public perceptions if they refuse. Secondly, even if some degree of choice is afforded to the charities, in that they can say what they don't want or have the capacity to handle, it remains the case that they are choosing the food to be consumed by others and thereby shaping the diets of individuals and families. Finally, the 'beneficiaries' are receiving products that may not be wanted or even consumed; they become the final arbiter over choice of disposal: consumption or discard.

Implications for local business and the construction of a de-commodified food system

The provision of large amounts of surplus food to charities also has serious implications for local businesses in a similar way to the experience of farmers in developing countries faced with the dumping of highly subsidised agricultural products from the EU or USA in their markets. How can a local business compete with free food and what does this mean for the challenge of promoting healthy and sustainable diets? Here in Cork, charitable organisations have made direct swaps between purchasing from local butchers and bread companies to using only meats and breads obtained through FoodCloud and FoodCloud Hub. In one case revealed through interview, this was a switch from buying 10 kilos of mince per week from a local butcher to feed clients to using surplus 'cooked' mince instead. Other organisations reported that in the absence of 'free food' they would not be distributing items such as potato crisps, biscuits and yogurts.

At the time of writing, Cork's first self-proclaimed food bank is preparing to open its doors on the 31st of May 2017 to begin a mission of "fighting food poverty in Cork", thanks to their partners which include Bord Bia (the Irish government's food promotion board), FoodCloud, Tesco and FEAD (Feed Cork, 2017). With new food banks joining the 'battle' against food waste, it is clear that charitable food is no longer an ad hoc arrangement for emergency provision; rather it has become an institutionalised mechanism for disposing of surpluses from a hyper-trophic food system that has the potential to threaten the viability of local food businesses. Ultimately, this further jeopardises the transition to a de-commodified, healthy, sustainable and just food system.

Conclusions

The global agri-food system has been lauded for its scale, efficiency and capacity to deliver more calories for a lower proportion of consumer spending than ever before, which has shaped the diets of a majority of the world's population. Yet, its capacity to produce more food than retail demand has led to a rising preoccupation with food waste and reinforced criticisms regarding the inherent unsustainability of this system (Sage, 2012). Cheap food has led to cheap lives – a cycle of impoverishment necessary to keep the current capitalist system afloat (Patel and Moore, 2017) – a forgotten aspect of the charitable food system where celebration keeps the status quo in place. The creation and growth of a 'secondary' food system designed to alleviate food poverty and eliminate waste has proven to be something of a breakthrough, for not only has it enabled a 'win-win' solution, it has also sanitised the role of Big Food companies and their grand philanthropic mission. The narrative power of charitable redistribution of commodified food is therefore the first significant obstacle to re-introducing food as a commons.

Our view is that food waste and diet-related ill-health are two sides of the same coin: both originating from, and facilitated by, the massification of food. Unfortunately, the logic of the global food system is to continue to expand production, justified by the apparent mainstream consensus that this is vital to feed a world of more than nine billion by 2050. As such, the likelihood that volumes of food surplus will also grow along with the associated poverty, intensification of production and the further disappearance of small-scale local producers and grocery stores. It is also likely that the number of food banks will increase – as will their cast of client beneficiaries – as there appears no other way that such volumes of food can be absorbed while enabling the food industry to reach targets such as having 'zero food waste' without actually addressing the root cause of that waste and the inherent injustices along the food chain. This also means a growing role for entrepreneurial intermediaries able to connect donors and recipients. High-tech, one-dimensional solutions to environmental or other problems generally receive an enthusiastic welcome in policy circles as they do not disturb the broader landscape of 'business-as-usual'. Not only does this allow governments to shirk their responsibilities, it makes the case for a multi-dimensional, multi-stakeholder, local-specific and resilience-enhancing 'food as a commons' approach more challenging.

Food redistribution charities are co-developing with, and are co-beneficiaries of, a broken food system and as such will act to deepen and hide the problem of inequality. The root problem is mass production and speculation, transforming food as a biological necessity, cultural artefact and public good replete with meanings (Vivero-Pol, 2017) into a commodity increasingly stripped of its nutritional, social and ecological value. Yet, responsibility for disposing of this material gets transferred from supermarkets and other businesses, where waste is a cost on the balance sheet, to charities, staffed by volunteers, and finally to beneficiaries who are now, seemingly, 'food poverty free', but ultimately second-class citizens fed with whatever is available and responsible for dealing with food waste. There are numerous unexplored implications arising from this arrangement that will repay further interrogation. Here, however, we simply seek to challenge the widespread and simplistic narrative that presents the diversion of food surplus as a solution to food poverty.

We maintain that food poverty is neither a logistical nor a food problem, but rather a symptom of a very unequal society. With an endless supply of surplus food, giving food as opposed to vouchers or money risks becoming the new norm, further etching away at the right to good food. By continuing this trajectory, inequalities are kept hidden, government responsibility is evaded and the social, ecological and health values of food are further discounted. In this way, accepting charitable food re-distribution as a solution raises profound moral questions regarding the privatisation, corporatisation and commodification of poverty and leads us further away from acknowledging food as a basic human right. It is clear that negotiating a progressive course in the contemporary era of austerity is an enormous political challenge. However, replacing the current two-tiered food system means working toward a multi-dimensional, multi-stakeholder, locally-specific, resilience-enhancing, sustainable system where food is regarded as a commons for all.

Acknowledgements

This research was funded by an Irish Research Council Government of Ireland Postgraduate Scholarship.

Notes

1 Long-term temporary workers that lack employee benefits and rights.
2 Ireland has the highest income inequality in Europe prior to social transfers and 29% of the population are classified as suffering from deprivation and poverty. Over the last 30 years, the top 10% of earners have increased their net wealth from 42% to 54% while the bottom 50% have decreased from 12% to 5% (Hearne and McMahon, 2016).
3 Ireland's response to homelessness was to accommodate families and individuals in hotels and B&Bs, costing the state €39 million in 2016 alone, more than double that of the previous year (Irish Times, 2017).
4 FoodCloud Hubs, formerly the Bia Food Initiative (BFI), is a social enterprise founded in 2012 to collect and store large volumes of surplus food for redistribution at minimal cost to the charity. In 2016, the BFI partnered with FoodCloud and became FoodCloud Hubs (FoodCloud, 2017)
5 In October 2017, Tesco Ireland launched the 'Community Big Chill' campaign to provide free freezers and fridges to charities (Tescoireland.ie, 2017).

References

Baker, P. and Friel, S. (2016). Food systems transformations, ultra-processed food markets and the nutrition transition in Asia. *Globalization and Health*, 12(1), p.80.

Basu, S. (2015). The transitional dynamics of caloric ecosystems: changes in the food supply around the world. *Critical Public Health*, 25(3), pp.248–264.

Berry, J. (1984). *Feeding hungry people*. New Brunswick: Rutgers University Press.

Booth, S. and Whelan, J. (2014). Hungry for change: the food banking industry in Australia. *British Food Journal*, 116(9), pp.1392–1404.

Bord Bia (2014). *Consumer desire for 'treats' driving bakery innovation.* [online] http://www.bordbia.ie/industry/manufacturers/insight/alerts/Pages/Consumerdesirefortreatsdrivingbakeryinnovation.aspx?year=2014&wk=19 [Accessed 10 June 2017].
Botsman, R. and Rogers, R. (2011). *What's mine is yours.* London: HarperCollins.
Brennan, C. (2017). *One in 10 Irish workers are earning the minimum wage or less.* [online] http://www.thejournal.ie/minimum-wage-ireland-3-3359255-Apr2017/ [Accessed 7 June 2017].
Butcher, L., Chester, M., Aberle, L., Bobongie, V., Davies, C., Godrich, S., Milligan, R., Tartaglia, J., Thorne, L. and Begley, A. (2014). Foodbank of Western Australia's healthy food for all. *British Food Journal*, 116(9), pp.1490–1505.
Campbell, H., Evans, D. and Murcott, A. (2017). Measurability, austerity and edibility: Introducing waste into food regime theory. *Journal of Rural Studies*, 51, pp.168–177.
Caraher, M. and Coveney, J. (Eds) (2016). *Food poverty and insecurity.* 1st ed. Basel: Springer.
Caraher, M. and Dowler, E. (2014) Food for Poorer People: Conventional and 'Alternative' Transgressions? In: Goodman, M. and Sage, C. (eds) *Food transgressions: Making sense of contemporary food politics.* Aldershot: Ashgate.
Caraher, M. and Furey, S. (2017). *Is it appropriate to use surplus food to feed people in hunger? Short-term Band-Aid to more deep-rooted problems of poverty.* London: Food Research Collaboration. [online] http://foodresearch.org.uk/wp-content/uploads/2017/01/Final-Using-food-surplus-hunger-FRC-briefing-paper-24-01-17-.pdf [Accessed 1 June 2017].
Carolan, M. (2011). *The real cost of cheap food.* 1st ed. London: Earthscan.
Clapp, J. and Scrinis, G. (2016). Big Food, nutritionism, and corporate power. *Globalizations*, 14(4), pp.578–595.
Collins, M. (2016). Earnings and low pay in the Republic of Ireland. *Journal of the Statistical and Social Inquiry Society of Ireland*, XLV, pp.146-176. [online] http://hdl.handle.net/2262/79413 [Accessed 2 June 2017].
Constitution of Ireland - Bunreacht Na hÉireann. (1937). [online] https://web.archive.org/web/20110721123433/http://www.constitution.ie/constitution-of-ireland/default.asp?UserLang=EN [Accessed 12 June 2017].
Crosscare (2017). *Crosscare - food services [home].* [online] http://www.crosscare.ie/index.php/food-provision-home [Accessed 24 May 2017].
De Schutter, O. (2012). End of mission statement: mission to Canada. Office of the United Nations High Commissioner for Human Rights. [online] http://www.srfood.org/images/stories/pdf/officialreports/201205_canadaprelim_en.pdf [Accessed 12 May 2017].
Dowler, E. and Lambie-Mumford, H. (2015). How can households eat in austerity? Challenges for social policy in the UK. *Social Policy and Society*, 14(03), pp.417–428.
DSP (2017). Social Inclusion Monitor 2015. Dublin: Department of Social Protection. [online] https://www.welfare.ie/en/downloads/SocialInclusionMonitor2015.pdf [Accessed 30 May 2017].
FEBA (2017). [online] https://www.eurofoodbank.org/ [Accessed 7 June 2017].
FareShare. (2017). *Tesco's food waste figures show 150% increase in charity redistribution.* [online] http://www.fareshare.org.uk/tescos-food-waste-figures-show-150-increase-in-charity-redistribution/ [Accessed 1 June 2017].
Feed Cork (2017). *Home.* [online] http://www.feedcork.com/ [Accessed 20 May 2017].
Fisher, A. (2017). *Big hunger.* Cambridge: The MIT Press.
Foodcloud (2016). *FoodClouds COO enjoys "Couples Volunteering" for our Food Rescue Project Tallaght.* [online] https://food.cloud/couples-volunteering/ [Accessed 23 May 2017].
Frenken, K. and Schor, J. (2017). Putting the sharing economy into perspective. *Environmental Innovation and Societal Transitions*, 23, pp.3-10.
Friel, S. and Conlon, C. (2004). *Policy response to food poverty in Ireland.* 1st ed. Dublin: Combat Poverty Agency.
FUSIONS (2016). *Estimates of European food waste levels.* Stockholm: European Commission. ISBN 978-91-88319-01-2.
Garthwaite, K., Collins, P. and Bambra, C. (2015). Food for thought: An ethnographic study of negotiating ill health and food insecurity in a UK foodbank. *Social Science & Medicine*, 132, pp.38–44.
Halpin, H. (2017). *'This will put farmers out of business' - Tesco criticised for low prices for lamb.* [online] http://www.thejournal.ie/icsa-hit-out-at-tescos-low-prices-3391634-May2017/ [Accessed 29 May 2017].
Hamari, J., Sjoklint, M. and Ukkonen, A. (2015). The sharing economy: Why people participate in collaborative consumption. *Journal of the Association for Information Science and Technology*, 67(9), pp.2047–2059.

Hearne, R. and McMahon, C. (2016). *Cherishing all equally: Gender, children and economic inequality in Ireland.* Dublin: TASC.

Heinrichs, H. (2013). Sharing Economy: A Potential New Pathway to Sustainability. *GAIA - Ecological Perspectives for Science and Society*, 22(4), pp.228–231.

Hill, S. (2015). *Raw deal.* 1st ed. New York: Palgrave Macmillan Trade.

Horton, J. (2004). Food banking: an international perspective. *Parity*, 17(3) pp 21–22.

GFN (2017). *Corporate and foundation partners - The Global FoodBanking Network.* [online] https://www.foodbanking.org/why-support-gfn/gfn-corporate-and-foundation-partners [Accessed 27 May 2017].

IFAN (2017). *Independent Food Aid Network UK.* [online] http://www.foodaidnetwork.org.uk/ [Accessed 10 June 2017].

Independent.ie. (2017). *Andrew Yaxley, CEO Tesco Ireland - My one big idea on climate change.* [online] http://www.independent.ie/irish-news/climate-change-and-you/andrew-yaxley-ceo-tesco-ireland-my-one-big-idea-on-climate-change-35079035.html [Accessed 24 May 2017].

IPES (2017). *Too big to feed: Exploring the impacts of mega-mergers, concentration, concentration of power in the agri-food sector.* Sustainable Food Systems [online] http://www.ipes-food.org/images/Reports/Concentration_FullReport.pdf [Accessed 11 December 2017].

IRSPEN (2017a). *Almost a third of Irish children are now overweight – study.* [online] http://www.irspen.ie/almost-a-third-of-irish-children-are-now-overweight-study/ [Accessed 9 December 2017].

IRSPEN (2017b). *HSE has lost €56M by not implementing national obesity treatment policy, experts warn.* [online] http://www.irspen.ie/hse-has-lost-e56m-by-not-implementing-national-obesity-treatment-policy-experts-warn/ [Accessed 1 June 2017].

Irish Tech News (2016). Social enterprise startup insights with Food Cloud, speaking at National Digital Week. [online] https://irishtechnews.ie/social-enterprise-startup-insights-with-food-cloud-speaking-at-national-digital-week/ [Accessed 6 June 2017].

Jonnet, I. (2017). *Ensuring that safe surplus food reaches those in greatest need.* [online] http://ec.europa.eu/health/newsletter/160/focus_newsletter_en.htm [Accessed 1 June 2017].

Kenny, T., Cronin, M. and Sage, C. (2018) A retrospective public health analysis of the Republic of Ireland's Food Harvest 2020 strategy: absence, avoidance and business as usual. *Critical Public Health*, 28(1) pp. 94–105.

Lambie-Mumford, H. and Dowler, E. (2014). Rising use of 'food aid' in the United Kingdom. *British Food Journal*, 116(9), pp.1418–1425.

Mann, J., Swinburn, B., Beaglehole, R., Mhurchu, C. and Jackson, R. (2015). Diverging global trends in heart disease and diabetes: implications for dietary guidelines. *The Lancet Diabetes & Endocrinology*, 3(8), pp.584–585.

Mansuy, J. and Ferrando, T. (2017). Food Waste at Time of Food Poverty: Are the EU and Its Member States Providing Sustainable Legal Solutions?. *SSRN Electronic Journal.*

Martin, C. (2016). The sharing economy: A pathway to sustainability or a nightmarish form of neoliberal capitalism?. *Ecological Economics*, 121, pp.149–159.

McGrane, M. (2015). *One year ago I joined a food revolution and started volunteering with FoodCloud. Here's what I've learned.* ChangeX Blog [online] https://www.changex.org/blog/one-year-ago-i-joined-a-food-revolution-and-started-volunteering-with-foodcloud-heres-what-ive-learned/ [Accessed 25 May 2017].

Monteiro, C., Moubarac, J., Cannon, G., Ng, S. and Popkin, B. (2013). Ultra-processed products are becoming dominant in the global food system. *Obesity Reviews*, 14, pp.21–28.

Milone. (2009). *Agriculture in transition. A neo-institutional analysis.* 1st ed. Assen: Van Gorcum & Comp. p.61.

Morozov, E. (2013). The 'sharing economy' undermines workers' rights. *Financial Times.* [online] https://www.ft.com/content/92c3021c-34c2-11e3-8148-00144feab7de [Accessed 16 May 2017].

Murphy, M. (2016). *FoodCloud Partnering with Bia Food Initiative.* Food For Thought. [online] http://www.fft.ie/foodcloud-partnering-with-bia-food-initiative/13889 [Accessed 23 May 2017].

Murphy, N. (2017). *VOOM Spotlight: FoodCloud: The Irish Continue To Lead The Way | Virgin Media Business Blog.* Virgin Media Business Blog. [online] http://vmbiz.ie/blog/voom-spotlight-food-cloud/ [Accessed 12 June 2017].

Nugent, C. (2017). *A time-series analysis of precarious work in the elementary professions in Ireland.* NERI WP 2017/No 43. [online] http://www.nerinstitute.net/download/pdf/a_timeseries_analysis_of_precarious_work_in_the_elementary_professions_in_ireland_may_17.pdf [Accessed 2 June 2017].

O'Halloran, M. (2018). Tesco contract terms criticised as Scargill joins picket. [online] https://www.irishtimes.com/news/ireland/irish-news/tesco-contract-terms-criticised-as-scargill-joins-picket-1.2980879 [Accessed 18 May 2017].

Patel, R. and Moore, J. (2017). *A history of the world in seven cheap things: A guide to capitalism, nature, and the future of the planet.* Oakland, CA: University of California Press.
Pearson-Stuttard, J., Bandosz, P., Rehm, C., Penalvo, J., Whitsel, L., Gaziano, T., Conrad, Z., Wilde, P., Micha, R., Lloyd-Williams, F., Capewell, S., Mozaffarian, D. and O'Flaherty, M. (2017). Reducing US cardiovascular disease burden and disparities through national and targeted dietary policies: A modelling study. *PLOS Medicine*, 14(6), p.e1002311.
Poppendieck, J. (1999). Sweet charity? Emergency food and the end of entitlement. Harmondsworth: Penguin.
Poppendieck, J. (2014). Food assistance, hunger, and the end of welfare. In: G. Riches and T. Silvasti (eds), *First world hunger revisited: Food charity or the right to food?*. 2nd ed. London: Palgrave-Macmillan, pp.176–190.
Purdam, K., Garratt, E. and Esmail, A. (2016). Hungry? Food Insecurity, Social Stigma and Embarrassment in the UK. *Sociology*, 50(6), pp.1072–1088.
Reeve, M. (2017). Feeding the nine billion. *Journal of the Institution of Environmental Sciences*, 26(1), p.57. [online] https://www.the-ies.org/sites/default/files/journals/es_feeding_nine_billion_feb_17.pdf [Accessed 7 June 2017].
Riches, G. (1997). *First World hunger.* 1st ed. New York: St. Martin's Press.
Riches, G. (2011). Thinking and acting outside the charitable food box: hunger and the right to food in rich societies. *Development in Practice*, 21(4-5), pp.768–775.
Riches, G. and Silvasti, T. (eds) (2014). *First world hunger revisited: Food charity or the right to food?*. 2nd ed. London: Palgrave-Macmillan.
Riches, G. and Tarasuk, V. (2014). Canada: Thirty Years of Food Charity and Public Policy Neglect. In: G. Riches and T. Silvasti (Eds), *First World Hunger Revisites*, 2nd ed. London: Palgrave-Macmillan.
Roberto, C., Swinburn, B., Hawkes, C., Huang, T., Costa, S., Ashe, M., Zwicker, L., Cawley, J. and Brownell, K. (2015). Patchy progress on obesity prevention: Emerging examples, entrenched barriers, and new thinking. *The Lancet*, 385(9985), pp.2400–2409.
Ronson, D. and Caraher, M. (2016). Big Society or Shunting Yards? Successful Failures. In: M. Caraher and J. Coveney (eds) *Food poverty and insecurity: International food inequalities.* Switzerland: Springer.
Ryan, O. (2017). *'One million tonnes of food are wasted in Ireland each year, people know that's wrong'.* [online] http://www.thejournal.ie/food-cloud-social-entrepreneurs-ireland-3299027-Mar2017/#comments [Accessed 23 May 2017].
Reilly, G. (2010). *Let them eat free cheese: Agriculture minister's Christmas gift to poor.* [online] http://www.thejournal.ie/let-them-eat-cheese-agriculture-ministers-christmas-gift-to-poor-43062-Nov2010/ [Accessed 9 December 2017].
Sage, C (2012) *Environment and Food.* London: Routledge.
SJI (2016). *Social Justice Ireland policy briefing: Poverty, deprivation and inequality.* Dublin: Social Justice Ireland. [online] https://www.socialjustice.ie/sites/default/files/attach/publication/4471/2016-07-04-sjipolicybriefingpoverty2016final2.pdf [Accessed 7 June 2017].
SJI (2017). *Median incomes increase but numbers in poverty still worrying.* [online] https://www.socialjustice.ie/content/policy-issues/median-incomes-increase-numbers-poverty-still-worrying [Accessed 7 June 2017].
Shannon, J. (2017). *Diabetes costs the State almost €70 million a year.* Irish Medical Times. [online] https://www.imt.ie/news/diabetes-costs-the-state-almost-e70-million-a-year-18-01-2017/ [Accessed 9 December 2017].
Stuckler, D. and Nestle, M. (2012). Big Food, food systems, and global health. *PLoS Medicine*, 9(6), p.e1001242.
Sundararajan, A. (2017). *The sharing economy.* 1st ed. Cambridge: The MIT Press.
Tarasuk, V. and Davis, B. (1996). Responses to food insecurity in the changing Canadian welfare state. *Journal of Nutrition Education*, 28(2), pp.71–75.
Tarasuk, V. and Eakin, J. (2003). Charitable food assistance as symbolic gesture: an ethnographic study of food banks in Ontario. *Social Science & Medicine*, 56(7), pp.1505–1515.
Tesco Ireland (2017). *Tesco Ireland – No good food will go to waste in Tesco Ireland by 2020.* [online] http://www.tescoireland.ie/news/news/article/no-good-food-will-go-to-waste-in-tesco-ireland-by-2020 [Accessed 11 December 2017].
van der Horst, H., Pascucci, S. and Bol, W. (2014). The "dark side" of food banks? Exploring emotional responses of food bank receivers in the Netherlands. *British Food Journal*, 116(9), pp.1506–1520.
Vivero-Pol, J. (2017). Food as commons or commodity? Exploring the links between normative valuations and agency in food transition. *Sustainability*, 9(3), p.442.

Ward, I. (2017). Ireland's forum on food waste. Food Waste Charter. [online] http://foodwasteforum.ie/wp-content/uploads/2017/04/Iseult-Ward-FoodCloud_Forum-on-Food-Waste-2017.pdf [Accessed 6 June 2017].
Welfare.ie. (2017). *FEAD - The Fund for European Aid to the most Deprived.* [online] https://www.welfare.ie/en/Pages/FEAD---EUROPEAN-AID-TO-THE-MOST-DEPRIVED.aspx [Accessed 12 May 2017].
Winson, A. (2004). Bringing political economy into the debate on the obesity epidemic. *Agriculture and Human Values*, 21(4), pp.299–312.
Wong, K. (2017). *Tackling food waste around the world: our top 10 apps. The Guardian.* [online] https://www.theguardian.com/sustainable-business/2017/feb/06/food-waste-apps-global-technology-leftovers-landfill [Accessed 25 May 2017].
Wood, Z. (2016). Tesco food waste rose to equivalent of 119m meals last year. *The Guardian.* [online] https://www.theguardian.com/business/2016/jun/15/tesco-food-waste-past-year-equivalent-119-million-meals [Accessed 31 May 2017].
Yaraghi, N. and Ravi, S. (2017). The Current and Future State of the Sharing Economy. IMPACT Series No. 032017. Brookings India. [online] https://www.brookings.edu/wp-content/uploads/2016/12/sharingeconomy_032017final.pdf [Accessed 29 May 2017].

19

COMMUNITY-BUILDING THROUGH FOOD SELF-PROVISIONING IN CENTRAL AND EASTERN EUROPE

An analysis through the food commons framework

Bálint Balázs

Introduction

'Grow-your-own' projects and community-based food self-provisioning (FSP) are nowadays regarded as vital components of alternative/local food systems (Balázs, 2012; Wiskerke and Viljoen, 2012). The growing popularity of community-based food self-provisioning has already attracted significant scholarship on various levels (national, regional, household, individual) on the topic, with the adoption of different perspectives (micro and macro) and disciplinary foci (Czegledy, 2002; Eyzaguirre and Bailey, 2009; Schmutz et al., 2014; De Hoop and Jehlička, 2017). This literature defines community-based food self-provisioning as the production and distribution of food by means other than buying and selling: in other words, a non-market distribution of local foods. It is accomplished primarily by gifting and bartering. It is informal, with no written agreements relating to the transactions. Beyond self-supply, barter and gifting of the surplus are central to FSP as practised by gardeners and non-farming households on private plots of land, fields or orchards (Jehlička and Smith, 2012).

In community-based food self-provisioning, the 'self' relates to the broader community, not the individual person or household; therefore, FSP is community-based in the sense that there is direct contact between producers and consumers. In contrast, when food is treated as a commodity, the two are separated by distance and by a number of intermediate agents. Producer-consumers in FSP remain outside of the food market, and food in this context is not a commodity and has no economic value as a commodity. Furthermore, FSP creates new and strengthens old communities that go beyond family, friends and neighbours. The producer-consumers rely on their own resources (time, money, organizational and learning capacities) to support such activities. As for motivation, these transactions may be inspired by the desire to diversify the diet, although the one key reason is the desire to compensate for the scarcity of healthy and fresh food at the local level (Balázs, 2016).

Political discourses citing the benefits of urban agriculture are gaining momentum and some governments are paying attention to community-based food self-provisioning (Balázs, 2016). Some green advocates and environmental organizations do not seem to value FSP (De Hoop and Jehlička, 2017). Its socio-economic significance is often ignored in the academic literature (Balázs, 2016). However, in contrast to most of the European Union, FSP in Central and Eastern Europe (CEE) remains an important component of customary food practices and accounts for a substantial share of the food supply (Alber and Kohler, 2008). About 40 percent of the non-farmers in CEE grow some of their own food, mostly vegetables and fruits, compared to just one-tenth of the non-farmers in Western European countries (Alber and Kohler, 2008; Jehlička and Smith, 2012).

The major purpose of this chapter is to show how FSP enables food as a commons and helps to build strong communities. The first part presents a detailed case study on the participants' motivations to undertake FSP, which is based on interviews conducted in Hungary and which is used to point out how FSP creates communities. The second part explores food as a commons from various perspectives.

FSP as unintended sustainability

Visions of sustainability do not neglect the topic of food; self-provisioning, local, fresh and seasonal food consumption seems to be the most exciting cross-cutting themes for a range of stakeholders (Carlsson-Kanyama et al., 2008; Király et al., 2013; Quist and Vergragt, 2006). However, it is only recently that FSP in post-socialist societies has been depicted as a socially inclusive practice compliant with the principles of sustainability. In general, the community and collective aspects of FSP – essentially a social and not an isolated individual activity – have been neglected in portrayals that emphasize the atomization of life in socialist and post-socialist CEE. The work of Petr Jehlička and his co-authors on FSP has demonstrated how gardeners contribute to sustainability without intentionally seeking to do so. Critically re-interpreting the study by Alber and Kohler (2008), Jehlička et al. (2012) contended that FSP in Central and Eastern Europe is a socially inclusive middle-class activity not generated by the lack of a sufficient food supply or as a survival strategy of the poor. Their research clearly shows how previous studies depicted 'these practices as backwards and contrast[ed] them with Western modernity. [...] Food self-provisioning is read as an index of path dependency, an economic coping strategy or as a faintly embarrassing cultural remnant' (Jehlička and Smith, 2012, pp. 79–80). Instead of being a survival strategy of the poor, a practice forced by necessity, they argued, FSP is performed by economically secure rather than insecure social groups as 'a complex bundle of practices that hold cultural, social significance that far outweighs economic explanation' (Jehlička and Smith, 2012, pp. 79–80).

Allotment gardens and urban agriculture can thus be seen as elements of a common urban food system based on the idea of food as a commons. These neo-traditional practices are significant and have strong potential: Sovová (2016), in her Brno Allotment Gardens study, points out that one-third of households' fruit and vegetable consumption can be covered from their urban gardens, though their environmental impact is unreported. The challenge was to determine the amount of food produced by gardeners and analysts used different methods, such as crowdsourcing data from citizen scientists (Gittleman et al., 2012) or extrapolating from the actual survey data (Sovová, 2016).

In their mixed methods research, Jehlička and Smith (2012) showed for the first time that the main motivation behind self-production is to access fresh and healthy food by enjoying a meaningful hobby and at the same time saving on food. As gardening experience is not independent of the knowledge of the current food system, allotments clearly compare their gardening achievements and production efficiency to smallholders on the market (Sovová, 2016). While environmental motivations have been very rarely reported, self-reported pesticide and fertilizer usage are very low; therefore, such sustainable food practices have been theorized as 'quiet sustainability' in a Central European context because of their non-intentional sustainable way of life (Smith and Jehlička, 2013). The same authors have defined FSP as:

> practices that result in beneficial environmental or social outcomes, that do not relate directly or indirectly to market transactions, and that are not represented by the practitioners as relating directly to environmental or sustainability goals ... Theirs is not a fulfilment of environmental obligations, an attempt to achieve 'resilience', or a response to limits, but the daily practice of a satisfying life. In other words, it is not just that the journey to sustainability is less difficult than is sometimes presented – large sections of humanity may already be on it without the need to proclaim the fact loudly.
>
> *(Smith and Jehlička, 2013)*

This definition highlights the fundamental importance of seeking a satisfying life. The key point that is missing from this is that people everywhere feel the need to do this together.

As the Central and Eastern European agri-food context is characterized by weak civic engagement and distrust among producers and consumers, FSP practices that renew knowledge about food can offer an adequate answer to global sustainability challenges, while at the same time building or rebuilding communal ties. To counterbalance the Western conceptualizations of FSP as an opportunity to expand food activism and foster social justice and environmental sustainability, De Hoop and Jehlička (2017) present post-socialist FSP as an environmentally friendly lifestyle completely undervalued by policy and environmental NGOs.

Motivations for regarding food as a commons in Hungary

To offer a concrete example of the general spatial-social context of household FSP practices informing the understanding of FSP as part of the food-as-a-commons discourse, we now turn to a mixed methods study from Hungary, a Central and Eastern European country still in its post-socialist cultural transformation phase. The case presented below explores the motivations for provisioning food from one's own produce and local food purchases as well as market and non-market exchanges as experienced by producer-consumers.

The analysis is based on a case study by the author exploring initial motivations for FSP (Balázs, 2016), incorporating data from a representative (omnibus) survey[1] of 1,200 personal interviews in November 2013 and qualitative interview data from 2012–2013 based on a snowball sampling of self-provisioners. The empirical evidence was collected and analysed in three stages. First, a total of ten semi-structured interviews with people active in FSP were analysed and examined in the light of two policy-maker and two expert interviews. Interview participants were identified through personal contacts and mirrored the heterogeneity of stakeholders in local food communities. Interviews lasted between 60 and 90 minutes, took place in urban settings and in the interviewees' own environment (Patton, 2005). All interviews were tape-recorded and transcribed verbatim before data analysis. The data analysis used thematic coding, a combination of meaning condensation, categorization and meaning interpretation of the relevant themes (Kvale, 1996). For further context and cross-checking the findings, data was collected through desk research (newspaper articles and advocacy literature) and further interviews with policy makers and experts. Finally, a national-level representative attitude survey was carried out in 2013 with 1,200 personal interviews. Data analysis used crosstabs to examine relationships within the data.

Several questions concerned food as a commons, such as: what kinds of socio-economic relationships emerge among people who share interests, resources, knowledge and values in household food production? How do practitioners experience benefits? What type of food community is created through these practices? How could the development of policies help gardeners to create strong bonds for food as a commons?

Collectivity and community-building in food self-provision

The following typical statements, taken verbatim from the interviews, illustrate the main value dimensions related to food commons.[2] The interviews show that FSP activities represent an integral part of these people's everyday lives and demonstrate the essentiality of food self-provisioning. Socio-economic benefits, individual motivations and collective values appeared as highly interlinked. FSP gardeners determine that they cannot afford to feed their families with certified organic food. On the contrary, the main inspiration behind FSP is keeping a family tradition, a practice that, according to some interviewees, brings related economic benefits and self-fulfilment:

> We have always been gardening since my childhood – I remember my parents and grandparents keeping animals, cultivating the land. This is a family tradition if one could say so. When we had a bigger garden, we could produce all kinds of fruits and vegetables so that we could preserve [them] and make marmalade. Altogether this resulted in economic benefits and the feeling that the food contains our work, which gave an extra emotional fulfilment. During the years, this has changed, since we have less land available, and we only keep the emotional aspect.(56-year-old social worker)

Beyond sharing the harvest, an important point for personal motivation is a reconnection with friends, family and the villagers:

> My mother is working like crazy, and we all follow her lead. However, when it is harvest time the whole family comes to help. … With the sour cherries, we even organize 'pick your own' campaigns. Moreover, sometimes when there is enough harvest we ask workers from the village to come. Grandma is picking cherries, even in her 83rd year. I usually climb the tree. And then somebody comes to carry it away, or we take it to the pick-up points…
>
> *(Marketing assistant from Budapest, aged 26)*

When relatives come to help, gardening helps to strengthen family bonds too:

> On weekends the grandchildren are coming to pick beans and apricots, they always come to help pick the fruit that they like... I could not do it alone, and they take the produce they gather... Boys do the spraying, and I do the hoeing. … I also keep chickens, there are seven left. I had 40 before, but I had to process and deep freeze [most of] them. My grandchildren also got some of this… formerly I had some pigs too...
>
> *(Pensioner, aged 72)*

Sharing or bartering the harvest creates a collective sense of fulfilment that goes beyond the family when friends and neighbours are invited to pick their own. Existing bonds are often strengthened not only by the exchange of advice but also the sharing of produce:

> We usually exchange sour cherries for strawberries. All this goes on in a friendly way as an exchange economy. As we have a large surplus…
>
> *(Budapest resident, aged 26)*

FSP also stimulates informal and experiential lifelong and life-wide learning such as, for example, effective means of weed control:

> I do not belong to any producer group… I do everything alone here… Routine advice from neighbours, and a lot of specialist reading. Everything is already documented quite well in books. There is a great professional literature, and sometimes I also have hobby gardening books. It is possible to learn these skills. Some amateurs are reporting on weed control… and now I have weeds everywhere. Imagine that I have to bend two million times to do the weeding. It is tough going.
>
> *(Besnyő resident, aged 80)*

> I am an urbanite – from Budapest. I spent much time in the countryside in my childhood. So I kept those memories, and also talked a lot with people and read the books. Much experience. Things were always changing a bit, but the basics stayed the same. In watering, we use the ancient method of digging a watering row next to the plants.
>
> *(Besnyő resident, aged 80)*

Through FSP, gardeners can create necessary reserves for the whole year:

> If we have a surplus, we first deep-freeze. Jams are nice, especially from apricot. Sour cherry also gives enough to preserve some. Now I am waiting for the blackberry and elderberry and will mix these three. Yum, it is so delicious.
>
> *(Szeged resident, aged 72)*

> I usually deep-freeze the beans and peas, as thawing frozen peas is good. All the rest, parsley, carrots and potatoes, go in the cold chamber to stay fresh. Onions and peppers go for pickles, as the family loves that and they usually take some home. Beetroot is pickled, being very healthy. Sour cherry goes for jam and preserves, likewise for strawberry if we have enough.
>
> *(Barcs resident, aged 78)*

> The surplus is preserved and frozen; every family member always has some in the fridge. We mostly consume fresh [produce] in huge quantities and cook and preserve. We store in all possible forms, including pálinka.
>
> *(Budapest resident, aged 26)*

> When I cannot use the fresh produce immediately, or it is not so good for preserving, I give it to the family members.
>
> *(Mosonmagyaróvár resident, aged 56)*

In multiple ways, food self-provision practices are producing food as a commons.

Community-based food self-provisioning is organized by households and closely tied to producer-consumer communities. The material element of the resource is the food, whereas the intangible elements are the symbols, skills, values and norms created around the use of such food. In what ways is food self-provisioning regarded as a commons? In this section, the six dimensions of food as a commons (after Vivero-Pol, 2017) and their intersections with FSP will be explored.

Food self-provisioning as production of essential material for survival

As a necessary *subsistence* for humans, food produced by gardeners for their own consumption, as well as for bartering or gifting, carries moral significance. This *materiality and essentiality of nourishment* has been highlighted by the economic development policy approach to FSP. Economic development policy in the 1980s represented FSP as a sign of scarcity. First of all, to illustrate the 'underdevelopment' of the economy, studies highlighted informality and the hidden character of such practices (e.g. Rose and Tikhomirov, 1993; Alber and Kohler, 2008). They also established that a high proportion of urban as well as rural residents were growing food, but, in the context of the Central and Eastern European transition to the market economy, such activities were considered as a purely economic practice, a backwards form of household livelihood strategy and a temporary survival mechanism under the stressful conditions of transition. Such Western-framed accounts formed the theoretical framework for constructing surveys and collecting data, especially in the work of Rose and Tikhomirov (1993) on urban peasantry. Surveys in post-Soviet Union Russia, Bulgaria, the Czech Republic, Slovakia and Poland concluded that farmers were

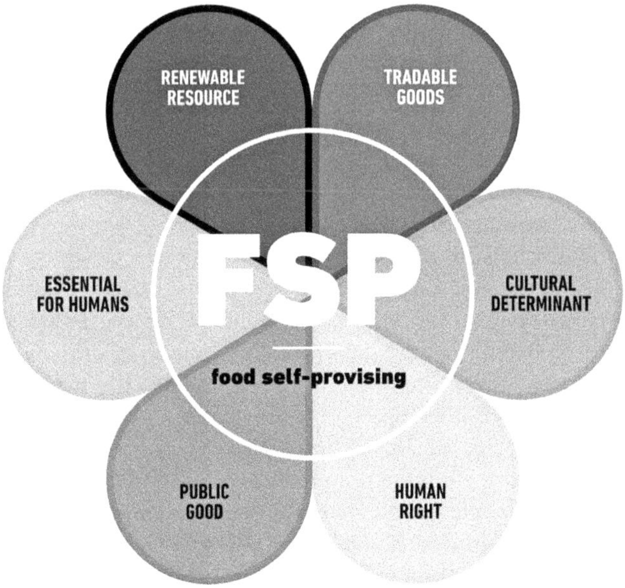

Figure 19.1 Factors of FSP enabling food as a commons.

Source: own figure based on Vivero-Pol (2017).

only a minority of all food producers. These studies also pointed to the high proportion of home growing in urban household plots and the liberal policies in generating sales of food from home as problems to be solved. Similarly, Seeth et al. (1998) dealt with food self-provisioners as victims of the economic transition, whereas Alber and Kohler (2008) coined FSP as a coping strategy in reaction to experienced difficulties in transition countries.

In sum, these studies interpreted FSP as an act of securing food supplies in response to economic hardship, a purely economic action determined by scarcity and deprived of any other non-commercial value. Such analyses used the influential interpretative framework of the Hungarian economist János Kornai as a basis for their approach in depicting frequent, intensive and chronic shortages of basic consumer goods (such as food) as overall systemic flaws of the centrally planned command economies (Kornai, 1980). Interestingly, research building on the economic development policy approach also identified a surprising element of FSP activities: social de-differentiation of urban and rural residents cultivating their households' plots. Still, the fact that FSP practices were non-market activities was interpreted as a sign of 'de-modernization' and associated with economically uncertain times. FSP was nothing but a household-level income-generating strategy that relied on work beyond a 'proper' first job (Bühler, 2004). Despite revealing the secret presence of such activities in official statistics, these economic studies could only perceive the traditional elements of employment, jobs creation and their market value. Although, they found associations (without causality) between family extension (birth of a second child) and food growing and income-generation activities, they did not see them as relevant, as they could not be represented in terms of exchange and Gross Domestic Product.

Food self-provision as culturally determined practice

The *cultural determinant* element of FSP can be traced in its diverse cultural meanings for gardeners that completely transgress market principles. Gardeners mostly start food provisioning activities through their access to small plots of land. FSP has turned out to be such an integral part of a convivial social life that it often leaves analysts puzzled. It is in this way that FSP allows community values to emerge around food and reinforces the meaning of good food for humans. In particular, FSP shapes the individual identity of self-provisioners by creating interpersonal, informal belonging to, or exclusion from, the produce network. One impact that has been observed is the strengthening of social bonds with relatives and friends and the building up of new relations among self-provisioners. These aspects go beyond market exchange have been best highlighted by views from *anthropology, ethnography* and *micro-sociology*.

In the case of FSP, studies on the motivational background have looked at how households created their economic sovereignty by relying on various strategies. One specific insight linked to the idea of food as a commons emerged from studies discussing FSP as attempts by the households to re-appropriate their food supply. For example, Smith (2003) analysed the symbolic meaning-making around home-produced food in Hungary as:

> It is considered and was considered better because the products of one's labour are made for the individual's satisfaction; they are not made for sale; therefore, they are not produced for money. They are something made at home, but are not so much required, but desired. (…) *Házi [homemade] food*, grown in the garden, raised in the yard, harvested, tended to, slaughtered, cooked, and served with special care and attention by family members is praised as a kind of an end in itself. It is functional only in that it brings people together in solidarity, enhances dignity, and produces individuality. It is not a way of organizing economic life like in the peasant cultures of the past, although

> it comes from a peasant mentality; it is an institution in opposition to the modernist state. The fact that many people make jam out of fruit, bacon out of their pigs, and pickled salads out of their vegetables to reduce their shopping budgets does not diminish the cultural meaning of *házi* foods.
>
> *(Smith, 2003, pp. 179–180)*

In this context, FSP appeared as a subcultural escapist move, an exit strategy from the communist regime. Nevertheless, the view from these anthropological studies also depicted FSP as a social practice related to the reaffirmation of (obligatory) family relations and cultural identity. Food in these studies was not only a good that moved around society but was characterized as food from *somewhere in particular*, a supplement for household food consumption.

Within this framework, Gibson-Graham (2006) divides commercial practices into market, alternative market (off-the-books, barter) and non-market (e.g. gift-giving, subsistence), while organizing labour practices into waged, alternative paid (cash-in-hand, reciprocity) and unpaid (family care, self-provisioning). Reciprocal exchange also proved to be the most accessible form to develop social networks (Sik, 1988). In post-Soviet Ukraine, Williams et al. (2012) used household surveys to map the persistence of this diversity of *everyday community economies* beyond market hegemony. Through research on post-socialist household work practices, authors have concluded that non-commodified financial practices were not some traditional, stagnant, declining, backwards and marginal sphere of business, but instead were alive and well, and even growing (Williams, 2005.; Williams et al., 2012). In particular, Czegledy's anthropological essay on urban peasants highlights the role of FSP as part of 'the changing set of demographies which link residence, employment, entrepreneurial behaviour and leisure activities with cultural notions of identity and self-provisioning ideas central to domestic life' (2002, p201). Similarly, Torsello's (2003) ethnographic study in Slovakia presents FSP as part of a 'subsistence economy,' a universal term created by classical anthropological literature on household economies. It serves only a portion of consumption, which gains special symbolic value through the food's origin, and contributes to strengthening social ties. FSP reflects people's relatedness, as it is underpinned by trust-based links which convey the rules of behaviour, obligation and expectations. Acheson (2007) studies household-level exchange networks in post-socialist Slovakia, highlighting the role of informal societal norms around caring for family members and related to extended and non-family members by gifting. Finally, Aistara (2015) shows how EU food-hygienic regulations restrict the processing and sale of homemade foods and thus contribute to the stigmatization of informal social networks and local practices as backwards.

Food self-provision from a human rights perspective

FSP also reinforces the concept of *culturally appropriate food* that challenges the commodification of food and therefore can be linked to the fulfilment of food as a *human right* (Kent 2008). Various modes of exchange, barter and gifting illustrate the various moral meanings that food can take for people, beyond being a commodity. Along these lines, *rural sociologists*, in their descriptions of agricultural restructuring in Central and Eastern European contexts, have adopted the notion of a duality of market and non-market production, as well as of the gradation of small-scale food production (Heinonen et al., 2007). In these reconstructions, farming families position themselves on the market as sellers or buyers or as outside the market, remaining self-sufficient. Analysts have pointed out how farming and non-farming households are embedded in their socio-economic networks, and how such social networks contribute to rural welfare, creating various path-making opportunities (e.g. in the non-farm economy, around environmental public goods). In rural Hungary, as Brown and Kulcsár (2001) show, household economic behaviour is dependent on food self-provisioning and sharing. Characterized by labour force attachment and heavy reliance on social welfare programmes, rural households derive a significant share of their food consumption from self-provisioning, and over half of the surveyed households produced and received at least one type of good or service outside of the market through inter-household exchange (Brown and Kulcsár, 2001).

From a human rights perspective, FSP is an elementary and empowering component of human development. It refocuses the debate onto the policies responsible for unjust food distributions and points to inequalities in our food system and to the entirely discriminatory practices of food markets. The term FSP is often used for self-sufficient farming households that are producers and consumers of their total agricultural production. Moreover, it is revealing that the concentration of such subsistence and semi-subsistence farming in Central and Eastern European countries is, in fact, outstanding; they account for four-fifths of semi-subsistence farms in the whole EU (Davidova et al., 2013). The high level of agricultural engagement of these societies has previously been studied by Kelemen and Kovách (2005), in a survey that same year, which found that around one-third of the Hungarian population is engaged in an agricultural activity or maintains access to a small piece of land one way or another. These producer-consumers are granted the possibility to farm thanks to this access to land, an insight that entered policy discussions in Hungary through the work of Davidova et al. (2013); however, even rural sociology faces some problems when looking at FSP, especially since adequate data is lacking; food production that takes place on plots that are smaller than the set size thresholds for data collection is excluded from official statistics. In essence, not much evidence is available about farming households' production activity or attitudes towards and motivations for food-self provisioning (Davidova et al., 2013). Megyesi (2016) presents estimations of farmers' self-provisioning by subtracting the number of farms that receive single area payments from the total number of farm units.[3] The difference is farms producing on less than 1 hectare – just enough for self-provisioning in the CEE context.

Food self-provisioning as a renewable resource for agri-food systems

As a contribution to a *natural and renewable resource*, FSP has created a unique eco-agri-food system for healthy plant cultivation in varied spatial settings and shaped by changing human dietary preferences. Agri-food studies' interest in FSP as part of an eco-agri-food system has led to the idea that FSP can reduce the negative or enhance the positive externalities of the food systems (Lang, 1999; Kortright and Wakefield, 2011). Much effort has been dedicated to transforming the farming sector through incentives, policies, subsidies and institutional-communicational measures towards an agricultural model that does not create externalities. At the same time, several environmental campaigns try to nudge consumers to turn towards sustainable lifestyles. All these solutions point to the growing acknowledgement of the current unsustainability of the agri-food system, a realization which would also value an approach that through unintentional practices can contribute to sustainability transition. However, little is said about the role of producers and the impact that non-market mechanisms have in improving the sustainability of the food system. Gardeners maintain agrobiodiversity by bringing back long-forgotten seed varieties and produce without environmental harm from pesticides or soil erosion. Through integrated management, home gardeners rely on local resources, enhance soil life, value ecosystem services and maintain agrobiodiversity (Birol et al., 2005; Eyzaguirre and Bailey, 2009; Schupp and Sharp, 2012; Calvet-Mir et al., 2016).

The socio-economic benefits can range from the capability for self-reliance to providing meaningful work for people who are less active in the job market (Landon-Lane, 2004). They also provide a healthy, active lifestyle and food for a broad range of individuals, decreasing the higher transaction costs of food provisioning in marginal price-sensitive communities and creating social cohesion and community well-being (Galhena et al., 2013; Schmutz et al., 2014).

Food self-provisioning as a public good

The *public good* dimension of FSP is linked to governmental/municipal public policies or citizens' collaborative practices, beyond individual, private interests. The broad spatial-social extent of FSP practices and the collective values around the self-production of food and sharing have inspired the recent studies by Jehlička and his co-authors, which point out the clear sustainability benefits of FSP and its relevance to alternative agri-food networks and food citizenship – in fact, they see it as a prerequisite to food democracy (Sage, 2014).

Growing one's food can also be described as a remarkable adaptive capacity, a component of *civic food systems* and a non-market source of local foods (Renting at al., 2012). Several studies position FSP in the context of *alternative (agri-)food networks* (AFNs, including market-based and community-based variants) and focus on how growing one's own food is typically a non-market-based activity closely tied to the consumption of fresh, seasonal, local food (Benedek and Balázs, 2016). In their exploration of the social bases of home gardening, Schupp and Sharp (2012) urge the positioning of FSP within the context of alternative food networks. At the household level, it is also apparent that local food advocates' well-intentioned slogans to 'buy local' or 'grow your own food' are not simple transactions; rather, such practices must be considered within the broader food-provisioning context (McIntyre and Rondeau, 2011). Whereas much of the work on AFNs deals with commodities valued by their price in the market (value-in-exchange), FSP cultivates value-in-use of food.

Also relevant here is *food security*, a perspective that analysts often adopt to emphasize that cities have differing degrees of food self-provisioning capacity, but rarely enough to satisfy their own food supply – especially in wealthy capital cities (Porter et al., 2014). In a county-level empirical comparative research study on Hungary, the Czech Republic, Scotland, the Netherlands and Germany, Vávra, et al. (2013) contend that resilience is a crucial concept in determining the sustainability achievements of FSP. From a food governance point of view, FSP activities are the essence of civic food networks (Renting et al., 2012). Self-sufficiency practices – whether obtaining healthy, fresh food, continuing a family tradition or maintaining a hobby – rely on a specific socio-economic basis. Similarly, seed exchange networking by non-monetized mechanisms has been promoted in the interests of self-reliance and alternative food provisioning (Balázs et al., 2015). Adaptations of such self-sufficiency practices and preferences for sustainable land management approaches extend to urban gardeners producing their food (Balázs, 2016).

Food self-provisioning as tradeable good

FSP creates solidarity and community as much as it goes beyond the market and trading. Food self-provisioners also rely on tradeable food commodities. Thus, analysts distinguish several types of FSP, based on their locations and the underlying motivations. There is an important distinction between traditional rural subsistence activities and more urbanized food self-providers. *Traditional* peasant gardens are cultivated in the backyards of farmers' houses to help meet food needs, while *leisure* farms are weekend houses or second homes of mainly urban dwellers who produce food more as a hobby than to meet serious needs (Kovách, 2012; Smith and Jehlička, 2012). *Subsistence farming*, which can be defined as marketing at most a minimal share of the produce, is getting more attention. Subsistence farmers have the most to lose through increased regulation, but the least power to influence agricultural and international trade policies (Davidova and Bailey, 2014; Fredriksson et al., 2016). A postmodern/neotraditional version of FSP, city allotment sites – also endangered by urbanization – have been studied by Sovová (2016) in the CEE context. The food sovereignty approach to FSP can be traced to the practice of *social cooperatives* and *municipal public work* in agriculture. Participants are highly motivated, but prioritized motivations differ: some prefer self-provisioning food; others to sell the surplus; while others are clearly aiming at market trade (Asztalos Morell, 2014).

Increasing policy attention to food quality, health and governance has triggered prolific discourses on sustainable consumption and diets. The reconceptualization of FSP has also profited from such *consumption research*, which previously recorded the shift of the consumer from the subject of the market to the promoter of meaningful social change through everyday ethics, learning and reflexivity. The concepts of 'post-consumerism' and 'prosumers' have been introduced to refer to a combination of everyday habituated production and consumption patterns, such as food self-provisioning, which are leading social transformation (Vergragt et al., 2014; Soper, 2000; Ritzer and Jurgenson, 2010). In FSP practices, analysts perceive a redefinition of boundaries between consumption and production; through their involvement in gardening, people develop skills and the distinction between amateurs and professionals is blurred. Values and norms of reciprocity that have been completely marginalized by the mainstream market emerge in certain FSP practices linked to barter, sharing, neighbouring, mutualization and cooperation, learning by doing agroecology (Schor and Fitzmaurice, 2015; Pellandini-Simányi, 2014). In sum, 'food-as-a-commons' does not negate the possibility of exchange (i.e. trading), but must be based on different values that are not exchange values.

Conclusions

The multiple dimensions of the food-as-commons concept (Vivero-Pol, 2017) are unique to the Central and Eastern European context, where the concentration of food retail and the relative underdevelopment of short food supply chains restrict the availability of fresh food (Kneafsey et al., 2013; Benedek and Balázs, 2016). FSP serves as an example of food as a commons primarily through the connections between food and community-building capabilities that clearly separate this practice from conventional food production and purchase for and through market mechanisms. Far from being an individual exotic hobby or a survival strategy, FSP in Hungary seems to be nurturing values necessary for individual, collective and social sustainability. Family and friendship networks are an important non-market source and a channel of local foods. Self-sufficiency practices are only partially explained by the need to save money or obtain healthy and fresh food. Collective values, such as maintaining family food traditions or sharing a hobby, appear to be equally, if not more important. Finally, production for one's own consumption and exchange is an expression of the democratic food system and a method to regain control over what people eat; it extends to all ages and income groups as well as to both rural and urban dwellers, and is neither centrally controlled nor defined (with some exceptions). In this way, food self-provisioning implements democratic changes into the food system without claiming a radical transformation (Sage, 2014; De Hoop and Jehlička, 2017). The various practices need to be further analysed to understand their potential for teaching about local, sustainable food and the prospects for more localized, alternative food futures.

Although mainstream macroeconomic theory suggests the growth paradigm is unquestionable and that market mechanisms are central to the construction of societies, food self-provisioning represents a contrary hope and has been shaping an alternative transition pathway towards sustainability of food systems for a long time. FSP practices can also very well be part of an innovative, utopian and just vision for the future (Balázs, 2016). As the above empirical examples reveal, there are entirely different motivations that support the self-provision of food that go well beyond the economic argument of scarcity and survival. In essence, FSP has both a resource/material component and a community/collective component that determines the rules of producing, sharing and consuming. FSP represents a radical act to shorten the space of the food chain from field to table and to socially extend the access to food from the household to the wider family and the community in which self-producers are embedded. The shared customary social organization of these practices is based on the values of caring, collectiveness, stewardship of the land and family.

Instead of individualizing environmental responsibility, FSP practices create an opportunity for community food growing, cooperation in manual labour, reciprocal exchange and collective consumption of produce. The challenge is how policy can open up such FSP spaces and create such communities where they do not currently exist. Sustain offers an interesting example (2014): a program that helps planning authorities and local governments introduce space for healthy food-growing communities when establishing new residential developments.

In the context of Central and Eastern European FSP, Sovová (2016) convincingly argues that to create sustainable urban food systems, a number of policy measures can help, the most relevant being those raising environmental awareness among gardeners on agroecology and mitigating the negative environmental impacts of conventional food production. In the context of Central and Eastern European countries, the possibility of highlighting the traditional character of this practice so as to improve the public image of allotments through narratives of change around food as a commons, responsible consumption and the construction of local, social and participative alternative food networks seems evident. In Hungary, several practical policy measures have been introduced under the auspices of the Rural Development Programme to facilitate particular types of FSP by means of protection and competition for home garden plots (Balázs et al., 2016). What is missing is symbolic support and high-level policy engagement with the fundamental norms of food sharing and the practices that reinforce them. Fortunately, scholars in recent years have turned to an alternative future-oriented narrative of FSP to revalue food as a commons, a social construct linked to multiple sustainability benefits (Vivero-Pol, 2013). FSP is gaining political significance as practices of gardeners have begun to govern their food system: a crucial step towards food democracy.

Acknowledgements

I wish to thank two of the editors of this volume, Jose Luis Vivero-Pol and Tomaso Ferrando, plus George Kent, another author of this volume, as well as all members of the Environmental Social Science Research Group (www.essrg.hu) for their helpful guidance to develop the main lines of the arguments in this chapter for a more sustainable food system.

Notes

1 An omnibus survey is a method of quantitative research where data collection is organized on a wide variety of subjects during the same interview. Usually, several researchers provide the content by paying to 'get on the omnibus'. Omnibus surveys thus use common demographic data collected from each respondent. The advantages to the research client include cost savings (because the sampling and

screening costs are shared across multiple clients) and timeliness (because omnibus samples are large and interviewing is ongoing). An omnibus survey generally uses a stratified sample.

2 The semi-structured interviews further explored the interlinkages of motivations and collective values. The interview excerpts are only illustrative and do not aspire to representativeness. Respondents of different genders, ages, educational levels and professional backgrounds who produce food for self-consumption (or in part for giving away/sharing) were chosen from both urban and rural areas, excluding professional farmers.

3 Due to limited administrative capacities and the absence of historical data, new member states (i.e. those that joined the European Union in 2004 and 2007) were granted the possibility of applying the single area payment scheme instead of applying the standard direct payment schemes. The single area payment scheme provides a flat-rate decoupled area payment for eligible agricultural land and replaces almost all payments granted in other than new member states. Under Regulation (EC) No 73/2009, the single area payment scheme was foreseen to expire. However, the 2013 reform of the Common Agricultural Policy permitted member states applying the single area payment scheme in 2014 to apply it until 2020. At present, the single area payment scheme is applied by all new member states except Slovenia, Malta and Croatia (Council Regulation (EC) No 73/2009).

References

Acheson, J. (2007) 'Household exchange networks in post-socialist Slovakia', *Human Organization*, vol 66, no 4, pp. 405–413

Aistara, G.A. (2015) 'Good, clean, fair … and illegal: paradoxes of food ethics in post-socialist Latvia', *Journal of Baltic Studies*, vol 46, no 3, pp. 283–298

Alber, J. and Kohler, U. (2008) 'Informal food production in the enlarged European Union', Social Indicators Paper 89, no 1, p. 113–127

Asztalos Morell, I. (2014) 'Workfare with a human face? Innovative utilizations of public work in rural municipalities in Hungary', *Metszetek*, vol 3, no 4, pp. 1–22

Balázs, B (2016) 'Food self-provisioning – the role of non-market exchanges in sustainable food supply', in A. Meybeck and S. Redfern (eds) *Sustainable value chains for sustainable food systems. A workshop of the FAO/UNEP Programme on Sustainable Food Systems*, FAO, Rome

Balázs, B. (2012) Local food system development in Hungary, *International Journal of Sociology of Agriculture and Food*, vol 19, no 3, pp. 403–421

Balázs, B., Smith, A., Aistara, G. and Bela, Gy. (2015) *Transnational Seed Exchange Networks*, TRANSIT: EU SSH.2013.3.2-1 Grant agreement no: 613169

Benedek, Zs. and Balázs, B. (2016) 'Current status and future prospect of local food production in Hungary: A spatial analysis', *European Planning Studies*, vol 24, no 3, pp. 607–624

Birol, E., Bela, Gy. and Smale, M. (2005) 'The Role of Home Gardens in Promoting Multi-Functional Agriculture in Hungary', *EuroChoices*, vol 4, no 3, pp. 14–21

Brown, D. L. and Kulcsár, L. (2001) 'Household economic behavior in post-socialist rural Hungary', *Rural Sociology*, vol 66, no 2, pp. 157–180

Bühler, C. (2004) 'Additional work, family agriculture, and the birth of a first or a second child in Russia at the beginning of the 1990s', *Population Research and Policy Review*, vol 23, no 3, pp. 259–289

Calvet-Mir, L., Riu-Bosoms, C., González-Puente, M., Ruiz-Mallén, I., Reyes-García, V. and Molina, J. L. (2016) 'The transmission of home garden knowledge: Safeguarding biocultural diversity and enhancing social–ecological resilience', *Society & Natural Resources*, vol 29, no 5, pp. 556–571

Carlsson-Kanyama, A., Dreborg, K. H., Moll, H. C. and Padovan, D. (2008). 'Participative backcasting: a tool for involving stakeholders in local sustainability planning', *Futures*, vol 40, no 1, pp. 34–46

Czegledy, A. (2002) 'Urban peasants in a post-socialist world: Small-scale agriculturalists in Hungary', in P. Leonard and D. Kaneff (eds) *Post-Socialist Peasant? Rural and Urban Constructions of Identity in Eastern Europe, East Asia and the Former Soviet Union*, Palgrave Macmillan, Basingstoke

Davidova, S. and Bailey, A. (2014) Roles of small and semi-subsistence farms in the EU, *EuroChoices*, vol 13, pp. 10–14

Davidova, S., Bailey, A., Dwyer, J., Erjavec, E., Gorton, M. and Thomson, K. (2013) *Semi-Subsistence Farming: Value and Directions of Development*, Brussels, Directorate General for Internal Policies. Policy Department B: Structural and Cohesion Policies

De Hoop, E. and Jehlička, P. (2017) 'Reluctant pioneers in the European periphery? Environmental activism, food consumption and "growing your own"', *Local Environment*, vol 22, no 7, pp. 1–16
Eyzaguirre, P. and Bailey, A. (2009) 'International case studies and tropical home gardens projects: Offering lessons for a new research agenda in Europe', in A. Bailey, P. Eyzaguirre and L. Maggioni (eds) Crop Genetic Resources in European Home Gardens: Proceedings of a Workshop, 3-4 October 2007, Ljubljana, Slovenia, Bioversity International, Rome
Fredriksson, L., Bailey, A., Davidova, S., Gorton, M. and Traikova, D. (2016), 'The commercialisation of subsistence farms: Evidence from the new member states of the EU', *Land Use Policy*, vol 60, pp. 37–47
Galhena, D. H., Freed, R. and Maredia, K. M. (2013). 'Home gardens: a promising approach to enhance household food security and wellbeing', *Agriculture & Food Security*, vol 2, no 1, p. 8
Gibson-Graham, J. K. (2006) *A Postcapitalist Politics*, Minneapolis, University of Minnesota Press
Gittleman, M., Kelli, J. and Brelsford, E. (2012) 'Using citizen science to quantify community garden crop yields', *Cities and the Environment (CATE)*, vol 5, no 1, article 4, http://digitalcommons.lmu.edu/cate/vol5/iss1/4
Heinonen, M., Nikula, J., Kopoteva, I., and Granberg, L. (2007) *Reflecting Transformation in Post-socialist Rural Areas*, Newcastle upon Tyne, Cambridge Scholars Publishing
Jehlička, P, Kostelecký, T. and Smith, J. (2012) 'Food self-provisioning in Czechia: Beyond coping strategy of the poor: A response to Alber and Kohler's "Informal food production in the enlarged European Union" (2008)', *Social Indicator Research* vol 111, no 1, pp. 219–234
Jehlička, P. and Smith, J., (2011) 'An unsustainable state: contrasting food practices and state policies in the Czech Republic', *Geoforum*, vol 42, no 3, pp. 362–372
Jehlička, P. and Smith, J., (2012) 'Sustainability and the "urban peasant": Rethinking the cultural politics of food self-provisioning in the Czech Republic', in P. Zahrádka and R. Sedláková (eds) *New Perspectives on Consumer Culture Theory and Research*, Cambridge Scholars, Newcastle upon Tyne
Kelemen, E. and Kovách, I. (2005) 'A vidéki népesség agrárintettsége', in I. Kovách (ed) *Vidék Mezőgazdaság, Európai Uniós csatlakozás – az értékek változása*, MTA Politikai Tudományok Intézete, Digitális Archívum, Budapest
Kent, G. (2008) *Global Obligations for the Right to Food*, Lanham, Rowman & Littlefield
Király, G., Pataki, Gy., Köves, A. and Balázs, B. (2013) 'Models of (future) society: Bringing social theories back in backcasting', *Futures*, vol 51, pp. 19–30
Kneafsey, M., Venn, L., Schmutz, U., Balázs, B., Trenchard, L., Eyden-Wood, T., Bos, E., Sutton, G. and Blackett, M. (2013) *Short Food Supply Chains and Local Food Systems in the EU. A State of Play of their Socio-Economic Characteristics*, Seville, JRC Scientific and Policy Reports, Joint Research Centre Institute for Prospective Technological Studies, European Commission
Kornai, J. (1980) *Economics of Shortage*, Amsterdam, North Holland
Kortright, R. and Wakefield, S. (2011) 'Edible backyards: a qualitative study of household food growing and its contributions to food security', *Agriculture and Human Values*, vol 28, no 1, pp. 39–53
Kovách, I. (2012) *A vidék az ezredfordulón: A jelenkori magyar vidéki társadalom szerkezeti és hatalmi változásai* Budapest, MTA Társadalomtudományi Kutatóközpont Szociológiai Intézet–Argumentum Kiadó
Kvale, S. (1996) *InterViews: An Introduction to Qualitative Research Interviewing*, Thousand Oaks, Sage Publications
Landon-Lane, C. (2004) *Livelihoods Grow in Gardens: Diversifying Rural Incomes through Home Gardens*, Rome, Food and Agriculture Organization of the United Nations (FAO)
Lang, T. (1999) 'The complexities of globalization: The UK as a case study of tensions within the food system and the challenge to food policy', *Agriculture and Human Values*, vol 16, no 2, pp. 169–185
McIntyre, L. and Rondeau, K. (2011) 'Individual consumer food localism: a review anchored in Canadian farmwomen's reflections', *Journal of Rural Studies*, vol 27, no 2, pp. 116–124
Megyesi, B. (2016) 'The changing role of local small-scale agri-food production in Hungary', in Proceedings – Conference on Democratizing Food Governance, Rome, 14 October 2016, pp. 99–102
Pellandini-Simányi, L. (2014) 'Bourdieu, ethics and symbolic power', *The Sociological Review*, vol 62, no 4, pp. 651–674
Patton, M. Q. (2005) *Qualitative Research. Encyclopedia of Statistics in Behavioral Science*, Wiley Online Library, http://onlinelibrary.wiley.com/doi/10.1002/0470013192.bsa514/full
Porter, J. R., Dyball, R., Dumaresq, D., Deutsch, L. and Matsuda, H. (2014) 'Feeding capitals: Urban food security and self-provisioning in Canberra, Copenhagen and Tokyo' *Global Food Security*, vol 3, no 1, pp. 1–7

Quist, J. and Vergragt, P. (2006) 'Past and future of backcasting: The shift to stakeholder participation and a proposal for a methodological framework', *Futures*, vol 38, no 9, pp. 1027–1045
Renting, H., Schermer, M. and Rossi, A. (2012) 'Building food democracy: Exploring civic food networks and newly emerging forms of food citizenship', *International Journal of Sociology of Agriculture and Food*, vol 19, no 3, pp. 289–307
Ritzer, G. and Jurgenson, N. (2010) Production, consumption, prosumption. The nature of capitalism in the age of the digital "prosumer"', *Journal of Consumer Culture*, vol 10, no 1, pp. 13–36
Rose, R. and Tikhomirov, Y. (1993) 'Who grows food in Russia and Eastern Europe?', *Post-Soviet Geography*, vol 34, no 2, pp. 111–126
Sage, C. (2014) 'The transition movement and food sovereignty: From local resilience to global engagement in food system transformation', *Journal of Consumer Culture*, vol 14, no 2, pp. 254–275
Schor, J. B. and Fitzmaurice, C. J. (2015) 'Collaborating and connecting: the emergence of the sharing economy' in L. Reisch and J. Thogersen (eds) *Handbook on Research on Sustainable Consumption*, Edward Elgar, Cheltenham
Schmutz, U., Lennartsson, M., Williams, S., Devereaux, M. and Davies, G. (2014) The Benefits of Gardening and Food Growing for Health and Wellbeing. London, Garden Organic; London, Sustain
Schupp, J. L. and Sharp, J. S. (2012) 'Exploring the social bases of home gardening', *Agriculture and Human Value*, vol 29, no 1, pp. 93–105
Seeth, H. T., Chachnov, S., Surinov, A. and Von Braun, J. (1998) 'Russian poverty: Muddling through economic transition with garden plots', *World Development*, vol 26, no 9, pp. 1611–1624
Sík, E. (1988) 'Reciprocal exchange of labour in Hungary' in R. E. Pahl (ed) *On Work: Historical, Comparative and Theoretical Approaches*, Blackwell, New York
Smith, J. (2003) 'From hazi to hyper market: Discourses on time, money, and food in Hungary', *Anthropology of East Europe Review*, vol 21, no 1, pp. 179–188
Smith, J. and Jehlička, P. (2013) 'Quiet sustainability: Fertile lessons from Europe's productive gardeners', *Journal of Rural Studies*, vol 32, pp 148–157
Soper, K. (2000) 'Other pleasures: The attractions of post-consumerism', *Socialist Register*, vol 36, pp 116–132
Sovová, L. (2016) 'Self-provisioning, sustainability and environmental consciousness in Brno allotment gardens', *Sociální studia/Social Studies*, vol 12, no 3, pp. 11–26
Sustain (2014) Planning sustainable cities for community food growing, London, Sustain
Torsello, D. (2003) *Trust, Property and Social Change in a Southern Slovakian Village*, LIT Verlag, Münster
Vávra, J., Cudlínová, E. and Lapka, M. (2013) 'Food and sustainability: Food self-provisioning and food shopping habits in the Czech Republic and selected EU countries', in *Proceedings of the International Scientific Conference - Region in the Development of Society* 2013, 10 October 2013, Brno, Czech Republic, Mendel University
Vergragt, P., Akenji, L. and Dewick, P. (2014) 'Sustainable production, consumption, and livelihoods: Global and regional research perspectives', *Journal of Cleaner Production*, vol 63, pp. 1–12
Vivero-Pol, J. L. (2013) 'Food as a Commons: Reframing the Narrative of the Food System', SSRN Working Paper Series, https://papers.ssrn.com/sol3/papers.cfm?abstract_id=2255447, Accessed on 6 December 2017
Vivero-Pol, J. L. (2017) 'Food as commons or commodity? Exploring the links between normative valuations and agency in food transition', *Sustainability*, vol 9, no 3, pp. 442
Williams, C., Nadin, S., Rodgers, P. and Round, J. (2012) 'Rethinking the nature of community economies: Some lessons from post-Soviet Ukraine', *Community Development Journal*, vol 47, no 2, pp. 216–231
Williams, C. C. (2005) 'Work organization in post-socialist societies', *Futures*, vol 37, no 10, pp. 1145–1157
Wiskerke, J. S. and Viljoen, A. (2012) 'Sustainable urban food provisioning: challenges for scientists, policymakers, planners and designers', in A. M. Viljoen and J. S. C. Wiskerke (eds) *Sustainable food planning: evolving theory and practice,* Wageningen, Wageningen Academic Publishers, pp. 19–35

PART V

Dialogue of alternative narratives of transition

20

CAN FOOD AS A COMMONS ADVANCE FOOD SOVEREIGNTY?

Eric Holt-Giménez and Ilja van Lammeren

Introduction

In 1996, La Vía Campesina raised the call for *food sovereignty*, the right of peoples to determine and control their own food systems. Food sovereignty challenged the concept of 'food security' and the privatization of land, seeds and water and advanced an array of environmentally sustainable and socially equitable alternatives to the destructive practices of the corporate food regime. The call became a global movement that today includes several hundred million small and medium food producers and hundreds of thousands of consumers. Food sovereignty is one of the pillars of the global counter-movements to neoliberalism.

The rejection of both industrial agriculture and of neoliberalism's tendency towards corporate privatization has led many food sovereignty advocates to embrace complementary concepts, like agroecology and the Right to Food. Parallel and overlapping notions have also emerged, like 'food justice,' the 'good food' movement, 'Slow Food' and 'food democracy'. These are often associated (or confused) with food sovereignty. Recently, some Right-to-Food scholars have introduced the notion of a 'food as a commons,' in which the commons is used both as a noun (shared resources) and as a verb (i.e., 'commoning' in which food is a common good) (Ferrando and Vivero-Pol, 2017).

At first glance, food sovereignty and the food commons have, well, much in common. Both are a rejection of the corporate food regime's tendency to manufacture food scarcity and both revive the histories and traditions of the Commons. Both reject the commodification of food and uphold food as a human right. Both propose transformative models for equitable and sustainable food systems as a basis for a new society.

But the etymologies of food sovereignty and food as a commons are not congruent. Food sovereignty is a call for resource equity emerging from the organized demands of peasants, pastoralists and fishers, while the food commons is a call for equal food access, emerging primarily from the professional community for human rights (mostly urban). The former has structural proposals addressing production (agrarian reform and farm parity, fair wages and decent working conditions for farm and food workers, the exemption of agriculture from free trade agreements and the dismantling of corporate power), while the latter focuses much more on normative valuations, distributive consumption and alternative nodes of governance. *La Vía Campesina*, whose 164 peasant organizations serve 200 million farmers around the world, is the social and cultural cornerstone of food sovereignty, while the food commons is known amongst a relatively small, but influential cadre of professionals and activist intellectuals (La Via Campesina, 2011).

Nonetheless, the two concepts are closely aligned and a clear political alliance between them might create a new lever for social change. Can the food commons help in the construction of food sovereignty? What are the conceptual, practical and political bases for such an alliance? Answering these questions requires assessing the notion of a food as a commons through the political-economic lens of food sovereignty and of the Commons itself.

The renewal of the Commons

Amidst the ongoing privatization of public, common pool and open access resources, the notion of the Commons is increasingly celebrated (Federici and Caffentzis, 2014). The renewed popularity of the Commons results from a widespread disaffection with the modern political and economic governance of both socialism and capitalism (Caffentzis, 2010). As an emergent key concept, the Commons symbolizes a platform to re-imagine access, use, community and values associated with governing resources (Cumbers, 2015). While the Commons may be celebrated for its unifying properties, the various forms of emerging commons that are being developed also demonstrate a confluence of political allegiances (Federici, 2011).

Whereas conventionally the Commons refers to both land-based resources and the rules, rights and obligations governing them, emerging commons range from local initiatives organized around principles of communal sharing to environmental commons that conceptualize ecological interdependencies at global scales of governance (Holder and Flessas, 2008).

More recently, the Commons have been broadened to include the more intangible aspects of social life, epitomized in the so-called 'creative,' 'cultural,' 'knowledge' and 'digital' commons (Bresnihan, 2015). Notably, the products of these immaterial commons are seen as continuously reproducible through recombinant technologies and human creativity, meaning one person's use does not subtract from somebody else's, but rather tends to increase the availability and productivity of common goods (De Angelis and Harvie, 2014). By presenting the act of sharing as a valued asset in and of itself, the commonwealth created in these immaterial commons offers a vision of society based on principles of abundance (Hardt and Negri, 2011). As such, these emerging commons act as an ideological counter-force to globalized markets, subverting the meaning and function of access and sharing within liberalized economies (Holder and Flessas, 2008; Caffentzis, 2010).

The call for 'food as a commons' integrates a variety of existing and emerging commons, from community gardens to recipes, under an overarching claim to the Right to Food (Ferrando and Vivero-Pol, 2017). Unlike traditional common property regimes, food commons asserts the right of individuals to not be excluded from resource use (Holder and Flessas, 2008). Deploying a language of rights, entitlement and justice, the vision of food as a commons is less defined by the specific actors and rules and more by the political claim to food as a common good (Vivero-Pol, 2017a). Significantly, arguments for a food commons question the capacity of both states and markets to ensure the right to food, and articulate a moral claim for food to exist outside of the capitalist mode of production and exchange (Vivero-Pol, 2013). While recognizing the lack of food access as the most pressing issue, the food as a commons argument departs from typical food security arguments based on an assumption of scarcity, asserting that 'the assumed private nature of food' is the root cause of hunger (Vivero-Pol, 2017d, p. 346).

This approach is novel—and bold—because previous work on the role of the Commons in securing food access mostly concerned land held in common, from which users derived multiple food and non-food products (Mosse, 1997). Proposing food *itself* as a commons does not ground the product to an actual place, but rather conceptualizes it at an aggregated, global supply scale. Taking the category of food as a whole (somewhat like water or air) provides conceptual space to develop the notion of food as a commons, to which human entitlement can appeal. Everyone should have the right to food. Can the right to food be secured through a commons framework? Possibly. But a number of questions need to be addressed regarding food and property regimes.

Food as property

Food is a physical good, embedded within property regimes used to govern its production and use. The materiality of food grounds its many meanings (e.g. as a commodity, a cultural expression, a basic need, etc.), but also sits objectively *outside* these meanings. Recalling Elinor Ostrom's measure for classifying the materiality of goods, private goods are characterized by the ease of excluding others and high subtractability of use—meaning gains obtained by one actor will reduce availability to others (Vatn, 2015). In comparison, public goods refer to resources for which individual behavior does not influence the shared resource base significantly and exclusion of actors is difficult, which may imply a certain level of detachment between users and governance of resources (Ostrom and Ostrom, 2015). Common-pool resources are those goods for which use is subtractable, but there is difficulty of exclusion—at least from within the site-specific group.

Food is a private good because the use of a food-substance reduces immediate availability to other actors and excludes others from consuming that particular food-product. Granted, when food supplies are abundant, the food-consumption by one individual does not significantly influence its availability to others and subtractability becomes low or insignificant (Vatn, 2015). This, however, does not change the materiality of food itself, but is the experience and aggregation of food-supplies in a given place and at a particular time.

Importantly, food as an essential material object embodies value (in the form of energy) from the sun, sea, sky and soil and labor (McMichael, 2013). The production of food thus inevitably utilizes a multitude of common-pool, public and private, ecologically interdependent resources, which may be governed under a wide variety of different property forms (Ostrom and Hess, 2007). Food also contains value as embodied *labor power*—the human effort used to collect or produce it. This is important because when food is exchanged, the *value* of this labor power is distributed among producers, middlemen, processors, retailers, distributers, shopkeepers and consumers. Even when food is shared freely, the value of labor power is shared (though it is socialized rather than privately appropriated). The different ways food is produced, distributed and consumed are, consciously or unconsciously, politically determined. Under modern capitalist regimes, private goods like food can be governed as commodities as well as public or common property.

While the material characteristics of a good do not determine the rules governing its use, the attributes of a given resource shape the relative benefits and costs associated with constructing and maintaining a given property form (Vatn, 2005). At the same time, governance of goods shapes the individual experience of degrees of excludability and subtractability of the good; on the one hand, human made technologies and legal constructs enable enclosures, on the other, political means can ensure access (Pistor and Schutter, 2015). As such, property regimes define whose interests are protected and provide means to coordinate competing uses (Vatn, 2015). While these rules are often so ubiquitous as to seem natural (e.g., the market), they are no more than socially constructed, normative instruments.

Relatively recent in human history, food scarcity became a function of the market, rather than of productive, infrastructural, or political conditions (Moore Lappé and Collins, 2015). The call for food as a commons recognizes this 'scarcity' of food under capitalism as a market fiction used to allocate otherwise abundant food supplies on the basis of economic demand, rather than need (Vivero-Pol, 2017d). It opposes the exclusions resulting from private property regimes in which access to food is made contingent upon purchasing power, as well as the added subtractability of food when food-products are used for more profitable, non-food purposes (Vivero-Pol, 2014). Accordingly, the call for food as a commons makes a moral argument for food to be considered as a common good, arguing that property regimes—as well as distinctions between private, public and common-pool resources—are social constructions (Vivero-Pol, 2017c).

This is true, and significant, but while individuals shouldn't be excluded from food, this doesn't mean that food itself is a non-excludable, non-rivalry good. While commodities are changeable human constructs, the material characteristics of food existed prior to those norms, rules and technical developments that accentuate, alter and capitalize on the physical attributes of food and remain crucial in the labor process, consumption and politics of food.

It is precisely the embedded materiality and systemic complexity of food that leads to its inescapable politics. Any proposal for resistance or transformation—whether ideational (like the food as a commons) or materially based (like food sovereignty)—will need to address the value, regimes and materiality of food. The social construction of the Commons (food or otherwise), and its relation to private property, is no less complex.

The Commons as Social Relations

Elinor Ostrom's ground-breaking analysis of common property regimes introduced the Commons as an alternative governance model to the state and the market and informed the design principles for the global commons, polycentric governance models and food as a commons (Ostrom, 1987; 1990; Ostrom et al., 1999). Ostrom's response to Garret Hardin's 'tragedy of the commons' identified common property as regimes in which members of a clearly defined group own the right to exclude non-members. This distinguished the Commons from 'open access' regimes that pose no limits on use. Her empirical work showed that the Commons operate by a set of social relations that mobilize the social differentiation among members, as well as between members and non-members, to regulate use and management of shared resource systems (Hardin, 1968; Ostrom, 1987, 1990).

Because common property regimes tend to harbor multiple, often conflicting uses, interdependencies and overlapping rules, rather than presuming attributes of governance based on property form, Ostrom tried to account for the 'bundle of rights and obligations' that governed resources. These she defined as the rights to access, to withdraw, to manage, to exclude and to alienate. Allocated differently among members of a commons, the bundle held by a given actor shapes their position to resources and to other actors (Ostrom and Hess, 2007). The 'right to access' regulates use by members and might be differentiated between areas, seasons or time of day; while the 'right to withdraw' specifies who can harvest what resource and for what purpose (Schlager and Ostrom, 1992). Further, there may be multiple, co-existing rules for a particular good. The 'right to management' then provides the right to regulate such use-patterns—decisions that potentially produce, increase or mitigate tensions between different users and uses of shared resources. Given the place-specific nature of these entitlements, restricting access by limiting members' 'right to alienation' (the right of a member to sell their entitlement to the Commons) and maintaining the ability to exclude non-members are often considered foundational to the sustainability of the Commons (Ostrom, 1987).

Ostrom's work countered dominant theories of human action. She showed that locally embedded institutions are complex systems of incentive structures amongst members—beyond the individual weighing of costs and benefits ascribed to open access and private property regimes (Ostrom, 2014). Desirable outcomes are not pre-determined but shaped by relationships between different members of a commons (Ostrom, 1990). Reciprocal expectations among actors, embodied through ethics, norms, trust and habit, create situations in which it becomes rational to cooperate, rather than compete, thus enabling sustainable use and management of the Commons (Ostrom, 1992).

Following on Ostrom's work, other researchers have pointed out that rules governing resources held in common do not merely arise as the preferable means to govern resources, but *emerge from existing structures of power* (Mosse, 1997). Members of a commons tend to occupy asymmetrical positions and the ways in which individuals relate to each other and common resources arise from existing social structures, shaped by divisions of labor and differential status based on gender, ethnicity, class and age (Nightingale, 2010). Members' respective bundle of rights are thus shaped by a wider set of social relations than those aimed at governing the Commons and may be better conceived as bundles of powers (Ribot and Peluso, 2003). The heterogeneity among actors' bundles of entitlements and responsibilities within a commons does not only reflect existing social structures, but inscribes these back upon the landscape in various ways, such as in the production of 'gendered crops' following gendered divisions of labor on commonly owned lands (Howard and Nabanoga, 2007). Actually, existing Commons thus defy simplistic notions of power as an outside force, disrupting and impinging upon an otherwise harmonious community; rather, the material resources and property regimes used to govern those resources constitute a domain through which *existing* relations of power are continuously enacted and transformed (Brown, 2007; Nightingale, 2014).

Common property regimes often govern resources where multiple activities are pursued under different property arrangements and products from commonly owned resource systems are privately appropriated and sold on markets (Ostrom and Hess, 2007). Common property regimes are rarely self-contained systems (Mosse, 1997; McCay, 2002). The resulting resource regimes do not resemble any set of property rights in their ideal form, but emerge through overlapping and combined property regimes (Smith, 2000; Fennell, 2010). Atavistic notions of a commons as an autonomous, rule-based system with clearly defined, relatively homogenous social groups organized around a circumscribed resource base, simply do not reflect the dynamic and ambiguous realities of resources held in common. Instead, the Commons often constitute a domain through which to navigate changing parameters of market conditions, the environment, social reproduction and the behavior of exogenous actors. We see this, for instance, by the way in which, just as value from 'rights to benefits' of members within a commons extend outwards through the sale of private goods and labor, the sustenance of the Commons may be sustained by the fulfillment of obligations with outside resources, such as immigrant remittances (Caffentzis, 2010).

While the limits and needs negotiated through the Commons are neither inherent nor abstract, understanding the dynamics of a commons requires locating user-groups within broader political and economic dynamics in which the Commons are embedded, and which shape long-term strategies and day to day decisions of its members (Turner, 2017). Rather than rules and norms operating autonomously, these are 'situated choices' and the evolving product of social and material realities. The Commons are not merely humanly constructed sets of rules nor a physical area of resources nor the perceptions actors have of these, but the way in which these are all combined and enacted. The 'situated interdependence' of actors and the Commons allows them to negotiate, in imperfect ways, changing, conflicting and uncertain needs, capacities and obligations (Bresnihan, 2015).

The Commons, capitalism and peasant resistance

The Roman Empire divided property into *res publicae, res privatae* and *res communes*: state, private and commons. Those things that couldn't be possessed and were available to all (open access) were *extra patrimonium* (Holt-Giménez, 2017). The Romans left some common property to people and villages because they understood its integral role to the production of an expropriable surplus and to the overall stability of the imperial regime. For thousands of years, the Commons ensured the food security that private property and government wouldn't or couldn't provide. In many ways and in many places around the world, the Commons still does.

The Commons can supplement food, fiber and other resource needs of small-scale farmers, pastoralists and fishers, lower their livelihood costs and allow them to sell their products cheaply, thus helping them compete with large-scale, capitalized production. However, this 'subsidy' from the Commons can cut both ways. If the Commons is used primarily to produce goods for market rather than for subsistence, low prices in the market can lead to the over-exploitation of the Commons. Also, when small-scale producers or their family members work for industrial wages, the Commons can enable industry to obtain peasant labor power more cheaply—essentially allowing industry to exploit workers by appropriating the food subsidy of the Commons. So, under certain conditions, the market and the private sector may indirectly benefit from the Commons. In recessionary or deflationary times capital may seek to privatize the Commons in order to put its wealth in land rather than hold it as money. Under periods of economic expansion, capital can use the power of the state to enclose the Commons in order to force smallholders to sell their land and move them into the labor market. Thus, while the Commons is a historic refuge for non-capitalist and non-state relations in the food system, it always has a relationship with capital and the state. At the same time, the Commons allows for increased agency, resiliency and room for maneuver among the peasantry, smallholders and villagers.

While it is fashionable to understand the Commons as a pre-capitalist sanctuary of un-commodified products and egalitarian social relations, this was not necessarily the case and the porous relation between the Commons and the changing modes of production and political organization in which it is embedded, is less often appreciated. With the capitalist mode of production and the industrialization of societies, the Commons and its livelihood functions were destroyed or increasingly marginalized. Nevertheless, the role of the Commons in the formation of capitalism is as important as the much more celebrated role of capitalism in the destruction of the Commons. The Commons was *pivotal* to both the destruction and the survival of the peasantry—a massive social force needed by capitalism for its labor, its nascent capacity to buy industrial products and its ability to fill the ranks of armies. Cheap food produced by the peasantry was not just important for the industrial take-off in cities, but also for the development of industrial agriculture in the countryside. Following the work of Kautsky, as late as the 1970s, Alain de Janvry characterized the peasant-capitalist relationship of Mexico's widely celebrated Green Revolution as 'functional dualism' (Janvry, 1981).

The persistence of the peasantry throughout the agrarian transitions of the seventeenth through the twentieth centuries cannot be understood simply through Kautsky's theory of self-exploitation of peasant family labor, but must also be understood through the role that the Commons played in the survival of peasant families (Kautsky, 1988). Whether in its reinstated forms (such as the Mexican *ejido*) or its many remnant forms (like the *acequia* system in the Southwestern United States), the Commons—however it was internally defined—has played an integral role in the interaction of the peasantry with whichever colonizing power was extracting their labor, food and resources. The function of the Commons in peasant society, since before the dawn of capitalism, has served both as a means of resistance *and* a means of exploitation. The balance between these two functions has depended on the correlation of forces between peasant production and capitalist production.

Just as the Commons has had a dynamic history with the dominant modes of production, it also has an extensive social and political history with hegemonic forms of state governance, notably with democracy. Capitalism and democracy evolved together. The particular form of democracy associated with capitalism is *liberal democracy,* based on the rights of property and the rights of the individual (or corporations that are treated as if they are individuals). The combination of capitalism and liberal democracy is inherently contradictory because the tendency of capital to concentrate wealth is antithetical to the distribution of power essential to democracy. This contradiction is managed by the *liberal nation-state* through forms of representative democracy.

Since the flowering of Greek democracy in 4 BC, private property has also been the basis for citizenship and power. But the protection of private property depends in no small degree on the existence of public property, that is, the state. Establishing and maintaining public property is not easy; the governed have to consent to be governed and although coercion can work for a while, unless there is a *social contract,* force is unsustainable in the long run. So the question is: how can the state reconcile the private ownership of the production of essential goods and services with the pursuit of the public good for all its citizens?

> [Private] property is continually in need of public justification—first, because it empowers individuals to make decisions about the use of scarce resource in a way that is not necessarily sensitive to others' needs or the public good; and second, because it does not merely permit that but deploys public force at public expense to uphold it.
>
> *(Waldron, 2004)*

Without the power of the state, individuals and corporations could not enforce their exclusive claims to property's uses and benefits. This is still the condition for private property. Public property—which theoretically belongs to all citizens—also requires the power of the state to ensure the production of public goods and services. Liberal democracy has a variety of legal structures for Commons protection and management, among them: trusts, reserves and forms of consuetudinary governance (e.g., *usos y costumbres*). The existence of Commons within liberal democracy depends on the liberal state and in no small measure on the regulatory frameworks for public goods. While private property, particularly under neoliberal regimes, is on one hand antithetical to the Commons, it is questionable whether the Commons could exist without the same nation-state that ensures the existence of private property (Holt-Giménez, 2017).

While the Commons is frequently described as a localized, democratic institution, it is not safe to project the conventions of liberal democracy onto the Commons. Not only were women frequently excluded from the Commons, the social relations of liberal democracy and the social relations of the Commons are historically and structurally distinct. In the face of the former's universality, the latter is specific to particular sets of social relations, geographies and cultures. The point is that the Commons is a social relation that is historically—and contentiously—embedded within a larger social relation. Today's liberal democracies and Commons are no strangers to each other and, while it is tempting to treat the Commons as something operating *outside* the current corporate food regime, it is and has been a subjugated part of our food systems for a very, very long time.

The call for a food as a commons

The call for food as a commons emanates not just as a reaction against neoliberal capitalism, but as a social expression both *within* and critical *of* liberal democracy. Indeed, the claim to the Right to Food itself is normatively possible thanks to liberal democracy's rights-based structure, while the very notion of a food commons defies the validity of liberal states as representative institutions. Can liberal democracy—an institution founded on private property and individual rights—promote a socialized property regime like the food commons?

The vision of a food commons notably differs from previous discourse on the Right to Food, not the least in its claim that capitalist market economies restrain the democratic capacity of liberal states and popular demands cannot be reconciled within the state-market duopoly (Vivero-Pol, 2014; Ferrando and Vivero-Pol, 2017). The food as a commons narrative scrutinizes the constitutional surface of the liberal state under which corporate actors are the primary architects of our food supply. Pointing to the limits of liberal multilateralism in which the World Trade Organization continues to overpower the United Nations' efforts to secure the Right to Food, the food commons declares the inherent tensions of liberal democracy as irresolvable through liberal constitutionalism alone.

As an alternative, the food commons envisions a broader 'communicative realm' not confined to liberal constitutionalism, but focused on the way 'discursive sources of order' can influence governance (Dryzek, 2000). The food commons thus proposes 'discursive democracy' in which civil actors democratize governance by contesting established conventions and influence decision-making bodies through rhetorical—rather than electoral—means.

The challenge taken up by food commons advocates then, is to somehow discursively construct a socialized and de-commodified food regime within the capitalist regime of commodity production—while working within a corporately compromised political structure designed for individual rights and private property. This is no trivial task, especially in countries where the constitution is routinely violated, where state violence reigns or where there is no social contract. Further, the history of the Commons has not been to alter hegemonic structures of capitalist or liberal-democratic dominance, but to provide a space of resistance for communities attempting to protect themselves from these very structures. The food commons as a political proposal rests on volatile and uncharted grounds.

The food commons narrative claims the reduction of food's multiple dimensions into a singular conception of food as a commodity is the 'discursive source of order' that drives hunger. Therefore, the proper valuation of food as an inalienable human right itself provides a weapon for transformation. Here, revalorizing food's multiple dimensions (and the rejection of food as a pure commodity) brings together a diversity of actors who, once enlightened with the rationale of food as a commons, assume agency as 'food citizens' (Vivero-Pol, 2017a).

The shift from liberal constitutionalism to the reflexive agency of civil actors exhibited in the vision of food as a commons is illuminated in the proposal for 'tricentric governance', in which 'self-regulated, civic collective actions for food' gain an increasing leverage over the state and market (Vivero-Pol, 2017d, p. 345). In essence, tricentric governance rebalances the political forces between government, economy and civil actions by re-appropriating public space from markets and the state, whereby civil actors engaging in collective action become an agency of their own.

Reaching below and beyond the surface of allegedly fractured states, the food commons ties in a diverse set of initiatives, movements and practices that preserve and rediscover commons-based solution as drivers of the transition to a food commons regime. While ranging widely, these are broadly classified as 'customary food movements' of indigenous communities;

subsistence and small food producers resisting enclosures; and 'contemporary collective innovations,' such as alternative food networks, urban gardening, etc. (Ferrando and Vivero-Pol, 2017). The concept of food as a commons further includes a variety of food-related things that may already be commonly understood as public goods (such as recipes and genetic resources) that are anticipated to provide a space for actors to converge around the multidimensional value of food.

Giving way to a 'mounting force of citizens' actions to reclaim food, the state is re-functioned from facilitator of the capitalist 'accumulation through enclosure' to regulator of capital interests and provider of enabling frameworks for food citizens (Vivero-Pol, 2014). The food commons is then tasked not only with socializing food, but also with transforming the liberal democratic state. This assumes the state is responsive to the demands of its citizens. For example, during the transition phase towards a new food regime, the state should provide incentives and enabling frameworks, such as basic food entitlement and food security floors, to support commoners as they initiate alternative means of food production and food-sharing at local scales and to 're-commonify' the structures that govern the food system, effectively pushing back the influence of both state and market (Vivero-Pol, 2013).

These collective actions for food are admittedly based in different traditions and localities. The forms of food production and food sharing vary and they draw from different civic, market and state sources. The notion of collective action in polycentric governance thus counters models of governance based on the aggregation of individual interests, operating instead as 'nodes' of connected but autonomous food-centers. Once such commons have gained ground, the thinking goes, this 'third pillar' of governance will evolve into self-regulated and diffuse collective action, replacing mechanisms of demand and supply with diverse, hybrid models of governance, effectively 'crowd-feeding the world' (Vivero-Pol, 2017d).

In a global food commons coordinated through a polycentric governance model, self-organized groups under self-negotiated rule develop 'food democracies', operating through 'nodes' of connected but autonomous food-centers, achieving—at last—a free association between producers and consumers (Caffentzis, 2010). At this point, the role of the state as regulator and provider of enabling frameworks can decrease and both markets and states are relegated to one of many ways of allocating resources.

The concept of tricentric and polycentric governance thus operates by a vision in which the valuation of food as a common good unites otherwise diverse actors within and across borders of nation-states and across neo-colonial world orders and elevates food above the capitalist mode of production in which it is currently embedded. The food as a commons narrative thus believes that *if* food could be *considered* as a commons, the global food system (and capitalism) *would* change (Vivero-Pol, 2017a). The extraordinary political agency placed in the proper valuation of food implies the proposal of food as a commons itself convinces both the rights-holder and beholder by "the unforced force of the better argument" (Habermas, 1996, p.306). Accordingly, the conception of food as a common good is assumed to morally bind existing legal frameworks for the Right to Food and enable new international treaties (presumably between liberal nation-states) to secure food's multiple valuations.

Unpacking food as a commons

The 'food as a commons' sweeping vision for a food-centric global transformation is perhaps more 'persuasive' than 'convincing' (Ferry, 2012). Either way, to consider its utility in the struggle for food sovereignty, a number of foundational assumptions need to be addressed. Foremost in the food as a commons narrative are the co-constitutive assumptions that food sharing is the material basis for the transformation of the capitalist food regime and that the ideational power of discursive democracy is a sufficient driver to refashion the regime's governance structure (Fuchs and Glaab, 2011). These basic assumptions give rise to a host of corollary suppositions. All should be interrogated.

First, in its critiques of capitalist fictions, the food as a commons discourse replicates a dis-embedded concept of food as well as a form of liberal politics that it presumes is 'common sense,' effectively removing these from the sphere of discursive struggle. Granted, the food-as-a-commons counter-movement deconstructs the hegemony of market liberalism and its representation of people as consumers, by which it recognizes the lack of food-access as socially constructed and incorporates alternative operating categories of human action and food (Ferrando and Vivero-Pol, 2017). Most obviously, its identification of food as a social construct with open and fluid boundaries serves to dislodge privileged notions of food as a commodity (Vivero-Pol, 2017b). Reacting against scarcity discourses and their legitimation of the industrial food-complex, the food-as-a-commons discourses poses food as an abundant, renewable resource to which all have sufficient access (Vivero-Pol, 2014). Likewise, the emphasis on shared values and cooperation amongst actors pushes against the rationale of individuals as merely self-interested, self-maximizing actors of competitive market structures. However, while food-supplies at the global scale may be continuously replenished, privileging 'the category of food as a whole' assumes the exchangeability and immediate availability of food (Vivero-Pol, 2017d, p. 340). This abstraction of food from its physical and geographical realities dis-embeds food from the labor process and from society (much as the capitalist food regime does) and arguably loses political and material ground that is historically essential to the Commons.

Second, in assuming that discursive, deliberative democracy will lead a transformation towards more sustainable and ethical forms of production, the political agent in the food commons narrative is the western 'eater' or food citizen. This seems like an uncomfortable projection of consumer politics upon the communities who produce most of the world's food. That food sharing—rather than ownership of the means of production or the redistribution of assets—be chosen as the lever for structural or transformation change is arguably an inherently liberal notion. And to the extent that the food commons narrative fails to address the roles of land and labor, it leaves much of capital's power intact (Turner, 2017).

While the *refusal* to consume food (boycotts, hunger strikes, etc.) is indeed a powerful economic and political tactic, this does not mean that a call for food as a commons in and of itself is necessarily a structural or even a strategic lever for transformation. Consumer calls for greener, healthier, less waste-producing food products, for example, are easily incorporated into the capitalist marketplace. Villagers living in poor rural communities throughout the developing world frequently prefer cheap, industrial products to their home-grown foodstuffs. In Cuba, where the state supports multiple, innovative forms of healthy food commoning, there is a growing desire for newly introduced industrial food commodities from the United States. None of this assumes that the majority of people *couldn't* prefer healthy, locally produced food over globalized, industrial food, but to assume that they do—and that this will drive systems transformation—is a heroic projection of values.

The food-as-a-commons literature makes risky assumptions about the nature of social change. The collective choice models applied portray institutional change as sequential, in which users' experiences and commitments generate expectations by which participants forge more beneficial arrangements (Johnson, 2004). This assumes property regimes are the product of an equitable selection process, in which stakeholders reciprocally construct the institutions best suited to the needs of the community. It also assumes a remarkable homogeneity of power within and between the nodes of polycentric governance.

Food—its production, consumption and means of exchange—has long been a tool of power, oppression and differentiation, sometimes through or within the Commons. Where commons and collective action models are regularly understood to require some level of trust, a shared past and expectations for the future, as well as a notion of mutual vulnerability, these attributes cannot be assumed to be common amongst all food citizens on the basis of being 'eaters' (McCay, 2002). To the extent that the idea of food as a commons provides a space for 'customary food movements' resisting ongoing enclosures and contemporary 'collective innovations' to converge around the multidimensional values of food, this kind of cross-sectoral convergence demands recognizing differences in position, how these came about and how they are sustained (Vivero-Pol 2017a). Any claim to commonality that obscures difference and embedded asymmetries between actors is likely to reproduce them.

In particular, the idea that social relationships found successful in one particular community can be transferred to higher scales of governance elides the implications of actors' embeddedness in a specific place and as *part of* larger political, economic and cultural structures (Mosse, 1997). While plurality and shared values around food may—potentially—be the basis for commoning *within* nodes in the polycentric mode of governance, the relations *between* different nodes will by no means be equal. If this is to be done at local scales (possibly with local, time-based currencies?), how will these communities equitably exchange goods with other localized food systems? What about states in which there is no social contract or limited fiscal capacity to secure public goods such as food baskets? These concerns are anathema to many governments, beyond the competencies of localized food nodes and certainly outside the political capacity of the United Nations system.

While there may be agreement on the essential nature of food as substance for life, there is little agreement on the different mechanisms of production and governance of food. In practice, 'self-regulated civic collective actions' decisions on 'appropriate' combinations of governance mechanisms inevitably draw on the existing structures of power and privilege (Johnson, 2004). The political ability and desirability to 'scale up' and 'scale out' local experiments to achieve the 'right combination' of collective action, government rules and incentives and private entrepreneurship are thus likely over-assumed. What may be a progressive collaboration between civil actors and innovative market enterprise to some, is exclusionary and symptomatic of oppressive market structures to others. Without adequate attention to privilege and power and their synergies with capital, the kind of polycentric governance proposed by the food commons runs the risk of advocating a kind of 'libertarian municipalism' (Cumbers, 2015).

Further, the notion that defining food as a public good provides sufficient ideational power to elevate it above the arena of capitalist market structures, is a belief that is poorly substantiated by past and contemporary examples of common and public goods and global public goods especially. Instead, we see public goods advanced through a rights-based structure of liberal constitutionalism and as such remain subject to the state's internal contradictions.

The main contradictions in the food-as-a-commons consumption-transformation proposal thus concern value, relations and scale. Accounting for these requires refraining from presuming any prior set of entitlements and foregrounding social and political-economic processes which substantiate, undermine or deny entitlement (Johnson, 2004). That is, sustained efforts aimed at democratizing food regimes demands a shift in focus from prescriptive 'right to benefit' to a descriptive understanding of the 'ability to benefit' (Ribot and Peluso, 2003). Constituted of various 'extra-legal mechanisms,' abilities to benefit from institutions and resources are shaped by class, social identity, access to capital, markets and technical means and networks which in turn shape the way food is produced, the flow of benefits and the capture of rents along value-chains (Ribot, 1998). To farmers, the ability to realize entitlements requires access to resources, infrastructure and markets as well as leverage over price. Because a proportionally larger amount of labor goes into producing food on smallholder farms compared to large industrial farms, recognizing the value of labor—an important aspect of food sovereignty—should be of special concern.

Finally, the food commons' discursive democracy approach to social change seems not to account for the global crisis of democracy, evident in the rise of right-wing populism and the extreme levels of material inequality in the world today (thanks to the neoliberal excesses of liberal democratic regimes). The masses are not clamoring for democracy (or equality), but political and economic *equity*. This is what is behind the revival of the Commons in the first place. But the conflation of discourse, democracy and equality within the food commons narrative prevents their strategic assessment. Any effort at 're-commonification' to construct a food commons, in fact, entails the de-commodification and dismantling of existing regimes of capital and power. Which is to say, anti-capitalist commons are, and will have to be, the product of class struggle (Federici and Caffentzis, 2014).

Food sovereignty, commoning, and class struggle

The notion of food sovereignty comes from a long history of peasant struggle that began, notably, with the struggle for the Commons in the face of enclosures pushed by large landowners and textile manufacturers beginning in the sixteenth century. The enclosures were bitterly contested by peasants, exploding in riots and rebellions in the face of the Enclosure Acts of the eighteenth and nineteenth centuries. The enclosures undermined the ability of people to feed themselves and created a destitute landless class that was obliged to work for wages. The battle over the Commons was at the core of a centuries long, anti-capitalist, class war. Any attempt to resist enclosures—or to reinstate a commons—is necessarily a chapter in this struggle.

The concept of food sovereignty is openly grounded in class. The 'sovereign' is the smallholder, struggling to assert control over productive land and resources, their own labor power and the value of their product: food. Because smallholders both consume and sell the food they produce, the call for food sovereignty addresses not only food as a right but food as a saleable good. Smallholders—like all workers—seek to be fairly compensated for the food they sell (parity).

As a component of class struggle, the Commons has taken many different forms over time. The de-commoditized role of food in revolutionary struggles has been significant, not only as a key component of resistance, but as a model for new social relations based on mutual aid. On the eve of the Spanish Civil War, peasants from the National Confederation of Labor (CNT) and the Federation of Land Workers pressured the new Republican government to implement land reform by carrying out massive occupations on half a million acres of farmland. When Nationalist forces led by General Francisco Franco rebelled against the Spanish Republic in July of 1936, anarchists formed worker-peasant militias to fight the fascists. As the Republic struggled to stay in power, in Aragon and in the Levante, peasant militias seized the property of the Church and large landowners, bringing them under collective rule. Tens of thousands of acres of municipal land were worked in common. Money was abolished. Food produced on the collectives was distributed freely and equally through ration cards. Labor, tools, goods and militia members flowed between collectives as needed. Commissions for the production and sale of dairy, livestock, rice, oranges, potatoes and other crops set quotas and arranged distribution. In the Levante alone, the Regional Federation of Agricultural Collectives managed five hundred collectives from fifty-four sub-cantonal federations and five provincial federations (Guerin, 1998). The CNT and the Land Workers Union paid wages to farmworkers and sold the surplus at cost to workers' unions in the cities. Similar experiences in agrarian collectivization took place in Castile and the Extremadura. The anarchist agrarian collectives established 'a new social order based on direct control of the country's productive resources by workers' committees and peasant assemblies' (Bookchin, 1994).

Food commoning was a deeply embedded political strategy constructed within a revolutionary movement for a new, classless society, governed on anarcho-syndicalist principles. While the collectivization of land and food was forged in the crucible of anti-capitalist revolution and an anti-fascist war, the social relations for the egalitarian administration of land, labor and the sharing of food built on longstanding peasant traditions (Bookchin, 1977). Two things become immediately clear; first, that the food as a commons was not the guiding principle, but a corollary system of organization of resources, and secondly, that the practice of food commoning was grounded in the practice of collectivization, itself part of traditional forms of mutual aid and village-scale Commons regimes.

Brazil's Landless Workers Movement is another example of anti-capitalist struggle in which the de-commodification of food plays a key role (Wright and Wolford, 2003). In 1984, a strong convergence of movements came together in 1984 to form the Brazilian Landless Workers' Movement (*Movimento dos Trabalhadores Rurais Sem Terra*, or MST). Taking advantage of the 1964 Constitutional Land Statute, in which land must serve a 'social purpose,' the MST organized hundreds of thousands of landless workers to occupy idle and socially unproductive lands belonging to the *latifundios* (large landed estates) (ibid.). With its roots in socialist activism, Liberation Theology and the popular education theories of Paulo Freire, the MST is at the forefront of agrarian reform. The MST identifies and occupies underutilized or empty lands to gain legal title and bring it into productive use using agroecology. Once underused land is successfully occupied, schools, cooperatives and credit unions are set up and the land is farmed to grow fruits, vegetables, grains, coffee and livestock. Present in 23 of Brazil's 25 states with over 1 million members, the movement has formed over 2,000 settlements, settling over 370,000 families with an estimated 80,000 more awaiting settlement. The MST has established a network of approximately 2,000 primary and secondary schools, partnered with 13 public universities, 160 rural cooperatives, 4 credit unions, food processing plants and retail outlets (Carter, 2015). Food, shared within and between settlements, is still sold in capitalist markets, however, and provides important income to MST farm families.

Like the Spanish collectives fifty years prior, the MST's food commoning meets members' immediate needs by socializing the means of production, as part of a larger political project for social transformation. It is strongly grounded in place (the settlement).

Today in Cuba, production cooperatives meet state production targets, sell in private markets and also donate a considerable surplus to schools, hospitals and retirement homes. They are able to do this because land has been de-commodified and is not a cost of production and because farmers enjoy state-enforced price floors that are close to parity. They also benefit from free health care, free education and highly subsidized housing, thus lowering their living costs. Food is one of many public goods that are regulated by the state and managed at local, regional and national scales. The peasant and cooperative sector was instrumental in overcoming Cuba's food shortages following the fall of the Soviet Union and the end of the massive food and petroleum subsidies. Food and food production are deeply embedded in socialist social relations in which unlimited access to land, water and markets, combined with free social services, protect and promote the smallholder farmer class. These conditions—addressed in other, non-state and intra-state ways by the examples of Spain and Brazil—may be *sine qua non* for the existence of a food commons.

There are many other examples—from the Commons in liberated Zapatista areas of Chiapas, Mexico to the Aymaran and Quechuan *ayllu* connected to indigenous resource struggles in the Andes—in which food and the Commons serve an integral role in the struggle for sovereignty.

Connecting the lessons and the politics of these existing experiences should be a central task for the food commons proposal.

Conclusion: Can the food commons advance food sovereignty?

The answer to this question rests, of course, with the social movements for food sovereignty. The central concerns from both a theoretical and historical perspective revolve around embeddedness, grounded-ness and political relevance. Is the notion of food as a commons embedded within the social relations of smallholder production? Is it grounded in material agrarian structures of food production? Does it serve a strategic political role in the construction of food sovereignty and liberation from the corporate food regime? In short, can food as a commons protect peasant, fisher and pastoralist livelihoods from the ravages of the market and from capitalist dispossession and can it help ensure their access to basic food-producing resources (land, water, forests) and can it build cross-sectoral solidarity, materially and politically?

These questions are admittedly less overarching than the proposals for a polycentrically governed global food regime based on food commoning. But in addressing them, we can begin to imagine how food commoning *could* help link rural and urban struggles fighting for land, social justice and political power. This exercise also releases the food as commons from the confines of its present meta-narrative and opens the possibility of food commoning as a strategy for the class, race and gender liberation struggles at the heart of the food sovereignty movement.

An embedded food commons should constitute, or at least reflect, the aspirational social relations of food sovereignty. One can imagine how food commoning could help link urban and rural struggles, strategically facilitating material and political alliances between agricultural and industrial sectors in non-exploitive ways that share costs, benefits and solidarity. It is less clear how this could happen at international scales. There is still much social learning to do prior to proposing a global food commons.

The strategic question for a possible alliance between the food sovereignty movement and food commons advocates is, how can the Commons—as a means rather than an end—help existing frontline struggles create a new, non-capitalist mode of production that is not built on the exploitation of labor, women and people of color? The viability of food as a commons will depend on whether or not actors from the front lines of struggles actually incorporate the food commons into their respective causes.

Bibliography

Bookchin, M. (1977) *The Spanish Anarchists: The Heroic Years 1868–1936*, Halper Colophon Books, New York

Bookchin, M. (1994) *To Remember Spain*, AK Press, San Francisco

Bresnihan, P. (2015) 'The More-Than-Human Commons: From Commons to Commoning', in S. Kirwan, L. Dawney and J. Brigstocke (eds) *Space, Power and the Commons: The Struggle for Alternative Futures,* Routledge, New York

Brown, K. M. (2007) 'Understanding the Materialities and Moralities of Property: Reworking Collective Claims to Land', *Transactions of the Institute of British Geographers*, vol 32, no 4, pp507–522

Caffentzis, G. (2010) 'The Future of "The Commons": Neoliberalism's "Plan B" or the Original Disaccumulation of Capital?', *New Formations*, vol 69, pp23–41

Carter, M. (ed.) (2015) *Challenging Social Inequality: The Landless Rural Workers Movement and Agrarian Reform in Brazil,* Duke University Press Books, Durham

Cumbers, A. (2015) 'Constructing a Global Commons in, against and beyond the State', *Space and Polity*, vol 19, no 1, pp62–75

De Angelis, M. and Harvie, D. (2014) 'The Commons', in M. Parker, G. Cheney, V. Fournier and C. Land (eds) *The Routledge Companion to Alternative Organization*, Routledge, New York
Dryzek, J. S. (2000) *Deliberative Democracy and Beyond: Liberals, Critics, Contestations*, Oxford University Press, New York
Dryzek, J. S. (2006) *Deliberative Global Politics: Discourse and Democracy in a Divided World*, Polity Press, Cambridge
Federici, S. (2011) 'Feminism and the Politics of the Commons', *The Commoner*, http://www.commoner.org.uk/?p=113, accessed 20 March 2017
Federici, S. and Caffentzis, G. (2014) 'Commons against and beyond Capitalism', *Community Development Journal*, vol 49, pp92–105
Fennell, L. A. (2011) 'Commons, Anticommons, Semicommons', in K. Ayotte and H. E. Smith (eds) *Research Handbook on Economics of Property Law*, Edward Elgar, Cheltenham
Ferrando, T. and Vivero-Pol, J. L. (2017) 'Commons and "Commoning": A "New" Old Narrative to Enrich the Food Sovereignty and Right to Food Claims', *Right to Food and Nutrition Watch*, pp50–56
Ferry, V. (2012) 'What is Habermas's "'Better Argument'" Good For?', *Argumentation and Advocacy*, vol 49, no 2, pp144–148
Fuchs, D. and Glaab, K. (2011) 'Material Power and Normative Conflict in Global and Local Agrifood Governance: The Lessons of "Golden Rice" in India', *Food Policy*, vol 36, no 6, pp729–735
Guerin, D. (1998) *No Gods, No Masters: An Anthology of Anarchism*, AK Press, San Francisco
Habermas, J. (1996) *Between Facts and Norms: Contributions to a Discourse Theory of Law and Democracy*, MIT Press, Cambridge
Hardin, G. (1968) 'The Tragedy of the Commons', *Science*, vol 162, no 13, pp1243–48
Hardt, M. and Negri, A. (2011) *Commonwealth*, Belknap Press, Cambridge
Holder, J. B. and Flessas, T. (2008) 'Emerging Commons', *Social & Legal Studies*, vol 17, no 3, pp299–310
Holt-Giménez, E. (2017) *A Foodie's Guide to Capitalism: Understanding the Political Economy of What We Eat*, Monthly Review Press, New York; Food First Books, Oakland
Holt-Gimenez, E., Patel, R. and Shattuck, A. (2009) *Food Rebellions: Crisis and the Hunger for Justice*, Pambazooka Press, London; Food First Books, Oakland
Howard, P. L. and Nabanoga, G. (2007) 'Are There Customary Tights to Plants? An Inquiry among the Baganda (Uganda), with Special Attention to Gender', *World Development*, vol 35, no 9, pp1542–1563
Janvry, A. (1981) *The Agrarian Question and Reformism in Latin America*, Johns Hopkins University Press, Baltimore
Johnson, C. (2004) 'Uncommon Ground: The "Poverty of History" in Common Property Discourse', *Development and Change*, vol 35, no 3, pp407–434
Kautsky, K. (1988) *The Agrarian Question*, Zwan Publications, London
Kirwan, S., Dawney, L. and Brigstocke, J. (2015) *Space, Power and the Commons: The Struggle for Alternative Futures*, Routledge, New York
La Via Campesina (2011) 'The International Peasant's Voice', *La Via Campesina: International Peasant's Movement*, https://www.viacampesina.org/en/index.php/organisation-mainmenu-44, accessed 13 March 2017
McCay, B. J. (2002) 'The Emergence of Institutions for the Commons: Context, Situations and Events', in E. Ostrom, T. Dietz, N. Dolsak, P. C. Stern, S. Stonich and E. U. Weber (eds) *The Drama of the Commons*, National Academies Press, Washington DC
McMichael, P. (2013) *Food Regimes and Agrarian Questions*, Fernwood Publishing, Nova Scotia
Moore Lappé, F. and Collins J. (2015) *World Hunger 10 Myths*, Food First Books, Oakland
Mosse, D. (1997) 'The Symbolic Making of a Common Property Resource: History, Ecology and Locality in a Tank-irrigated Landscape in South India', *Development and Change*, vol 28, no 3, pp467–504
Nightingale, A. J. (2010) 'Beyond Design Principles: Subjectivity, Emotion, and the (Ir)Rational Commons', *Society & Natural Resources*, vol 24, no 2, pp119–132
Nightingale, A. J. (2014) 'Questioning Commoning', *Social & Cultural Geography*, vol 15, no 8, pp 980–982
Ostrom, E. (1987) 'Institutional Arrangements for Resolving the Commons Dilemma: Some Contending Appraches', in B. McCay and J. Acheson (eds) *The Question of the Commons: The Culture and Ecology of Communal Resources*, University of Arizona, Tucson
Ostrom, E. (1990) *Governing the Commons: The Evolution of Institutions for Collective Action*, Cambridge University Press, Cambridge
Ostrom, E. (1992) 'Community and the Endogenous Solution of Commons Problems', *Journal of Theoretical Politics*, vol 4, no 3, pp343–351

Ostrom, E. et al. (1999) 'Revisiting the Commons: Local Lessons, Global Challenges', *Science*, vol 284, no 5412, pp278–282
Ostrom, E. (2010) 'Beyond Markets and States: Polycentric Governance of Complex Economic Systems', *Transnational Corporations Review*, vol 2, no 2, pp1–12
Ostrom, E. (2014) 'Collective Action and the Evolution of Social Norms', *Journal of Natural Resources Policy Research*, vol 6, no 4, pp235–252
Ostrom, E. and Hess, C. (2007) 'Private and Common Property Rights' *Social Science Research Network*, https://papers.ssrn.com/abstract=1936062, accessed 1 January 2017
Ostrom, V. and Ostrom, E. (2015) 'Public Goods and Public Choices', in D. H. Cole and M. D. McGinnis, *Elinor Ostrom and the Bloomington School of Political Economy: Resource Governance*, Lexington Books, Washington, DC
Pistor, K. and Schutter, O. D. (2015) *Governing Access to Essential Resources*, Columbia University Press, New York
Ribot, J. C. (1998) 'Theorizing Access: Forest Profits along Senegal's Charcoal Commodity Chain', *Development and Change*, vol 29, no 2, pp307–341
Ribot, J. C. and Peluso, N. L. 'A Theory of Access', *Rural Sociology*, vol 68, no 2, pp153–181
Schlager, E. and Ostrom, E. (1992) 'Property-Rights Regimes and Natural Resources: A Conceptual Analysis', *Land Economics*, vol 68, no 3, pp249–262
Smith, H. E. (2000) 'Semicommon Property Rights and Scattering in the Open Fields', *The Journal of Legal Studies*, vol 29, no 1, pp131–169
Turner, M. D. (2017) 'Political Ecology III: The Commons and Commoning', *Progress in Human Geography*, vol 41, no 6, pp795–802
Vatn, A. (2005) *Institutions and the Environment*, Edward Elgar Publishing, Cheltenham
Vatn, A. (2015) *Environmental Governance: Institutions, Policies and Actions*, Edward Elgar Publishing, Cheltenham
Vivero-Pol, J. L. (2013) 'Food as a Commons: Reframing the Narrative of the Food System', *Social Science Research Network*, https://papers.ssrn.com/abstract=2255447, accessed 1 January 2017
Vivero-Pol, J. L. (2014) 'The Food Commons Transition: Collective Action for Food Security', *The Broker*, http://www.thebrokeronline.eu/Articles/The-food-commons-transition, accessed 2 January 2017
Vivero-Pol, J. L. (2017a) 'Food as Commons or Commodity? Exploring the Links between Normative Valuations and Agency in Food Transition', *Sustainability*, vol 9, no 3, p442–445
Vivero-Pol, J. L. (2017b) 'How Do People Value Food? Systematic, Heuristic and Normative Approaches to Narratives of Transition in Food Systems', PhD thesis, Université Catholique de Louvain, Belgium
Vivero-Pol, J. L. (2017c) 'The Idea of Food as Commons or Commodity in Academia. A Systematic Review of English Scholarly Texts', *Journal of Rural Studies*, vol 53, pp182–201
Vivero-Pol, J.L. (2017d) 'Transition towards a Food Commons Regime: Re-Commoning Food to Crowd-Feed the World', in G. Ruivenkamp & A. Hilton (eds.). *Perspectives on Commoning*, Zed Books, London
Waldron, J. (2004) 'Property and Ownership', *Stanford Encyclopedia of Philosophy*, http://plato.stanford.edu/archives/spr2012/entries/property/, accessed 12 January 2017
Wright, A. L. and Wolford, W. (2003) *To Inherit the Earth: The Landless Movement and the Struggle for a New Brazil*, Food First Books, Oakland

21

LAND AS A COMMONS

Examples from the UK and Italy

Chris Maughan and Tomaso Ferrando

Land is at the root of the food system

We assume that if you chose to read this book it is because you share with us the idea that the fundamentals of the contemporary food system must be rethought. Food poverty, non-communicable diseases, environmental degradation, social injustice and resource depletion are only a few of the visible consequences of a food system intended to produce and distribute food not on the basis of needs and rights, but on the demands of the market.

For such a fundamentalist market logic to prevail, food must be defined and treated like a commodity, no differently than a mobile phone or crude oil. In an era of ubiquitous online purchasing and the 24-hour drive-through, it is even easier for food to be thought of this way, divorced from its social and ecological contexts. When food is commodified, David Harvey reminds us (2003), the complexity of social and productive relationships becomes hidden – a fetish is created that covers the imbalances of power on which such processes rely, as well as their negative ecological impacts.

The shift from food as a commodity to food as a commons that is advocated in this book, we argue, must thus begin with the mapping and identification of those struggles and political choices that are hidden behind the act of consumption. Of all the elements, from labour to seeds, from water to the means of production, we believe that particular attention must be paid to the de-commoditization of the land and the soil that make food possible. In other words, we believe that there cannot be a true *commoning* of the food system as long as land (and water and all the other elements that are essential to the generation of food) continue to be defined through the prism of absolute proprietary regimes. Indeed, the political choices around the allocation of rights and limits concerning land and its fertility lie at the core of any society, not only of its food system, and a transformation of the latter requires a reconfiguration of what land is and the role that it plays in social and ecological reproduction.

Much of this will be familiar to readers with a basic understanding of the commons and its histories. Indeed, the enclosure of agricultural lands is often identified as the basis on which modernity and the western economic system were created (Wallerstein, 1974; Arrighi, 1998). The land of First Nations in North America; the commons in the United Kingdom; the forests and pastoral land communally managed throughout the Global South: all have been seized and fenced, subordinated to the logic of private ownership, capital accumulation and economic growth. Once agricultural land is reduced to its exchange value, it is also easily stripped of its social and cultural meanings, which are context-specific and epistemically determined (Diaz et al., 2018). Its living complexity is simplified to a substrate upon which to build limitless private wealth, insensible to any resultant damage to the social and ecological fabric. Land becomes a homogeneous, standardized and objectified *asset* (Scott, 1999).[1] Any attempt to change the mainstream food system in line with ecological and justice concerns must therefore begin with recognizing land and soil as a common resource, one that must be collectively, sustainably and democratically managed, generating benefits that are accessible and shared, especially among those most in need.

In dealing with land as a commons for a commons-based food system, this chapter does not engage substantially with the history of enclosures nor with the long history of resistance to such activities. Rather, we decided to focus on contemporary forms of mobilization that combine the idea of the commons with land and soil as key components in the creation of an ecological and democratic food system. However, given the space constraints and the abundance of examples from all over the world, we had to make a decision based on our intellectual interest and our geographical backgrounds. Firstly, we constructed our contribution around the idea that the commons most certainly continue to be an issue of great political significance, but not without its contradictions and inconsistencies. Following Laforge et al. (2016), we thus offer a framework which acknowledges this diverse field of action, from the regime-oriented (or 'dominant') through the reformist to the most radical of proponents. Secondly, as two citizens of the Global North who are involved in local movements in our own geographies, we considered it appropriate to speak about those examples and realities that we know best. This is not intended to diminish the many stories of land struggles and resistance in the name of the commons originating from the Global South. We hope the reader will not dismiss our choice as mere academic Eurocentrism; on the contrary, we hope you will be curious to know more about the struggles for land and the commons that are taking place at 'the core' of the global capitalist project.

With this in mind, the following section introduces a range of examples of 'food commoning' in the United Kingdom, including those that provide evidence of the effort to 'co-opt' the power of the commons and contain any political fallout that might arise from this; those that we call 'collaborators' or reformists, i.e. those attempting to find a space of compromise within the circuits of capitalism, to carve out a legitimate niche within the enclosure; and finally, those groups who position themselves in a stance of 'contestation' with mainstream food regimes, calling for a radical overhaul of the current system. The next section moves to Italy and examines the case of 'Mondeggi as a Common Farm without Masters' to offer another case of 'contestation'. This example reflects on the possibility of linking the theory and practice of the commons with the struggle for land and food sovereignty (Borras et al., 2015) and imagining a post-capitalist property regime where public, private and common coexist (Ireland and Meng, 2017). By way of conclusion, we offer a reflexive appeal for all who read this to engage directly with these rare and ephemeral examples of the commons – to learn from and join those currently fighting to build and sustain such spaces and to ensure that stories and actions of commoning continue to be told and experienced.

Commoning at the heart of the enclosures: co-optation, collaboration and contestation

From one perspective, it is a miracle that 'the commons' has any political currency in the UK at all. As the title of this subsection indicates, Britain constituted the epicentre of the capitalist accumulation project. From Thomas More's (2016 [1516]) disquieting account of sheep that 'devastate and depopulate fields [...leaving] no land free for the plough' to Wallenstein's unflinchingly detailed World-Systems volumes (1974), the annals of British enclosure offer a grim, 700-year-cavalcade of misery[2]. And that's not all; history has not been kind to the commons as a concept – whether brutally (but unfairly) disavowed in the now infamous 'Tragedy of the Commons' (Hardin, 1968) or expertly unravelled by the brilliant Shared Assets (2015), to many 'the commons' has become a problematic term at best – it never really worked, or even existed at all, and it never will. But if you look to the grassroots today, the opposite is irrefutable. The commons has, against all odds, retained the power of a clarion call, capable of rallying the vestiges of an ancient agrarian struggle for land justice, keeping alive a vision of food and farming for the many, not the few.

The most recent chapter in the 700-year-history of the erosion of common land in the UK has been the much-documented disappearance of public spaces (Grolle, 2008). In food growing terms, this has been seen most extensively in the loss of municipal allotments, which have fallen by more than 75% since the 1950s (NWCAA, 2017). In this context, one might welcome any effort to halt this trend and (re)-open any and all spaces for publicly accessible, community food-growing purposes. Indeed, recent years have seen an unprecedented spike of interest in the development of various forms of community food production, such as community supported agriculture (CSAs), community gardens and *guerilla* gardening (Guitart et al., 2012; Tornaghi, 2017).

Increasingly, and in the terms of Laforge et al. (2016), this popularity has also inevitably been the target of 'co-optation', in many instances creating insidiously negative outcomes for the communities purported to benefit from their implementation. This confounding phenomenon has even coined a delightful neologism: 'Green LULUs'; that is, 'locally unwanted land uses'. Evidence is emerging that the placement of green infrastructures (e.g. community gardens), far from being desired by, or of obvious benefit to, proximate low-income residents, can have the adverse effect of increasing rents in surrounding areas (i.e. 'gentrifying' them), pushing out working class or otherwise low-income communities who have historically resided there. Research done in this area has consistently emphasized the deleterious role played by corporate backers of such Green LULUs, particularly where decision-making around their placement and implementation is taken out of the hands of local residents (Anguelovski et al., 2017).

A recent example in the UK is the 'Bags of Help' project run by Tesco supermarket. The scheme, intended to fund community projects (many of which have a food-growing focus), rather predictably selected projects based on their proximity to Tesco stores and awarded its grants on condition of strict adherence to 'acknowledge [...] the support of Tesco in any published documents' (Groundwork, 2017). As such, the scheme is clearly not only run-of-the-mill 'green wash', but also a fairly elaborate extension of Tesco's marketing strategy, complete with 'signposting' to the nearest outlet. The argument here is not that the 'Bags of Help' scheme is particularly inimical – such behaviour is, unfortunately, only to be expected from a corporate mainstay like Tesco – instead, the point to grasp here is that community or 'commons' represents spaces and processes that are being steadily appropriated and that their transformative potential will only work if they are truly designed and determined by the communities who use them. 'Bags for Help' is first and foremost about Tesco, not the enrichment of civic and communal cultures, ecological food alternatives, or collective assets.

The jury is still out on the impact (negative or otherwise) of Tesco's 'Bags of Help' scheme (at the time of writing, the project is only part way through its second round of awards), but it is undeniably problematic to have community initiatives determined not by local needs, but ultimately by the marketing strategy of a corporate superpower. Indeed, put this way, who can ignore the existential threat Tesco and its global supply chains pose to attempts to reclaim and develop food commons? In what ways, we must ask, are the political potentialities of such community schemes being distorted and subverted by their supposed benefactors? Such counterfactuals are, sadly, almost impossible to determine.

Thankfully, in the UK at least, such corporate co-optation of food communities and food commons is relatively unusual (for another example, however, see the Heineken Orchard Blossom[3]). Much more well-known are attempts by organizations to 'play the game' of private ownership in order to create commons-like spaces where food and food-related knowledges can be influenced by, and be of benefit to, the majority. Over the last two decades a number of such initiatives have emerged using a variety of community-based platforms. CSAs are perhaps the most prominent community-based platform (Urgenci, 2016), but there are also land trusts (e.g. Kindling Trust in Manchester, the Bristol Land Trust), cooperatives (e.g. the Ecological Land Cooperative, Organic Lea), knowledge sharing platforms (e.g. Farm Hack, Soil Hack) and innovative right-to-food activism (e.g. The Real Junk Food Project), all of which make an effort to establish legal grounds on which to redistribute land, food and agroecological knowledge and build a food system around a decommoditized and commoning paradigm.

Perhaps the most widely publicised example of such 'collaborative' food commoning is 'Incredible Edible' (IET), based out of Todmorden in Yorkshire, which aims to convert public assets, especially ornamental flowerbeds, into food-producing spaces. IET boasts many things, including a rapidly expanding network of affiliated projects (over 100 in the UK and more than 700 worldwide (Peart, 2015)), beneficial impacts on wildlife, increases in healthy eating practices, community cohesion and flourishing local food economies (IET, 2017b). IET is notable for its use of radical language, such as its deliberately playful neologism for *guerrilla* gardening – 'propaganda gardening'. IET's aim is nakedly to 'revolutionise local food systems' through the 'power of small actions' (Russi, 2015) and explicit about its intention to 'repurpos[e] the commons for open source food and agricultural biodiversity' (Paull, 2013). IET has been singularly successful (in the UK at least) in reinvigorating and exporting an idea of the urban food commons in which 'there is something for everyone' (IET, 2017a).

That said, IET's collaborative or 'big tent' approach (see McMichael, 2008) often means that it treads a difficult line between its radical objectives and some of its more mainstream collaborative aspirations. Indeed, certain of IET's actions might appear incongruous in the context of an explicit attempt to revolutionize and re-democratize local food systems. Most famously perhaps is the instance of repurposing a flower bed outside the town police station, which on the surface looked like a daringly subversive act of guerrilla gardening, but which was in fact welcomed by local police, to the great delight of IET. A similar level of incongruity is achieved by IET's focus on tourism and local enterprise. Tourism is pushed particularly hard by IET, who declare triumphantly (and with tongue firmly in cheek) 'Welcome to Todmorden: the only town in Britain to attract vegetable tourists' (IET website 2017). Again, this is not a 'gotcha' point intended to undermine the success of IET; IET has been successful where many others have failed in popularizing the idea of community-led urban food growing. It is simply to point out that IET's strategy is one circumscribed by the logic of private capital, not one which seeks to directly dismantle it and organize society around different axes. Indeed, it begs the question, what would food commoning look like if it did take seriously systemic change, particularly a movement away from the capitalist-logic of contemporary food systems?

Radical or 'contesting' models of food commoning still remain a relatively fringe phenomenon in the UK; that said, they often punch well above their weight, being routinely led by small, committed groups of activists able to gain traction by exposing the fatal logic of modern land ownership and food production. The examples below are thus united by their preparedness not only to create (or 'reclaim') commons-like spaces where food and food-related knowledge sharing can be influenced by, and be of benefit to, all (as with previous examples), but to signal a preparedness to risk direct confrontation (and sometimes arrest and imprisonment) to achieve those ends. This preparedness is not, it must be stressed, an unthinking opposition, but rather a necessary consequence of the deeply ingrained injustices of the contemporary food system and the need to take decisive action to construct alternatives.

Land occupations have a long history in the UK, often in response to spikes in acts of enclosure and appropriation. In more recent years, perhaps the best-known examples are the road protests of the late 1980s and early 1990s, the successes of which have been well documented (Plows, 2008). One less well-discussed outcome of these protests, however, was the influence such actions had on food-focused land protests of the following decades. The influence, for example, on Reclaim the Fields, Grow Heathrow, Yorkley Court and anti-fracking protests in West Sussex, North Yorkshire and Lancashire during the 2010s, can be directly connected to these earlier struggles. A key moment in the contemporary radical food commoning movement was undoubtedly Grow Heathrow, a land occupation in the outskirts of west London. Though the explicit purpose of this occupation was (and remains) to prevent the development of a third runway at Heathrow, the group also signalled the role of food, not only in their name, but also by declaring their aim 'to develop and promote community and resource autonomy to support long-term community resilience' (Transition Heathrow, 2017).

The site Grow Heathrow chose to occupy – a piece of semi-derelict land – was also once used to cultivate exotic fruit for sale in London, before such production was outsourced to the developing world and to transnational commodity traders (Patel, 2007). This (un)happy coincidence allowed Grow Heathrow occupiers to immediately begin growing food in the tumbledown skeletons of the old polytunnels that remained on site, in pointed opposition to the neo-colonial logic of the modern industrial food system. Added to this, Grow Heathrow has worked hard to foster a culture of continuous communal living on the site by organizing regular events, especially those based around the sharing of food either grown locally, foraged or salvaged from the local urban waste stream. Grow Heathrow offers a 'live' example of a community based around the values of social justice, environmental sustainability and frugal abundance.

The success of Grow Heathrow is unprecedented; in March 2018 they will enter their 9th year of continuous occupation; and yet, Grow's success can be seen not only in their enduring presence and numerous positive impacts on the local area (MacKinnon and Derickson, 2013), but also in their function as a hub or catalyst for other forms of food commons activism. Reclaim the Fields (est. 2011), Yorkley Court Community Farm (2012–2016) and later the Land Justice Network (previously 'Land for What?' est. 2016) have all directly interacted with the site. Over the last eight years, Grow Heathrow has used their space to convene an ambitiously wide range of political 'issues', including resource scarcity, democratic governance, climate change, sustainable food systems and housing shortages (to name but a few intersections). Indeed, if IET has 'something for everyone', Grow Heathrow might be said to offer *radical action for everyone*, with food and land at the centre of a just and ecological reorganization of British social and political life.

Another site of note is, of course, Yorkley Court (YCCF) itself, which was an occupation of a farm 'where the land ownership is contested and where the soil is abused by industrial agriculture' (Reclaim the Fields, n.d.). YCCF's vision was to cultivate the land 'according to the principles of agroecology in any aspect of managing of the land and the principles of Food Sovereignty when deciding upon the future development of the farm' (YCCF, 2015). As justification for their actions Yorkley Court cited the '1000 years of struggle' over which period 'the ownership of fields, forest and commons has been progressively concentrated into the hands of a few powerful landowners' (ibid).

Though Yorkley Court occupation was eventually ended in March 2016, along with Grow Heathrow (and other intentional and public-facing land projects too numerous to mention here) their mere existence continues to fuel ongoing food commons activism in the UK and beyond. Of particular note in recent years has been the development of the Land Justice Network (LJN), whose focus on the commons is explicit. In their manifesto, 'Our Common Ground', LJN states that 'Land is the main uniting factor underpinning most of our struggles for social and environmental justice, whether for genuinely affordable housing or food growing, for preserving nature or community space' (LJN, 2017). Indeed, what makes LJN so exciting is their retention of a radical, 'contesting' programme within a 'big tent' frame. To have the Landworkers' Alliance, The Radical Housing Network, The New Economics Foundation and Just Space all around the same table is exhilarating and a sign of the enduring and broad appeal of radical land reform and the commons.

The next few years promise to be an exciting time for a grassroots contestation, re-construction and re-articulation of what Eleanor Ostrom (2007) called a 'new commons'; that is, working out where the soil, land and food production fit within a dizzying constellation of emergent commons (especially digital- and knowledge-related). In a time of unprecedented social and environmental challenges, it will be these new spaces (not a revivalist commons based on a medieval fantasy) which will provide space for robust networks of solidarity to develop, vital in the ongoing fight for social and ecological justice.

Regaining and redistributing control: Mondeggi Bene Comune – Farm without Masters

Moving our attention to Italy, the story of land as a commons can be tracked back to 2011, a schizophrenic year for the Italian commons. In June, more than seventeen million citizens voted to support the idea of water as a common good (*Acqua bene commune*) and defeated the legislative attempt to further privatize this essential resource (Carrozza and Fantini, 2013; Mattei and Quarta, 2014). The referendum represented the political and legal victory of hundreds of local water committees and a thousand testimonies that water is an essential element of life and cannot be appropriated by the few. It was a moment of rediscovery of the commons and great debate, with the multiplication of platforms for intellectual engagement and spaces of practical action (Quarta and Ferrando, 2015). However, in less than six months the country was dragged into the direst phase of post-crisis austerity and commodification. Under the government of Prime Minister Mario Monti (a former Goldman Sachs and EU competition commissioner) the rhetoric of guilt, urgency, catastrophe and inevitability was translated into market-based solutions, cuts in public expenditures and large-scale privatization.

As with the historical enclosure of the British commons, law became the tool to impose a socio-economic transformation through privatization and exclusion. The combination of public authority and the rhetoric of the state of emergency of public finances (Agamben, 2003) legitimized the implementation of a radical (i.e. going to the roots) bio-political project of socio-economic discipline. The separation between the Italian government and the people could not be stronger. On the one hand, common-sense was undergoing a powerful re-construction from the bottom-up with a referendum attended by more than seventeen million people. On the other hand, the executive was inflicting on the country a new round of the neoliberal shock doctrine (Klein, 2009). As the name of the decree suggests ('*Decreto Salva Italia*', i.e. Save Italy),[4] the legal measure was introduced with urgency as an exceptional intervention justified by the context of international crisis and the need to find assets that could be sold and contribute to the financial stability of the public balance. In a reprise of the '*There Is*

No Alternative' (TINA) mantra of the '80s, the decree raised the retirement age, reduced the budget of the national health system and public transportation and used public guarantees to save private banks (Art. 8). Public land was specifically addressed by Article 66, which institutionalized the sale of public agricultural land to generate income and revitalize the national agri-food sector.

Although Article 66 also contained a clause designed to support farmers under the age of 35, it did not take long before the movements that had challenged the privatization of water recognized the very essence of the decree: privatize public land not for the many but to support businesses and entrepreneurs with enough capital to afford it. The response was immediate and was framed in the terms of the commons. On the ground, local and national movements concerned by the access to land joined forces with those campaigning around water as a commons and launched the '*Campagna Terra Bene Comune*' (Campaign for Land as a Commons) to oppose the decree. Similar to the water campaign that had preceded the referendum, the language of the commons was used to build a framework for resistance and bottom-up contestation that was sufficiently broad to welcome people and organizations with diverse backgrounds and trajectories. Given the silence from the public administration and the urgency of the situation, it did not take long before local collectives decided to assume a proactive attitude and, on the wave of the occupations of private and public urban spaces that had taken place in the last years in Italy, begin occupations of their own.

As in the UK, radical or 'contesting' models of food commoning are a relatively fringe phenomenon in Italy. However, their theoretical and practical relevance is increasing, and their experiences are particularly interesting from the point of view of interconnecting struggles and bottom-up construction of post-capitalist proprietary forms. Of all the experiences that emerged from the *Terra Bene Comune* movement, we decided to introduce the example of the *Mondeggi Bene Comune – Fattoria Senza Padrone* ('Mondeggi as a Commons - Farm Without Master') and its legal and political implications. However, we are aware that several realities have emerged in connection with *Campi Aperti* (Open Fields) and *Genuino Clandestino* (Genuine Clandestine), two national networks that developed a national profile and that have provided the intellectual and material context in which the struggle for land in Italy has been unfolding (Angrisi, 2011).

The reasons why we chose to examine the case of Mondeggi are three-fold: first of all, Mondeggi has prioritized combining the struggle for land with the construction of a just, democratic and ecological food system. Secondly, the commoners have been practicing the idea of land as a commons by constructing a form of property that is based both on the static idea of property over something (e.g. land, means of production, some animals) and the dynamic notion of communal land management (commoning the land as a way of living on the land in community and sharing, caring, and producing for more than mere individual return).[5] Thirdly, one of the authors had the opportunity to visit the farm, meet some of the Mondeggi commoners and spend time listening to their stories, hopes and concerns. It was a unique opportunity to learn about the complexity of commoning the land.

This experience has raised multiple questions that cannot be fully addressed here and that are linked with the multidimensional nature of the project, which is simultaneously an act of resistance (an occupation against the privatization of a historical farm nearby Florence), an agroecological project (with a permaculture vegetable garden), a platform to engage with the surrounding communities and bring them back to the land (with a project of collective stewardship of olive trees), a venue for creativity and art performances, an opportunity for adults and children alike to reconnect with nature, a laboratory to preserve agricultural traditions and give them new life, a home, and an example of what it means to live and resist together.

The next two sub-sections offer a short introduction to the history of Mondeggi and a brief reflection on the way in which this experience utilizes vocabulary and aspirations of the commons to keep together food sovereignty, land sovereignty and agroecology.

Brief history of Mondeggi Bene Comune

Mondeggi as a community of resistance cannot be detached from those they struggle against, i.e. the Province of Florence and, later, the Metropolitan City of Florence. Legally speaking, Mondeggi is, in fact, a farm of over 200 hectares owned by a bankrupt public corporation, with the Province of Florence as the sole shareholder (later transferred to the City of Florence). In November 2013, after a fruitless attempt to engage in a conversation with the Province to obtain a concession on the area, the first act of trespass was organized on an abandoned olive grove. For the trespassers, some of whom then became occupiers, the action was not only about the exclusionary nature of private ownership, but about demanding that the public authority recognize the nature of the land as a collective good and facilitate the establishment of a new local food system around the land of Mondeggi. As a sign of their desire to bridge the gap between land and community, the olives were transformed into oil and distributed for free during local markets. The message was clear: what is public belongs to everyone and its benefits should be distributed rather than accumulated. In their words, priority should be given to the intrinsic connection between the 'free access to land and the right to a genuine, accessible, locally produced food' (Mondeggi Bene Comune, 2017a).

At the end of 2013, after a long silence and the decision by the commoners to occupy two houses and a warehouse, the Province of Florence announced an auction for the land, hoping to find a buyer. Finding a private owner of a public land became the way in which the Province could walk away from any responsibility and any role in the management of the land. The threat of eviction suddenly became more real. However, it also increased the solidarity and the interest around the farm. When no one made an offer for the farm and the Metropolitan City of Florence replaced the Province as owner of the farm, a diverse group of activists and citizens had already gathered 'around the defence of Mondeggi and intervened on its state of abandonment, intensifying the organization of events opened to the whole population to foster local awareness and sociality, take care of buildings, and develop numerous agricultural, social, and cultural projects' (Mondeggi Bene Comune, 2017a). Comforted by the events, the commoners issued a Declaration of Principles and Intents to establish the trajectories that would have guided the conversion 'of abandoned public goods into commons self-managed by and accessible to the community' (Mondeggi Bene Comune, 2017b). Food and land sovereignty became, together with agroecology and social justice, the new horizon to follow.

Food and land sovereignty under the cover of the commons

Since its inception, the experience of *Mondeggi Bene Comune* was aligned with the national campaign for Land as a Commons. It aimed to oppose the privatization of productive land and soil and to build a new proprietary regime that rejected the idea of bourgeoise property and its vision of 'sole and sole despotic dominion which one man claims and exercises over the external things of the world, in total exclusion of the right of any other individual' (Blackstone, 2016). To this objectifying and appropriative vision, the commoners aimed to oppose a bottom-up, participatory and autonomous proprietary form that went beyond the public/private division and focused on democratic management, just redistribution of utilities and respect of the ecological limits of the planet. If crises are a moment of forced choice, the occupiers use this moment to define themselves against the Monti government and to be practically and theoretically involved in the creation of a commons-based alternative aimed at supporting 'peasant and small-scale agriculture, constructing territorial communities and a participatory system of stewardship' (Mondeggi Bene Comune, 2017a) where individual and socialized properties would coexist because the most important element was not the right over something, but the duty to redistribute to those most in need.

If Hardin defines the land commons as an inevitable expression of individualism, competition, over-exploitation and lack of coordination, Mondeggi moves in the opposite direction. Mondeggi offers an example of how communities can gather around land and soil, not only in order to demonstrate an alternative way of relating to the land, but also to connect the struggle for democratic land control with a 'strategic rebooting of the broader agricultural and food system' so that 'such democratisation [will not] fizzle out and revert back to older or trigger newer forms of land monopoly' (Borras et al, 2015). What is most striking about the experience of Mondeggi is that the commoners utilize the vocabulary and practice of the commons to combine the struggles for land and food sovereignty. The case of Mondeggi offers a concrete example where the political ideal of food as a commons dialogues and reinforces the paradigm of food sovereignty: decommodification of food and land is presented as the necessary prerequisite of a democratic, just and autonomous food system.

In order to achieve this double goal of land and food sovereignty, the structure of governance reinforces and consolidates the sense of belonging and responsibility, but also recognizes the fact that different members may have different weight in the decision-making process, depending on their level of commitment and participation.[6] In addition, the common nature of the land also means that Mondeggi is recognized as part of a broader social context and of a planet with limited resources. As a consequence, a stewardship project was created to invite the local population to take care of olive trees and share the benefit of the harvest[7] and local markets represent a preferred venue for the sale of the farm's products. Furthermore, the principles of agroecology are adopted as binding requirements for any agricultural practice, so that nothing in Mondeggi happens without consideration of the ecological implications of human activities. In this way, the circle is closed; under the umbrella of the commons, land sovereignty is actively practised in order to reach food sovereignty through agroecological practices, in an inclusive way, as long as the principles, values and objectives of a just and democratic food system are respected by all who participate.

Despite the public title over the land, Mondeggi and its *commoners* the daily threat of eviction. They are not alone. Between 2013 and 2017, the Italian Ministry of Finance and the Ministry of Agriculture have identified 1500 hectares of public land to be put on the market (sold or individually rented) without considering the possibility of pursuing alternative forms of collective management or titling, such as concessions or land trusts (Terrevive, 2017). Although it is easy to be caught by a wave of optimism when visiting Mondeggi, it is important to remember that the struggle for the recognition of land and food as a commons stretches out before us, full of obstacles. The next five years will tell us whether Mondeggi can become a significant coordinate in the movement for food sovereignty and land reform or, on the contrary, just another missed opportunity to think differently about land and the food system.

What next?

As Philip McMichael (2008, p. 219) suggests, 'the significance of the food sovereignty movement is that, in the narrative of capitalist modernity, its project is virtually unthinkable'. In the examples we have discussed, especially Mondeggi and other instances of 'contesting' food commoning in the UK, a similar dynamic can be seen. While they remain, in the face of the juggernaut of capitalist development, almost 'unthinkable', they offer deep veins of inspiration and hope in a world with apparently diminishing supplies of both. Indeed, it is the dynamic between these two poles which often get their proponents so excited. But we must also continue to bridge this gap.

The question remains then – where do we go from here? How do we ensure that these rare and ephemeral examples of the commons do not simply evaporate or remain the exception to the rule? As we have seen, what often seems to characterize and unite the most striking examples of food commoning is their ability to bring agricultural land back into collective control. Conversely, examples which, while powerful, arguably manifest troubling (or even fatal) contradictions, are those which are not truly owned, determined and directed from the 'bottom-up', but are, in short, circumscribed by the logic of capital, an anathema to the commons.

By way of conclusion, we offer two key observations on this apparent dichotomy. The first is an appeal: to ensure the continuation of these rare and ephemeral examples of the commons, we need to act, to join those currently fighting to build and sustain such spaces, and ensure that stories of commoning continue to be told. This appeal is, of course, offered reflexively; it is much easier to call others to radical action than offer a practical guide on how to do it well. The dynamics of such spaces are always in flux and when and where they emerge is fiendishly hard to predict. So, in supplement to this appeal, we also suggest renewed effort in thinking through the processes that keep such spaces in collective ownership and management.

In this regard, food sovereignty is once again of value. The food sovereignty movement offers numerous examples of collective management of land, from the now legendary MST and La Via Campesina (Holt-Giménez and Van Lammeren, this volume; Barbosa, 2016; Martinez-Torres and Rossett, 2014) to emergent examples in the Global North. In the UK, Canada and Australia, for example, a spate of 'people-centred food policy' processes have been gaining momentum (see Laforge et al., 2016), not least due to their successful innovation of certain forms of civic participation, which have brought democratic legitimacy to grassroots alternatives to the corporate food regime. This promising movement is already making explicit links to food commons (e.g. The Peoples Food Policy, 2017; Esteva, 2014) and may also provide ways into radical land reform for those less comfortable with direct action. In the Global South, examples like the civic struggles to claim fair access to water, energy and food in South Africa (Chapter 15 in this volume) show the possibility to tinker with the paradigm of commons as an old/new political claim for the dispossessed and those made disposable by the capitalist system.

Food (as has been stated many times) is eaten by everyone and therefore has incredible convening power; however, meaningful engagement with food governance and alternative food paradigms is done only by a select few. In this chapter we have told stories of people who have decided to dedicate their lives to tackle a central problem in the Western food system: the commodification and enclosure of land. In the United Kingdom, in Italy and elsewhere, younger and older generations are thinking, imagining and practising new food systems starting at its core: the soil that nourishes the crops, the farm that binds communities, and the very land on which we stand. We believe the exciting forms of civil-society-led processes mentioned above will contribute to the co-production of a new vision and provide important connective tissue between the radical outliers of food commoning, progressive property advocates, and the broad base of support for food systems which nourish the collective, rather than enriching the few. Watch this space (and get engaged)!

Notes

1 The homogenization of land and the loss of its diversity is often the outcome of agrarian reforms based on individual titles and absolute property rights. As a consequence, this kind of political measure is increasingly opposed by bottom-up organizations like the landless movement (*Movimento Sem Terra*) in Brazil.

2 http://www.thelandmagazine.org.uk/articles/short-history-enclosure-britain

3 Another example is offered by Helping Blossom Britain, a partnership between the alcohol corporation Heineken, The Urban Orchard Project and The Bulmer Foundation. According to the project's website, '[t]he aim is to create sustainable, long-term orchards planted and managed by the community, for the community.' The choice of orchards was determined by the possibility of transforming apples into cider, i.e. by the connection between nature and the activity of the corporate 'philanthropist'. Communities will gather around urban orchards and create new spaces, but these spaces have been identified by an external actor and are intrinsically connected with a business, that of alcoholic beverages, that is often criticized for the appropriation of water and staples, but also because it is increasingly concentrated and uniform. For further info, see https://www.heineken.co.uk/article.php?article=34011413964597.

4 Law Decree n. 201/111 of 6 December 2011, converted by Parliament into law on 22 December 2011 (law 2014/2011) and entered into force on 1 January 2012.

5 We are aware of the other experiences of bottom-up and spontaneous communing of public spaces that have been taking place throughout the country, such as the Valle Theatre in Rome, the Cavallerizza in Turin and the Colorificio Occupato in Pisa. Another experience that rotates around land is that of Caicocci, a publicly owned farmland near Perugia that was left abandoned until some members of the national network *Genuino Clandestino* took custody of it.

6 Art 10 of the Declaration identifies five categories of community members, each one with different rights and duties. Those who garrison (*presidianti*), the guardians (custodi), the guests (osptii), the wayfarers (viandanti) and the beneficiaries (fruitori). See Mondeggi (2017b)

7 For example, the project *Mondeggi Terreni Autogestiti* (Mo.TA, Mondeggi Self-Managed Land) was launched, inviting citizens and organizations to take care of a parcel of olive trees (each one of seventeen trees), manage them, clear the land around them, maintain the broader ecological balance, harvest their fruits and receive a percentage of all the oil that is produced with all the olives, including those of other parcels and other trees. Altogether, Mo.TA is about caring of the commons, creating community and strengthening their connection with the land.

References

A People's Food Policy (2017) A People's Food Policy. Retrieved December 14, 2017, from https://www.peoplesfoodpolicy.org.

Agamben, G. (2003) *State of Exception*, University of Chicago Press.

Angirsano, N. (2011) Genuino Clandestino: Movimento di Resistenze Contadine, InsuTv Documentary, available in Italian at http://vimeo.com/34322825, accessed 13 December 2017.

Anguelovski, I., Connolly, J. J. T., Masip, L. and Pearsall, H. (2017) Assessing green gentrification in historically disenfranchised neighborhoods: A longitudinal and spatial analysis of Barcelona. *Urban Geography*, 39(3), 1–34. https://doi.org/10.1080/02723638.2017.1349987

Arrighi, G. (1998) Capitalism and the modern world-system: Rethinking the nondebates of the 1970's. *Fernand Braudel Center Review*, 21(1), 113–29.

Barbosa, L. P. (2016) Educação do Campo [Education for and by the countryside] as a political project in the context of the struggle for land in Brazil. *The Journal of Peasant Studies*, 1–26.

Borras, S., Franco J. and Monsalve-Suárez, S. (2015) Land and food sovereignty, *Third World Quarterly*, 36(3), 600–617.

Carrozza C. and Fantini, E., eds. (2013) *Si scrive acqua.. Attori, pratiche e discorsi nel movimento italiano per l'acqua bene comune*, Accademia University Press.

Diaz, S. et al. (2018). Assessing nature's contributions to people. *Science*, 359(6373), 270–272.

Esteva, G. (2014) Commoning in the new society. *Community Development Journal*, 49, 144–159.

Grolle, S. (2008) *Privatisation of public space*, GRIN Verlag.

Groundwork (2017) Groundwork | Terms and Conditions of Grant. Retrieved October 13, 2017, from https://www.groundwork.org.uk/Sites/tescocommunityscheme/pages/terms-and-conditions-of-grant-tes.

Guitart, D., Pickering, C., and Byrne, J. (2012) Past results and future directions in urban community gardens research. *Urban Forestry and Urban Greening*, 11(4), 364–373. https://doi.org/10.1016/j.ufug.2012.06.007.

Hardin, G. (1968) The Tragedy of the Commons, *Sciences*, 162(3859), 1243–1248.

Hess, C. (2008) Mapping the New Commons. Retrieved October 12, 2018, from http://dlc.dlib.indiana.edu/dlc/bitstream/handle/10535/304/Mapping_the_NewCommons.pdf

IET (2017a) Propagating success. An evaluation of the social, environmental and economic impacts of the Incredible Edible Todmorden Initiative. Retrieved October 12, 2018, from https://www.incredible-edible-todmorden.co.uk/file_download/333/IET+Evaluation+Summary+Report+FINAL.compressed.pdf.

IET (2017b) Seeing is believing, but some folk like the facts! Incredible Edible Todmorden | News. Retrieved October 13, 2017, from https://www.incredible-edible-todmorden.co.uk/news/seeing-is-believing-but-some-folk-like-the-facts.

Ireland, P. and Meng, G. (2017) Post-capitalist property. *Economy and Society*, 46(3–4), 1–28.

Klein, N. (2009) *The Shock Doctrine*, Random House of Canada.

Laforge, J. M. L., Anderson, C. R., and McLachlan, S. M. (2016) Governments, grassroots, and the struggle for local food systems: Containing, coopting, contesting and collaborating. *Agriculture and Human Values*. https://doi.org/10.1007/s10460-016-9765-5

MacKinnon, D., and Derickson, K. D. (2013) From resilience to resourcefulness: A critique of resilience policy and activism. *Progress in Human Geography*, 37(2), 253–270.

Martínez-Torres, M. E., and Rosset, P. M. (2014) Diálogo de saberes in La Vía Campesina: food sovereignty and agroecology. *The Journal of Peasant Studies*, 41, 979–997. https://doi.org/10.1080/03066150.2013.872632.

Mattei U. and A. Quarta (2014) *L'acqua e il suo diritto*, Ediesse.

Mondeggi Bene Comune (2017a) Three years of Popular Stewardship. Retrieved September 13, 2017 from https://mondeggibenecomune.noblogs.org/la-nostra-storia/dossier-three-years-of-popular-stewardship/.

Mondeggi Bene Comune (2017b) Dichiarazione di Gestione Civica di un Bene Comune Mondeggi Bene Comune – Fattoria Senza Padrone. Retrieved September 14, 2017 from https://mondeggibenecomune.noblogs.org/mondeggi-bene-comune/dichiarazione-di-gestione-civica/.

Moore, J.M. (2015) *Capitalism in the Web of Life*, Verso.

More, T. (2016 [1516]) *Utopia*. Cambridge University Press.

NWCAA. (n.d.). How many allotment sites are there. Retrieved October 13, 2017, from http://www.nwcallotmentassociation.com/how-many-allotment-sites-are-there.html

Patel, R. (2007) *Stuffed and Starved: Markets, Power and the Hidden Battle for the World Food System*, Black Inc, Melbourne.

Paull, J. (2013) 'Please Pick Me' – How Incredible Edible Todmorden is repurposing the commons for open source food and agrobiodiversity. In: Franzo, J.; Hunter, D.; Borelli, T. and Mattei, F. (eds.) *Diversifying Foods and Diets: Using Agricultural Biodiversity to Improve Nutrition and Health*. Earthscan, Routledge, Oxford, 336–345.

Peart, A. (2015) Incredible Beginnings, TED Conferences LLC, available at http:/incredibleediblenetwork.org.uk/incredible-beginnings, accessed 13 October 2017.

Plows, A. (2008) Towards an analysis of the 'success' of UK green protests. *British Politics*, 3(1), 92–109.

Quarta A. and T. Ferrando (2015) Italian property outlaws: From the theory of the commons to the praxis of occupation. *Global Jurist* 3, 261–290.

Reclaim the Fields (n.d.) A brief history of Reclaim the Fields UK. Retrieved October 13, 2017, from https://www.reclaimthefields.org.uk/about/a-brief-history-of-reclaim-the-fields-uk/.

Russi, L. (2015) *Everything Gardens and other Stories: Growing Transition Culture*, University of Plymouth Press, Plymouth.

Scott, J.C. (1999) *Seeing Like a State*, Yale University Press, New Haven.

Shared Assets (2015) Working towards a 21st Century Commons. Retrieved October 13, 2017, from http://www.sharedassets.org.uk/policy/working-towards-a-21st-century-commons/.

Terrevive (2017) Online gli elenchi dei terreni agricoli in vendita o in affitto nel corso del 2017. Retrieved December 12, 2017 from http://www.agenziademanio.it/opencms/it/notizia/Terrevive-online-gli-elenchi-dei-terreni-agricoli-in-vendita-o-in-affitto-nel-corso-del-2017/.

The Land Justice Network (2017) Our Common Ground statement. Retrieved September 7, 2017, from https://www.landjustice.uk/our-common-ground-statement/

Tornaghi, C. (2017) Urban agriculture in the food-disabling city: (Re) defining urban food justice, reimagining a politics of empowerment. *Antipode*, 49(3), 781–801.
Transition Heathrow (n.d.) Transition Heathrow: About Us. Retrieved October 13, 2017, from http://www.transitionheathrow.com/about-us/.
Urgenci (2016) Overview of Community Supported Agriculture in Europe. Retrieved October 12, 2018, from http://urgenci.net/wp-content/uploads/2016/05/Overview-of-Community-Supported-Agriculture-in-Europe-F.pdf.
Wallerstein, I. (1974) *Capitalist Agriculture and the Origins of the European World-Economy in the Sixteenth Century*, University of California Press, Berkeley.
YCCF (2015) Why we are here. Retrieved October 13, 2017, from https://yorkleycourt.wordpress.com/why-we-are-here/.

22

THE CENTRALITY OF FOOD FOR SOCIAL EMANCIPATION

Civic food networks as real utopias projects[1]

Maria Fonte and Ivan Cucco

Introduction: the transformative potential of food

In activist narratives and academic debates, food has become an agent of change. Rather than considering it as a simple commodity, food, articulated as 'local food' or even better as 'civic food', has been linked to transformative projects and visions of a more equitable, ethical and sustainable economy and society. Its alternativeness rests on the connection to the territory, i.e. to the agro-ecological and social context in which food is produced and consumed. As for the agro-ecological conditions, the local food discourse promotes respect for the environment and nature. As for the social context, it emphasizes the inclusion of people who are currently marginalized by the corporate and increasingly financialized food production system. This movement implies, on the one hand, the protection of small farms and their tacit knowledge; on the other hand, a concern for the poor voiced in requests for their right to a healthy, culturally sound diet.

A variety of practices and movements have gathered under the 'local food' umbrella in both the global North and South. Beyond the widespread diffusion of farmers' markets, prominent examples of local food initiatives in the global North include Community Supported Agriculture (CSA), Association pour le Maintien d'une Agriculture Paysanne (AMAP) and Solidarity Purchasing Groups and Food Teams. In Eastern Asia, the globalization process brought about by the Trans-Pacific Partnership (TPP) agreement is acting as a catalyst for experiences related to local food, organic agriculture and Alternative Food Networks (AFNs) (Hisano, 2015). In the global South, as well as in many Eastern European and Mediterranean countries, local food initiatives not only propose an alternative to the corporate food system, but also aim at valorizing local markets, traditional foods and peasant and subsistence agriculture as forms of 'quiet sustainability' (Smith and Jehlička, 2013). Especially in Latin America, local food initiatives are contributing to the elaboration and consolidation of new practices linked to a novel model of agro-ecological development, based on the valorization of farmers' knowledge and experimentation in opposition to the expert-led model of the science- and technology-intensive GMO agriculture (Altieri and Nicholls, 2010).

The transformative practices and academic discourses that deploy the concept of local food appear far from homogeneous. Different initiatives mobilize the concept of the 'local' for various transformative aims: environmental sustainability (Feenstra, 1997; Pretty et al., 2005;

Pirog et al., 2001, 2003; Garnett, 2007); the revitalization of rural economies and the reconnection of urban consumers to agriculture and nature (Renting et al., 2003; Marsden et al., 2000; Fonte 2008; Brunori et al., 2012); the agro-ecological revolution, especially in Latin America (Altieri and Toledo, 2011); or advancing land entitlements for marginalized groups in the face of encroaching dispossession in the Global South (McMichael, 2014). Local food is thought of also as a crucial element in protectionist and neo-ruralist ideologies promoting bounded, defensive spatial strategies (Hinrichs, 2000; Born and Purcell, 2006).

These contradictions point out to the limited heuristic value of 'local food' as an analytical category, particularly when decoupled from an explicit attention to the political and power dimension of the local, considered in its own right as well as in its multi-scalar interactions. The contribution of a 'reflexive localism' (Goodman et al., 2012; Fonte, 2013b) and the opening up of food studies to insights from other disciplines (e.g. critical human geography) highlight the importance of *politics in places* (Hinrichs, 2015). In this perspective the 'local' becomes the space for the enactment of a political agenda and the nexus of a political reflexion involving the construction of a more equitable and more sustainable food economy (Harris, 2009), an emancipatory project leading the way to a transition toward sustainability (Hinrichs, 2014).

Building upon these considerations, in this chapter we explicitly focus on the political and transformative dimensions of local food projects and propose to read local food initiatives as 'real utopias' projects, whose central aim is transforming the food economy in the direction of environmental sustainability, social justice, democracy and solidarity. Looking at local food discourses and movements through the framework of a transition theory offers the advantage of connecting local food practices to other proposals for emancipatory responses to the disruptive effects of neoliberal capitalism, such as participatory budgeting and empowered participatory governance, the social economy of welfare provision, the experience of transition cities, the sharing economy and open-source software.

In the next section we will locate the Real Utopias Project in the realm of transition theories, highlighting similarities and differences with the multi-level perspective on socio-technical regime transition (Geels, 2005; Geels and Schott, 2007) and social practice theory (Reckwitz, 2002; Schatzki,1996). In the remainder of the article, we will analyse local food as a 'real utopia', articulating our reading of the literature on local food and civic food networks with the three core dimensions proposed by Wright (2006, 2010): a critique of the dominant food system; a prefiguration of the desirable future food economy; and a proposal of strategies for getting there. A discussion will follow, aimed at highlighting how only embracing the multi-dimensional value of food as a commons (Vivero-Pol 2017a) may free up its full transformative potential in the direction of a more just and sustainable society.

Transition theories and the Real Utopias Project

Our interpretation of local food as a heuristic for the transition to sustainability draws upon different theories: the multi-level perspective (MLP), social practice theory (PT), and the Real Utopias Project. Both MLP and social practice theory deal with the problem of how to advance towards more sustainable consumption and production systems; the Real Utopias perspective is instead fundamentally interested in visions (and practices) that aim at building new institutional architectures that can enhance social emancipation and deepen democratic participation.

The MLP (Geels, 2005; Geels and Schott, 2007) has to date offered the most articulated theory of transition towards environmental sustainability of the economic system. It relies on the concept of 'strategic niche management' and sees the achievability of the transformation objectives as a progression from interstitial or symbiotic strategies to the reconfiguration of the entire system in the direction of sustainability. Innovative ideas and solutions emerge and are developed in spaces protected from market competition; in these innovation niches, learning processes are set in motion and new forms of economic and social organization are experimented. The resulting innovations may be scaled up to system-wide proportions or may be co-opted by the dominant system through adequate governance mechanisms, so as to provide a basis for new socio-technical regimes that reconfigure the system as a whole. In some way, interstitial niche innovation may obtain different results as to the degree of system transformation, depending on the pressure deriving from landscape (structural, exogenous) variables and the capacity of the new social-technical (niche) solutions to resolve the economic and social contradictions of the dominant regime. Policy recommendations stemming from MLP literature are primarily related to the *governance* of the socio-technical regime change, through a mix of tools that may go from networks governance in an early phase of niche stimulation, to regulation, standards, tax and subsidies in later phases, aimed at widespread uptake of the innovation.

Practice theory is concerned not so much with how to govern the necessary changes in the socio-technical structure (which may represent the production side of a sustainable economy), but with how to change routinized social practices, especially the social practice of consumption, so as to reach sustainability goals (Reckvitz, 2002; Schatzki, 1996; Shove et al., 2012). Practice theory is a 'social theory' in so much as its unity of analysis is not the single firm or the individual, but specific social practices, routinized types of behaviour constituted by a nexus of interrelated elements of different nature, both material (such as objects and things) and immaterial (competences and meanings). Theorizing about the transition to an enhanced sustainable food system may be seen as a challenge to dominant routinized food practices and as the foreshadowing of a new practice, based on new norms and understandings, a new material infrastructure and a new agency (Warde, 2005; Halkier ,2009; Crivitis and Paredis, 2013; Fonte, 2013a). In this case, the transformation theory is directed at devising policy interventions addressing the systemic challenges of changing social consumption practices in their meanings, material structures and ways of understandings.

Neither MLP nor social practice theory is concerned with a theory of the new possible institutional architecture of the relationships between the State and economic and social powers, which may guarantee a more just economy and society. Their main interest lies in the modification of social practices or in the transformation of the socio-technical system to enhance sustainability and address environmental problems. In the words of Geels et al. (2015: 6):

> Transformation towards new transport, electricity heat or agro-food system and practices are more radical than the solution in the reformist position, but do not necessarily presume the abandonment of capitalism, economic growth or the embrace of frugality.

In this respect, Geels et al. (2015) differentiate the 'reconfiguration position' of MLP and PT proponents from the revolutionary sustainable consumption and production position associated with 'new economics' (Jackson, 2009), the 'de-growth approach' (Demaria et al., 2013) or the 'sharing economy' (Rifkin, 2014). The reconfiguration position, while accepting that "'green' innovation or practices should not only be environmentally sustainable, but also economically viable and socially acceptable" (Geels et al., 2015: 7), does not aim to simultaneously solve problems of poverty, inequality and democratic accountability. The revolutionary position, on the contrary, suggests that addressing environmental problems necessarily requires more fundamental changes in the values basis of the economic system and a shift to a more egalitarian and equitable society.

In contrast to both MLP and PT, the Real Utopias Project is an explicitly emancipatory social project: 'emancipatory' because it centres around a moral purpose (ending social oppression and the creation of conditions for human flourishing); and 'social' because it implies a belief that emancipation depends on the transformation of the social world, not only of the individual nor the economy (Wright, 2006).

According to Fung and Wright (2003) and Wright (2006; 2010) the Real Utopias Project embraces the tension between dreams and practice. 'Utopia' implies developing visions of alternatives to dominant institutions that embody aspirations for a better world, while 'real' refers to proposing desired alternatives that are viable and achievable. Real Utopias is a theory of social change animated by a normative foundation in the values of equality and fairness, democracy and freedom, and solidarity and communities, the same values that constitute the normative basis of the approaches to food as a commons (Ruivenkamp and Hilton, 2017)).

The exploration of Real Utopias initiatives is an integral part of an emancipatory social science and its core tasks are threefold:

- Explain *why we want to change* the present system, i.e. elaborate a diagnosis and critique of existing institutions, according to specified moral principles.
- Explain *where we want to go,* i.e. envisage viable alternatives to the present social structure.
- Propose *how to get where we want to go*, i.e. develop a theory of transformation for realizing the desirable alternatives.

Proposals for transformation need to be desirable, viable and achievable. The achievability of a viable alternative depends "upon the extent to which coherent, compelling strategies can be formulated which both help to create the conditions for implementing alternatives in the future, and have the potential to mobilize the necessary social forces to support that alternative when such conditions occur" (Wright, 2006: 99).

Next we will articulate our reading of 'local food' as a Real Utopias Project by highlighting how the 'local food' concept is used: (1) as a diagnosis and a critique of the mainstream food system; (2) as the prefiguration of a more sustainable, democratic and just food economy; (3) as a starting point in order to identify strategies for realizing the desirable alternatives.

Local food as a critique of the present

"The food system is broken" (Vivero-Pol 2017a: 2). While the 'food as commons' perspective poses at the centre stage of its transformative project and analysis the role played by the commodification of food (Vivero-Pol 2017b and 2017d), the 'local food' movement conveys its critique to the dominant food system through the concept of space. Friedmann (1992) was among the first to identify the main problems of the industrial food system (or better, in her words, the Fordist food regime) in terms of 'distance' and 'durability'. Distance has to do with *space*, intended as both the space of production and the space of consumption. Kloppenburg et al. (1996: 38) also refer to distancing as the main challenge in the food system: "If the mitigation of the deleterious effects of distancing is one of the central challenges posed by the operation of the global food system, then greater attention to proximity — to that which is relatively near — should be an appropriate response." The pioneering concept of a 'foodshed' is conceived exactly as a response able to 'repair and fix' the food system. In a recent paper, Hinrichs (2015: 1) writes: "ideas of 'local' and 'place' have been pressed into service to fix food and link it to a more meaningful somewhere, redressing harms associated with a global neoliberal food and agricultural system". Actually, 'distancing disempowers' according to Kloppenburg et al. (1996: 36);

so, due to the physical and social distancing that characterize the global food system, control passes to those who know how to act at a distance: the big corporations and multi-nationals.

What, then, are the harms that the idea of 'proximity' should fix and repair? In the literature on local food, distance has been analysed in a double meaning: geographical and social; these two aspects are different, but interconnected. In its *geographical meaning*, local food is intended as a critique of the long physical distance food travels 'from farm to fork' in industrialized, consolidated global value chains. Attention to food miles links concerns over food to environmental preoccupations with climate change and emissions of carbon dioxide and other greenhouse gases from transport.

A sharp contrast is drawn between the short chain for local food and the long distance food is required to travel in the conventional, centralized, industrialized food system. In the United States, Pirog and Rasmussen (2008) analysed the transport arrangements for 28 fruits and vegetables to Iowa markets via local and conventional food distribution systems and calculated that produce in the conventional system travelled an average of 1,546 miles (about 2,500 kilometres) while by contrast locally sourced food travelled an average of just 44.6 miles (72 kilometres). In the European Union, food consumption's contribution to GHG emission is calculated to be 31 per cent of total consumption (Tukker et al., 2006).

From an *economic geography perspective*, the relation between urban and rural is also questioned. The rural has always been conceptualized as a provider of labour, goods and services functional to the necessities of the city. From here stems the 'urban bias' (Lipton, 1977) of any development vision, which sees agriculture and countryside as subordinated to urban needs. This dichotomized vision of space has now been overcome as food has been legitimated among the 'urban questions', not only through the notion and initiatives of urban agriculture but also through a new awareness of the importance of food for the 'hungry city' (Steel, 2013). A new conceptualization of the rural-urban partnership through the concepts of foodshed or the food city region aims at overcoming this dichotomy (Morgan et al., 1996; Kloppenburg et al., 1996).

From *a social perspective*, production processes in the agro-industrial food complex are de-territorialized, placeless and centred around the commodification of food (food from nowhere). For many scholars, the reduction of food dimensions to one of a commodity is at the roots of the failure of the global food system (Magdoff and Tokar, 2010; Zerbe 2009). For example, according to Patel (2007), food as a commodity is traded through global value chain based on unfair exchange relations which favour big intermediaries (above all the retailing industry) against the interests of agricultural producers and final consumers, with the former not earning a living income and the latter paying too much for food. This leads to great social paradoxes, like the simultaneous presence of overproduction, obesity and food shortages: at the global scale, in 2014 750 million people were suffering from hunger (FAO, 2015), while 1.9 billion were overweight, of whom 670 million were obese and at risk of illnesses, such as diabetes and cardio-vascular diseases.

Attempts to transform the mainstream food system were already in place in the 1960s; organic agriculture was certainly the most important and most widely known among such attempts[2]. But in the 1990s a widespread perception that the organic movement had dropped its alternative/environmental ideological baggage grew up. The organic movement was seen as seduced by multi-national retailing firms with the prospect of a mass market (Blythman, 2005). Furthermore, organic certification was seen as encouraging non-local food consumption, with consequent increases in costs for producers and prices for local consumers. The logic of input substitution leading to conventionalization was considered to be an increasing trend in the organic movement and became part of the problems that needed to be 'fixed' (Guthman, 2003; Buck et al., 1997; Darnhofer et al., 2010). The 'post-organic' (Moore, 2006) local food movement shifted its focus, therefore, to the necessity of bringing consumers, and the cultural value of food, back into the food system, re-establishing a connection between the producer and the consumer and addressing the sustainability not only at the point of production, but also in the distribution system of the food chain (Fonte, 2010: 6).

Local food as the prefiguration of where we want to go: from local agriculture to civic food networks

The transformative horizon of localness is predicated on the re-connection of food to the territory in which it is produced and consumed. Localness articulates different dimensions of proximity: geographical, social, cultural and ecological. It is associated with space and short distance, but also with place, regions and territories: small-scale farms, multi-functional agriculture, quality food, rural livelihoods and sustainable community agriculture. Local food is ultimately about much more than short distance:

> The local foods movement is about an ethic of food that values reviving small scale, ecological, place-based, and relationship-based food systems...Large corporations peddling junk food are the exact opposite of what this is about.
>
> *(Severson 2009)*

From a sociological viewpoint, local food is considered place-embedded, the opposite of the placeless commodity of industrial agriculture. This concept of *embeddedness* imparts social meaning to notions of place, social meaning to be elaborated by the rural communities inhabiting the 'places' in question. The shortness of local food chains makes it possible to trace the food almost personally to the individual farmer who produced it, enabling relations of *trust* to be established in the local society. Food production is re-contextualized within the formal and informal social relationships that constitute the basis for community life. Geographical proximity is important because it implies and favours *social proximity*, i.e. face-to-face interactions between producers and consumers. It can foster a vision and a practice of 'food as a commons', governed by cooperation, sharing, stewardship, sustainability and direct democracy (Vivero-Pol, 2017d).

Social interactions have a significant impact on rural community life. Local food becomes part of a political project for *keeping rural communities alive* and constructing local economies which respect natural resources, give attention to cultural and biological diversity, defend the economic sustainability for small farmers and promote social justice and food sovereignty. The *place-embeddedness* of food may thus be conceived of as local society's *resistance* strategy against globalization and neo-liberalism (Polanyi, 1944) and it comes forward as a cultural, individual and collective societal response to the commodification of everything (Strassen, 2003), especially food, driven by the neo-liberal economy.

Place-embeddedness is, however, not strictly identic to 'localness'. Actually, the relation between 'localness' and 'sustainability' has been problematized, both in theory and in practice. The environmental impact of the food economy does not depend only on the distance 'from farm to fork', but also on how the food is transported, grown, transformed and prepared. Only a life-cycle analysis of food can yield an accurate assessment of the total volume of gas emissions linked to its production and distribution. The difficulty of establishing well-defined boundaries for the notion of 'locality', taking into account the conditions for the entire life-cycle of production, appears to undermine the usefulness of 'localness' as a category for the analysis of food systems sustainability (Garnett, 2007).

Other authors question the relation between localness and 're-embedding'. To use the words of DeLind (2011), we must ask ourselves: 'is local food taking us where we want to go?' Hinrichs (2015) invites us to look at how the distribution of interests and power across different groups of farmers and consumers, as well as across varied organizations and institutions, serves to concentrate or spread the benefits and risks when fastening food to a locality. Hinrichs also suggests to explore more seriously the effect of fastening food to a specific locality in historical circumstances when flexibility is needed in order to respond to emerging environmental, climate change or health challenges. The concept of 'local trap' (Born and Purcell, 2006) wants to highlight the risks implicit in assuming that place proximity always results in benefit or repair for environmental impact and social justice.

Finally, localness is a descriptive concept and its limited heuristic value is evident when we want to distinguish a progressive versus a defensive localism or reconcile localism with 'a sense of planet' (Heise, 2008) or with a 'global sense of place' (Massey, 1994). From these critiques and from the quest for a more reflexive localism, the need has emerged to assume more explicitly the concept of 'civic agriculture' (Lyson, 2004) and civic values into the conceptualization of local food.

Renting et al. (2012) propose Civic Food Networks (CFNs) as a complementary category to concepts such as 'short food supply chains' and 'local(ized) food systems'. CFNs may better express the processes of change in the agri-food governance mechanisms, showing the increasingly important role of civil society (and to some extent of local and regional administrations) compared to market forces and to the (national) State; they imply a new conception of food citizenship and food democracy and the regeneration of food governance mechanisms. CFNs refer to the network of all actors involved in the local food system that, as ecological citizens, partake in responsibility for the sustainability of the food economy and endorse the value of food as a commons and a right. In this respect CFNs refer to food commoning as a different form of governance than market allocation or state governance (Bloemen and Hammerstein, 2015) and aim to guarantee food as a right, i.e. guarantee access (both physical and economic) to sustainable food to all people, individuals and communities. There is no transition to sustainability if sustainable food is for the elite; as a member of the GAS movement in Rome said, a system that guarantees sustainable food only for an elite, exactly as the industrial food system does, is unjust, and cannot lead to sustainability (Fonte, 2013a).

Finally sustainability, social justice, solidarity and food democracy are the challenges that CFNs will have to face. The 'utopian' food economy towards which the CFNs' vision aims is a local-based food system, which can endorse civic values like sustainability, but also social justice and solidarity: in this vision, food cannot be a simple commodity; it is the basis on which to construct an emancipated society in which basic rights are acquired and fairness, solidarity and democracy may prevail.

How we get there: transformative strategies

The local food movement is too heterogeneous to express a single, defined political theory for the transformation of the food system. Different stances within the movement could be enrolled into a reformist, a revolutionary or a reconfiguration position (Geels et al., 2015). Furthermore the 'buy local' prescription is, in itself, not sufficient to trigger a transformation in the desired directions of the food system, let alone of the whole society.

In any case, the local food movement is not based on an individual consumer strategy of buying local, as much as on collective strategies for the transformation of the food system. According to the Real Utopias Project, the collective strategies that can move a system in the direction of social emancipation can be classified in three broad categories according to their relationship to State institutions: ruptural, interstitial and symbiotic. These categories do not represent heterogeneous positions regarding the ultimate goal of a social emancipatory movement ("where we want to go"), but they rather identify differentiated strategies for achieving these goals ("how to get there"). Different segments of the local food movement can therefore be differentiated according to the nature of the collective strategies they propose and adopt for setting in motion the desired transformation of the food system.

'Ruptural transformations' (or "smash the State strategy", Wright, 2006: 122) aim at creating new institutions of social empowerment through a complete break with existing forms of social structures, based on the idea that confrontation and political struggle will create a radical disjuncture with existing institutions. Ruptural transformations may be partial, rather than total, as they can involve a subset of institutions rather than the foundations of a social system: "The unifying idea is of sharp discontinuity and rapid change, rather than metamorphosis over an extended period of time" (Wright, 2006: 122).

'Interstitial transformations' ("ignore the State strategy") initiatives have in common the idea of building alternative institutions and deliberately fostering new forms of social relations that embody emancipatory ideals through direct action. As also theorized in MLP, interstitial transformations operate in niches at the margins of capitalist society, where they are often not perceived as an immediate threat to dominant classes and elites. Yet, cumulatively, such initiatives create enlarged spaces for non-commodified, non-capitalist, 'diverse' (Gibson-Graham, 2006) economic and social relations, which "can not only make a real difference in people's lives, but potentially constitute a key component of enlarging the transformative scope for social empowerment in the society as a whole" (Wright, 2006:122).

'Symbiotic transformations' ("use the State strategy") "involve strategies in which extending and deepening the institutional forms of popular social empowerment also solves certain practical problems faced by dominant classes and elites" (Wright 2006: 122). Symbiotic transformations have a contradictory character, often taking advantage of a tension between the short- and long-term effects of institutional change. In the short term, symbiotic forms of social empowerment are in the interests of elites and dominant classes; in the long term they can shift the balance of power towards broader social empowerment. It may also be thought that advances in bottom-up social empowerment will be most stable and defendable when social empowerment helps solve also real problems faced by capitalists and other elites. Positive compromises between different interests may be realized (producers and consumers; rural and urban economies) through collaborative problem-solving processes. Examples of symbiotic transformative processes may range from civic renewal movements to food councils, watershed councils and territorial development pacts.

Local food movements seem to consider interstitial and symbiotic strategies as the most appropriate to their emancipatory project, instead of ruptural or oppositional strategies.[3] It would also appear that the choice of the transformation strategy is processual, i.e. it changes according to the different stages of the movement life-cycle but also according to different contextual contingencies. For example, the Fair Trade and organic movements were considered to operate according to an interstitial strategy in their early stages of development, but the consolidation of their initiatives and the strengthening of their economic realities were accompanied by a switch to a symbiotic strategy (Renard, 1999; Smith, 2006).

Interstitial activities have in common the strategy of building alternative institutions and fostering new forms of social relations through direct action rather than through the support of the State. This approach seems to characterize large portions of the local food movement. Among the Italian alternative food movement, the most prominent proponents of an interstitial strategy are

Solidarity Purchasing Groups, which look with suspicion at State interference in their initiatives. At the same time, other segments of Italian alternative agriculture like Slow Food, the Campagna Amica Farmers' Markets network and the Italian Association of Organic Agriculture (AIAB) seems to opt for a symbiotic strategy (Fonte and Cucco, 2015).

Food councils and public procurement policies intended to promote local, organic, healthy food also adopt a symbiotic strategy in pursuing the transformation of the food system, although the symbiotic relationship is generally established with the local institutions rather than with the central State. Municipal authorities and representatives of different local interests collaborate with civil society organizations in designing and implementing policies that foster the interests of local agriculture and local communities of food. By working with local institutions, especially at the municipal or regional level, CFNs are implementing initiatives that better consolidate new experiences of food democracy and food justice. In all these initiatives, the role of local actors is important in terms of agency. The contribution of CFNs has been particularly relevant in this regard, since they have stressed the transformative potential of the 'consumer' intended not as the neo-liberal agent operating to make markets work better, but as the collective of '*ecological citizens*' (Seyfang, 2006) ready to assume responsibility for local and global problems and to organize collective democratic action in order to advance toward a sustainable solution.

Discussion: The multi-dimensional value of food as basis for its transformative potential

The paradigm shift in the global food economy envisaged in the Real Utopias Project depends on food regaining its multi-dimensional value, beyond commodification. Vivero-Pol (2017a) highlights six fundamental dimensions of food: food as public good, human right, cultural determinant, essential for humans, renewable resource and tradable good. Food is perceived as a commons when the different dimensions are all properly valued and the tradeable dimension does not obscure the non-economic ones. If we look at alternative agriculture and civic food networks in Italy as an example, we find that all of them (Slow Food, Campagna Amica, organic agriculture, welfare agriculture and Solidarity Purchasing Groups) have been inspired, motivated and operate according to a vision of food as a common social concern (Fonte and Cucco, 2015).

In the 1990s, traditional actors like Coldiretti – the biggest farmers' federation in Italy – have abandoned the previous productivist approach and adopted a vision in which the multi-functional role of agriculture is at the centre of new social relations, linking the common interests of farmers, restaurateurs and citizens. Farmers market and Campagna Amica (a foundation established by Coldiretti in 2008 in order to coordinate and promote farmers' direct selling initiatives) shops become poles of renewal and revitalization and food the expression of a community of people linked to their territory.

For the organic movement the renewal of the agro-food regime is predicated on a new relation between the farmer, the consumer and nature. The health of the soil and the health of consumers are at the core of new techniques, contributing to the diffusion of a new form of knowledge different from the 'industrial-based' knowledge of private and public extensionists. We can see here the ambition of organic agriculture to recuperate and reaffirm the principles of traditional agriculture and traditional knowledge as a commons, to anchor them in participative forms of scientific knowledge and elaborate standards as an aid to communication with consumers. Going back and valorizing traditional knowledge as a commons is fundamental in the emergence and development of organic agriculture (see Reyes-García et al. in this volume).

Slow Food approach is, since its start, a cultural approach to food. The 1989 Slow Food's Paris Manifesto interpreted the many European food controversies of the 1980s and 1990s as the crisis of the modernization model of the food system, based on industrial standardization, homogenization and commodification of food. It proposed to react to the crisis through "the political appropriation of food as symbol of collective or contested national identity", implying "the protection of threatened food and the diversity of cultural landscapes" (Leitch, 2003: 441). Since the beginning, Slow Food implements its 'politics of pleasure' through the *convivia*, groups of territorial associates who promote the common experimentation of dinners, food and wine tasting events as a form of new social relations based on pleasure. But the originality of the movement lies in the development of a *new*, post-hedonist politics, in which ecosystem diversity is central to taste and slowness and "becomes a metaphor for a politics of place, concerned with local cultural heritage, regional landscapes and idiosyncratic material cultures of production, as well as international biodiversity and cosmopolitanism"(Leitch, 2003: 453).

Civic food networks take initiative for something more than the quality of the product: local, traditional, typical, endangered – or simply good – food. Food embodies new social relations, is the expression of ecological citizenship, democracy, justice and solidarity. Solidarity Purchasing Groups (Gruppi di Acquisto Solidale, GAS), who represent the most innovative experience of civic food networks, state in their policy document: "The objective of GAS is to provide for the purchase of goods and services whilst attempting to realize a more human vision of the economy, that is an economy closer to the real needs of people and the environment, expressing an ethos of critical consumption that unites people instead of dividing them..." (Retegas, 1999: 4). Food is not a mere commodity in this vision. It is vested with political power to make a better world (Fonte, 2013a):

> starting from food, we can change many things. We start from a little as a step towards changing everything: the way people eat or drink or use the environment... it is our small contribution to making a better world.
>
> *(GAS151, in Fonte, 2013a)*

Food is a universal, fundamental need with the potential to generate intimate relations between people. For GAS members, food is at the centre of a community of relations and the re-appropriation of social spaces that has been lost.

Conclusions

In the last decades, and not only in Italy, food has emerged as a key theme for mass mobilization and theoretical elaborations for a post-capitalist economy (Gibson-Graham, 2006). Differently from the oppositional movements of the Fordist era, which relied on a unified class pursuing ruptural strategies aimed at 'smashing the state', civic food networks adopt interstitial and symbiotic strategies with the aim of building alternative consumption/production practices and social relations in the food economies. For this reason, local food movements have attracted critiques of being functional to a neo-liberal politics of discarding the State (Guthman, 2007).

When viewed through the lens of transition theories, especially the Real Utopias framework, the strategies adopted by local/civic food movements appear however to have a significant, even if not fully developed, transformative potential. Innovative food niches are mobilizing non-ruptural strategies in the service not only of sustainability, but also of a more general social emancipatory project based on fairness, solidarity and participative democracy, which recognize a role both in the State and the civil society.

The foundation of such an emancipatory project is in the overcoming of the reductionist vision of food as a commodity and in the embracing of its complexity and multi-dimensional value of food as a commons. Like water, participatory budgeting, empowered participatory forms of governance, the social economy of welfare provision, the experience of transition cities, the sharing economy and open-source software, food coalition of collective actors reclaim and posit an enlarged space of public domain and commons as the basis for a more sustainable, just and democratic society.

Acknowledgements

Thanks to the editors of this volume, who have involved us in this important project of a common reflexion on food as a commons.

Notes

1 In a slightly different form, this article has been published in the online journal *socio.hu* special issue in English # 3, 2015, *The social meaning of food* with the title 'Local food and civic food networks as a real utopias project'. We thank the editors of this journal for the permission to reprint the article.
2 This is not to say that the organic movement was born in the 1960s. It started earlier and at different times in different countries (Reed 2010).
3 This does not want to exclude that the objective of the local food movement is 'ruptural', i.e. wants to create a radical disjuncture with existing institutions, but only to indicate that the adopted strategies are not openly oppositional.

References

Altieri, M.A. and Nicholls, C.I. (2010) 'Agroecologia: potenciando la agricultura campesina para reverir el hambre y la inseguridad alimentarias en el mundo', *Revista de Economia Critica* 10, 62–74.

Altieri, M.A. and Toledo, V.M. (2011) 'The agroecological revolution in Latin America', *Journal of Peasant Studies* 38, 587–612.

Bloemen, S. and D. Hammerstein. (2015) The EU and the commons: A commons approach to European knowledge policy. Berlin and Brussels: Commons Network and Heinrich Böll Stiftung; http://commonsnetwork.eu/wp-content/uploads/2014/02/June17_The-EU-and-the-Commons-A-Commons-Approach-to-European-Knowledge-Policy-true-final.pdf, accessed on 5 February 2018.

Blythman, J. (2005) 'Organic food is not necessarily the automatic choice for the ethical consumer', *The Ecologist*, 17 June 2005.

Born, B. and Purcell, M. (2006) 'Avoiding the local trap: scale and food system in planning research', *Journal of Planning Education and Research* 26, 195–207.

Brunori, G., Rossi, A. and Guidi F. (2012) 'On the new social relations around and beyond food. Analysing consumers' role and action in Gruppi di Acquisto Solidale (Solidarity Purchasing Groups)', *Sociologia Ruralis*, 52(1), 1–30.

Buck, D., Getz, C. and Guthman J. (1997) 'From farm to table: the organic vegetable commodity chain of northern California', *Sociologia Ruralis*, 37(1), 1–20.

Crivitis, M. and Paredis, E. (2013) 'Designing an explanatory practice framework: Local food systems as a case', *Journal of Consumer Culture*, 13(3), 306–336.

Darnhofer, I., Lindenthal, T., Bartel-Kratochvil, R. and Zollitsch, W. (2010) 'Conventionalisation of organic farming practices: from structural criteria towards an assessment based on organic principles. A review', *Agronomy for Sustainable Development*, 30(1), 67–81.

DeLind, L.B. (2011) 'Are local food and the local food movement taking us where we want to go? Or are we hitching our wagons to the wrong stars?' *Agriculture and Human Values*, 28(2), 273–283.

Demaria, F. Schneider, F., Sekulova, F. and Martinez-Alier, J. (2013) 'What is Degrowth? From an activist slogan to a social movement', *Environmental Values*, 22, 191–215.

FAO (2015) *The State of Food and Agriculture*, Rome.

Feenstra, G. (1997) 'Local food systems and sustainable communities', *Americaan Journal of Alternative Agriculture*, 12, 28–36.

Fonte M. and Cucco I. (2015) 'The Political Economy of Alternative Agriculture in Italy'. In A. Bonanno and L. Busch (eds.) *Handbook of the International Political Economy of Agriculture and Food*. Elgar Publishing: Cheltenham, UK, 264–294.
Fonte, M. (2008) 'Knowledge, Food and Place. A way of producing, a way of knowing'. *Sociologia Ruralis*, 48(3), 220–222.
Fonte, M. (2013a) 'Food consumption as social practice: Solidarity Purchasing groups in Rome', *Journal of Rural Studies*, 32, 230–239.
Fonte, M. (2013b) 'Reflexive localism: Toward a theoretical foundation of an integrative food politics', *International Journal of Sociology of Agriculture & Food*, 20(3), 397–402.
Fonte, M. (2010) 'Introduction: food relocalisation and knowledge dynamics for sustainability in rural areas'. In M. Fonte and A.G. Papadopoulos (eds.) *Naming Food After Places, Food Relocalisation and Knowledge Dynamics in Rural Development*. Ashgate Publishing, Ltd: Farnham, UK, 1–34.
Friedmann, H. (1992) 'Distance and durability: Shaky foundations of the world food economy', *Third World Quarterly* 13(2), 371–383.
Fung, A. and E.O. Wright (eds.) (2003) *Deepening democracy*. Verso: London, UK.
Garnett, T. (2007) 'Animal Feed, Livestock and Greenhouse Gas Emissions: What are the Issues?' Paper presented to the Society of Animal Feed Technologists, Coventry, 25 January 2007.
Geels, F. W. (2005) 'Processes and patterns in transitions and system innovations: Refining the co-evolutionary multi-level perspective', *Technological forecasting and social change* 72(6), 681–696.
Geels, F.W. and Schott, J. (2007) 'Typology of socio-technical transition pathways', *Research Policy* 36, 399–417.
Geels, F.W., McMeekin, A., Mylan J. and Southerton, D. (2015) 'A critical appraisal of Sustainable Consumption and Production research: The reformist, revolutionary and reconfiguration positions', *Global Environmental Change* 34, 1–12.
Gibson-Graham, J.K. (2006) *A postcapitalist politics*. University of Minnesota Press: Minneapolis, USA.
Goodman, D., DuPuis, E.M. and Goodman, M.K. (2012) *Alternative Food Networks*. Routledge: London, UK.
Guthman, J. (2003) 'The trouble with 'organic lite' in California: a rejoinder to the 'conventionalisation' debate', *Sociologia Ruralis* 44(3), 301–316.
Guthman, J. (2007) 'Why I am fed up with Michael Pollan et al.' *Agriculture and Human Values* 24, 261–264.
Halkier, B. (2009) 'A practice theoretical perspective on everyday dealings with environmental challenges of food consumption'. *Anthropology of Food* (Online), 5 September 2009. http://aof.revues.org/6405, accessed 18 May 2017.
Harris, E. M. (2009) 'Neoliberal subjectivities or a politics of the possible? Reading for difference in alternative food networks', *Area* 41(1), 55–63.
Heise, U. K. (2008) *Sense of Place and Sense of Planet: The Environmental Imagination of the Global*, Oxford University Press: Oxford, UK.
Hinrichs C. (2000) 'Embeddedness and local food systems: notes on two types of direct agricultural market', *Journal of Rural Studies* 16, 295–303.
Hinrichs, C. (2014) 'Transitions to sustainability: a change in thinking about food systems change?' *Agriculture and Human Values* 31, 143–155.
Hinrichs, C. (2015) 'Fixing food with ideas of "local" and "place"' *Journal of Environmental Studies and Science*, 6(4), 759–764.
Hisano, S. (2015) Food security politics and alternative agri-food initiatives in Japan; Working Paper 131. Graduate School of Economics, Kyoto University, 1–32. http://www.econ.kyoto-u.ac.jp/~chousa/WP/131.pdf, accessed 5 February 2018.
Jackson, T. (2009) *Prosperity without Growth*. Earthscan: London, UK.
Kloppenburg J. R., Hendrickson J. and Stevenson, G.W. (1996) 'Coming into the foodshed' *Agriculture and Human Values* 13, 33–42.
Leitch, Alison. (2003). 'Slow Food and the politics of pork fat: Italian food and European identity' *Ethnos* 68(4), 437–462.
Lipton, M. (1977) *Why Poor People Stay Poor: Urban Bias in World Development*. Harvard University Press: Cambridge, USA.
Lyson T.A. (2004) *Civic Agriculture: Reconnecting Farm, Food and Community*, University Press of New England: Lebanon, USA.
Magdoff, F. and B. Tokar, (eds.) (2010) *Agriculture and Food in Crisis: Conflict, Resistance, and Renewal*. Monthly Review Press: New York, USA.

Marsden T. K., Banks, J. and Bristow, G. (2000) 'Food supply chain approaches: exploring their role in rural development' *Sociologia Ruralis* 40, 424–438
Massey, D. (1994) *Space, Place and Gender*. Polity Press: Cambridge, UK.
McMichael, P. (2014) 'Historicizing food sovereignty', *The Journal of Peasant Studies* 41(6), 933–957.
Moore, O. (2006) 'Farmers' markets, and what they say about the perpetual post-organic movement in Ireland'. In G. Holt and M. Reed (eds.) *Sociological Perspectives of Organic Agriculture: From Pioneer to Policy*, CABI Publishing: Oxfordshire, UK, 18–36.
Morgan, K., Marsden, T. and Murdoch, J. (2006) *Worlds of Food: Place, power and Provenance in the Food Chain*. Oxford University Press: Oxford, UK.
Patel, R. (2007) *Stuffed and Starved: Markets, Power and the Hidden Battle for the World's Food System*. Melville House Publishing: Brooklyn, USA.
Pirog, R ., Van Pelt, T., Enshayan, K. and Cook, E. (2001). *Food, Fuel, and Freeways: An Iowa Perspective on How Far Food Travels, Fuel Usage, and Greenhouse Gas Emissions*. Leopold Center for Sustainable Agriculture. http://lib.dr.iastate.edu/leopold_pubspapers/3/, accessed 23 May 2017.
Pirog, R. and Benjamin, A. (2003) *Checking the Food Odometer: Comparing Food Miles for Local Versus Conventional Produce Sales to Iowa Institutions*, Leopold Center for Sustainable Agriculture. http://lib.dr.iastate.edu/leopold_pubspapers/130/, accessed 23 May 2017.
Pirog, R. and Rasmussen, R. (2008) *Food, Fuel and the Future: Consumer Perceptions of Local Food, Food Safety and Climate Change in the Context of Rising Prices*. Leopold Center for Sustainable Agriculture. http://lib.dr.iastate.edu/leopold_pubspapers/126/, accessed 23 May 2017.
Polanyi, K. (1944) *The Great Transformation. The political and economic origins of our time*. Farrar & Rinehart: New York, USA.
Pretty, J. N., Ball, A. S. Lang, T.andMorison, J. I. L. (2005) 'Farm cost and food miles: An assessment of the full cost of the UK weekly food basket' *Food Policy* 30, 1–19.
Reckwitz, A. (2002) 'Towards a theory of social practices'. A development in culturalist theorizing. *European Journal of Social Theory* 5, 243–263.
Reed, M. (2010) *Rebels for the Soil: The Rise of the Global Organic Food and Farming Movement*. Earthscan: London, UK.
Renard, M. C. (1999) 'The interstices of globalization: The example of fair coffee' *Sociologia Ruralis* 39(4), 484–500.
Renting, H., Schermer, M. andRossi, A. (2012) 'Building food democracy: Exploring civic food networks and newly emerging forms of food citizenship' *International Journal of Sociology of Agriculture and Food,* 19(3), 289–307.
Renting, H., Marsden, T. and Banks, J. (2003) 'Understanding alternative food networks: exploring the role of short food supply chains in rural development' *Environment and Planning A* 35, 393–411.
Retegas, 1999. Documento Base dei GAS; http://www.retegas.org/upload/dl/doc/GASDocumentoBase.PDF, accessed 26 May 2017.
Rifkin, J. (2014) *The zero marginal cost society*. Palgrave Macmillan: Basingstoke, UK.
Ruivenkamp G. and A. Hilton (2017) *Perspectives on Commoning: Autonomist Principles and Practices*. Zed Books: London, UK.
Schatzki, T. (1996) *Social Practices. A Wittgenstein Approach to Human Activity and the Social*. Cambridge University Press: Cambridge, UK.
Severson, K. (2009) 'When "local" makes it big'. *The New York Times* (Online), 12 May 2009. http://www.nytimes.com/2009/05/13/dining/13local.html? pagewanted=all&_r=0, accessed 23 May 2017.
Seyfang, G. (2006) 'Ecological citizenship and sustainable consumption: Examining local organic food networks' *Journal of Rural Studies* 22(4), 383–305.
Shove, E., Pantzar, M. and Watson, M. (2012) *The Dynamics of Social Practice. Everyday Life and How It Changes*. Sage: London, UK.
Smith, A. (2006) 'Green niches in sustainable development: The case of organic food in the United Kingdom' *Environment and Planning C* 24(3), 439–458.
Smith, J. and Jehlička, P. (2013) 'Quiet sustainability: Fertile lessons from Europe's productive gardeners' *Journal of Rural Studies* 32, 148–157.
Steel, C. (2013) *Hungry City: How Food Shapes Our Lives*. Vintage: London, UK.
Strassen, S. (2003) *Commodifying Everything: Relationships of the Market*. Routledge: London, UK.
Tukker, A., Huppes, G., Guinée, J., Heijungs, R., de Koning, A., van Oers, L., Suh, S., Geerken, T., Van Holderbeke, M., Jansen, B. and Nielsen, P. (2006) Analysis of the life cycle environmental impacts related to the final consumption of the EU-25; http://ec.europa.eu/environment/ipp/pdf/eipro_report.pdf, accessed 23 May 2017.

Vivero-Pol, J. L. (2017a) 'Food as commons or commodity? Exploring the links between normative valuations and agency in food transition' *Sustainability* 9(3), 442.
Vivero-Pol, J. L. (2017b) 'The food commons transition: collective actions for food and nutrition security'. In: G. Ruivenkamp and A. Hilton (eds.) *Perspectives on Commoning: Autonomist Principles and Practices*. Zed Books: London, UK, 325–379.
Vivero-Pol, J. L. (2017c) 'The idea of food as commons or commodity in academia. A systematic review of English scholarly texts' *Journal of Rural Studies* 53, 182–201.
Vivero-Pol, J. L. (2017d) 'How do people value food? Systematic, heuristic and normative approaches to narratives of transition in food systems' PhD Thesis, August 2017, Université Catholique de Louvain.
Warde, A. (2005) 'Consumption and theory of practice' *Journal of Consumer Culture* 5, 131–153.
Wright, E.O. (2006) 'Compass Points. Towards a Socialist Alternative' *New Left Review* 41, 93–124.
Wright, E.O. (2010) *Envisioning Real Utopias*. Verso: London, UK.
Zerbe, N. (2009) 'Setting the global dinner table. Exploring the limits of the marketization of food security'. In J. Clapp and M.J. Cohen(eds.) *The Global Food Crisis. Governance Challenges and Opportunities*. The Centre for International Governance Innovation & Wilfrid Laurier University Press: Waterloo, Canada, 161–175.

23

CLIMATE CHANGE, THE FOOD COMMONS AND HUMAN HEALTH

Cristina Tirado-von der Pahlen

Introduction

Climate change, poverty, waste, inequity, pollution, the loss of biodiversity, the breakdown of society: these 21st century challenges are symptoms of a planet in distress, a planet driven by industrialism to the threshold of collapse. Climate change and environmental degradation impact water, land, biodiversity, and other natural resources, and the socio-economic systems and structures upon which the food commons depend. This degradation will exacerbate the existing undernutrition problem and, at the same time, undernutrition undermines climate resilience and the climate adaptation capacity of vulnerable populations. In concrete terms, undernutrition mortgages the possibility of building a food commons regime.

On the other hand, the current global food system is one of the main contributors to climate change, environmental degradation and to diet-related adverse health impacts. The global food system is responsible for up to 29% of all human caused greenhouse gas (GHG) emissions, 70% of freshwater use and two thirds of terrestrial biodiversity loss, with animal-based foods being major contributors to these environmental changes (Vermulen et al. 2012; Tubiello et al. 2015). One-third of food produced for human consumption is wasted and the emissions from food waste are a major contributor of climate change (FAO 2013).

Although the current global food system delivers a diverse and healthy diet to those who can afford it, it also leaves 795 million people hungry, two billion micronutrient-deficient, two billion overweight and over 600 million people obese (IFPRI, 2016). Malnutrition in all its forms (undernutrition, overweight and obesity) has become a universal challenge and all countries in the world suffer from one or more forms of malnutrition. Undernutrition is one of the world's most serious but least addressed socioeconomic and health problems, affecting the poorest the most (UNSCN 2014). This issue raises social justice, equity and ethical considerations (Tirado et al. 2018).

At the same time, current diets with high intakes of meat, fat and sugar, represent a major risk to health, social systems and environmental life support systems (Aleksandrowicz et al. 2016, GPAFSN 2016). Unhealthy diets and resulting increases in body mass indices (overweight) are major contributors to Non-Communicable Diseases (NCDs) (especially type II diabetes, coronary heart disease and some cancers) through the inadequate intake of fruits, vegetables, nuts, seeds and dietary fiber and through the high consumption of red and processed meats (Sabate and Soret 2014).

The alarming pace of climate and environmental changes and the consequent impacts on food systems resilience, nutrition and humans' health require the reshaping of food systems and diets with a focus on sustainability and health. We need to re-examine our relationship with food and re-organize our food systems with an equity, justice and ethical perspective (Tirado et al., 2018) as a food commons which will serve future generations.

This perspective calls for a new food paradigm and that is where the food commons comes in. This approach to food is inherently oriented towards the preservation and reproduction of natural resources, equity and the prioritization of human needs rather than maximizing profit-making. According to Vivero-Pol (2017a) the food commons are compounded of: 1) the natural and non-material resources (foods, recipes, traditional food and agricultural knowledge, etc.), 2) the communities who share the resources (local, national or global), 3) the commoning practices people use to produce, transform and eat food and 4) the moral narrative that sustains the main purpose of the food system (i.e. to sustainably produce healthy food for all).

In order to address the increasing nutritional, health and climate change challenges, our food system could be re-imagined as a commons to ensure food and nutrition security and sustainability for present and future generations. As several contributions to this edited volume discuss, the unprecedented nature and complexity of these challenges require an integrated and holistic approach to find a solution. This food commons system would nourish communities around the world, improving the availability, access and utilization of food resources over time through self-regulated collective actions, shared knowledge and sustainability solutions.

This chapter analyzes the inter-linkages of climate change, sustainable and healthy food systems and diets, human nutrition and health using the food commons framework described by Vivero-Pol (in this volume) and considering the multiple dimensions of food. The chapter starts by analyzing the impacts of climate change on the multiple dimensions of the food commons and it then looks at how food demand and dietary choices affect climate change. The chapter concludes proposing the food commons as a framework for the transition to sustainable and healthy diets in order to address nutrition and health challenges and reduce climate risk to the global food system.

Climate change impacts the multiple dimensions of the food commons

Climate change is already disrupting the global food system and these disturbances are projected to increase in number and intensity in coming years. Climate change, variability and extreme weather conditions amplify environmental degradation and threaten the already disadvantaged socio-economic systems that are required to support the transition towards a food commons system. This environmental degradation results in adverse impacts on food and nutrition security, health and the human, eco-systems and planetary resilience capabilities.

In this section, we want to analyze how climate change and variability pose a major risk to the multiple dimensions of food as a commons defined by Vivero-Pol (2017b), including: food as a human right, as an essential human resource for life and health, as a renewable natural resource, as a public good and as a cultural determinant. In addition, this section looks at how a food commons system could be capable of connecting with food and nutrition security, public health promotion and sustainability.

Food as a human right

The paradigm of a food commons is constructed around the notion that access to safe and healthy food – either self-produced, purchased, or shared – is a basic human right. Climate change amplifies all threats to the food supply and therefore challenges society's ability to provide the right to food. The right to food is the right to have regular, permanent and unrestricted access, either directly or by means of financial purchases, to quantitatively and qualitatively adequate and sufficient food corresponding to the cultural traditions of the people to which the consumer belongs (Elver 2015).

The Inter-Governmental Panel on Climate Change 5th Assessment Report (IPCC-5AR) concluded that global temperature increases of ~4°C or more above late-20th-century levels, combined with increasing food demand, would pose large risks to food security globally and regionally (IPCC 2014). Unless action is taken to reduce global emissions, it is predicted that climate change could cut the projected improvement in food availability by about one-third by 2050 (Springmann et al. 2016a). Changes in temperature and precipitation, without considering effects of CO2, will very likely contribute to increased global food prices by 2050, with estimated increases ranging from 3% to 84% (Porter et al. 2014). These increased prices will contribute to food insecurity and undernutrition.

Worst-case projections based on high GHG concentrations, high population growth and low economic growth estimate that the global number of people at risk of undernutrition would increase by as much as 175 million – above today's level – by 2080 (Brown et al. 2015). Undernutrition interacts with infectious disease, causing an estimated 3.5 million preventable maternal and child deaths annually, threatening the right to health and life. In particular, undernutrition during pregnancy and early childhood is very often the beginning of a vicious cycle; a deficient child's development very often leads to social and economic vulnerabilities later in life, particularly in women, which then perpetuates undernutrition in the next generation.

Climate change may have an impact on rates of severe stunting in children, which is estimated to increase by up to 23% in central sub-Saharan Africa and 62% in South Asia (Lloyd et al., 2011). The millions of people who have experienced under-nutrition early in life face many challenges and are often not able to make a full contribution to the social and economic development of their households, communities and nations when they become adults. The resulting impacts in terms of lost national productivity and economic growth are huge (IFPRI, 2016) and prevent the realization of the human rights to food, to health and to the environment. From a sustainability perspective, the right to food and to a healthy natural environment are inextricably related, since environmental degradation jeopardizes the planet's capacity to meet rising food needs (Von Braun and Brown, 2003). Actually, climate change effects will diminish the availability and reliability of total food produced.

Essential resource for humans

Adequate nutrition is a critical part of human health, life and development. It is related to improved infant, child and maternal health, stronger immune systems, safer pregnancy and childbirth, lower risk of non-communicable diseases (e.g. diabetes and cardiovascular disease) and longevity. People with adequate nutrition can create opportunities to gradually break the cycles of poverty and hunger (World Bank 2006).

Over two billion people suffer from micronutrient deficiencies, in particular vitamin A, iodine, iron and zinc, among others. Climate-induced food price spikes increase food insecurity (Hertel et al. 2010) and are likely to result in additional micronutrient deficiencies, as low-income people may reduce their consumption of micronutrient-rich foods (e.g. animal products, fruit and vegetables) in an effort to maintain consumption of increasingly expensive staples.

Climate-related extreme weather events can further exacerbate seasonal food shortages, thus impacting food security, undernutrition and resulting in large population displacements.

Displaced people are susceptible to undernutrition and micronutrient deficiencies because they frequently depend on food aid that may be insufficient. The number of natural disasters has increased due to more frequent extreme weather events – more than 80% of natural disasters are climate related – and the impacts on humanitarian assistance are substantial (UNISDR 2015). With increasing risks of climate-related disasters, there is a need to better protect populations who are already food- and nutrition-insecure by developing community based nutrition-sensitive disaster risk reduction strategies and management practices that strengthen the food commons (See Kent in this volume).

In addition, increasing concentrations of CO2 in the atmosphere (the same phenomenon that drives climate change) results in a reduction in protein concentration and other nutrients (e.g. calcium, iron, zinc and vitamins) in many human plant crops (Taub et al. 2008, Taub 2010). Analysis of food balance sheets from the Food and Agriculture Organization (FAO) shows that in 2010 roughly 667 million people were living in countries whose populations received at least 60% of their dietary zinc and iron from grains (e.g. wheat and rice or legumes). Similarly, 1.9 billion people who lived in these countries received at least 70% of one or both of these nutrients from these crops (Myers et al. 2014). Increasing concentrations of CO2 reduce the nutritional value of important crops by reducing the overall mineral concentrations by 8% and increasing the total non-structural carbohydrates (mainly starch and sugars) in crops (e.g. wheat, rice and soybeans) and this reduction in nutritional value may contribute to the prevalence of micronutrient deficiencies and obesity problems (Loladze 2014).

Public good

Climate change and climate-related disasters exacerbate many socio-economic factors that determine public health and nutrition (e.g. insufficient access to adequate food, resources and education) and threaten the functioning of institutions that are critical for human health and well-being, including public health and nutrition services, and social protection systems.

Climate change impacts on water, sanitation, and energy availability have major implications for food access and utilization and may affect undernutrition and health outcomes (Porter et al. 2014; Smith et al. 2014). Populations in water-scarce regions are likely to face decreased water availability, particularly in the sub-tropics, with implications for the consumption of safe food and drinking water. In other areas, flooding and increased precipitation are likely to contribute to increased incidences of infectious and diarrheal diseases, which are a major public health concern. Most of the projected climate-related human disease burden will result from increases in diarrheal diseases and malnutrition. Diarrheal diseases particularly affect nutrient absorption and food utilization. Climate change is projected to increase the burden of diarrheal diseases in low-income regions by approximately 2–5% in 2020 and will impact low-income populations already experiencing a large burden of disease. Climate change and variability can impact the occurrence of food safety hazards at various stages of the food chain, from primary production to consumption. Food safety has implications on public health and affects the food as a public good. Temperature increases and changes in rainfall patterns have an impact on the persistence and patterns of occurrence of bacteria, viruses, parasites and mycotoxins and the patterns of their corresponding food-borne diseases, affecting health and nutrition (Tirado et al. 2010).

Cultural determinant

Traditional (or customary) food systems based on non-commodified valuations of food (or commons-based food producing resources), can contribute to nutrition security, health, sustainable use of resources and resilience. However, climate change is leading to biodiversity loss and can have a major impact on traditional food systems of different communities, in particular those of indigenous peoples, traditional fisher folks, subsistence farmers and gatherers, pastoral and nomadic communities, etc.

The Intergovernmental Panel on Climate Change (IPCC) has identified key risks of climate change across sectors and regions including, among others: i) loss of marine and coastal ecosystems biodiversity and the goods and services they provide for coastal livelihoods, especially for fishing communities in the tropics and the Arctic; ii) loss of terrestrial and inland water ecosystems, biodiversity and the ecosystem and the goods, functions, and services they provide for livelihoods and iii) food insecurity and the breakdown of food systems linked to warming, drought, flooding and precipitation variability and extremes, particularly for poorer populations (IPCC 2014). Biodiversity and natural resources loss can pose major risks to indigenous diets and traditional medicines with adverse consequences on nutrition and health and traditional knowledge and cultural heritage of indigenous communities. Traditional food systems are further threatened because of the increasing loss of indigenous peoples' traditional territories due to climate change mitigation measures (e.g. carbon sinks and renewable energy projects) (FAO 2008).

Indigenous peoples rely on natural resources and the food commons for the provision of traditional foods, fuel and medicines and will be particularly affected by the impacts of climate on ecosystems, biodiversity and the environment. Indigenous peoples are often the most marginalized and disadvantaged groups in terms of receiving the resources needed for well-being, including food and health care. Extreme poverty and food insecurity often characterizes these groups. For example, approximately 10% of the circumpolar population that is indigenous is particularly vulnerable to climate change. Factors contributing to their vulnerability include their close relationship with the land, the location of communities in coastal regions, reliance on the local environment and on the food commons for key aspects of their diet and economy, and socio-economic factors. The traditional diet of populations in circumpolar regions is likely to be negatively affected by changes in animal migrations and distribution, and human access to them, partly because of the impacts of increasing temperatures on snow and ice timing and distribution (Conflanileri 2007, Smith et al. 2014). For indigenous communities, the protection of their natural resources and food commons is essential for their lives, ecosystems and cultural heritage.

Strengthening and encouraging engagement in local food traditions is an important strategy to improve diets and reverse negative food-related health outcomes. Approaches promoting the food commons eco-systems conservation and agro-ecological food systems can contribute to food security, nutrition and health in a sustainable manner and have a key role in protecting bio-cultural knowledge and heritage and can provide novel opportunities to address climate challenges.

Renewable resource

Although rarely comprehended or managed as a renewable resource, food systems are a special kind of human-dominated ecosystem (called an agro-ecosystem) that can be managed to produce the organic matter and material cycling required for ecosystem health. A defining feature of modern food systems is the uncoupling of crop from livestock production and from food consumption in order to achieve narrowly defined efficiencies of land, labor and capital (Lyson 2004). This uncoupling of solar energy flow and material recycling in the agro-ecosystem creates a need for fossil fuel energy inputs and produces the wastes responsible for much of the environmental, social and economic degradation attributed to industrial agriculture and food systems. The absence of a holistic and agro-ecological approach to food system is thus among the responsible for the production of GHGs that drive global warming, the creation of dead zones at the mouth of the world's major rivers, declining food quality, the increase in rural poverty and the extraordinary waste of food.

Food losses and waste impact both food security and nutrition and the sustainability of food systems in their capacities to ensure good quality and adequate food for this generation and for future generations (HLPE 2014). In particular, food waste that is left rotten in landfills has adverse effects on climate change and on the effective enjoyment of human rights, particularly the right to food.

According to FAO, nearly one-third of food produced for human consumption, or approximately 1.3 billion tonnes per year is either lost or wasted globally. Without accounting for GHG emissions from land use change, the carbon footprint of food produced and not eaten is estimated to be 3.3 Gtonnes of CO2 equivalent: as such, food wastage (i.e. food waste and loss) ranks as the third top emitter after USA and China (FAO 2013). A commons food system could change the approach to food waste and loss; considering food as a commons is such that distribution does not happen anymore on the basis of purchasing power, therefore there is no more food thrown away because it cannot be bought. More importantly, a food commons requires a consideration of the true cost of production, so that quantity is replaced with quality and there is a reduction of the amount of food circulating and being funneled towards consumers.

Global food demand

The trend towards an increased food demand is expected to continue for the next decades with a 70% increase in total food demand between 2005/2007 and 2050 (FAO 2013). At the same time, dietary patterns are shifting towards the consumption of more foods from animal origins, including fish. Actually, global demand for beef is projected to increase by 95%, and animal-based foods, in general, by 80%, between 2006 and 2050 (Ranganathan et al. 2016). This growth is expected to be concentrated in urban areas of emerging economies, particularly China and India.

Increasing global trends in meat consumption are expected to increase the GHG emissions related to food from 30 to 80% by 2050 and can have profound long-term impacts on the availability and pricing of certain basic foods and access to nutritionally diverse food sources (Friel et al. 2009). Predictive studies show that if global diets change in an income-dependent way (i.e. tending to have more animal protein), global-average per capita dietary GHG emissions from crop and livestock production would increase 32% from 2009 to 2050 (Tilman and Clarke 2014). These dietary changes present a set of complex challenges for climate change mitigation, agriculture and nutrition and require a shift towards more sustainable and healthy food consumption patterns that protect the food commons in the coming decades.

Sustainable and healthy diets to reduce global warming

To ensure that 9 billion people will have access to a nutritious and healthy diet that is produced in a sustainable manner, meeting climate change targets by 2050 is a global challenge. Many of the world's food systems are exceeding or approaching planetary limits and are compromising the capacity of the planet to produce food in the future.

It is estimated that the current food system is responsible for more than 30% of the total emissions (Niles et al. 2018). By 2050, GHG emissions from food and agriculture could rise by as much as 80% due to the increased consumption of animal products. Food-related GHG emissions could account for half of all emissions allowed by targets for keeping the global rise in temperature to less than 2°C by 2050 and could exceed total permissible levels by 2070 (Hedenus et al. 2014; Springmann et al. 2016b). The Paris Climate Agreement aims to keep the global temperature rise this century well below 2°C above pre-industrial levels and to pursue efforts to limit the increase even further, to 1.5°C, which is a significantly better line of defence against the worst impacts of the changing climate on health and food (IPCC 2018).

The link between a food commons system that is sustainable, healthy and climate-friendly may be represented by the notion of sustainable diets as those that take into consideration how the food system influences health and environmental outcomes and vice versa. Sustainable diets have been defined as those "diets with low environmental impacts which contribute to food and nutrition security and to healthy life for present and future generations" (FAO and Bioversity 2010). These diets are protective and respectful of biodiversity and ecosystems, culturally acceptable, accessible, economically fair and affordable, nutritionally adequate and safe and healthy, while optimizing natural and human resources (FAO and Bioversity 2010). The make-up of a diversified, balanced and healthy diet varies depending on individual needs (e.g. age, gender, lifestyle, degree of physical activity), cultural context, locally available foods and dietary customs.

Sustainable and healthy diets can improve public health and nutritional outcomes while contributing to the reduction in GHG emissions and climate change mitigation goals (HLPE 2012, Tilman and Clark 2014, Springmann et al. 2016b). A general transition to more nutritious and diverse diets (with fewer processed foods and more fruits and vegetables) frequently results in reduced GHG emissions as well as a likely reduction in non-communicable diseases (Green et al. 2015, Milner et al. 2015). At the global level, it is estimated that transitioning towards more plant-based diets that are in line with WHO global dietary guidelines on healthy eating (WHO 2015) and on recommendations of the World Cancer Research Fund (WCRF 2007), could reduce global mortality by 6–10% and food-related greenhouse gas emissions by 29–70% compared with a reference scenario in 2050 (Springmann et al. 2016b).

Strategies to reduce the production and consumption of animal-based foods among wealthier populations could help limit GHG emissions from food production and from land use for agricultural expansion, as well as benefiting the health of adults in groups consuming high amounts of foods from animal origins (Monsivais et al. 2015, Ranganathan et al. 2016). Dietary shifts towards reduced over-consumption of protein by reducing consumption of animal-based foods, particularly beef, could close the food gap by up to 30% globally, while substantially reducing agricultural resource use and global environmental impacts (Ranganathan et al. 2016). This potential to close the food gap suggests the importance of a paradigmatic shift towards producing and eating less meat in high- and medium-income societies, in order to leave enough "planetary environmental space" to allow for low income and undernourished populations around the world to increase their protein production to meet their dietary needs. This shift calls for a different vision of the global food system as a global food commons system that is shared by the entirety of humanity and which should be protected in order to provide food for all. It is not just about my own food preferences, or my own country's food needs, but the entire planet's possibilities.

Food as medicine: a critical food dimension

The relevance of diets to human health leads us to add to the multiple dimensions of the food commons described by Vivero-Pol (2017b) the consideration of food as a medicine to enrich the multidimensional approach. Food has been considered fundamental for health since Hippocrates studied the medicinal properties of foods in Ancient Greece (Kleisiaris et al. 2014). Many traditional medicine systems practiced throughout history, including Traditional Chinese Medicine and Ayurvedic Medicine, consider food to be medicine and hold the belief that a healthy diet is necessary for protecting health.

A healthy diet is essential for health promotion and helps protect against malnutrition in all its forms, as well as non-communicable diseases (NCDs), including diabetes, heart disease, stroke and cancer (WHO 2015). According to the World Health Organization (WHO), consuming at least 400g (5 portions) of fruits and vegetables per day can save 2.7 million lives (WHO 2009) by preventing NCDs. In this context, as explained above, healthy diets shall also be sustainable since this is fundamental to maintain healthy consumption patterns worldwide. Particularly, food is considered as medicine for HIV-positive and Type 2 Diabetes patients (Palar et al. 2017). Good nutrition is critical for the millions of people living with HIV/AIDS since nutrition can postpone HIV/AIDS-related illnesses and prolong life. Food insecurity negatively impacts health through nutritional (e.g., weight, diet quality), mental health (e.g., depression, stress), and behavioral pathways (e.g., medication adherence and disease self-management) (Weiser et al. 2015).

Medically appropriate food assistance to food-insecure individuals improves health outcomes in resource-rich settings, is associated with improved HIV and Diabetes health, and improves medication adherence and mental health (Palar et al. 2017).

Nutrition policies can provide incentives for improving diets, for strengthening the nutritional focus of health services (particularly in the context of anti-retroviral therapy and home-based care) and for ensuring nutritionally balanced food aid as a safety net for people who are acutely food-insecure or at risk of becoming so (e.g. households hosting AIDS orphans) (Gillespie and Kadiyala 2005).

Many of the geographical areas highly affected by the HIV/AIDS epidemic are at the same time impacted by climate change. HIV/AIDS-affected populations in developing countries often depend on the food commons for good nutrition. Therefore, it is essential to promote the climate resilience of the food commons system in affected communities, including the establishment of appropriate nutrition-sensitive safety nets, community-governed school feeding programs and others. In this context, the state has a key role as the facilitator (or as an enabling state as described by Pazaitis and Bauwens in this volume) of specific dynamics of the food commons through their support of nutrition-sensitive safety nets.

Food commons and transformative approaches towards sustainable and healthy diets

The current patterns of food production and consumption having food valued and exchanged as a pure for-profit commodity are not sustainable and they bring as consequences the high GHG emissions attributed to agriculture, the triggering of unhealthy eating habits that cause the rise of NCDs, the unabated hunger and the rising obesity pandemic. The paradigm of the food commons should be constructed around the notion that access to food and to healthy environments are basic human rights which, at the same time, determine the right to health.

A food commons system should guarantee good nutrition while ensuring sustainable and healthy diets for all.

The definition of a sustainable diet will differ depending on the communities that are adopting them. There are key principles of nutrition-sensitive and health-promoting sustainable diets, including:

- Enhancing the quantity, nutritional quality and diversity of local nutrient-rich food production for the promotion of local and seasonal foods and diverse diets.
- Respecting socio-cultural contexts and identities: Strategies must be suitable for the microclimate, local and community needs, and the community's socio-cultural values and heritage and local or regional identity.

- Promoting dietary patterns with less animal and processed products and more fruits, vegetables, grains and pulses, which have a huge transformative potential for public health and climate change mitigation.
- Targeting the most vulnerable groups and ensuring social inclusion and resilience: Social protection through increasing households' income, strengthening rural and urban services and investing in sustainable agriculture (e.g. agro-ecology) is critical so that households become less exposed, less sensitive, more adaptive, and more resilient to a range of climate shocks.
- Ensuring gender equality engagement: Women serve as agents of social change and development through their unique roles in the family and in child care, household food provisioning, food choices and health and natural resource stewardship.
- Adopting a multi-sectoral approach and good governance: A number of policy, institutional and governance solutions are necessary for the establishment of healthy and sustainable food systems and diets. These solutions require the engagement of consumers, producers, processors, retailers, food services, local and national governments, public health groups, educators, chefs, civil society, local policy councils, etc.

Approaches to accelerating a shift towards less GHG emission-intensive, more sustainable and healthier diets should evolve from particular socio-economic and cultural contexts with conditions of the food systems in focus. For example, effective approaches to sustainable and healthy diets have been implemented through centuries (since the times of Ancient Greece) by Mediterranean cultures and communities. The Traditional Mediterranean diet is considered healthy (Martinez-Gonzalez 2016) and sustainable because of its high reliance on plant-based foods and biodiversity which ensures nutritional quality, a variety of food preparation techniques, respect for seasonality, commitment to local culture and traditions, diversity of agro-ecological landscapes and low environmental impacts due to low consumption of animal products (FAO and Bioversity 2010). For example, a recent study shows that Spain (a country with a Mediterranean diet) is predicted to have the longest forecast life expectancy, exceeding 85 years for both sexes in 2040 (Foreman et al. 2018). In the Mediterranean basin meals are usually shared with family and friends for the communal pleasure of eating together. Quality family time and social relationships are also determinants of wellbeing, health and longevity (Yang et al. 2016).

The Nordic Council of Ministers Food Policy Lab has just launched a "Solutions Menu" for sustainable food policies covering nutrition, food culture and identity, public food and meals, food waste and sustainable diets (Halloran et al. 2018). These solutions include the promotion of traditional and local foods, biodiversity, gastronomy, artisanal foods, cooking programs, local and regional identity, rural development and tourism, networks that bring producers and consumers closer together, school gardens and public meals and establishing healthy and sustainable institutional food procurement and urban food services (with low meat content). For example, the national school meal programmes in the Nordic Region, which manage food as a public concern and not just as a commodity, date back to the 1940s and now focus on nutritional equality, healthy eating habits and the use of food as a pedagogical tool. The 'Copenhagen model' demonstrates that it is possible to increase the availability and affordability of organic food in public institutions and support the retail market for organic products through the push-pull mechanism of public procurement (Nordic Council of Ministers 2014).

There are other drivers for shifts to sustainable and healthy diets currently under study that include economic interventions. For example, diverting public subsidies from food commodities (e.g. soybean, corn, oil palm, sugar), which are meant to create ultra-processed food or biofuels, to fresh fruits and vegetables can contribute to promoting sustainable and healthy diets. Subsidies on healthier substitutes in retail settings are effective in influencing consumption, leading to a higher ratio of expenditure on healthy foods. The inclusion of healthy and sustainable foods in nutrition assistance programs – that is, the scheme for the most deprived or food stamps programs – can also encourage consumer shifts.

Food based dietary guidelines are also key drivers for changing dietary patterns towards healthy and sustainable and low climate-altering emission diets. In general, the sustainable dietary guidance focuses on eating local and seasonal, decreasing meat consumption, eating more fruits, vegetables and plant-based protein, reducing energy intake and reducing food waste. Brazil's dietary guidelines also address the social equity and economic aspects of sustainability and urge people to avoid ultra-processed foods that damage traditional food cultures and health, while simultaneously promoting locally available meals (not food products) and encouraging home-made cooking and eating together.

Rights-based mechanisms, which focus on vulnerable groups, especially social protection programs, like food and cash transfers and universal school feeding programs, can help vulnerable households become less exposed, less sensitive and more adaptive to a range of climate-related shocks and, at the same time, promote healthy diets. Universal school feeding programs, understood as a public policy to promote a public good (healthy diets for children), can provide opportunities for sustainable and local food procurement and education on healthy diets. For example, Brazil has one of the largest school feeding programmes in the world and provides free meals in all public schools which are usually cooked on school premises. Free meals are valued and governed as a human right, which every Brazilian child is entitled to enjoy. The program emphasizes the use of locally produced fruits and vegetables and tries to purchase food from local smallholders as much as possible, in order to bolster small farmers' incomes as well as school enrolments and the nutrition of school-aged children. The spin-off effects of this program go beyond the supply of free daily meals to kids, having economic and social impacts on the local communities where those schools are placed. These programs could be further linked to public purchases targeting local organic producers and to community education programs to avoid food waste and promote healthy eating habits. The fact that the meals are cooked daily on school premises requires a nutritionist to supervise the special dietary requirements of children and the 2–3 communal meals offer a space where healthy food habits are taught and practised.

Public procurement can have an important role in establishing healthy and sustainable food systems based on the idea of food as a commons. Public procurement refers to the process by which public authorities (e.g. government departments, regional and local authorities or bodies governed by public law), purchase works, goods or services from private companies or directly from the producers. The European Commission Green Public Procurement is an instrument that uses purchasing power to choose environmentally friendly products can make an important contribution to sustainable consumption and production and this may include food purchasing. This idea is grounded in the understanding that public procurement is an instrument of the State to achieve a public good, by influencing market supply and demand. There is no invisible hand to match the lowest offer with the demand, but a purportedly hand that chooses healthy, local or organic food producers. Local Food Policy Councils or local communities or institutions (e.g. schools, cooperatives or agencies) can have a major role in promoting sustainable local procurement policies. For example, the Oakland Unified School District (California, USA) has encouraged changes in their purchasing guidelines to include a minimum amount of sustainably and ethically produced plant-based menu options.

Conclusions

Developing a global food system to deliver healthy diets for a growing population, while reducing environmental impacts and climate change, is one of greatest challenges of our world. A global shift towards environmentally sustainable and healthy food systems and diets is urgently needed to avoid the worst impacts of climate change. Along those lines, to sustain that shift, it is compulsory to revalue food dimensions that are non-economical, because they cannot be valued in monetary terms, but they are essential for human beings (e.g., dimensions such as food as a human right, cultural determinant or its essentialness for human survival). How much does something cost that is vital for everybody's life? Moreover, the dimension of valuing food as a medicine is also very relevant in this context whereby industrial food harms people's health and obesogenic food is legally accepted in all countries.

The current food system, re-imagined as a commons, could provide collective and state-driven solutions to address nutritional challenges, health and climate risks, based on the knowledge of the crowd and customary solutions that have been working for centuries, and contribute to support sustainability. As discussed in this chapter, a sustainable and fair food system based on the idea of food as a commons would deliver food security and nutrition for all in such a way that the economic, social and environmental bases to generate food security and nutrition for future generations are not compromised (HLPE 2014). A system where profit-maximization is not the only driving ethos, where an enabling state could divert subsidies to civic actions for food, to small-scale organic producers and to guarantee a minimum access to food for all, and where self-regulated collective actions can thrive to support their own communities. This food commons system (or network of systems) would nourish the world while improving the availability, access and utilization of food resources over time, all of this while bearing in mind the food security, nutrition and health of the coming generations.

A sustainable food system should not only minimize negative impacts to our planet (e.g. land degradation, water pollution and deforestation) but should also integrate agriculture, climate action and biodiversity conservation in order to contribute to agro-ecological resilience and to positive human nutrition and health outcomes. It should also incorporate cultural preferences and build on traditional knowledge and protect cultural heritage. In a nutshell, it should value adequately and incorporate in the policy-making the different food dimensions that are relevant to humans.

Nutrition-sensitive and sustainable food systems are fundamental to reducing undernutrition and improving nutrition-security in a changing climate (Tirado et al. 2013). Furthermore, food systems need to be repositioned from just supplying food to providing high-quality diets for all (GPAFSN 2016). A paradigm shift from industrial agriculture (where food is just valued and governed as a commodity) to diversified agro-ecological systems (where food is valued as a multi-dimensional commons) should be put at the centre of any political project aimed to address food security, environmental protection, nutritional adequacy and social equity. Social equity, justice and ethical considerations should be fundamental values of sustainable food systems (Tirado et al. 2018).

Encouraging a global transition to culturally acceptable, sustainable and healthy diets that are nationally and locally appropriate can be a major contribution to both a reduction in climate emissions and improved public health and nutrition. For example, shifting consumption away from animal products, especially meat from ruminant sources that are raised in confined allotments in high-meat consumption societies, towards less emission-intensive diets (either plants-based or with extensive-grazing livestock), would result in reduced GHG emissions as well as likely reductions in diet-related non-communicable diseases. A bottom-up approach to this transition is critical and there are some strategies that communities, food policy councils and alternative food networks could begin to explore at their local, state or regional levels (e.g. reducing contributions to climate change (mitigation) by facilitating a shift to more plant-based diets through promoting sustainable local and State Procurement Policies).

The paradigm of food as a commons and its holistic approach to ecology, food systems and society make it possible to recognize that climate mitigation strategies, sustainable diets and health are intertwined and require integrated solutions and coherent policies and those policies cannot just be based on economic rationalities. We need to understand that internal motivations to do things differently for climate, for eating habits and for health are multi-faceted, including drivers that are culturally rooted, based on emotions, influenced by media or mediated by social pressures. Recognizing the importance of the non-economic dimension of food can enable stakeholders and communities to prioritize multiple impacts, opportunities and interests, forging a coherent, consensual approach where the notion of food commons (as opposed to commoditized food) is utilized to shape and sustain sustainable and healthy diets that contribute to nutrition, health, environmental and climate solutions. Policies that include, but are not limited to, agriculture, health, food and nutrition, dietary guidance, gender equity, environment, water, energy, waste, trade, transportation, economics and trade, need to be integrated through a multi-stakeholder process to promote sustainable and healthy food systems and diets. At the same time, it will be essential to be aware of the potential negative implications that different mitigation options may have on nutrition and food security. The production of biofuels and the loss of edible crops for the sake of maximizing profit per hectare at any cost, for example, calls for increased policy coherence and institutional and cross-sectoral collaboration at the local, national and international level. Similarly, food waste and its donation without the provision of systemic solutions represent a major attempt against the food commons approach, food justice and equity and raise serious ethical considerations that should be analyzed (Kenny and Sage, this volume).

The nutrition and health communities and civil society should engage in multi-sectoral decision-making processes that support nutrition-sensitive climate adaptation and mitigation initiatives promoting sustainable and healthy food systems and diets among others. Nutrition and sustainable diets should be considered in national processes related to climate action under the United Nations Framework Convention on Climate Change (UNFCC), (e.g. National Adaptation Plans (NAPs), National Determined Commitments (NDCs), Nationally Appropriate Mitigation Actions (NAMAs), among others). The nutrition community should contribute to IPCC assessments and to the work on adaptation and health (including malnutrition) under the Nairobi Work Programme (NWP) on impacts, vulnerability and adaptation to climate change. Stakeholders involved in the UNFCC work related to agriculture and food security should draw on support from international institutions (e.g. the Committee on World Food Security (CFS)) to integrate the right to food and other human rights and the idea of the food commons, as guiding principles for climate action.

Agreement on shared rights-based principles of sustainability in promoting healthy diets is needed. Integrated policies implemented through collaborative action to reduce the speed and impacts of climate change on the Global Food System while underwriting better outcomes in nutrition and health are critical for the implementation of a food commons system. Effective transformative approaches to sustainable and healthy diets should be evidence-based, socio-culturally and geographically appropriate, sustainable, inclusive, democratic, integrated, equitable, ethical, inspiring and embraced by the communities and the institutions that have shaped them.

Placing the multi-dimensionality of food as a commons at the center of the development of policies to adapt to and mitigate the impacts of climate change and to advance public health and nutrition would be essential to promote a food commons approach within the sustainable development agenda. Effective engagement of local communities and civil society in the governance of the food commons is necessary to create a transparent process and a policy environment where climate, food and health policies can be envisaged and integrated.

References

Aleksandrowicz, L., Green R., Joy E.J.M., Smith, P. and Haines, A. 2016. The Impacts of Dietary Change on Greenhouse Gas Emissions, Land Use, Water Use, and Health: A Systematic Review. PLoS ONE 11(11): e0165797.

Brown, M.E., Antle, J.M., Backlund, P., Carr, E.R., Easterling, W.E. et al. 2015. Climate Change, Global Food Security, and the U.S. Food System. Washington D.C., USA: USDA, UCAR and NCAR. http://www.usda.gov/oce/climate_change/FoodSecurity2015Assessment/FullAssessment.pdf (Accessed on 15 June 2018).

Confalonieri, U., Menne, B., Akhtar, R., Ebi, K.L., Hauengue, M., et al., eds. 2007. Climate change 2007: impacts, adaptation and vulnerability. Contribution of Working Group II to the Fourth Assessment Report of the Intergovernmental Panel on Climate Change. Cambridge and New York: Cambridge University Press: 391–431.

Elver, H. 2015. Interim report of the Special Rapporteur on the right to food submitted to the 70th General Assembly, 70/287 United Nations Human Rights, Office of the High Commissioner.

FAO. 2008. Climate Change and Food Security: A Framework Document. Rome, Italy: United Nations Food and Agriculture Organization (FAO).

FAO. 2013. The State of Food and Agriculture. Rome, Italy: United Nations Food and Agriculture Organization (FAO). http://www.fao.org/publications/sofa/2013/en/ (Accessed on 15 June 2018).

FAO and Bioversity International. 2010. Sustainable Diets and Biodiversity: Directions and solutions for policy, research and action. Rome, Italy: FAO and Bioversity International. http://www.fao.org/docrep/016/i3004e/i3004e00.htm (Accessed on 15 June 2018).

FAO. 2015. FAO success stories on climate smart agriculture. Rome, Italy: FAO. http://www.fao.org/3/a-i3817e.pdf (Accessed on 15 June 2018).

Foreman, K.J., Marquez, N., Dolgert, A. et al. (2018). Forecasting life expectancy, years of life lost, and all-cause and cause-specific mortality for 250 causes of death: reference and alternative scenarios for 2016–40 for 195 countries and territories. Lancet.

Friel, S., Dangour, A.D., Garnett, T., Lock, K., Chalabi, Z., Roberts, D. 2009. Public health benefits of strategies to reduce greenhouse-gas emissions: food and agriculture. Lancet, 374: 2016–2025.

Gillespie, S. and Kadiyala, S., 2005. HIV/AIDS and food and nutrition security: interactions and response. American Journal of Agricultural Economics, 87(5): 1282–1288.

GPAFSN. 2016. Food systems and diets: Facing the challenges of the 21st century. London, UK: Global Panel on Agriculture and Food Systems for Nutrition.

Green, R., Milner, J., Dangour, A.D., Haines, A., Chalabi, Z. et al. 2015. The potential to reduce greenhouse gas emissions in the UK through healthy and realistic dietary change. Climate Change, 129: 253–265.

Halloran, A., Fischer-Møller, M.F., Persson, M. and Skylare, E. 2018. Solutions Menu: A Nordic guide to sustainable food policy. Copenhagen, Denmark: Nordic Council of Ministers. https://www.diva-portal.org/smash/get/diva2:1214792/FULLTEXT01.pdf (Accessed on 15 June 2018).

HLPE. 2012. Food security and climate change. A report by the High Level Panel of Experts on Food Security and Nutrition of the Committee on World Food Security. Committee on World Food Security. Rome, Italy: FAO.

HLPE. 2014. Food losses and waste in the context of sustainable food systems. A report by the High Level Panel of Experts on Food Security and Nutrition. Committee on World Food Security. Rome: FAO.

IFPRI. 2016. Global Nutrition Report 2016: From Promise to Impact: Ending Malnutrition by 2030. Washington, DC: International Food Policy Research Institute.

IPCC. 2014. Summary for policymakers. In Field, C.B., Barros, V.R., Dokken, D.J., Mach, K.J., Mastrandrea, M.D. et al., eds. Climate Change 2014: Impacts, Adaptation and Vulnerability. Part A: Global and Sectoral Aspects. Contribution of Working Group II to the Fifth Assessment Report of the Intergovernmental Panel on Climate Change. Cambridge and New York: Cambridge University Press. 1–32.

IPCC. 2018. Global Warming of 1.5 °C. IPCC special report on the impacts of global warming of 1.5 °C above pre-industrial levels and related global greenhouse gas emission pathways, in the context of strengthening the global response to the threat of climate change, sustainable development, and efforts to eradicate poverty.

Kleisiaris, C.F., Sfakianakis, C. and Papathanasiou, I.V. 2014. Health care practices in ancient Greece: The Hippocratic ideal. Journal of Medical Ethics and History of Medicine, 7: 6.

Lloyd, S.J., Kovats, S. and Chalabi, Z. 2011. Climate change, crop yields, and undernutrition: development of a model to quantify the impact of climate scenarios on child undernutrition. Environ Health Perspectives, 119: 1817–1823.

Loladze, I. 2014. Hidden shift of the ionome of plants exposed to elevated CO2 depletes minerals at the base of human nutrition. eLife, 3: e02245.

Lyson, T.A. 2006. Civic agriculture: Reconnecting farm, food and community. Medford, USA: Tufts University Press.

Martínez-González, M. A. 2016. Benefits of the Mediterranean diet beyond the Mediterranean Sea and beyond food patterns. *BMC Medicine*, 14: 157.

Milner, J., Green, R., Dangour, A.D. et al. 2015. Health effects of adopting low greenhouse gas emission diets in the UK. BMJ Open 5: e007364. doi:10.1136/bmjopen-2014-007364.

Monsivais, P., Scarborough, P., Lloyd, T., Mizdrak, A., Luben, R. et al. 2015. Greater accordance with the Dietary Approaches to Stop Hypertension dietary pattern is associated with lower diet-related greenhouse gas production but higher dietary costs in the United Kingdom. American Journal of Clinical Nutrition, 102(1): 138–145. doi:10.3945/ajcn.114.090639.

Myers, S.S., Zanobetti, A., Kloog, I., Huybers, P., Leakey, A.D. et al. 2014. Increasing CO2 threatens human nutrition. Nature, 510(7503): 139.

Niles, M., Ahuja, R., Barker, T., Esquivel, J., Gutterman, S., Heller, M., Mango, N., Portner, D., Raimond, R., Tirado, C. and Vermeulen, S. 2018. Climate change mitigation beyond agriculture: A review of food system opportunities and implications. *Renewable Agriculture and Food Systems*, 33(3): 297–308.

Nordic Council of Ministers 2014. Nordic Nutrition Recommendations 2012: Integrating nutrition and physical activity. Copenhagen, Denmark: Nordic Council of Ministers Secretariat. http://norden.diva-portal.org/smash/get/diva2:704251/FULLTEXT01.pdf (Accessed on 15 June 2018).

Palar, K., Napoles, T., Hufstedler, L.L. et al. 2017. Comprehensive and medically appropriate food support is associated with improved HIV and diabetes health. Journal of Urban Health, 94: 87.

Porter, J.R., Xie, L., Challinor, A.J., Cochrane, K., Howden, S.M. et al. (eds.). 2014. Climate change 2014: impacts, adaptation, and vulnerability. Part A: global and sectoral aspects. New York, USA: Cambridge University Press. 485-533.

Ranganathan, J., Vennard, D., Waite, R., Dumas, P., Lipinski, B. et al. 2016. Shifting Diets for a Sustainable Food Future: Creating a Sustainable Food Future. Washington D.C., USA: World Resource Institute. http://www.wri.org/sites/default/files/Shifting_Diets_for_a_Sustainable_Food_Future_0.pdf (Accessed on 15 June 2018).

Sabaté, J. and Soret, S. 2014. Sustainability of plant-based diets: back to the future. The American Journal of Clinical Nutrition, 100: 476S–482S.

Smith P., et al. 2014. Agriculture, Forestry and Other Land Use (AFOLU). In Edenhofer, O. et al. (eds.). Climate Change 2014: Mitigation of Climate Change. Contribution of Working Group III to the Fifth Assessment Report of the Intergovernmental Panel on Climate Change Cambridge and New York: Cambridge University Press.

Springmann, M., Mason-D'Croz, D., Robinson, S., Garnett, T., Godfray, H.C.J. et al. 2016a. Global and regional health effects of future food production under climate change: a modelling study. The Lancet, 387: 1937–1946.

Springmann, M., Godfray, H.C.J., Rayner, M. and Scarborough, P. 2016b. Analysis and valuation of the health and climate change cobenefits of dietary change. Proceedings of the National Academy of Sciences of the United States, 113(15): 4146–4151.

Taub, D.R., Miller, B. and Allen, H. 2008. Effects of elevated CO2 on the protein concentration of food crops: a meta-analysis. Global Change Biology, 14(3): 565–575.

Taub, D. (2010). Effects of rising atmospheric concentrations of carbon dioxide on plants. Nature Education Knowledge, 3(10): 21.

Tilman, D. and Clark, M. 2014. Global diets link environmental sustainability and human health. Nature, 515(7528): 518–522.

Tirado, M.C., Clarke, R., Jaykus, L.A., McQuatters-Gollop, A. and Frank, J.M. 2010. Climate change and food safety: a review. Food Research International, 43(7): 1745–1765.

Tirado, M.C., Crahay, P., Mahy, L., Zanev, C., Neira, M. et al. 2013. Climate change and nutrition: creating a climate for nutrition security. *Food and Nutrition Bulletin*, 34(4): 533–547.

Tirado-von der Pahlen, M.C., Arias, D., Comim, F. and Sassi F. 2018. Social equity, justice and ethics: missing links in eco-agri-food systems. In UNEP. TEEB for Agriculture and Food: Scientific and Economic Foundations. Chapter 5. Geneva, Switzerland: UN Environment Programme.

Tubiello, F.N., Salvatore, M., Ferrara, A.F., House, J., Federici, S. et al. 2015. The contribution of agriculture, forestry and other land use activities to global warming, 1990–2012. *Global Change Biology*, 21(7): 2655–2660.

UNSCN, 2014. Nutrition and the Post-2015 Sustainable Development Goals: A technical paper. Rome, Italy: United Nations System Standing Committee on Nutrition.

UNISDR. 2015. Science for Disaster Risk Reduction: Case studies. Geneva, Switzerland: UNISDR. https://www.unisdr.org/partners/academia-research/case-studies. (Accessed on 15 June 2018).

Vermeulen, S.J., Campbell, B.M. and Ingram, J.S., 2012. Climate change and food systems. *Annual Review of Environment and Resources*, 37: 195–222.

Vivero-Pol, J.L. 2017a. The idea of food as commons or commodity in academia. A systematic review of English scholarly texts. Journal of Rural Studies, 53: 182–201.

Vivero-Pol, J.L. 2017b. Food as Commons or Commodity? Exploring the links between normative valuations and agency in food transition. Sustainability, 9(3): 442.

Von Braun, J. and Brown, M.A. 2003. Ethical questions of equitable worldwide food production systems. *Plant Physiology*, 133(3): 1040–1045.

Weiser, S.D., Palar, K., Hatcher, A., Young, S., Frongillo, E.A. and Laraia, B. 2015. Food Insecurity and Health: A Conceptual Framework. In Ivers, L.C., ed., Food Insecurity and Public Health. Boca Raton, USA: CRC Press. 23–50.

World Bank. 2006. World Development Report 2006: Equity and Development. Washington, D.C., USA: World Bank.

WCRF. 2007. Food, Nutrition, Physical Activity, and the Prevention of Cancer: a Global Perspective. Washington D.C., USA: World Cancer Research Fund American Institute for Cancer Research.

WHO. 2009. Protecting health from climate change. Connecting science, policy and people. Geneva, Switzerland: World Health Organization. http://www.who.int/globalchange/publications/reports/9789241598880/en/ (Accessed on 15 June 2018).

WHO. 2015. Healthy diet fact sheet N°394. Geneva, Switzerland: World Health Organization. http://www.who.int/en/news-room/fact-sheets/detail/healthy-diet (Accessed on 15 June 2018).

PART VI
Conclusions

24

FOOD AS COMMONS

Towards a new relationship between the public, the civic and the private

Olivier De Schutter, Ugo Mattei, Jose Luis Vivero-Pol and Tomaso Ferrando

Why this book?

This book was motivated by the need to approach with a fresh look what we regard as perhaps the most embarrassing predicament of the Anthropocene/Capitalocene (Capra and Mattei, 2015, Altvater et al., 2016, Moore, 2017). We live in an era with roughly the same number (about one billion) of over-fed people and of people lacking access to nutritious food (which means that do not know in the morning if they will be able to feed themselves and their children during the day). Our era also stands out by the remarkable amount of food that is wasted in some parts of the world and by the unprecedented number of livestock that populates this planet (Patel and Moore, 2017). Moreover, in the current phase of neoliberal capitalism that dominates in the Anthropocene/Capitalocene, the ecological footprint is out of control; some rich people (the majority in the Global North and the elite in the Global South) can enjoy every day food shipped from thousands of miles away on gas gulping aircrafts and boats that pollute the environment beyond imagination. Such luxury, the result of the worldwide colonization of diets, would be impossible without a very significant environmental subsidy; if all the externalities had to be internalized, eating Nile Perch would be unaffordable to most people everywhere. The subsidy is ultimately paid by the poor in the South and, in general, will certainly be paid by future generations. Unless we deal with and avoid the hidden social and environmental costs that are so far unaccounted for in the hegemonic food system (TEEB, 2018)

While the rich eat healthy fruit salads of organic mangoes in Stockholm and Palo Alto, poor people (in the North and in the South) are fed, when at all, with industrially made high-calorie junk food, such as chicken nuggets or cheeseburgers, that produce, among other things, resistance to antibiotics that are widely given to the unlucky martyrs of this capitalist food chains. This ultra-processed food (Monteiro et al., 2018) produced by the Big Food industry is basically made up of food components from cash crops (e.g. maize, wheat, soybean) that are extracted, ultra-processed and re-amalgamated again or that feed industrially-raised animals that are then slaughtered in the millions after having suffered beyond imagination and produced massive amounts of greenhouse gas. These horrors are not the response to the needs of the billions of people who live in the Anthropocene/Capitalocene. They are the necessary and inevitable imperatives of capitalist extraction in a context where food is understood not as the fundamental depository of use value, but as a commodity considered in its exchange value only. Food as a commodity is exclusively represented through a quantitative approach which has its own scientific measure system: the calories, which fuel both cars and human bodies to ensure they are fit for work. Quality is expelled from this approach and so is any non-mechanistic systemic vision of the food chain. Introducing the commons as a way of thinking is an attempt to introduce quality and holistic and systemic thinking as well as a phenomenological, rather than positivistic, approach in addressing the catastrophic consequences of capitalist logic applied to food. The capitalist food industry in its full reach is actually the horror story this book has tried to tell (though incompletely) in order to bankrupt the current system and seek new avenues. This book is thus part of a broader search for sustainable institutional arrangements that, as humankind, we urgently need to develop before it is too late for us and for the planet (Capra and Mattei, 2015).

This collection of essays explores the consequences of the commodification of food, and it seeks to identify whether food can instead be redefined as a "commons", and what this redefinition might entail. This may sound like a provocative thesis. First, the commodification of food has such a long and hegemonic history that it has become difficult to even imagine that it could be conceptualized differently; recent archaeological discoveries suggest that the production and storage of food at the individual household level, rather than at the level of the larger social group – the tribe or the community –, started some 6,000 years ago, with the first urban settlements in Mesopotamia, which corresponds to modern-day Iraq (Shepperson, 2017). Obviously, capitalism, and corporate-capitalism in particular, has imposed the logic of exchange value to such an extent that these remote episodes of commodification may seem of mere archaeological interest in a phenomenology of the present; yet they do illustrate, at the same time as the historical weight of the dominant paradigm, the possibility of alternatives (Polanyi, 1957). Secondly, food is par excellence a good that is both excludable and rival; access can easily be fenced off from food as a material object and what one person eats is not left to be consumed by others. Therefore, following the classic typology of Paul Samuelson (1954), food is a private good and there would, in principle, be no reason to believe that market mechanisms would not be appropriate to ensure optimal allocation of resources for its production.

These are precisely the misunderstandings that the essays on this book aim to dissipate. Redefining food as a "commons" is not to negate that either food itself (the calories and micronutrients we ingest) or most of the factors required for its production (including in particular land and human labour) can be privatized and that they have been so historically. It is to suggest, rather, that treating food *as a mere commodity*, while ignoring its multiple dimensions and the various other functions it fulfils, stifles our institutional imagination as to how its production and consumption could be governed. That would not only ignore the various social innovations that today, often at the grassroots level, are challenging this approach, at least implicitly. It would also be problematic on efficiency and equity grounds if the purpose is to save capitalism from itself, and on structural grounds in a full-fledged critique based on the material conditions of the present. In these conclusions, we explore the problems associated with treating food as a commodity to be distributed by market mechanisms. We then ask whether turning to state bureaucracy offers a better solution. Next, we turn to the idea that framing food as a commons may be an alternative to both. We conclude with a proposal to redefine the relationships between the state, the community and the market (or the respective logics of the public, the civic and the private) in order to create the space necessary for commoning practices to expand.

Why food cannot be treated as a mere commodity

Treating food as a mere commodity implies that the choices about how much to produce and what to produce are meant to respond to demand. In the perfectly functioning markets that neoclassical economic models still rely on, supply follows price signals; higher prices caused by a rise in demand therefore should lead to increase production, in turn bringing prices down to an "equilibrium" price (which Smith called the "natural" price). However, this idealized view presents a number of problems.

Food markets' imperfections

First, it is entirely unrealistic. Food markets are, in fact, imperfect to the point of caricature. Concentration of power has significantly increased at various segments of food chains, both as a result of the industrialization of agricultural production and of food processing and as a result of globalization. In the mid-twentieth century, the problem of concentration concerned primarily the big commodity traders – ADM, Bunge, Cargill or Louis Dreyfus – dominating the networks of international trade, particularly for the major cereals. Today, concentration has increased significantly not just in the middle segment, but also at the two ends of the chain. On the side of the input providers, the 130 billion USD-valued merger between the US agro-chemical giants Dow and DuPont Pioneer (now Corteva), combined with Bayer's buyout of Monsanto for 67 billion USD and ChemChina's acquisition of Syngenta for 43 billion USD (and the planned merger with Sinochem) will result in 70% of the total agrochemical industry being in the hands of only three megacorporations (IPES-Food, 2017). On the side of retailers, global retailers, using their superior logistical abilities and bargaining power in upstream markets, now increasingly supply not only rich consumers – ten supermarkets supply half the food in the European Union, according to recent estimates (Oxfam, 2018) – but also the urban middle class in emerging economies (Reardon and Berdegué, 2002; Reardon et al., 2012).

Moreover, concentration at one segment of the chain leads to concentration at the other segments. Large retailers tend to prefer to source from large wholesalers and large processing firms; this allows them to reduce transaction costs and to have access to a diversity of product types in a "one-stop shop" ; the invoicing system is formalized, allowing the retailers to discharge their accounting obligations for value-added tax accounting and product liability; and the packaging and branding of products is superior to that which smaller processors or wholesalers would be able to achieve. This leads to what some authors have called a "mutually reinforcing dual consolidation": the more large retailers dominate consumers' markets, the more large commodity buyers tend to dominate the upstream markets.

Imperfect markets are not a new phenomenon, of course; economist Joan Robinson had already conceptualized such imperfections in the 1930s. But the positive feedback loops (or self-reinforcing mechanisms) that now exist between the ability of the largest and more powerful players of the food systems to control the logistics and strengthen their dominant position (as buyers) by extracting favourable conditions from their suppliers or (as sellers) from their clients, are now threatening to put the system off balance (IPES-Food, 2017). Indeed, this process leads to a race towards the bottom: it results in lower wages for farmworkers and in a lower remuneration for independent agricultural producers that supply raw materials. Large buyers can obtain from sellers a number of concessions that reflect their dominant buyer power, such as discounts from the market price that correspond to the savings made by the seller due to increased production or passing on to the seller certain costs associated with functions normally carried out by the buyer, such as grading of livestock or stocking of shelves. These concessions only make it more attractive for the retailers to source from these dominant buyers, since they may benefit from this superior buyer power that such larger suppliers have. It also further strengthens the position of the dominant buyers, who can acquire a competitive advantage over less dominant buyers in downstream markets, leading to acquisition by larger agribusiness firms of dominance on both the buying and selling markets.

The end consumer may benefit from these trends, both because of the economies of scale achieved by the dominant players and because the abuse of buyer power may lead to lower prices at the end of the chain. But such gains are purely quantitative and arguably are paired with important qualitative losses. As to small food producers, they systematically lose. These farmers buy their inputs at retail prices and they sell their produce at wholesale prices. Moreover, as a narrow set of large firms increasingly act as gate-keepers to the high-value markets of rich countries, small-scale farmers find it increasingly difficult to join these supply chains and the gap is growing between large and small producers in a context in which both categories of producers compete for access to resources, credit and political influence. Larger producers have easier access to capital and thus to non-land farm assets such as storage, greenhouses or irrigation systems. They can more easily comply with the volumes and standards requirements that the agrifood companies – the commodity buyers, the processors and the retailers, depending on which sources directly from the producer of raw materials – seek to impose. Small farmers can only compensate for these disadvantages by their lower labor costs, or because they are a less risky sourcing option to the buyers, since the larger farmers have more market options and thus can be less reliable. The disturbing consequence is that small farmers pay a high entry fee into global supply chains; because of these structural obstacles they face, they can only compete by a form of self-exploitation for instance by agreeing to low wages for those (often family members) working on the farm and by agreeing to be locked into a situation of high dependency towards the buyer. This is one major reason why undernutrition persists in many parts of the developing world, often as a result of extreme deprivation in rural areas. Only a small segment of the farming population still manages to thrive in an increasingly competitive environment, in which farmers can survive only by achieving economies of scale. They must get big or they must get out; many stay small and barely survive.

The idealized picture of well-functioning agricultural and food markets driven by price signals is unrealistic in another way; most food producers don't respond easily. Often price increases lead them to make production choices that, six or eight months later, lead to oversupplying the markets, which generates the price slumps that eventually result in a loss of income for them, since all farmers respond to the same price signals. Rather than fixing the issue and improving the market, pure price signals degenerate in the "hog cycle" well known to agricultural economists (Haas and Ezekiel, 1926, Hanau, 1928, Coase and Fowler, 1935). Or, instead, farmers cannot adapt. When the prices of coffee or cocoa go down on international markets, for instance, the small producers of coffee or cocoa beans do not reduce production: they increase it in order to compensate for the resulting losses and because they have to meet a number of expenses – for education, healthcare or housing – that cannot be reduced. Often, switching to something else is simply not an option, since they depend on the soil, weather conditions, access to knowledge and seeds and markets, which are fixed. While the "hog cycle" is more a characteristic of commodity markets dominated by large and relatively highly capitalized producers and the second problem (called the "commodity problem") is more usual for tropical crops and for relatively smaller-sized farms, what both cases have in common is that prices cannot efficiently direct production. To believe they could do so entirely ignores both the agronomic and the economic *modus operandi* that are typical of agricultural production.

Obliterating inequalities

A second problem with this idealized view that sees food as a commodity that should be produced and allocated on the basis of price mechanisms alone is that such a view is blind to the impacts of inequalities. Inequalities within countries are reaching levels unheard of since the 1930s, and despite all the talk about globalization allowing nations to catch up, even inequalities between countries remain high (Atkinson, 2015; Bourguignon, 2015; Stiglitz, 2015). However, as long as food supply is driven by market demand (and that is what drives food production since food is treated as a commodity), it is the purchasing power of the rich, not the essential needs of the poor, that directs how resources are used – which foodstuffs are produced, under which conditions and for which markets. Food prices do not necessarily reflect human needs. Rather, prices are an indicator of demand, as expressed by those with purchasing power; the richer you are, the more votes you have in influencing the allocation of resources. As noted by Scitovsky, the representation of the world into prices and its commodification means that the marketplace is analogous to a plutocracy; it is "the rule of the rich", he wrote, "where each consumer's influence on what gets produced depends on how much he spends" (Scitovsky, 1992: 8). This is perfectly illustrated by the surface of land required in Argentina or Brazil to produce soy or maize for animal feed in industrial livestock processes or by the deforestation resulting from oil palm plantations in Indonesia or Malaysia to compensate for the diversion of colza or sugar beet for energy in Europe. In a globalized world in which food is traded across borders and investments flow freely, the poor may be priced out from the use of resources, while the purchasing power of the rich may guide the direction of agricultural development.

Such distortions are also the price we pay for inequality. Indeed, the reason that large areas of farmland can be dedicated to producing feedstock to satisfy the overconsumption of meat in affluent societies or to fuel their cars, is because consumers in rich countries can command the resources that will allow their lifestyles to continue unchallenged. This fundamental inequity is reflected in the fundamental structure of property and contract law, though these areas of law generally obfuscate their complicity in achieving such a result (Mattei and Quarta, 2018). Similarly, the huge amounts of retail and consumer waste in rich countries is correlated with the fact that, as incomes have grown, the proportion of the household budget spent on food has diminished.[1] This correlation highlights the limits of the reasoning according to which the expansion of trade in agricultural commodities leads to efficiency gains by encouraging a division of labor according to comparative advantage; in fact, the expansion of trade has also resulted in the luxury tastes of the richest parts of the world being allowed to compete against the satisfaction of the basic needs of the poor (Lambin and Meyfroidt, 2011). This trend is particularly concerning in terms of competition for natural resources needed for food production, particularly land. While it is true that a purely Malthusian view of land as finite oversimplifies the issue of competition for scarce resources, as the productivity of land can be increased to a certain extent and as some land can still be brought into production, recent research has highlighted the considerable social and ecological costs of doing so. Once these tradeoffs are taken into account, this research shows, there is significantly less cropland available for future expansion than has been traditionally assumed in most scenarios (Lambin, 2012; Lambin et al., 2013). Here, again, the structure of property rights on land facilitates and legitimizes these impacts, in the name of its apparent neutrality, as already noted by J.R. Commons in 1927 in his *Reasonable Value: A Theory of Volitional Economics* (see Hiroyuki, 2018; see also De Schutter, 2015).

Dismissing the planetary boundaries

A third problem with treating food as a mere commodity is that the economic logic guiding production choices – including the allocation of resources and research and development programs of large firms – entirely ignores the ecological logic. Agricultural production that maintains soil health and resilience in the face of a changing climate should prioritize diversity through mixed cultures and frequent rotations, biological control of pests (rather than reliance on pesticides) and minimize the use of external (non organic) inputs (IPES-Food, 2016).

But the markets command to do the exact opposite; the negative externalities resulting from unsustainable forms of agricultural production and from a heavy reliance on fossil energy are not counted in the costs of production and farmers are pressured to become providers of cheap raw materials to the food manufacturing industry, since that's how profits are made for shareholders – by "adding value". The result is industrialized farming on large areas of land to allow for the mechanization of production.

The environmental impacts are now well understood, but they have not led to questioning the mechanism through which production choices are guided by an exclusively profit-maximizing logic, incentivized by legal and financial systems (Clapp and Isakson, 2018). The spread of monocultures, though they allowed and were allowed by mechanization, resulted in a significant loss of agrobiodiversity; crop species, such as indigenous leafy vegetables, small-grained African cereals, legumes, wild fruits and tree crops, are now gradually disappearing as they are displaced by the production of rice, maize and wheat (Jacobsen et al., 2013). Indeed, biodiversity loss, for which the spread of industrial agriculture is chiefly responsible (FAO, 1997: 33, FAO, 2010: 15–16), is the domain in which the world has moved furthest beyond the "safe operating space" for humanity (Steffen et al., 2015). Largely as a result of unsustainable farming practices, an estimated 33% of soils worldwide are moderately to highly degraded due to erosion, nutrient depletion, loss of organic matter, acidification, salinization, compaction and chemical pollution (FAO, 2015). The resulting loss of natural soil fertility forced an ever-greater reliance on chemical (nitrogen-based) fertilizers to maintain yields (Loveland and Webb, 2003), but this in turn polluted freshwater (Parris, 2011) and as phosphate and nitrogen water pollution reaches the oceans, natural fertilization processes are stimulated spurring algae growth that absorbs the dissolved oxygen required to sustain fish stocks (Paerl and Huisman, 2012, Chislock et al., 2013).

Moreover, mainstream food systems are now a chief contributor to the growth of greenhouse gas emissions. In 2005, it was estimated that agriculture accounted for approximately 10-12% of total man-made greenhouse gas emissions, in the form of nitrous oxide from the use of fertilizers, methane from flooded rice fields and livestock and carbon dioxide from the loss of soil organic carbon in croplands and, due to intensified grazing, on pastures (Smith et al., 2007). But it is not agricultural production alone, it is food production more broadly – food processing and packaging and the logistics of food distribution – that consumes large amounts of energy; approximately 2,000 litres per year in oil equivalents are required to supply food for each American, which accounts for about 19% of the total energy used in the United States (Pimentel et al., 2008). Indeed, the production of fertilizers, herbicides and pesticides, the tillage, irrigation and fertilization and the transport, packaging and conservation of food all require considerable amounts of energy, so that as much as one-third of greenhouse gases from human activity are linked to how food systems developed (HLPE, 2012; Vermeulen et al., 2012; FAO, 2017). Not only is food production itself threatened by the pressures it exercises on ecosystems, including by the apparently uncontrollable growth of emissions responsible for global warming, but it also has developed a huge dependency on fossil energies – the gas needed for the production of fertilizers and the oil needed for machinery and the processing and transport of food –, which makes it unsustainable.

The statist alternative

Food, therefore, cannot be treated as a mere commodity. Indeed, as abundantly shown by the materials in this book, accepting such a purely quantitative (and reductionist) framing of food produces catastrophic results on efficiency, equity and sustainability grounds. This is not to imply, however, that such governance should be centralized in hands of a state bureaucracy. In the commons approach, in fact, the state, just like the capitalist market, is more part of the problem than of the solution. Indeed, state bureaucracies are notoriously ill-equipped to ensure an effective coordination of complex systems, such as food systems. Although states can and must fulfil a number of roles that are essential both for the smooth functioning of markets and for social justice – from providing infrastructures and enforcing food safety regulations and from combating cartels and abuses of dominant positions to ensuring redistribution through taxes and subsidies –, their centralized knowledge cannot be a substitute for the dispersed knowledge and practices of the actors of food systems, whose relationship to the context in which they operate and ability to adapt to changing local circumstances are an essential element of the resilience of the system; for all the naivete of their belief in markets, this, at least, was a key insight of libertarians such as Hayek or (Michael) Polanyi, whose ideas about the virtues of "spontaneous" or "polycentric" orders (Hayek, 1960; Polanyi, 1951) are now echoed in references to polycentric governance in Elinor Ostrom's work (Ostrom, 1990) or in Yochai Benkler's discussion of commons-based peer production (Benkler, 2006). As much as these authors disagree on essential aspects of how societies should be governed, they share the core intuition at the basis of the Hayek-Polanyi line of libertarianism; that the pretence of the state to have privileged access to the kind of knowledge that is required to steer society is similar in kind to, and hardly less preposterous than, the pretence of divine monarchs of the past to have a direct relationship with God. States are not gods and they do not have god-like powers to know what each individual should be ordered to do in order for the common good to emerge.

Nor is this all. The claim of state bureaucracies to command the course of societies also fails on empirical grounds. Recent scholarship has highlighted the extent to which political elites are attentive to the expectations of economic elites, rather than to those of ordinary citizens. These "political elites" include, in particular, elected representatives and not just the technocrats populating public administrations; the problem, it seems, is not so much (or not only) that administrations have too much power and that they escape the control of politicians, but that the politicians themselves are captured (Gilens, 2012; Gilens and Page, 2014).[2] The disproportionate influence which corporations exercise in the political system is only marginally related to the selling out of politicians, whether in the extreme form of corruption or in the more subtle and socially acceptable forms of electoral campaign financing and revolving doors. It has to do, primarily, with the ability of corporations to serve politicians with a convincing narrative, which portrays them as champions of the 'low-cost' economy, in which efficiency gains associated with economies of scale and the division of labour across different jurisdictions in global supply chains allow them to serve mass consumption, provided they are left free to organize production and are not forced to internalize the full range of negative externalities caused by their operations.

Political elites are also poorly equipped to respond to concerns about long-term ecological impacts of industrial production models. In part, this is due to the short-termism associated with electoral cycles; it is this "democracy failure", this gap between the myopic preoccupations of electoral politics and the requirements of intergenerational justice, which had led a number of authors to propose that democracy be "circumscribed" or made more "reflexive" by introducing a longer-term perspective in the form, for instance, of the establishment of a "higher chamber" of non-elected experts with a veto power that they are to exercise as representatives of future generations (Bourg and Whiteside, 2010). Perhaps even more importantly, recent research in cognitive psychology examining risk perception has highlighted that elites are hard-wired to ignore environmental risks that could question their individualistic and meritocratic worldview – one which, as they occupy a dominant position within society, they are likely to adhere to. Indeed, this research shows that information about risks associated with unsustainable types of food production, particularly environmental risks, should be processed by various groups of the population in order to reduce the cognitive dissonance between the data with which they are provided and the cultural values they entertain (Kahan et al., 2007). The so-called "White male effect" in risk perception is such that, for those occupying a dominant position in society, more and better information improving their scientific literacy or their numeracy does not lead them to assess risk in accordance with the warnings of scientists in areas, such as climate change, where the risk tends to be underestimated: quite to the contrary, it seems that they would rather use this gain to strengthen their scepticism, as if unwilling to recognize information that runs counter to their interests (Kahan et al., 2011). While this may be comforting to the individual (and may be said, therefore, to be "rational" from the individual's point of view, as the discomfort of cognitive dissonance is minimized), the consequences for society are clearly sub-optimal, presenting us with a collective action problem which Kahan et al. (2011) have called the "tragedy of the risk-perception commons"; it calls for a science of communication about risk that can take into account the fact that we live in culturally diverse and pluralistic societies and that each sub-group within society may have to be addressed differently for social norms within the sub-group to evolve (Higgs, 2015).

A final limitation of reliance on state action to transform food systems is that, as anticipated many years ago by what social psychologists labelled "reactance" theory (Wicklund, 1974), individuals resist adopting conduct that they perceive to be imposed on them from without. In response, researchers now insist on a shift of attention from extrinsic to intrinsic motivations. The work of Richard Ryan and Edward Deci provides perhaps the most explicit attempt to demonstrate the importance of intrinsic motivations in explaining individual behavior (Ryan and Deci, 2000a and 2000b; Moller et al., 2006). The so-called "self-determination theory" they pioneered emphasizes that lasting behavioural change depends on individuals acting on the basis of their own values and deeply held beliefs, rather than external rewards or penalties. Interventions "from above", whether in the form of top-down regulation or in the form of economic incentives, may be insufficient to disrupt routines in eating behaviour and to bring about change at the desired scale. State bureaucracies, using tools of regulatory injunctions or economic incentives, treat individuals like objects, rather than as subjects of their own history (Arendt, 1958) and members of a broad and universal community of citizens. Individuals on whom rules are imposed, to whom subsidies are promised or who are threatened with having to pay taxes will comply with the rule, capture the subsidy or avoid paying the tax – but they will otherwise pursue their own life objectives, deviating as little as possible from such objectives that they have set for themselves. In contrast, behavioural changes that rely on the *intrinsic* motivations of the individual shall be resilient; because they are based on the individual's identity or self-image or on the values that the individual treats as his/her own, such changes will persist in time, even though the context (and the external incentives it provides) may have evolved. To change the food system is is thus essential to work against the fragmentation of individual identities and strengthen the imaginary of community as a source of reward and space of emancipation.

Food as commons

It is thus that we arrive at the conclusion that, because of the failures of both "the market" and "the state", a different dimension, that of the commons, should be introduced to approach the predicaments of the Anthropocene/Capitalocene that the global food catastrophe shows in all its immediate drama. As already shown in the introduction and unfolded in multiple chapters, the commons present a complex epistemology. Considering the commons as a sort of civic sector, capable of coexisting with the fundamental capitalist institutions, already betrays a political choice that is questionable and indeed has been questioned (Hardt and Negri, 2009; Mattei, 2011;

Dardot Laval, 2015; De Angelis, 2017). Moreover, a number of understandings co-exist as to what is implied by reframing food as a commons. While all these understandings share the view that neither unaccountable markets nor state bureaucracies are appropriate mechanisms to exclusively guide production choices, to allocate resources or to ensure equitable access to food, they range from the least to the most radical versions. The revival of commoning practices, indeed, has two distinct sources. Some stem from social innovations, which emerged in a relatively recent past, primarily in urban settings and at the initiative of upper-middle class activists. Others develop in traditional communities that have hitherto resisted the full commodification of food, and maintained customary forms of governance of natural resources as well as non-commoditised means of food production and distribution despite the expansion of the capitalist frontier. It would go beyond the ambition of this chapter to deepen the inquiry as to how to subsume the practice (customary and contemporary) of commoning into a broader theory of the commons. However, we do offer a few hints in the concluding section.

Food as commons and the right to food

One approach – the least radical – is to insist on the need to guarantee the right to food as a human right, whereby an absolute human need (see Chapter 7 in this volume) is framed as a mandatory entitlement. The right to food is the right of every individual, alone or in a community with others, to have physical and economic access at all times to sufficient, adequate and culturally acceptable food that is produced and consumed sustainably, preserving access to food for future generations (Committee on Economic, Social and Cultural Rights, 1999: paras. 6-7). Thus understood, the right to food can be secured (i) by obtaining incomes from employment or self-employment; (ii) by social transfers from the state or from family and community networks; or (iii) by own production, for individuals who have access to land and other productive resources. Through these channels, that often operate in combination with one another, the right to food should ensure that each person has access to a diet that, "as a whole … contains a mix of nutrients for physical and mental growth, development and maintenance, and physical activity that are in compliance with human physiological needs at all stages throughout the life cycle and according to gender and occupation" (id., para. 9). Thus, a convenient way to summarize the normative content of the right to food is by referring to the requirements of availability, accessibility, adequacy and sustainability, all of which must be built into legal entitlements and secured through accountability mechanisms.

Such an entitlements approach suggests that states be held accountable in a range of areas which influence the ability for each individual to have access to a healthy diet – from minimum wage legislation to social protection floors and from access to productive resources for food producers to the provision of subsidized food items in support of poor consumers or directly providing food to those who cannot afford it through food banks or food pantries. The right to food thus may serve to control state bureaucracies as well as to ensure that markets are not allowed to deprive people of access to at least the amount of food essential to lead active and healthy lives. However, it also has a number of limitations. It does not question the commodification of food, the production of which still relies on market mechanisms (and on price signals in particular) and access to which still depends on purchasing power; although, in a perspective built on the right to food, food buyers shall be supported in having access to food by being guaranteed decent wages or social protection, the dominant paradigm for food production and access remaining unchallenged. This approach can thus perfectly co-exist with capitalist markets (Ferrando, 2016).

Moreover, the language of human rights and the mechanisms through which human rights are monitored and enforced remain specialized, sometimes perceived as neglectful of social and economic justice (Moyn, 2018) and Western-centric; the language is that of human rights experts and it is experts who populate the mechanisms with all the political limits of professionalism (Sarfatti-Larson, 1977). Even the most robust legal and policy frameworks designed to support the right to food may lack any self-instituting dimension; they tend to proritize individual above collective rights, they impose obligations mostly on states and not on other individuals or corporate actors and they require the prior approval of states to become operational (it is in that sense that they remain state-centric) (Charvet and Kaczynska-Nay, 2008; Claeys, 2014). For all these reasons, human rights, although they ensure an indispensable protective function, may also be seen in other respects as disempowering – as robbing rights-holders from their ability to define the objectives of their struggles for justice and to define the pathways towards realizing them (Kennedy, 2002, 2012; Hopgood, 2013).

Food as commons and customary forms of tenure

But may human rights reinvent themselves? Are we now witnessing a form of "re-commonification" at work in this field of law? Indeed, the revival of the "commons" based on the recognition of customary forms of tenure and governance of resources that traditional communities have developed, forms a remarkable development in recent international law. Indigenous or traditional communities have, until now, lived at the external limits of the market society. Their livelihoods have been supported by the organisation of solidarity and care within their community (see, for example, Chapter 8 in this volume) and their livelihoods exhibit a non-instrumental relationship to Nature, one that recognizes our interdependence with Nature and does not see it simply as a resource to exploit or as a free dumping site (Chapter 4 in this volume). The common property regimes that these groups rely on are increasingly referred to in recent instruments, with a view to ensuring that such regimes shall be protected from encroachment. They are spectacularly endorsed, for instance, in the *Voluntary Guidelines on the Responsible Governance of Tenure of Land, Fisheries and Forests in the Context of National Food Security (VGGT)*, adopted in 2012 by the Committee on World Food Security (CFS). These guidelines note that "there are publicly-owned land, fisheries and forests that are collectively used and managed (in some national contexts referred to as commons)", and that "States should, where applicable, recognize and protect such publicly-owned land, fisheries and forests and their related systems of collective use and management, including in processes of allocation by the State" (guideline 8.3).

Two years after the adoption by the CFS of the VGGT, another intergovernmental committee of the FAO, the Committee on Fisheries (COFI), adopted the *Voluntary Guidelines for Securing Sustainable Small-Scale Fisheries in the Context of Food Security and Poverty Eradication* (SSF guidelines). The guidelines, the outcome of a three-year-long participatory process conducted between 2010 and 2013, encourage states to recognize forms of co-management of fisheries based on customary forms of tenure: "Local norms and practices, as well as customary or otherwise preferential access to fishery resources and land by small-scale fishing communities including indigenous peoples and ethnic minorities, should be recognized, respected and protected in ways that are consistent with international human rights law" (guideline 5.4). This illustrates a significant shift in perspective. In contrast to the dominant view, according to which a clarification and strengthening of access rights, including by the use of transferable fishing quotas, would increase economic efficiency and avoid overfishing (see, for instance, World Bank, 2004 or Cunningham et al., 2009), the SSF guidelines suggest that priority should go to improving access to fishing rights to the communities who need it most – and who could be best placed to manage the common-pool resources concerned and to monitor catches at the local level.

The shift to co-management of fisheries vesting collective rights with fishing communities is also based on the broadly positive assessment made of the establishment of fishing zones (whether in lakes or coastal areas in seas), reserving fishing in these areas for local communities, allowing them to manage access rights (Sharma, 2011; Special Rapporteur on the right to food, 2012, para. 58; Ratner et al., 2014). A 2011 study comparing 130 co-management schemes (covering 44 developed and developing countries) thus demonstrated how local communities have often been able to develop legitimate institutions of self-governance and establish sustainable approaches to managing fishing intensity and ecosystems impacts, provided strong community leaders emerge and robust social capital exists to monitor compliance with individual or community quotas (Gutiérrez et al., 2011). Other studies highlighted that co-management schemes could be successful, provided certain conditions are present (Townsend, 2008; Béné et al., 2009), including, in particular, an enabling institutional enviroment at the national level (Nielsen et al., 2004; Lewins et al., 2014) and a tradition of cooperation within the community (Jamu et al., 2011). The SSF guidelines express this new consensus.

Finally and most recently, the revival of the commons in international human rights law can be seen in the proposal for a *Declaration on the rights of peasants and other people working in rural areas*, initially submitted by Bolivia within the Human Rights Council, and strongly inspired by the *Via Campesina*, the transnational network of small-scale food producers (for background of this attempt, see Claeys, 2015 and Golay, 2015). At the time of writing, the proposal was under negotiation within an Intergovernmental Working Group of the Council. Article 5(1) of the draft Declaration refers to the right of peasants and other people working in rural areas 'to have access to and to use the natural resources present in their communities that are required to enjoy adequate living conditions' and to their right 'to participate in the management of these resources and to enjoy the benefits of their development and conservation in their communities'. Under the heading 'Right to land and other natural resources', Article 17(1) provides that 'Peasants and other people living in rural areas have the right, *individually and collectively*, to the lands, water bodies, coastal seas, fisheries, pastures and forests that they need to achieve an adequate standard of living, to have a place to live in security, peace and dignity and to develop their cultures' (emphasis added); and in wording clearly inspired by the 2012 VGGT, Article 17(3) adds:

> States shall provide legal recognition for land tenure rights, including customary land tenure rights, not currently protected by law. All forms of tenure, including tenancy, must provide all persons with a degree of tenure security that guarantees legal protection against forced evictions. States shall recognize and protect the natural commons and their related systems of collective use and management.[3]

Thus, a counter-movement is emerging within international human rights law, as a reaction to the push towards privatization and enclosures of which, for centuries, international law has been an instrument (see also Bakker, 2007). Needless to say, the long history of plunder (Mattei and Nader, 2008) and of constantly changing legal lingo deployed for centuries to hide its establishment and continuation, should warn us against the risks of patronizing and essentializing the "local communities" by old and new carriers of the white man's burden. It is a fact that the corporate-driven transformation of land tenure systems from commons institutions (serving the long-term interest of a community that includes future generations) into capitalist institutions (serving the imperatives of capital accumulation for private individuals in this generation) has erased our capacity to study and understand the most resilient institutional alternatives (Ferrando, 2017). The regeneration of commons institutions of land tenure may be the most crucial prerequisite of a meaningful institutionalization of the idea of food as commons (Chapter 21 in this volume). The establishment of this conversation in international law is therefore welcomed; though in the absence of a systemic critique and a comprehensive reconfiguration of the premises of the international legal system, it still seems highly insufficient.

Food as commons and social innovations

Contemporary developments within human rights support, therefore, some form of re-commonification. But they still do so largely in a defensive mode, rather than in a mode that empowers. In contrast, the more radical approach to "re-commoning" insists on the self-instituting dimension that human rights lack. In all world regions, a range of social innovations are developing that re-invent how food is produced, processed, distributed and valued, developing alternatives to the dominant paradigm that sees (and treats) food as a commodity. These innovations include community gardens maintained by neighbors, community-supported agriculture, in which the risk is shared between the farmer and the eaters and in which the eaters express solidarity with the farmer; vegetables grown for the community members to pick, referred to as "incredible edibles"; cooperative supermarkets or social groceries. Such social practices challenge not only the dominant representation of food as a commodity, but also the representation of the individual that accompanies it – a representation both implicit and powerful, indeed, powerful precisely *because* often left implicit. These social practices refuse to see individuals in Hobbesian terms, as guided by appetites and aversions, and constantly searching to satisfy the former and to avoid the latter, for the selfish maximization of their utility. C.B. Macpherson perfectly summarized the relationship between that view of the individual and the bourgeois society that was emerging as Hobbes was writing the *Leviathan* in the mid-seventeenth century:

> We live in a market society. Our behaviour, our values, are largely shaped, directly or indirectly, by the requirements of the market. We are bourgeois men. So were the men Hobbes described and analysed. ... His scientific method required him to build up a model of man and of society, and the models he constructed were bourgeois models. Since the main body of his science was produced by deduction from these models, it is a science of bourgeois society.
>
> *(Macpherson, 1968: 11–12)*

This science of society we inherited from Hobbes is as much normative as it is positive; by becoming the dominant reconstruction of how we (i.e. some people in Western civilization at a specific moment of history) have come to act with one another, it also prescribes how we should behave.

Instead, these social innovations introduce a diagonal dimension between the horizontality of market relationships mediated by prices and the verticality of the state imposing regulations on the individual; on this diagonal axis, individuals relate to one another not as market agents seeking to conclude a transaction, nor as competitors, but as members of a community collaborating in pursuing a collective action for the commonwealth. Food is a "commons" in that its production and distribution responds to a logic of solidarity and mutual help, rather than to a logic of competition and exclusion, because people recognize mutual neediness and the essentiality of eating (again Chapter 7 in this volume). Even more importantly, the rules governing food production and allocation are set by the individuals who are involved in this collective action: as such, it is not only the state bureaucracy that is democratized and held accountable, it is society itself that is democratized, as the rules resulting from self-governance replace the rules of the market (see for example Chapter 14, Chapter 15 or Chapter 19 in this volume). Thus private law, after the parenthesis of modernity, is recognized again as the *law of the privates* that, outside of the reductionism of possessive individualism, self-regulate their transacting in the interest of nature and community (see Capra and Mattei, 2015). This bottom-up generated law of the commons serves a crucial counter-hegemonic function (Mattei and Quarta, 2018).

This perspective overlaps only partly with the analysis provided by Elinor Ostrom of "common pool resources" in her seminal contribution of 1990 (Ostrom, 1990). For what we have in mind when referring to food as a commons is not a physical entity exploited according to rules a community has chosen for itself. More precisely, this is not the *only* form that "commoning" may take; the term encompasses a much broader set of social innovations by which relationships around food are redefined, by the social actors themselves, who occupy a space between the state and the market, escaping both bureaucratization and commodification, a space that these actors seek to democratize by inventing the rules by which it shall be governed (Dardot and Laval, 2015; Hardt and Negri, 2009). Although the re-writing of the rules carried out by innovative civic food actions and alternative imaginations can include food being traded for profit (therefore still working as a tradeable good, but not only and not primarily) (Vivero-Pol, 2017a), the commodification of food erases any trace of commonality, as philosopher Michael Sandel (2013) explained for other goods. As a phenomenological expression, food plays multiple functions depending on specific circumstances and commodification is incompatible with this plurality.

Food as commons moves us far away from the positivistic approach that dominates Ostrom's work as well as that of modern social science. Indeed, like water, food entirely defeats the neat separation between the subject and the object, the positive and the normative and the objective scientific description and the subjective political preferences. We are the food we eat as well as the water we drink. We cannot observe food from a perspective detached by the material conditions in which we operate. Similarly, food and the food system are co-defined by natural circumstances, human decisions and anthropocentric practices. The commons approach shows us that there is no resource (private, public or collective) that becomes such without human inputs and therefore dependent on political conditions and choices (food as a social construct). There is no abstract subject separated from an abstract object. The commons allow us to observe food, as everything else, in an ecological way that connects the dots of the interdependent material conditions that compose the web of life. There is no luxury without starvation, no starvation without luxury; that is the challenge of the commons to the transcendent and naturalizing vision of positive science.

A prospective scenario: the tri-centric scheme to steer a different transition pathway

Deconstructing food as a commodity and reconstructing it as a commons requires establishing a tri-centric governance system recombining market rules, public regulations and self-regulated collective actions, re-arranged to maximize the potential of each. Food would be produced, consumed and distributed by agreements and initiatives formed by state institutions, private producers and companies, together with self-organized groups under self-negotiated rules. Those agreements would include not only private–public partnerships (PPPs), but also public-commons partnerships (PCPs), a new institution that deserves to be further explored (Piron and Cogolati 2016) and that the city of Turin (Italy) and its *administrazione condivisa* exemplifies

(Bottiglieri et al. 2016). Those governing agreements serve commoning practices by enabling access and promoting food through a multiplicity of open structures and peer-to-peer practices enabling sharing and co-producing food-related knowledge and edible products. The development of this tricentric governance would combine civic collective actions for food, an enabling state and socially-responsible private enterprises (see Figure 24.1).

The civic collective actions for food are already happening, with people producing food by themselves or getting organized in food buying groups, community-supported agriculture or sharing meals clubs. The fast-growing constituency involved in this transition can value food as a commons (Chapter 22 in this volume). It calls for a different kind of state, with different duties and skills to steer that transition. Such a state would aim at partnering and at stimulating social and just innovation, rather than at steering by command-and-control via policies, subsidies, regulations and the use of coercion. That would be a "partner state" acting as an enabling supervisor and considering food as a public good (Chapter 5 in this volume). The private sector itself should be supported in its diversity, as there is a need to count on a different breed of private enterprises in order to satisfy the needs unmet by collective actions and state guarantees. This private sector shall be driven by a different ethos while making profit, bound by the requirements to contributing to social aims and the satisfaction of the needs of the community rather than moved by the goal of profit-maximization at any cost. In that sense, the market would be seen as a means towards an end (wellbeing, happiness, social good) with a primacy of labor and natural resources over capital. We will expand on these three roles below.

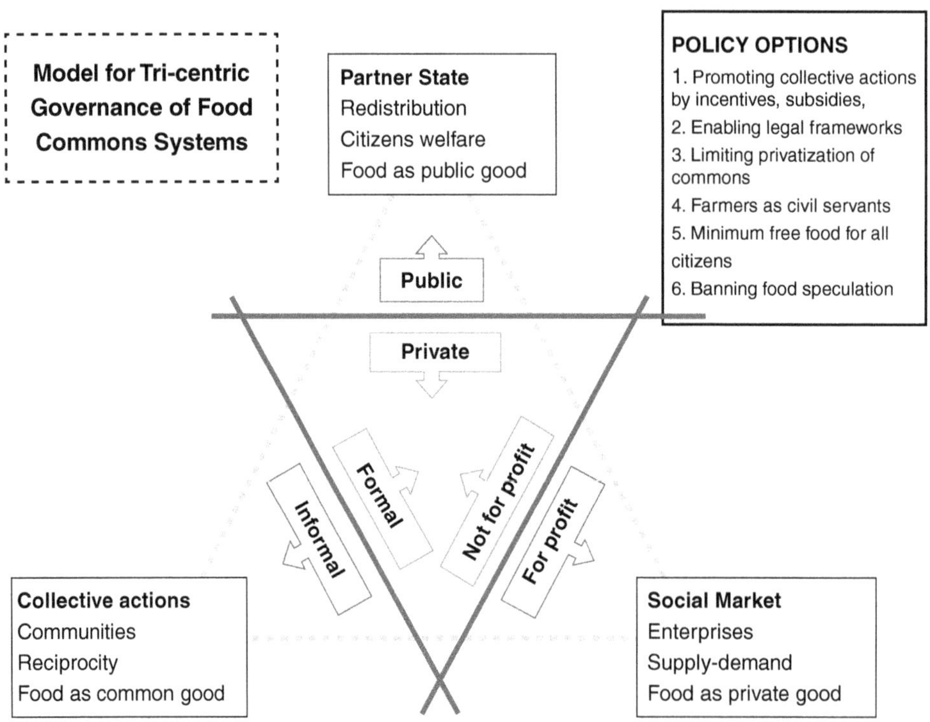

Figure 24.1 The ideational tri-centric governance model for transition in food systems.
Source: Vivero-Pol (2017b).

Civic collective actions to govern food as commons

Civic food networks generally emerge at the local level and aim to preserve and regenerate the commons that are important for the community. There have been two streams of civic collective actions for food running in parallel. In rural areas, small-scale food producers, relying on low-input or agroecological production, seek to develop types of farming (as well as local processing and marketing) that evade the constraints of long supply chains controlled by large food manufacturers or retailers. In urban and peri-urban areas, alternative food networks (AFNs) are emerging, for instance in the form of community-supported agriculture (CSAs) or urban agriculture. These AFNs are led, on the one hand, by concerned food consumers who want to reduce their food footprint, produce (some of) their own food, improve the quality of their diets and free themselves from corporate-retail control and, on the other, by the urban poor and migrants motivated by a combination of economic necessity and cultural attachments. Over the last 20 years, these two transition paths have been growing in parallel, both increasingly relying on the vocabularies of food sovereignty and agroecology. They remain to a certain extent disconnected, however, divided by geographical and social boundaries. But the maturity of their technical and political proposals and reconstruction of "rururban" connections have paved the way for a convergence of interests, goals and struggles. Large-scale societal change requires broad, cross-sector coordination. It is to be expected that the movement for the revival of peasant agriculture and AFNs will continue (and need) to grow together, beyond individual organisations, to knit a new, more finely meshed and wider food commons capable of confronting the industrial food system for the common good (Ferrando and Vivero-Pol, 2017).

The partner state to govern food as a public good

The state has an essential role to play in providing an enabling framework for re-commonification. The transition towards a food commons regime will need a different kind of state, however, with different duties and skills to steer that transition. These desirable functions are shaped by partnering and innovation rather than command-and-control via policies, subsidies, regulations and the use of force. This enabling state plays a role as shaper and creator of markets and facilitator for civic collective actions to flourish. This state presents characteristics that borrow both from the partner state (as theorized by Kostakis and Bauwens, 2014) and from the Entrepreneurial State (as described by Mazzucato, 2013). The partner state has democratic public authorities playing a sustaining role (enabling and empowering) in the direct creation by civil society of common value for the common good. Unlike the Leviathan paradigm of top-down enforcement, this type of state sustains and promotes commons-based peer-to-peer production. Among the duties of the partner state, Silke Helfrich[4] mentioned the prevention of enclosures, triggering of the production/construction of new commons, co-management of complex resource systems that are not limited to local boundaries or specific communities, oversight of rules and charts, care for the commons (as mediator or judge) and initiator or provider of incentives and enabling legal frameworks for commoners governing their commons. The entrepreneurial state, meanwhile, fosters and funds social and technological innovations that benefit humanity as public ideas that shape markets (e.g., in recent years, the Internet, Wi-Fi, GPS), funding the scaling up of sustainable consumption (like the Big Lottery Fund supporting innovative community food enterprises that are driving a sustainable food transition in the UK) and developing open material and non-material resources (knowledge) for the common good of human societies. Public authorities will need to play a leading role in support of existing commons and the creation of new commons for their societal value.

The non-profit maximizer private sector

The private sector presents a wide array of entrepreneurial institutions, encompassing family farming with just a few employees (FAO, 2017), for-profit social enterprises engaged in commercial activities for the common good with limited dividend distribution (Defourny and Nyssens, 2006) and transnational 'too-big-to-fail' corporations that exert near-monopolistic hegemony on large segments of the global food supply chain (van der Ploeg, 2010). The latter are managed on behalf of unknown (or difficult to track) shareholders whose main goal is to maximize their (short-term) profits rather than equitably produce and distribute sufficient, healthy and culturally appropriate food to people everywhere. The challenge for the private sector is to adjust direction, to be driven by a different ethos while making profit – keeping, indeed, an entrepreneurial spirit, but also focusing much more on social aims and satisfying needs. Or, put the other way around, the private sector role within this tri-centric governance will operate primarily to satisfy the food needs unmet by collective actions and state guarantees. The market will be seen as a means towards an end (wellbeing, happiness, social good, climate change mitigation, etc.) with a primacy of labour and natural resources over capital. Thus, this food commons transition does not rule out markets as one of several mechanisms for food distribution; it does reject market hegemony over our food supplies, however, emphasizing that other avenues are available.

Local transitions towards the organisation of local, sustainable food production and consumption are taking place today across the globe (e.g. Ghent in Belgium,[5] Torino in Italy,[6] Toronto in Canada,[7] Fresno in the US[8]). Inspired, often unwillingly, by principles along the lines of Elinor Ostrom's (1990, 2009) poly-centric governance, food is being produced, consumed and distributed by agreements and initiatives formed by state institutions, private producers and companies, together with self-organised groups under self-negotiated rules that tend to have a commoning function by enabling access to and promoting food in all its dimensions through a multiplicity of open structures and peer-to-peer practices aimed at sharing and co-producing food-related knowledge and items. The combined failure of state fundamentalism (in 1989) and so-called 'free market' ideology (in 2008), coupled with the emergence of these practices of the commons, has put this tri-centric mode of governance back on the agenda.

The transition period for this regime and paradigm shift should be expected to last for several decades, a period where we will witness a range of evolving hybrid management systems for food similar to those already working for universal health/education systems. The era of a homogenized, one-size-fits-all global food system will be replaced by a diversified network of regional foodsheds designed to meet local needs and re-instate culture and values back into our food system (The Food Commons, 2011). The Big Food corporations will not, of course, allow their power to be quietly diminished, and they will, inevitably, fight back by continuing to do what has enabled them to reach such a dominant position today: legally (and illegally) lobbying governments to lower corporate tax rates and raise business subsidies, combating the adoption of restrictive legal frameworks (related to GMO labelling, TV food advertising, local seed landraces, etc.) and generally using the various powers at their disposal to counter alternative food networks and food producing systems. To emphasize, the confrontation is likely to continue over decades: paralleling and reversing the industrialization and commodification path that led us to this point is not a matter of days. Imagination and motivation are essential, along with a self-reflective attitude that allows the movement to get stronger and more aware.

Conclusions

This book has a double purpose with regard to the valuation of food and the exploration of alternatives. The authors and editors aimed to get free from the mental straitjacket imposed on thinkers, practitioners and activists by the dominant narrative of capitalism, founded on neoclassical economics and absolute propietary schemes, which has done so much to delay the search for alternatives to the multiple crises our societies are facing at the begininng of 21st century. By proposing that food could be valued, enacted, produced, distributed and governed as a commons, we want to reject the historical and present commodification of food. We are tired of the practice of simplification, expropriation and appropriation that deprives such an essential resource of its multiple non-economic dimensions and that treats it according to an economic logic in disregard of its ecological impacts. Secondly, we want to lay the foundations for an alternative normative imagination of food and food systems: a vision deeply rooted in history – though disruptive and innovative – at the same time that enables the exploration of other policy options to build just, sustainable and democratic food systems.

As the British economist John Maynard Keynes once wrote in what has since become a cliché in the social sciences, "[t]he difficulty lies not so much in developing new ideas as in escaping from old ones" (Keynes, [1936] 2016). In that sense, we have tried to get rid of the entrenched understanding of food as a commodity (and as a private good in economic terms). In doing so, we have posited that commons are at the same time a very ancient and rather innovative framework to govern natural resources that are essential to human survival. Thus, food commons can easily be perceived as an emancipatory alternative, a knowledge carrying a moral purpose to combat exclusion and create conditions for human flourishing (Wright, 2013) or a disruptive narrative that challenges the power relations in the industrial food system and deepens food democracy.

The challenge for the future is to develop political, cultural and institutional conditions that allow the aggregate of predicaments addressed by the label "*food as commons*" to imagine and materialize solutions capable of making the global (dis)order regain a modicum of decency and legitimacy. The narrative of food as commons may unlock practices and policies that were discarded, forgotten or simply not permitted by the hegemonic mainstream because they were not aligned to the capitalist mantra. Such practices and policies have been generally dismissed as too radical, too naïve or too expensive.

Modernity, through the ideological and repressive apparatus of the State (Althusser, 1973) made of law, education and scientific professionalism, has systematically transformed commons into capital in a very successful way; it achieved this transformation successfully not only for food itself, but also for land, seeds and labour, and increasingly for knowledge, water and even, with the introduction of "rights to pollute", clean air. Individualization and competition have been the outcome of and the fuel for this process. Food, an absolute and universal need for humans, has shared the fate of every other commons (assuming that we can disentangle them from each other). It has been transformed into capital to be understood as an abstract naturalized entity, disembedded from its social relations, places of production and consumption and separated from the material conditions of life that can be produced and reproduced only within the capitalist logic.

This logic has divided the world into two spheres, the public (state) and the private (market), while insisting that these spheres are of an ontologically different nature and that their various combinations exhaust the repertoire of institutional possibilities. Where, however, there is neither state nor private property (which, however, always requires the state even to exist), there are the commons. The commons were described by Garrett Hardin as places of no law and then rescued, a generation later, by Elinor Ostrom through the idea of material common-pool resources better run by communities, as something worth scientific inquiry. Both Hardin and Ostrom, however, share the same positivistic logic aimed at an objective ontology of the commons. For them, the commons were ontological; for us, the commons are phenomenological; they are located in place and time and continuously re-created and co-constructed by the inextricable ecology of the web of life.

The binary logic of modernity was easy to challenge when it became abundantly clear that the private and the public sectors do share the very same logic of plunder through mechanisms, such as revolving doors, regulatory capture and the military industrial complex already denounced by President Eisenhower in 1955 (Mattei and Nader, 2008). The insurrection of the Zapatistas in Chiapas and the struggle against the privatization of the provision of water services in Cochabamba opened the way for the global challenge of the conspiracy between political institutions and global transnational corporations that victimize the commons. Significantly, food and water have long been priority issues in the Global South. The last round of extraction and commodification that we are experiencing in its full-fledged forms today (i.e. land and water grabbing in the Global South (De Schutter, 2011), fish stocks overexploitation, pollution of the sea with plastic and the atmosphere with industrial by-products) started after the global crisis of 2008 and is becoming truly global in its victimization, also involving the weak inhabitants of the Global North plundered of every previous conquest of democracy and public welfare. In this grim global scenario, the commons provide an inspirational and aspirational tool of resistance and a theoretical challenge to the binary vision of the state and market duopoly. The theoretical challenge must be morphed into institutions, policies and laws to provide a generative alternative to the current extractive scenario (Capra and Mattei, 2015). Practices of collective food production, sharing and consumption, customary and contemporary, stemming from subsistance agriculture as well as from the new possibilities offered by the internet-based circular economy have to be promoted and developed as a robust counter-narrative.

Again, there are different levels of ambition. Some of the civic alternatives that are in the process of being developed as local forms of direct participation in the administration of certain cities (e.g. Naples, Ghent, Barcelona, Brussels, Turin, Porto Alegre, most of them being termed as commons alternatives) are already tailored to food predicaments in order to make the civic sector more robust. Think about the possibility to apply the participatory logic developed in the case of the Naples acqueduct to food supply (private or public) or to the case of the Mondeggi Bene Comune and the establishment of a commonly managed farm based on the principles of agroecology, accessibility, anti-patriarchy and decolonization (Chapter 21 in this volume). In this case, an institutionalized "civic sector" could complement rather than substitute the public and the private, perhaps ultimately providing some desirable transformations. It is, however, crucial to understand that a thriving social solidarity economy, if it is not supported by a strong educational effort, is at high risk of being itself co-opted by the state or the market, as it happened to ideas such as the Green Economy, organic food or the distribution of surplus food through partnerships between large retailers and charities. Capitalism has a fantastic mimetic power and unless confronted with permanent political pressure from social movements, is very unlikely to allow genuine alternatives to coexist with its loyal cronies.

A more ambitious move is therefore to deploy the *fundamental need* of food as a powerful counter-hegemonic weapon in an international war of narratives of the Gramscian type. Politically speaking, this is not a reformist but rather an openly revolutionary approach (either counter- or alter-hegemonic). The civic constituencies in both the Global North and the Global South might be mobilized on an issue as universal, direct and dramatic as hunger. Most revolutions started on issues of food scarcity and food rebellions are still taking place in recent years throughout the world (Holt-Gimenez and Patel, 2012). Such a war of narratives, positions and cultural hegemony requires subjectivities endowed with an aspirational vision and an inspirational plan. The vision is not abstract but rather embeddeed in the material conditions of the Anthropocene/Capitalocene, whose structural imperatives are leading to a catastrophe which appears increasingly inevitable; four out of nine planetary boundaries have already been surpassed (Steffen et al., 2015) and an increasing number of people live unbearable lives of deprivation and domination. Since the early times of the Huguenot constitutional theory, a right/duty to resist is triggered in front of a model of sovereignty leading to the destruction of the political community (Menkel-Meadow, 2012). The ecological thinking embedded in the commons provides a standard of evaluation of law and policy in the political choice that is open to everybody: obey or resist. Resistance against ruinous laws or policies, if it remains an individual act, may at best give us a hero often dead or imprisoned. Collective resistance can change the world, which makes it crucial to work hard in a process of collective subjectivization based on eco-literacy and systemic thought, the only way to even see through the fog of the capitalist propaganda. It is our hope that the alternative imaginations of food and food systems offered in this book may contribute to the kind of vision and horizon that are needed to subvert the deadly logic of global state-corporate capitalism.

Notes

1 A 2011 study prepared at the request of the FAO estimates that 1.3 billion tonnes of food produced for human consumption -- about one third of the total -- are lost or wasted (Gustavsson et al. 2011). The levels of waste per capita of households are much higher in rich countries than in developing countries; while consumers in Europe and North America waste 95–115 kg/year, this figure is only 6–11 kg/year in Sub-Saharan Africa and South/Southeast Asia.
2 Public choice theory provides, in this regard, a description of politics that is not without foundation (Buchanan and Tullock 1958, Stigler 1971) and should call for a reaction precisely from those, the believers in the strength of democracy, who are suspicious of its normative prescriptions.
3 Article 21(3) uses a similar formula with respect to access to water: 'States shall respect, protect and ensure access to water, including in customary and community-based water management systems ....'
4 In a text quoting her to be found here: http://wiki.p2pfoundation.net/Partner_State (accessed on 27 July 2018).
5 https://stad.gent/smartcity-en/news-events/expert-michel-bauwens-researches-ghent-%E2%80%98commons-city-future%E2%80%99 (accessed on 21 August 2018).
6 https://iucfood.wordpress.com/2017/08/06/making-sustainable-food-policies-a-reality-first-ipes-food-local-lab/ (accessed on 21 August 2017).
7 http://tfpc.to/ (accessed on 21 August 2017).
8 http://www.thefoodcommons.org/ (accessed on 21 August 2017).

Bibliography

Althusser, L. 1973. *Lo Stato e i suoi apparati*. Editori Riuniti, Roma.
Altvater, E., Crist, E., Haraway, D., Hartley, D., Parenti, C. and McBrien, J., 2016. *Anthropocene or Capitalocene?: Nature, History, and the Crisis of Capitalism*. Pm Press, Oakland.
Arendt, H. 1958. *The Human Condition*. Chicago: University of Chicago Press.
Atkinson, A.B. 2015. *Inequality. What Can Be Done?*., Harvard University Press, Cambridge.
Bakker, K. 2007. The 'Commons' versus the 'Commodity': Alter-globalization, Anti-privatization and the Human Right to Water in the Global South. *Antipode*, 39 (3), 430–455.
Béné, C., Belal, E., Ousman Baba, M., Ovie, S., Raji, A., Malasha, I., Njaya, F., Na Andi, M., Russell, A. and Neiland, A. 2009. Power Struggle, Dispute and Alliance Over Local Resources: Analyzing 'Democratic' Decentralization of Natural Resources through the Lenses of Africa Inland Fisheries. *World Development*, 37 (12), 1935–1950.
Benkler, Y. 2006. *The Wealth of Networks: How Social Production Transforms Markets and Freedom*, Yale University Press, New Haven.

Bottiglieri, M., Pettenati, G. and Toldo, A. 2016. *Toward the Turin Food Policy. Good Practices and Visions.* Milan: FrancoAngeli.
Bourg, D. and Whiteside, K. 2010. *Vers une démocratie écologique. Le citoyen, le savant et le politique*, Seuil, Paris.
Bourguignon, F. 2015. *The Globalization of Inequality.*, Princeton University Press, Princeton.
Buchanan, J. and Tullock, G. 1958. *The Calculus of Consent: Logical Foundations of Constitutional Democracy.* Liberty Fund, Indianapolis.
Capra, F. and Mattei, U. 2015. *The Ecology of Law.* Berrett Khoeler, Oakland.
Charvet, J., and Kaczynska-Nay, E. 2008. *The Liberal Project and Human Rights: The Theory and Practice of a New World Order.* Cambridge University Press, Cambridge.
Chislock, M.F., Doster, E., Zitomer, R.A., and Wilson, A.E. 2013. Eutrophication: Causes, Consequences, and Controls in Aquatic Ecosystems. *Nature Education Knowledge* 4 (4), 10.
Claeys, P. 2014. Vía Campesina's Struggle for the Right to Food Sovereignty: From Above or from Below? In *Rethinking Food systems. Structural Challenges, New Strategies and the Law*, eds. N.C.S. Lambek, P. Claeys, A. Wong, and L. Brilmayer, 29–52. Dordrecht, Heidelberg, New York, London: Springer.
Claeys, P. 2015. Food Sovereignty and the Recognition of New Rights for Peasants at the UN: A Critical Overview of La Via Campesina's Rights Claims over the Last 20 Years. *Globalizations*, 12 (4), 452–465.
Clapp, J. and Isakson, R. 2018. *Speculative Harvests. Financialization, Food, and Agriculture.* Halifax, Nova Scotia: Fernwood Publishing.
Coase, R. and Fowler, R.F. 1935. Bacon Production and the Pig-Cycle in Great Britain. *Economica*, 2 (6), 142–167.
Committee on Economic, Social and Cultural Rights. 1999. *General Comment No. 12: the Right to Food.* UN doc. E/C.12/1999/5.
Committee on World Food Security. 2009. *Reform of the Committee on World Food Security.* U.N. Doc. CFS:2009/2Rev. 2.
Cunningham, S., Neiland, A.E., Arbuckle, M. and Bostock, T. 2009. Wealth-Based Fisheries Management: Using Fisheries Wealth to Orchestrate Sound Fisheries Policy in Practice. *Marine Resource Economics*, 24 (3), 271–287.
Dardot, P. and Laval, C. 2015. *Commun. Essai sur la révolution au XXIème siècle.* La Découverte, Paris.
De Angelis, D.M., 2017. *Omnia sunt communia: On the Commons and the Transformation to Postcapitalism.* Zed Books Ltd, London
De Schutter, O. 2011. The Green Rush: The Race for Farmland and the Rights of Land Users. *Harvard International Law Journal*, 52 (2), 503–559
De Schutter, O. 2015. The Role of Property Rights in the Debate on Large-Scale Land Acquisitions. In *Large-Scale Land Acquisitions: Focus on South-East Asia, International Development Policy series No. 6*, eds. C. Gironde, C. Golay, and P. Messerli, 53-77. Geneva: Graduate Institute Publications, Boston: Brill-Nijhoff.
Defourny, J. and Nyssens, M. 2006. Defining social enterprise. In *Social Enterprise. At the Crossroads of Market, Public Policies and Civil Society*, ed. M. Nyssens, 3–26. London: Routledge.
FAO. 1997. *The State of the World's Plant Genetic Resources for Food and Agriculture.* Food and Agriculture Organization of the United Nations, Rome.
FAO. 2010. *Second Report on the State of the World's Plant Genetic Resources for Food and Agriculture.* Food and Agriculture Organization of the United Nations, Rome.
FAO. 2015. *Status of the World's Soil Resources.* Food and Agriculture Organization of the United Nations, Rome.
FAO. 2017. *The Future of Food and Agriculture. Trends and Challenges.* Food and Agriculture Organization of the United Nations, Rome.
Ferrando, T. 2016. Il Sistema Cibo Bene Comune. In *Beni Comuni 2.0. Contro-egemonia e nuove istituzioni*, eds. A. Quarta and M. Spanò. Milan: Mimesis.
Ferrando, T. 2017. Land Rights at the Time of Global Production: Multi-Spatiality and 'Legal Chokeholds'. *Business and Human Rights Journal*, 2 (2), 275–295.
Ferrando, T. and Vivero-Pol, J.L. 2017. Commons and 'Commoning': a 'New' Old Narrative to Enrich the Food Sovereignty and Right to Food Claims. In *Right to Food and Nutrition Watch 2017*, ed. FIAN International, 50–56. https://www.righttofoodandnutrition.org/commons-and-commoning-new-old-narrative-enrich-food-sovereignty-and-right-food-claims (Accessed on August 21, 2018).
Gilens, M. 2012. *Affluence & Influence: Economic Inequality and Political Power in America.* Princeton University Press, Princeton.

Gilens, M. and Page, B. 2014. Testing Theories of American Politics: Elites, Interest Groups and Average Citizens. *Perspectives on Politics*, 12 (3), 564–581.

Golay, G. 2015. *Academy In-Brief No. 5: Negotiation of a United Nations Declaration on the Rights of Peasants and Other People Working in Rural Areas*. Geneva Academy of International Humanitarian Law and Human Rights, Geneva.

Gustavsson, J. et al. 2011. *Global Food losses and food waste. Extent, causes, and prevention*. Swedish Institute for Food and Biotechnology (SIK) and FAO, Rome.

Gutiérrez, N. L., Hilborn, R. and Defeo, O. 2011. Leadership, Social Capital and Incentives Promote Successful Fisheries. *Nature*, 470 (7334), 386–389.

Haas, G.C. and Ezekiel, M. 1926. *Factors Affecting the Price of Hogs*. US Department of Agriculture, Washington DC.

Hanau, A. 1928. Die Prognose der Schweinepreise. *Vierteljahrshefte zur Konjunkturforschung/Sonderhefte*, 18 (1930), 46.

Hardt, M. and Negri, A. 2009. *Commonwealth*. Harvard University Press, Cambridge.

Hayek, F. 1948 (1937). Economics and Knowledge. In *Individualism and Economic Order,* 33-46. Chicago: Chicago University Press.

Hayek, F. 1960. *The Constitution of Liberty*. Chicago University Press, Chicago.

Higgs, S. 2015. Social Norms and Their Influence on Eating Behaviours. *Appetite*, 86, 38–44.

HLPE. 2012. *Food Security and Climate Change*. HLPE Report No. 3. High-level Panel of Experts on Food Security and Nutrition Rome. Committee on World Food Security.

Hiroyuki, U. 2018. John R. Commons's Criticism of Classical Economics. *Journal of Economic Issues*, 52 (2), 396–404.

Holt-Gimenez, E. and Patel, R. eds., 2012. *Food Rebellions: Crisis and the Hunger for Justice*. Food First Books, Oakland.

Honneth, A. 2017 (2015). *L'idée du socialisme* (orig. *Die Idee des Socialismus*). Gallimard, Paris.

Hopgood, S. 2013. *The Endtimes of Human Rights.*: Cornell University Press, Ithaca.

Horkheimer, M. 1930 (1974). *Débuts de la philosophie bourgeoise de l'histoire* (orig. *Anfänge der bürgerlichen Geschichtsphilosophie*). Payot, Paris.

Human Rights Council. 2017. *Draft Declaration on the Rights of Peasants and Other People Working in Rural Areas*. Presented by the Chair-Rapporteur of the Working Group, U.N. Doc. A/HRC/WG.15/4/2.

IPES-Food. 2016. *From Uniformity to Diversity. A Paradigm Shift from Industrial Agriculture to Diversified Agroecological Systems*. International Panel of Experts on Sustainable Food Systems, Brussels.

IPES-Food. 2017. *Too Big to Feed. Exploring the Impacts of Mega-Mergers, Consolidation and Concentration of Power in the Agri-Food Sector*. International Panel of Experts on Sustainable Food Systems, Brussels.

Jacobsen, SE., Sørensen, M., Pedersen, S.M. and Weiner, J. 2013. Feeding the World: Genetically Modified Crops versus Agricultural Biodiversity. *Agronomy for Sustainable Development*, 33 (4), 651–662.

Jamu, D., Banda, M., Njaya, F. and Hecky, R.E. 2011. Challenges to Sustainable Management of the Lakes of Malawi. *Journal of Great Lakes Research*, 37 (1), 3–14.

Kahan, D.M., Braman, D., Gastil, J., Slovic, P. and Mertz, C.K. 2007. Culture and Identity-Protective Cognition: Explaining the White Male Effect in Risk Perception. *Journal of Empirical Legal Studies*, 4 (3), 465–505.

Kahan, D.M., Wittlin, M., Peters, E., Slovic, P., Ouellette, L.L., Braman, D. and Mandel, G. 2011. *The Tragedy of the Risk-Perception Commons: Culture Conflict, Rationality Conflict, and Climate Change*. Cultural Cognition Project Working Paper No. 89.

Kennedy, D. 2002. The International Human Rights Movement: Part of the Problem? *Harvard Human Rights Journal*, 15, 101.

Kennedy, D. 2012. The International Human Rights Regime: Still Part of the Problem? In *Examining Critical Perspectives on Human Rights*, eds. Dickinson, R., Katselli, E. Murray, C. and Pedersen, O.W., 19–34. Cambridge: Cambridge University Press.

Keynes, J.M. (1936) 2016. *General Theory of Employment, Interest and Money*. Atlantic Publishers & Dist, New Delhi.

Kostakis, V. and Bauwens. M. 2014. *Network Society and Future Scenarios for a Collaborative Economy*. London: Palgrave MacMillan.

Lambin, E. and Meyfroidt, P. 2011. Global Land Use Change, Economic Globalization, and the Looming Land Scarcity. *Proceedings of the National Academy of Sciences*, 108, 3465–3472.

Lambin, E. 2012. Global Land Availability: Malthus versus Ricardo. *Global Food Security*, 1, 83–87.

Lambin, E., Gibbs, H.K., Ferreira, L., Grau, R., Mayaux, P., Meyfroidt, P., Morton, D.C., Rudel, T.K., Gasparri, I. and Munger, J. 2013. Estimating the World's Potentially Available Cropland Using a Bottom-Up Approach. *Global Environmental Change*, 23 (5), 892–901

Lewins, R., Béné, C., Baba, M.O., Belal, E., Donda, S., Lamine, A.M., Makadassou, A., Mamane Tahir Na., A., Neiland, A.E., Njaya, F., Ovie, S., and Raji, A. 2014. African Inland Fisheries: Experiences with Co-Management and Policies of Decentralization. *Society and Natural Resources*, 27 (4), 405–420.

Lind, D. and Barham, E., 2004. The Social life of the Tortilla: Food, Cultural Politics, and Contested Commodification. *Agriculture and Human Values*, 21 (1), 47–60.

Loveland, P.J and Webb, J. 2003. Is There a Critical Level of Organic Matter in the Agricultural Soils of Temperate Regions a Review. *Soil and Tillage Research*, 70, 1–18.

Macpherson, C.B. 1968. Introduction. In *Leviathan*, Hobbes, T. London: Penguin Books.

Mattei, U. and Nader, L. 2008. *Plunder: When the Rule of Law is Illegal*. Blackwell, Oxford.

Mattei, U. 2011. *Beni Comuni. Un Manifesto*. Laterza, Bari.

Mattei, U. and Bailey, S. 2013. Social Movements as Constituent Power. The Italian Struggle for the Commons. *Indiana Journal of Global Legal Studies*, 20 (2), 965–1013

Mattei, U.and Quarta, A. 2018. *The Turning Point in Privale Law. Ecology, Technology and the Commons*. Edward Elgar, Cheltenham.

Mazzucato, M. 2013. *The Entrepreneurial State. Debunking Public vs. Private Sector Myths*. London: Anthem Press.

Menkel-Meadow, C. 2012. Introduction and Coda: International Dispute Resolution. In *Complex Dispute Resolution: Volume III*, ed. Carrie Menkel-Meadow, #. Farnham: Ashgate Press.

Moyn S. 2018. *Not Enough: Human Rights in an Unequal World*. Harvard University Press, Cambridge.

Moller, A.C., Ryan, R.M. and Deci, E. 2006. Self-Determination Theory and Public Policy: Improving the Quality of Consumer Decisions Without Using Coercion. *Journal of Public Policy and Marketing*, 25 (1), 104–116.

Moore, J.W. 2017. The Capitalocene, Part I: On the Nature and Origins of Our Ecological Crisis. *The Journal of Peasant Studies* 44 (3), 594–630.

Monteiro, C.A., Cannon, G., Moubarac, J.C., Levy, R.B., Louzada, M.L.C. and Jaime, P.C. 2018. Ultra-Processing. An Odd 'Appraisal'. *Public Health Nutrition* 21 (3), 497–501.

Nielsen, J.R., Degnbol, P., Viswanathan, K.K., Ahmed, M., Hara, M., Raja Abdullah, N.M. 2004. Fisheries Co-Management—an Institutional Innovation? Lessons from South East Asia and Southern Africa. *Marine Policy*, 28 (2), 151–-160.

Ostrom, E. 1990. *Governing the Commons: The Evolution of Institutions for Collective Action*. New York: Cambridge University Press.

Ostrom, E. 2009. *A Polycentric Approach for Coping with Climate Change*. Policy Research working paper WPS 5095. Washington, DC: World Bank. https://openknowledge.worldbank.org/bitstream/handle/10986/9034/WPS5095_WDR2010_0021.pdf (Accesed on August 12, 2018).

Oxfam. 2018. *Ripe for Change. Ending Human Suffering in Supermarket Supply Chains*. OXFAM. https://policy-practice.oxfam.org.uk/publications/ripe-for-change-ending-human-suffering-in-supermarket-supply-chains-620418.

Paerl, H.W. and Huisman, J. 2012. Climate Change: Links to Global Expansion of Harmful Cyanobacteria. *Water Research*, 46, 1349–1363.

Parris, K. 2011. Impact of Agriculture on Water Pollution in OECD Countries: Recent Trends and Future Prospects. *International Journal of Water Resources Development*, 27, 33–52.

Patel, R. and Moore, J.W. 2017. *A History of the World in Seven Cheap Things: A Guide to Capitalism, Nature, and the Future of the Planet*. Univ of California Press, Berkeley.

Pimentel, D., Williamson, S., Alexander, C., Gonzalez-Pagan, O., Kontak, C. and. Mulkey, S. 2008. Reducing Energy Inputs in the US Food System. *Human Ecology*, 36, 459–471.

Piron, J. and Cogolati, S. 2017. *Vers des partenariats publics-communs*. June 2017. http://www.etopia.be/spip.php?article3209 (Accessed on August 21, 2018).

Polanyi, M. 1951. *The Logic of Liberty*. Liberty Fund, London. Reprinted by Routledge.

Polanyi, K. 1957. The Economy as Instituted Process. In *Trade and Market in the Early Empires: Economies in History and Theory*, eds. K. Polanyi, C. Arendberg and H. Pearson, 244. Free Press, New York.

Ratner, B.D., Åsgård, B. and Allison, E.H. 2014. Fishing for Justice: Human Rights, Development, and Fisheries Sector Reform. *Global Environmental Change*, 27, 120–130.

Reardon, T. and Berdegué, J.A. 2002. The Rapid Rise of Supermarkets in Latin America. Challenges and Opportunities for Development. *Development Policy Review*, 20 (4), 317–334.

Reardon, T., Timmer, C.P. and Minten, B. 2012. Supermarket Revolution in Asia and Emerging Development Strategies to Include Small Farmers. *Proceedings of the National Academy of Sciences*, 109 (31), 12332–123337.
Ryan, R.M. and Deci, E.L. 2000a. Internal and External Motivations: Classic Definitions and New Directions. *Contemporary Educational Psychology*, 25 (1), 54–67.
Ryan, R.M. and Deci, E. 2000b. Self-Determination Theory and the Facilitation of Intrinsic Motivation, Social Development, and Well-Being. *American psychologist*, 55 (1), 68–78.
Samuelson, P.A. 1954. The Pure Theory of Public Expenditure. *Review of Economics and Statistics*, 36 (4), 387–389.
Sandel, M.J. 2013. *What Money Can't Buy: The Moral Limits of Markets*. Farrar, Straus and Giroud, New York.
Sarfatti-Larson, M. 1977. *The Rise of Professionalism: A Sociological Analysis*. University of California Press, Berkeley.
Scitovski, T. 1992. *The Joyless Economy: The Psychology of Human Satisfaction*. Oxford: Oxford University Press.
Sharma, C. 2011. Securing Economic, Social and Cultural Rights of Small-Scale and Artisanal Fishworkers and Fishing Communities. *Journal of Maritime Studies*, 10 (2), 41–62.
Shepperson, M. 2017. *How Ancient Lentils Reveal the Origins of Social Inequality*. The Guardian, 11 October 2017.
Smith, P., D. Martino, Z. Cai, D. Gwary, H. Janzen, P. Kumar, B. McCarl, S. Ogle, F. O'Mara, C. Rice, B. Scholes and O. Sirotenko. 2007. Agriculture. In *Climate Change 2007: Mitigation. Contribution of Working Group III to the Fourth Assessment Report of the Intergovernmental Panel on Climate Change*, eds. B. Metz, O.R. Davidson, P.R. Bosch, R. Dave and L.A. Meyer, #. Cambridge and New York: Cambridge University Press.
Special Rapporteur on the right to food. 2012. *The Fisheries Sector and the Right to Food*. Interim Report of the Special Rapporteur on the right to food, Olivier De Schutter, to the 67th session of the General Assembly, UN doc. A/67/268.
Steffen, W., Richardson, K., Rockström, J., Cornell, S.E., Fetzer, I., Bennett, M.E., Biggs, R., Carpenter, S.R., de Vries, W., de Wit, C.A., Folke, C., Gerten, D., Heinke, J., Mace, G.M., Persson, L.M., Ramanathan, V., Reyers, B. and Sörlin, S. 2015. Planetary Boundaries: Guiding human development on a changing planet. *Science*, 347 (6223), DOI:10.1126/science.1259855
Stigler, G.J. 1971. The Theory of Economic Regulation. *The Bell Journal of Economics and Management Science*, 2 (1), 3–21.
Stiglitz, J. 2015. *The Great Divide*. W.W. Norton and Company, New York.
TEEB. 2018. *Measuring what matters in agriculture and food systems: a synthesis of the results and recommendations of TEEB for Agriculture and Food's Scientific and Economic Foundations report for The Economics of Ecosystems and Biodiversity*. UN Environment, Geneva.
The Food Commons. 2011. *The Food Commons 2.0. Imagine, Design, Build*. October 2011. http://www.thefoodcommons.org/images/FoodCommons_2-0.pdf (accessed on August 21, 2018).
Townsend, R., Shotton, R. and Uchida, H., eds. 2008. *Case Studies in Fisheries Self-Governance*. FAO Fisheries Technical Paper No. 504. Food and Agriculture Organization, Rome.
van der Ploeg, J.D. 2010. The Food Crisis, Industrialized Farming and the Imperial Regime. *Journal of Agrarian Change* 2(10): 98–106.
Vermeulen, S.J., Campbell, B.M. and Ingram, J.S.I. 2012. Climate Change and Food Systems. *Annual Review of Environmental Resources*, 37, 195–222.
Vivero-Pol, J.L. 2017a. Food as Commons or Commodity? Exploring the Links between Normative Valuations and Agency in Food Transition. *Sustainability*, 9 (3), 442.
Vivero-Pol, J.L. 2017b. *How Do People Value Food? Systematic, Heuristic and Normative Approaches to Narratives of Transition in Food Systems*. PhD Thesis. October 2017. Faculty of Bioengineers. Universite catholique de Louvain, Belgium. https://dial.uclouvain.be/pr/boreal/object/boreal:191763
Wicklund, R.A. 1974. *Freedom and Reactance*. Oxford: Lawrence Erlbaum.
World Bank. 2004. *Saving Fish and Fishers: Toward Sustainable and Equitable Governance of the Global Fishing Sector*. World Bank, Washington, DC.
Wright E.O. 2013. Transforming Capitalism through Real Utopias. 2011 Presidential Address. *American Sociologist Review*, 78, 1–25.

INDEX

For Product Safety Concerns and Information please contact our EU
representative GPSR@taylorandfrancis.com
Taylor & Francis Verlag GmbH, Kaufingerstraße 24, 80331 München, Germany

www.ingramcontent.com/pod-product-compliance
Ingram Content Group UK Ltd.
Pitfield, Milton Keynes, MK11 3LW, UK
UKHW050053200926
475702UK00019B/2814
* 9 7 8 0 3 6 7 6 2 8 5 6 7 *